TM 9-2320-272-10
Operator Manual
5 Ton 6x6 Truck
M939 series

Includes Change 2

September 2004

The M939 series truck is a 5 ton (10,000 pound) 6x6 US Military truck that was developed to replace the M39 and M809 series of trucks. It is capable of carrying it's design load over all terrain and in all weather. A number of specialty bodies were produced. An estimated 44,590 units were produced. This is the operator manual. It is being reprinted as a convenience to operators, enthusiasts and private owners.

Should you have suggestions or feedback on ways to improve this book please send email to Books@OcotilloPress.com

---

Edited 2021 Ocotillo Press
ISBN 978-1-954285-60-6

Ocotillo Press
Houston, TX 77017
Books@OcotilloPress.com

ARMY TM 9-2320-272-10
AIR FORCE TO 36A12-1C-441

TECHNICAL MANUAL

OPERATOR'S MANUAL

FOR

TRUCK, 5-TON, 6X6, M939, M939A1, AND M939A2 SERIES TRUCKS (DIESEL)

TRUCK, CARGO: 5-TON, 6X6 DROPSIDE, M923 (2320-01-050-2084) (EIC: BRY); M923A1 (2320-01-206-4087) (EIC: BSS); M923A2 (2320-01-230-0307) (EIC: BS7); M925 (2320-01-047-8769) (EIC: BRT); M925A1 (2320-01-206-4088) (EIC: BST); M925A2 (2320-01-230-0308) (EIC: BS8);

TRUCK, CARGO: 5-TON, 6X6 XLWB, M927 (2320-01-047-8771) (EIC: BRV); M927A1 (2320-01-206-4089), (EIC: BSW); M927A2 (2320-01-230-0309) (EIC: BS9); M928 (2320-01-047-8770) (EIC: BRU); M928A1 (2320-01-206-4090) (EIC: BSX); M928A2 (2320-01-230-0310) (EIC: BTM);

TRUCK, DUMP: 5-TON, 6X6, M929 (2320-01-047-8756) (EIC: BTH); M929A1 (2320-01-206-4079) (EIC: BSY); M929A2 (2320-01-230-0305) (EIC: BTN); M930 (2320-01-047-8755) (EIC: BTG); M930A1 (2320-01-206-4080), (EIC: BSZ); M930A2 (2320-01-230-0306) (EIC: BTO);

TRUCK, TRACTOR: 5-TON, 6X6, M931 (2320-01-047-8753) (EIC: BTE); M931A1 (2320-01-206-4077) (EIC: BS2); M931A2 (2320-01-230-0302) (EIC: BTP); M932 (2320-01-047-8752) (EIC: BTD); M932A1 (2320-01-205-2684) (EIC: BS5); M932A2 (2320-01-230-0303) (EIC: BTQ);

TRUCK, VAN, EXPANSIBLE: 5-TON, 6X6, M934 (2320-01-047-8750) (EIC: BTB); M934A1 (2320-01-205-2682) (EIC: BS4); M934A2 (2320-01-230-0300) (EIC: BTR);

TRUCK, MEDIUM WRECKER: 5-TON, 6X6, M936 (2320-01-047-8754) (EIC: BTF); M936A1 (2320-01-206-4078) (EIC: BS6); M936A2 (2320-01-230-0304) (EIC: BTT).

HEADQUARTERS, DEPARTMENTS OF THE ARMY AND THE AIR FORCE

WARNING

EXHAUST GASES CAN KILL

1. DO NOT operate your vehicle engine in enclosed area.
2. DO NOT idle vehicle engine with vehicle windows closed.
3. DO NOT drive vehicle with inspection plates or cover plates removed.
4. BE ALERT at all times for exhaust odors.
5. BE ALERT for exhaust poisoning symptoms. They are:
   - Headache
   - Dizziness
   - Sleepiness
   - Loss of muscular control
6. IF YOU SEE another person with exhaust poisoning symptoms:
   - Remove person from area
   - Expose to open air
   - Keep person warm
   - Do not permit person to move
   - Administer artificial respiration* or CPR, if necessary

*For artificial respiration, refer to FM 21-11.

7. BE AWARE, the field protective mask for Nuclear-Biological-Chemical (NBC) protection will not protect you from carbon monoxide poisoning.

THE BEST DEFENSE AGAINST EXHAUST POISONING IS ADEQUATE VENTILATION.

WARNING

HIGH INTENSITY NOISE

Hearing protection is required for all personnel working in and around this vehicle while the engine is running (reference AR 40-5 and TB MED 501).

WARNING SUMMARY

- If Nuclear, Biological, or Chemical (NBC) exposure is suspected, all air filter media should be handled by personnel wearing protective equipment. Consult your unit NBC officer or NBC NCO for appropriate handling or disposal instructions.
- Extreme care should be taken when removing surge tank filler cap if temperature gauge reads above 175°F (79°C). Steam or hot coolant under pressure will cause injury.
- Pump brakes gradually when slowing or stopping vehicle on ice, snow, or wet pavement. Sudden stop will cause vehicle wheels to lock, engine to stall, and loss of power steering. Failure to do this will result in injury or death.
- Ground spike must be driven into ground 18-24 inches (46-61 centimeters) and spike cable connected to the chassis before power can be taken from outside source. Failure to do this will result in electrical damage, injury, or death.
- Ensure spike cable ring terminal makes good contact with bare metal. If necessary, scrape contact area clean of dirt, paint, or rust. Failure to do this will result in electrical damage, injury, or death.
- Vehicle will become charged with electricity if A-frame contacts or breaks high voltage wire. Do not leave vehicle while high voltage line is in contact with A-frame or vehicle. Failure to do this will result in injury or death. Notify nearby personnel to have electrical power turned off.
- Do not put vehicle in motion until warning light is extinguished, and alarm (buzzer) stops sounding. If air pressure gauges indicate less than 90 psi (621 kPa), turn ignition and battery switches to OFF positions, and notify unit maintenance. Failure to do this will result in injury or death.
- Alcohol used in alcohol evaporator is flammable, poisonous, and explosive. Do not smoke when adding fluid and do not drink fluid. Failure to do this will result in injury or death.
- Stay clear of dump body and cab protector at all times during loading and unloading operations. Dump body can unexpectedly raise when a heavy load is dropped into dump body and will cause injury or death.
- Do not smoke, have open flames, or make sparks around the batteries, especially if the caps are off. They can explode and cause injury or death.
- Do not lower load without a ground guide. Direct all personnel to stand clear of lifting operations. Swinging loads will cause injury or death.
- Do not perform fuel system checks or inspection while smoking or near fire, flames, or sparks. Fuel could ignite, causing damage to vehicle, injury, or death.
- This vehicle has been designed to operate safely and efficiently within the limits specified in this TM. Operation beyond these limits is prohibited IAW AR 70-1 without written approval from the Commander, U.S. Army Tank-automotive and Armaments Command, ATTN: AMCPEO-CM-S, Warren, MI 48397-5000.
- Block vehicle wheels if operating site is on a grade, no matter how slight. Failure to do this will result in injury or death.

WARNING SUMMARY (Cont'd)

- Do not use hand throttle while driving. The hand throttle will not disengage when brakes are applied. Failure to do this will result in injury or death.
- Open van door slowly. Personnel may be on ladder. Use caution when using ladder.
- Use ground guide when backing up to parked semitrailer. Failure to do this will result in damage to vehicle, injury, or death.
- Do not pull tractor forward beyond approach ramps until all air lines are disconnected. Failure to do this will result in injury or death.
- Death or serious injury to soldiers, or damage to army equipment will occur if the instructions in this procedure are not followed.
- Liquid surge results from liquid's movement in partially-filled tanks, which may result in loosing control of 5,000 gallon fuel tankers causing vehicle damage, injury or death to personnel.
- On cross-country terrain, payload is limited to 3,000 gallons of fuel if the prime mover is an M931/A1/A2 or M932/A1/A2 series 5-ton tractor. Failure to comply may result in damage to vehicle or injury or death to personnel.
- Stopping distance is generally increased by ABS technology. ABS technology is designed to perform a conventional braking technique called "stab" braking automatically using wheel speed sensors. Drivers must understand they should not pump the brakes on an ABS-equipped vehicle, as this will deactivate the ABS. Drivers must also understand that by removing pressure from the brake pedal, drivers can also deactivate the ABS. Failure to comply may result in damage to vehicle or injury or death to personnel.
- When the ABS senses impending wheel lockup, the ECU will modulate the relays which will repeat a "release and recharge" cycle of air in the brake chambers. Unlike a car's ABS, where you can feel this modulation on the brake pedal, you will not feel any modulation of the brake pedal on an air brake system. When the ABS does modulate, you will feel a jerking sensation of the vehicle as the brakes rapidly release and lock. Failure to comply may result in damage to vehicle or injury or death to personnel.
- M939 series vehicles have a conventional air brake system, which is very sensitive. Drivers of these vehicles must be well-trained in operating tactical vehicles with air brakes. Air brakes are unique because braking force is proportional to pedal travel, but the driver does not experience resistance from the brake pedal. An inexperienced driver may respond to lack of resistance by applying too much force to brake pedal. Operators can be confident that M939 series trucks equipped w/ABS brakes have more than adequate brake capacity for safe mountain terrain operations.
- Do not drive too fast for total weight of vehicle, amount of fuel in tanker, length and angle of grade, road conditions, and weather. Failure to do so may result in damage to vehicle or injury or death to personnel.
- Comply with warning signs indicating length and angle of grade. Failure to comply may result in damage to equipment or injury or death to personnel.

WARNING SUMMARY (Cont'd)

- Ensure vehicle is moving 10–15 mph (16–24 km/h) slower than posted ramp speed for entrance or exit ramps. Failure to comply may result in vehicle rollover, causing damage to vehicle or injury or death to personnel.
- ABS technology is designed to maintain rolling traction and steering. The rolling action may produce longer stopping distances on some surfaces, such as freshly fallen snow or loose gravel. The ABS steering advantage outweighs any braking disadvantage on these surfaces. Evasive steering techniques are designed to allow the driver to steer the vehicle clear of damage. By maintaining a speed reduction without wheel lockup, ABS increases steerability of the vehicle. The driver should use just enough steering movement to adjust the vehicle to a clear space on the roadway.
- Avoid steering more than necessary to clear an obstacle. Oversteering may cause a skid, jackknife, or rollover. Failure to comply may cause damage to vehicle or injury or death to personnel.
- Do not drive faster than road or weather conditions permit. Maximum safe speed limit for normal highway driving is 55 mph (88 km/h).
- Stopping can be adversely affected by poor road/weather conditions. Drive at a slow, safe speed in poor conditions to avoid excessive braking. Failure to comply may result in damage to equipment or serious injury or death to personnel.
- Do not pump brakes that are locking on a vehicle equipped with ABS when stopping. ABS will automatically release wheels that are locking and apply pressure to the other wheels. Failure to do so may result in damage to vehicle or injury or death to personnel.

ARMY TM 9-2320-272-10
AIR FORCE TO 36A12-1C-441
C2

TECHNICAL MANUAL
NO. 9-2320-272-10
TECHNICAL ORDER
NO. 36A12-1C-441

CHANGE
NO. 2

HEADQUARTERS,
DEPARTMENT OF THE ARMY
AND AIR FORCE
WASHINGTON, D.C., 30 SEPTEMBER 2004

OPERATOR'S MANUAL
FOR
TRUCK, 5-TON, 6X6, M939, M939A1, and M939A2 SERIES TRUCKS (DIESEL)

TRUCK, CARGO: 5-TON, 6X6 DROPSIDE, M923 (2320-01-050-2084) (EIC: BRY); M923A1 (2320-01-206-4087) (EIC: BSS; M923A2 (2320-01-230-0307) (EIC: BS7); M925 (2320-01-047-8769) (EIC: BRT); M925A1 (2320-01-206-4088) (EIC: BST); M925A2 (2320-01-230-0308) (EIC: BS8); TRUCK, CARGO: 5-TON, 6X6 XLWB, M927 (2320-01-047-8771) (EIC: BRV); M927A1 (2320-01-206-4089), (EIC: BSW); M927A2 (2320-01-230-0309) (EIC: BS9); M928 (2320-01-047-8770) (EIC: BRU) M928A1 (2320-01-206-4090) (EIC: BSX); M928A2 (2320-01-230-0310) (EIC: BTM);

TRUCK, DUMP: 5-TON, 6X6, M929 (2320-01-047-8756) (EIC: BTH); M929A1 (2320-01-206-4079) (EIC: BSY); M929A2 (2320-01-230-0305) (EIC: BTN); M930 (2320-01-047-8755) (EIC: BTG); M930A1 (2320-01-206-4080); (EIC: BSZ); M930A2 (2320-01-230-0306) (EIC: BTO); TRUCK, TRACTOR: 5-TON, 6X6, M931 (2320-01-047-8753) (EIC: BTE); M931A1 (2320-01-206-4077) (EIC: BS2); M931A2 (2320-01-230-0302) (EIC: BTP);

M932 (2320-01-047-8752) (EIC: BTD); M932A1 (2320-01-205-2684); EIC: BS5); M932A2 (2320-01-230-0303) (EIC: BTQ); TRUCK, VAN, EXPANSIBLE: 5-TON, 6X6, M934 (2320-01-047-8750) (EIC: BTB); M934A1 (2320-01-205-2682); (EIC: BS4); M934A2 (2320-01-230-0300) (EIC: BTR); TRUCK, MEDIUM WRECKER: 5-TON, 6X6, M936 (2320-01-047-8754) (EIC: BTF); M936A1 (2320-01-206-4078) (EIC: BS6); M936A2 (2320-01-230-0304

TM 9-2320-272-10, August 1996, is changed as follows:

1. Remove old pages and insert new pages as indicated below.
2. New or changed material is indicated by a vertical bar in the margin of the page.

Remove page	**Insert page**
Warning a and Warning b	Warning a through Warning d
None	A and B
i through iii/(iv blank)	i through iii/(iv blank)
1-5 through 1-8	1-5 through 1-8
1-15 and 1-16	1-15 and 1-16
1-19 through 1-26	1-19 through 1-26
1-49 and 1-50	1-49 and 1-50

Remove page	Insert page
1-57 through 1-68	1-57 through 1-68
1-71 and 1-72	1-71 and 1-72
2-1 through 2-4	2-1 through 2-4
2-35 through 2-50	2-35 through 2-50
2-53 and 2-54	2-53 and 2-54
2-57 through 2-66	2-57 through 2-66
2-73 and 2-74	2-73 and 2-74
2-77 through 2-84	2-76.1 through 2-84
2-87 through 2-90	2-87 through 2-90
2-97 through 2-102	2-97 through 2-102
2-105 through 2-110	2-105 through 2-110
2-119 and 2-120	2-119 and 2-120
2-133 and 2-134	2-133 and 2-134
2-141 through 2-144	2-141 through 2-144
2-155 through 2-158	2-155 through 2-158
2-173 and 2-174	2-173 and 2-174
2-181 and 2-182	2-181 and 2-182
2-203 and 2-204	2-203 and 2-204
2-207 and 2-208	2-207 and 2-208
3-1 and 3-2	3-1 and 3-2
3-7 through 3-14	3-7 through 3-14
3-17 and 3-18	3-17 and 3-18
3-25 through 3-28	3-25 through 3-28
3-61/(3-62 blank)	3-61/(3-62 blank)
A-1 and A-2	A-1 and A-2
B-3 through B-6	B-3 through B-6
B-9 through B-24	B-9 through B-24
C-1 through C-5/(C-6 blank)	C-1 through C-5/(C-6 blank)
D-1 through D-4	D-1 through D-4
INDEX 1 through INDEX 18	INDEX 1 through INDEX 23 (INDEX 24 blank)
DA Form 2028-2	DA Form 2028

3. File these change sheets in front of the publication for reference purposes.

By Order of the Secretary of the Army:

PETER J. SCHOOMAKER
General, United States Army
Chief of Staff

Official:

JOEL B. HUDSON
Administrative Assistant to the
Secretary of the Army
0407602

By Order of the Secretary of the Air Force:

JOHN P. JUMPER
General, United States Air Force
Chief of Staff

Official:

GREGORY S. MARTIN
General, United States Air Force
Commander, Air Force Materiel Command

Distribution:

To be distributed in accordance with the initial distribution number (IDN) XXXXXXX, requirements for TM 9-2320-272-10.

ARMY TM 9-2320-272-10
AIR FORCE TO 36A12-1C-441
C1

TECHNICAL MANUAL
NO. 9-2320-272-10

HEADQUARTERS,
DEPARTMENT OF THE ARMY
AND THE AIR FORCE
WASHINGTON, D.C., *22 Feb 1999*

OPERATOR'S MANUAL FOR TRUCK, 5-TON, 6X6, M939, M939A1 AND M939A2 SERIES (DIESEL)

TRUCK, CARGO: 5-TON, 6X6, DROPSIDE
M923 (2320-01-050-2084), M923A1 (2320-01-206-4087), M923A2 (2320-01-230-0307), M925 (2320-01-047-8769), M925A1 (2320-01-206-4088), M925A2 (2320-01-230-0308);
TRUCK, CARGO: 5-TON, 6X6,
M924 (2320-01-047-8773), M924A1 (2329-01-205-2692), M926 (2320-01-047-8772), M926A1 (2320-01-205-2693);
TRUCK, CARGO: 5-TON, 6X6, XLWB,
M927 (2320-01-047-8756), M927A1 (2320-01-206-4089), M927A2 (2320-01-230-0309), M928 (2320-01-047-8770), M928A1 (2320-01-206-4090), M928A2 (2320-01-230-0310);
TRUCK, DUMP: 5-TON 6X6,
M929 (2320-01-047-8756), M929A1 (2320-01-206-4079), M929A2 (2320-01-230-0305), M930 (2320-01-047-8755), M930A1 (2320-01-206-4080), M929A2 (2320-01-230-0306);
TRUCK, TRACTOR: 5-TON 6X6,
M931 (2320-01-047-8753), M931A1 (2320-01-206-4077), M931A2 (2320-01-230-0302), M932 (2320-01-047-8752), M932A1 (2320-01-205-2684), M932A2 (2320-01-230-0303);
TRUCK, VAN, EXPANSIBLE: 5-TON 6X6,
M934 (2320-01-047-8750), M934A1 (2320-01-205-2682), M934A2 (2320-01-230-0300), M935 (2320-01-047-8751), M935A1 (2320-01-205-2683) M935A2 (2320-01-230-0301);
TRUCK, MEDIUM WRECKER: 5-TON, 6X6,
M936 (2320-01-047-8754), M936A1 (2320-01-206-4078), M936A2 (2320-01-230-0304).

This change is issued in conjunction with Safety of Use Message (SUOM), TACOM Control No. 98-07. It emphasizes safe driving procedures to be followed by M939 vehicle operators.

TM 9-2320-272-10, August 1996, is changed as follows:

Remove page	Insert page
1-23 and 1-24	1-23 and 1-24
2-105 and 2-106	2-105 and 2-106

By Order of the Secretary of the Army:

DENNIS J. REIMER
General, United States Army
Chief of Staff

Official:

JOEL B. HUDSON
Administrative Assistant to the
Secretary of the Army
05676

By Order of the Secretary of the Air Force:

RONALD R. FOGLEMAN
General, United States Air Force
Chief of Staff

Official:

HENRY VICCELLIO, JR.
General, United States Air Force
Commander, Air Force Materiel Command

DISTRIBUTION:

To be distributed in accordance with the initial distribution number (IDN) 380385, requirements for TM 9-2320-272-10.

LIST OF EFFECTIVE PAGES

NOTE: The portion of the text affected by the changes is indicated by a vertical line in the outer margins of the page.

Dates of issue for original and changed pages are:
Original .... 0 .... 15 August 1996
Change .... 1 ...... 22 February 1999
Change .... 2 ..... **30 September 2004**

TOTAL NUMBER OF PAGES IN THIS PUBLICATION IS 490.
CONSISTING OF THE FOLLOWING:

Page No.	*Change No.
warning a	0
warning b	2
warning c Added	2
warning d Added	2
A - B Added	2
i	0
ii - iii	2
iv blank	0
1-1 - 1-4	0
1-5 - 1-7	2
1-8 - 1-15	0
1-16	2
1-17 - 1-19	0
1-20 - 1-25	2
1-26 - 1-48	0
1-49	2
1-50 - 1-56	0
1-57 - 1-59	2
1-60 - 1-61	0
1-62 - 1-63	2
1-64 - 1-65	0
1-66 - 1-67	2
1-68 - 1-71	0
1-72	2
1-73 - 1-82	0
2-1	0
2-2 - 2-4	2
2-5 - 2-35	0
2-36 - 2-37	2
2-38 - 2-39	0
2-40 - 2-41	2
2-42 - 2-43	0
2-44 - 2-48	2
2-49	0
2-50	2
2-51 - 2-52	0
2-53 - 2-54	2
2-55 - 2-57	0
2-58 - 2-60	2
2-61	0
2-62	2
2-63	0
2-64 - 2-65	2
2-66 - 2-73	0
2-74	2
2-75 - 2-76	0
2-76.1 - 2-76.2 Added	2
2-77	2
2-78	0
2-79	2
2-80	0
2-81 - 2-84	2
2-85 - 2-86	0
2-87	2
2-88	0
2-89	2
2-90 - 2-97	0
2-98 - 2-101	2
2-102 - 2-104	0
2-105	2
2-106	0
2-107 - 2-110	2
2-111 - 2-119	0
2-120	2
2-121 - 2-133	0
2-134	2
2-135 - 2-141	0
2-142	2
2-143 Deleted	2
2-144 - 2-155	0
2-156	2
2-156.1 - 2-156.8 Added	2
2-157	2
2-158 - 2-172	0
2-173	2
2-174 - 2-180	0
2-181	2
2-182 - 2-202	0
2-203	2
2-204 - 2-207	0
2-208	2
2-209 - 2-240	0
3-1	2
3-2 - 3-6	0
3-7	2
3-8 - 3-9	0
3-10 - 3-14	2
3-15 - 3-16	0
3-17	2
3-18 - 3-24	0
3-25	2
3-26	0
3-27	2
3-28 - 3-60	0
3-61	2
3-62 blank	0
A-1 - A-2	2
B-1 - B-3	0
B-4 - B-6	2
B-7 - B-8	0
B-9 - B-10	2
B-11	0
B-12 - B-17	2
B-18	0
B-19	2
B-20 - B-21	0

*Zero in this column indicates original page.

LIST OF EFFECTIVE PAGES (Cont'd)

Page No.	*Change No.
B-22 - B-23	2
B-24	0
C-1	0
C-2 - C-5	2
C-6 blank	0
D-1	0
D-2 - D-4	2
E-1 - E-3	0
E-4 blank	0
Index 1 - Index 23	2
Index 24 blank	2
DA Form 2028	2

*Zero in this column indicates original page.

ARMY TM 9-2320-272-10
AIR FORCE TO 36A12-1C-441

TECHNICAL MANUAL
NO. 9-2320-272-10

TECHNICAL ORDER
NO. 36A12-1C441

HEADQUARTERS,
DEPARTMENT OF THE ARMY
AND THE AIR FORCE
Washington DC, 15 August 1996

OPERATOR'S MANUAL
FOR
TRUCK, 5-TON, 6X6, M939, M939A1, AND
M939A2 SERIES TRUCKS (DIESEL)

This change supersedes a portion of LO 9-2320-272-12, dated June 1990.

REPORTING ERRORS AND RECOMMENDING IMPROVEMENTS

You can help improve this publication. If you find any mistakes or if you know of a way to improve the procedures, please let us know. Submit your DA Form 2028 (Recommended Changes to Equipment Technical Publications), through the Internet, on the Army Electronic Product Support (AEPS) website. The Internet address is http://aeps.ria.army.mil. If you need a password, scroll down and click on "ACCESS REQUEST FORM." The DA Form 2028 is located in the ONLINE FORMS PROCESSING section of the AEPS. Fill out the form and click on SUBMIT. Using this form on the AEPS will enable us to respond quicker to your comments and better manage the DA Form 2028 program. You may also mail, fax or email your letter or DA Form 2028 direct to: AMSTA-LC-CI Tech Pubs, TACOM-RI, 1 Rock Island Arsenal, Rock Island, IL 61299-7630. The email address is TACOM-TECH-PUBS@ria.army.mil. The fax number is DSN 793-0726 or Commercial (309) 782-0726.

HOW TO USE THIS MANUAL

1. The information contained in this manual can be accessed in several ways.

a. If you know what area you are looking for, use the front cover index. Find the appropriate box and match it to the blackened pages on the side of the book, or use the page number listed in the box.

b. If you are looking for a specific paragraph, refer to the index at the back of the manual.

2. This manual consists of:

a. Chapter 1, Introduction – provides information for completing forms and records, gives a familiarization of equipment, and a functional and physical description of major equipment.

b. Chapter 2, Operating Instructions – provides information needed to use or operate the vehicle.

c. Chapter 3, Maintenance Instructions – provides information covering lubrication, troubleshooting, and corrective maintenance.

3. Types of notations

a. Warnings are posted immediately prior to text covering any area that would present a situation that may result in injury or death. Compliance is mandatory.

b. Cautions will be found on the same page and preceding the text covering any area that would present a situation that may result in damage to equipment.

c. Notes will precede text covering an area with the intent to alter normal procedures for unique situations or equipment, or point out areas of special concern.

4. Appendices have been provided to:

a. Basic Issue Items (BII) list – specify the minimum essential items to place and maintain the M939, M939A1, and M939A2 (M939/A1/A2) series vehicles in operation, and are maintained with the vehicle.

b. Additional Authorization List (AAL) – specify items not shipped with vehicles, but are authorized for support of the M939/A1/A2 series vehicles.

c. Expendable/Durable Supplies and Materials List – specify expendable/durable supplies and materials you will need to operate and maintain M939/A1/A2 series vehicles.

CHAPTER 1
INTRODUCTION

Section I. GENERAL INFORMATION

M923A2 AND M925A2

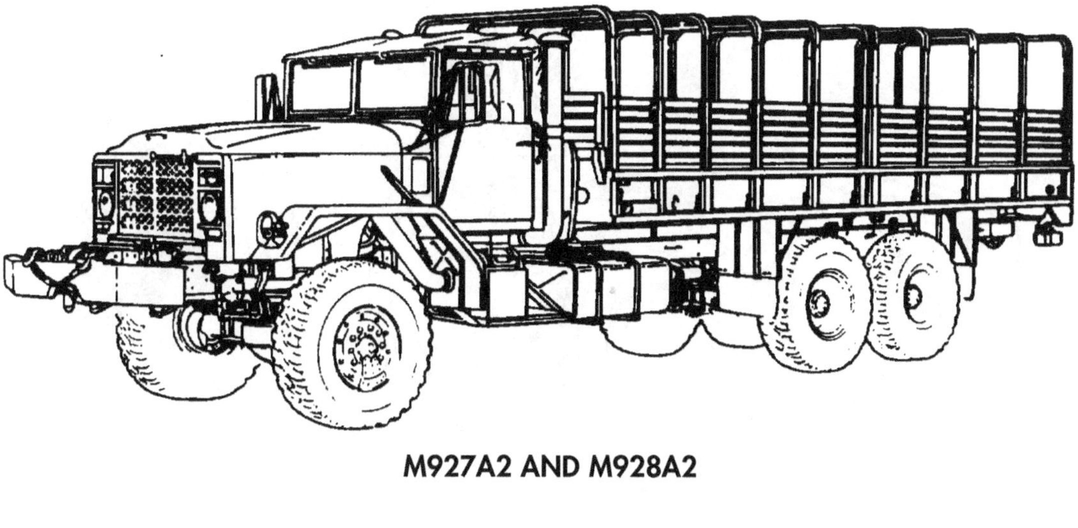

M927A2 AND M928A2

M929A2 AND M930A2

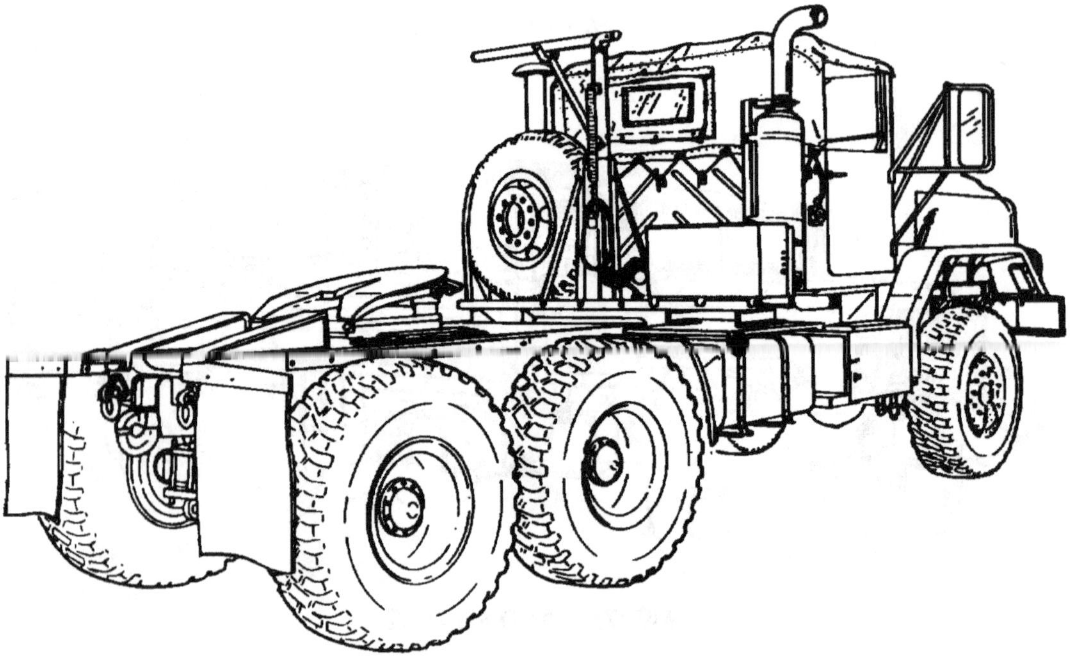

M931A2 AND M932A2

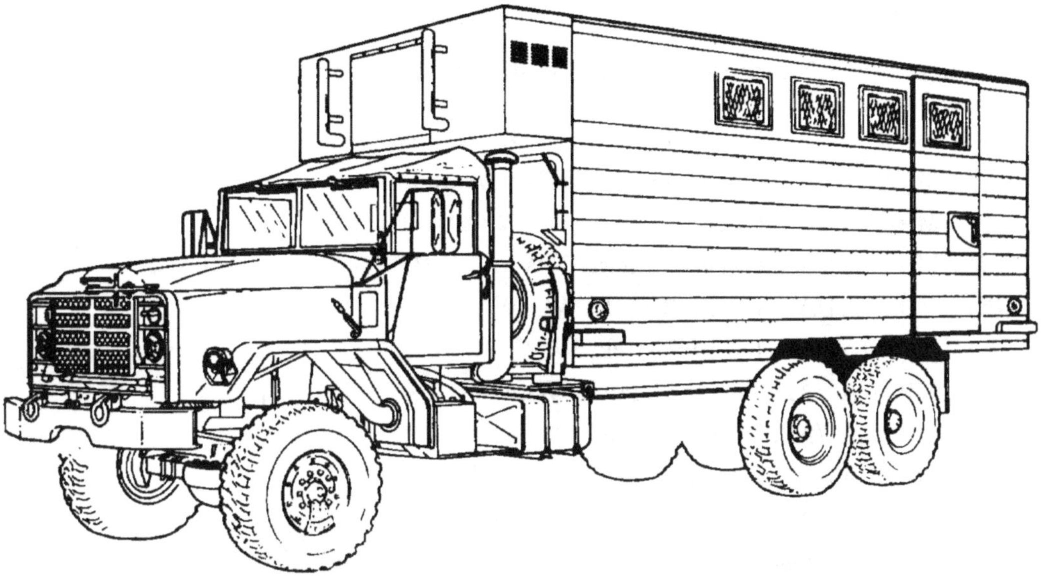

M934A2

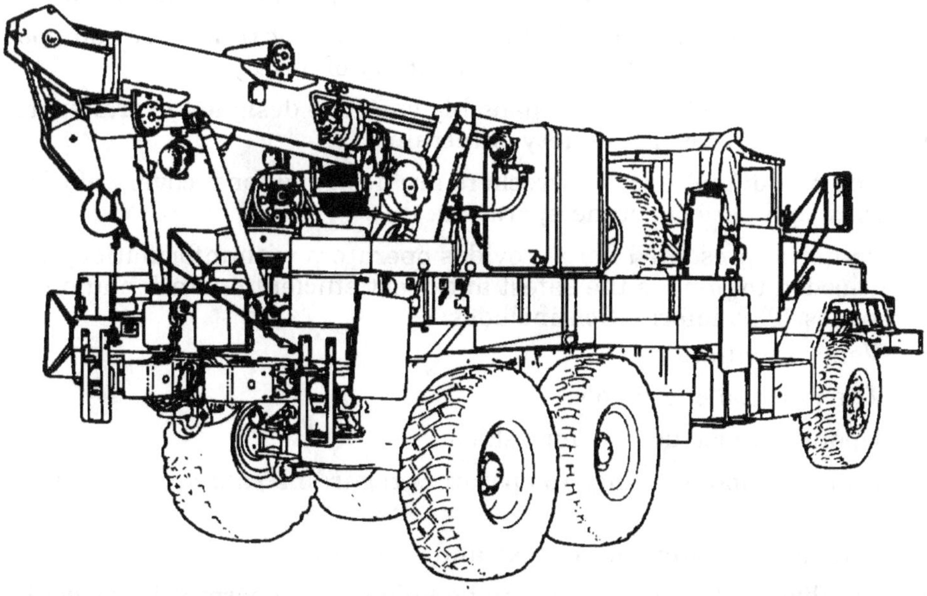

M936A2

1-1. SCOPE

a. This operator's manual contains instructions for operating and servicing the following M939/A1/A2 series vehicles:

(1) M923/A1/A2, Cargo Truck, WO/W (Dropside)

(2) M925/A1/A2, Cargo Truck, W/W (Dropside)

(3) M927/A1/A2, Cargo Truck, WO/W (XLWB)

(4) M928/A1/A2, Cargo Truck, W/W (XLWB)

(5) M929/A1/A2, Dump Truck, WO/W

(6) M930/A1/A2, Dump Truck, W/W

(7) M931/A1/A2, Tractor Truck, WO/W

(8) M932/A1/A2, Tractor Truck, W/W

(9) M934/A1/A2, Expansible Van, WO/W

(10) M936/A1/A2, Medium Wrecker, W/W

b. Vehicles' purpose.

(1) The M923/A1/A2, M925/A1/A2, M927/A1/A2, and M928/A1/A2 series cargo trucks provide transportation of personnel or equipment over a variety of terrain and climate conditions.

(2) The M929/A1/A2 and M930/A1/A2 series dump trucks are used to transport various materials over a variety of terrains. Each vehicle can be equipped with troop seat, and tarpaulin and bow kits for troop transport operations.

(3) The M931/A1/A2 and M932/A1/A2 series tractor trucks are equipped with a fifth wheel used to haul a semitrailer over a variety of terrain.

(4) The M934/A1/A2 series expansible vans are designed to transport electronic base stations over a variety of terrain.

(5) The M936/A1/A2 series wreckers are designed for recovery of disabled or mired vehicles, and perform crane operation.

c. The material presented here provides operators with information and procedures needed to provide the safest and most efficient operation and servicing of these vehicles. This information includes:

(1) Vehicle limitations.

(2) The function of controls.

(3) Operation instructions for vehicle.

(4) Cautions and warnings to operators regarding safety to personnel and equipment.

(5) Operator maintenance checks and services.

(6) Troubleshooting procedures to be followed by operator if the vehicle malfunctions.

(7) Operator forms and records.

1-2. MAINTENANCE FORMS

a. Vehicle Maintenance Forms and Records. Department of the Army forms and procedures used for equipment maintenance will be those prescribed by DA Pam 738-750 as contained in the maintenance management update.

b. Hand Receipt Manual This manual has a companion document with a TM number followed by -HR (which stands for Hand Receipt). The TM 9-2320-272-10-HR consists of preprinted hand receipts (DA Form 2062) that list end item related equipment (i.e., COEI, BII, and AAL) you must account for. As an aid to property accountability, additional -HR manuals may be requisitioned from the following source in accordance with procedures in chapter 3, AR 310-2:

The U.S. Army Adjutant General Publications Center
2800 Eastern Blvd.
Baltimore, MD 21220

1-3. CORROSION PREVENTION AND CONTROL (CPC)

Corrosion Preventive and Control (CPC) of Army materiel is a continuing concern. It is important that any corrosion problems with this item be reported so that the problems can be corrected and improvements can be made to prevent problems in the future items.

While corrosion is typically associated with rusting of metals, it can also include deterioration of other materials, such as rubber and plastic. Unusual cracking, softening, swelling, or breaking of these materials may be a corrosion problem.

If a corrosion problem is identified, it can be reported using Standard Form 368, Product Quality Deficiency Report. Use of key words such as "corrosion", "rust", "deterioration", or "cracking" will ensure that the information is identified as a CPC problem.

The form should be submitted to the address specified in DA Pam 738-750.

1-4. DESTRUCTION OF ARMY MATERIEL TO PREVENT ENEMY USE

Procedures for destruction of Army materiel to prevent enemy use can be found in TM 750-244-6.

1-5. REPORTING EQUIPMENT IMPROVEMENT RECOMMENDATIONS (EIR'S)

If your vehicle needs improvement, let us know. Send us an EIR. You, the user, are the only one who can tell us what you don't like about your equipment. Let us know why you don't like the design or performance. The preferred method for submitting QDRs is through the Army Electronic Product Support (AEPS) website under the Electronic Deficiency Reporting System (EDRS). The web address is: https://aeps.ria.army.mil. This is a secured site requiring a password which can be applied for on the front page of the website. If the above method is not available to you, put it on an SF 368, Product Quality Deficiency Report (PQDR), and mail it to us at: U.S. Army Tank-automotive and Armaments Command, ATTN: AMSTA-TR-E/PQDR MS 267, 6501 E. 11 Mile Road, Warren, MI 48397-5000. We'll send you a reply.

1-6. WARRANTY INFORMATION

Deleted

1-7. NOMENCLATURE CROSS-REFERENCE LIST

The following is an alphabetical list of commonly used military terms that appear in this manual. This list is cross-referenced to commonly understood terms used in everyday speech that mean the same thing.

Engine Coolant	antifreeze/water
Exhaust Stack	tailpipe
Failsafe Unit	warning buzzer
Fording	crossing through water
Grade	steepness of hill
Hydraulics	operated by oil pressure
Inclement Weather	bad weather (rain, snow, high winds)
Indicators	gauges, warning lights, etc.
Mired	stuck in mud or snow
Operation	task
Operator	driver
Slaving	jump starting
Splash Shields	mud flaps
Transport	to carry
Turning Radius	distance needed to make a U-turn
Usual Conditions	good roads, good weather

1-8. LIST OF ABBREVIATIONS

All abbreviations that appear in this manual are listed below.

AAL	Additional Authorization List
ABS	Antilock Brake System
AC	Alternating Current
BII	Basic Issue Items
B.O.	blackout
BRT	bright
CC	cross-country
°C	degree Celsius
cm	centimeter
Contd	continued
CPC	Corrosion Prevention Control
CTIS	Central Tire Inflation System
cu ft	cubic feet
CW	chain (and) wire rope (lubricating oil)
DA	Department of Army
DC	Direct Current
DFA	diesel fuel (arctic)

1-8. LIST OF ABBREVIATIONS (Contd)

drv	drive
ECU	Electronic Control Unit
EIR	Equipment Improvement Recommendation
emer	emergency
°F	degree Fahrenheit
ft	feet/foot
g	gram
GAA	grease, automotive, and artillery
gal.	gallon
GO	gear oil
GOS	gear oil (sub-zero)
Hwy	highway
in.	inch
kg	kilogram
km	kilometer
km/h	kilometers per hour
kPa	kilopascal
l	liter
lb	pound
lb-ft	pound-feet
lg	long
LO	Lubrication Order
m	meter
MAC	Maintenance Allocation Chart
mi	mile
mpg	miles per gallon
mph	miles per hour
N•m	Newton meter
NBC	Nuclear, Biological, or Chemical
NSN	National Stock Number
OE/HDO	oil, engine/heavy duty oil
OEA	oil, engine (arctic)
oz	ounce
para.	paragraph
pg	page
PMCS	Preventive Maintenance Checks and Services
pr	pair
psi	pounds per square inch
pt	pint
PTO	power takeoff
qt	quart
rpm	revolutions (turns) per minute
TM	technical manual
w/hlg	with hydraulic liftgate
w/o	without
wo/w	without winch
w	with
w/w	with winch
xlwb	extra long wheelbase
yd	yard

1-9. GLOSSARY

The following list shows definitions of military terms that appear in this manual. Other terms in this manual are defined in the paragraph where they first appear.

Angle of Approach – Angle between front tires and front bumper

Angle of Departure – Angle between rear tires and rear bumper

Fording – Crossing through water

Grade – Steepness of terrain

Hydraulic – Operated by oil pressure

Operator – Driver of vehicle

Tarpaulin – Canvas cover

Slaving – Jump starting

6X6 – Each vehicle has six axle ends and all six are capable of driving

Torque – Measurable force or power required to do work

Torqued – Requirement for a specified force to be applied to ensure proper seating or seal

Section II. EQUIPMENT DESCRIPTION AND DATA

1-10. EQUIPMENT DESCRIPTION INDEX

1-11. EQUIPMENT CHARACTERISTICS, CAPABILITIES, AND FEATURES

a. The 5-ton, 6x6, M939, M939A1, and M939A2 series vehicles are designed for use on all types of roads, highways, and cross-country terrain. These vehicles also operate in extreme temperatures such as arctic weather conditions.

b. The M939 series vehicles are an improved version of the M809 series. The improvements make M939 series vehicles more reliable and easier to operate. The major improvements are:

WARNING

This vehicle has been designed to operate safely and efficiently within the limits specified in this TM. Operation beyond these limits is prohibited IAW AR 70-1 without written approval from the Commander, U.S. Army Tank-automotive and Armaments Command, ATTN: AMCPEO-CM-S, Warren, MI 48397-5000.

(1) Automatic Transmission

(2) Improved Power Steering System

(3) Complete Airbrake System

(4) Improved Cooling System

(5) Improved Electrical System

(6) Three-Crew Member Cab

(7) Tilt Hood

(8) Hydraulically Powered Front Winch

c. The M939 series vehicles use 11:00 x R20 tires and have rear tandem duals, while the M939A1 and M939A2 series vehicles use 14:00 x R20 super sized tires and rear tandem singles. The mounted tires and spare on each vehicle are non-directional in design and use.

d. The M939 and M939A1 series vehicles employ the Cummings (NHC 250) 250 horsepowered engine, while the M939A2 series vehicles use a smaller, turbocharged Cummings (6CTA8.3) 240 horsepowered engine. The M939A2 series vehicles additionally have the Central Tires Inflation System (CTIS) which allows for greater tactical mobility.

e. All M939/A1/A2 series vehicles utilize the same automatic transmission, are equipped with a spare tire mount at the rear of the cab, have removable canvas cab tops, and are supplied with pintle hooks and air connections used for towing.

f. The 5-ton load limit rating of M939/A1/A2 series vehicles does not mean these vehicles are limited to 5-ton payloads. A vehicle rating only indicates the maximum amount of cargo weight the vehicle axles and frame can withstand when operating under the worst cross-country conditions.

1-12. LOCATION AND DESCRIPTION OF MAJOR COMPONENTS

a. This paragraph contains information regarding the major components that makeup the specific models of the M939, M939A1, and M939A2 series vehicles. These are:

(1) Chassis

(2) Cab

(3) Body

(4) Engine

(5) Fuel Tank

b. Chassis Types:

(1) M927/A1/A2, M928/A1/A2, and M934/A1/A2 model vehicles utilize the 215 in. (546.1 cm) Extra Long Wheelbase (XLWB) chassis. This facilitates transporting of large and awkward loads.

(2) M923/A1/A2, M925/A1/A2, and M936/A1/A2 model vehicles utilize the 179 in. (454.7 cm) wheelbase.

(3) M929/A1/A2, M930/A1/A2, M931/A1/A2, and M932/A1/A2 model vehicles utilize the 167 in. (424.2 cm) wheelbase.

c. Cab Assembly: all models use the same cap assembly even though minor changes are made to accommodate options for specific models.

d. Body Assembly:

(1) M923/A1/A2 and M925/A1/A2 model vehicles use the same cargo body.

(2) M927/A1/A2 and M928/A1/A2 model vehicles uses the same extended cargo body.

(3) M929/A1/A2 and M930/A1/A2 model vehicles use the dump body.

(4) M931/A1/A2 and M932/A1/A2 model vehicles use the fifth wheel approach plate and check plate.

(5) M934 and M939/A1/A2 model vehicles use the van body.

(6) M936 and M939/A1/A2 model vehicles use the crane and body assembly.

e. Engine Type: All M939/A1 series vehicles employ the Cummins – NHC 250 engine, while M939A2 series vehicles use the Cummins – 6CTA8.3 engine.

f. Fuel Tank Types:

(1) Initial issue tanks used on M939/A1 series vehicles are top fill, but replacement tanks may be substituted with M939A2 series vehicle side fill fuel.

(2) Fuel tank quantities and capacities differ between models. Refer to table 1-6 for specific information on your model.

1-12. LOCATION AND DESCRIPTION OF MAJOR COMPONENTS (Contd)

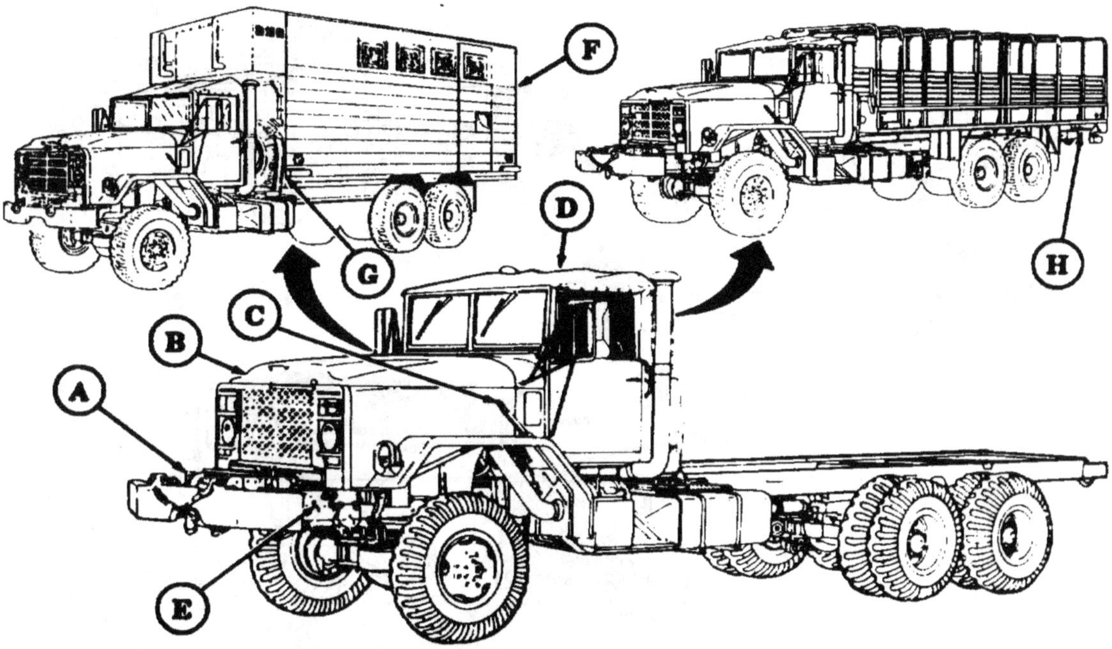
F
D
G
H
C
B
A
E

215 IN. (546.1 CM) EXTRA LONG WHEELBASE (XLWB) CHASSIS

(A) **FRONT WINCH** – Used on M925/A1/A2, M928/A1/A2, M930/A1/A2, M932/A1/A2, and M936/A1/A2 model vehicles to recover mired vehicles and in conjunction with A-frame kit for lifting operations.

(B) **TILT HOOD** – Tilts forward and can be locked open to gain access to engine components.

(C) **HOOD LATCHES** – Hold hood down when closed.

(D) **CANVAS CAB ROOF** – Can be folded down to reduce overall height and facilitate use of machine gun mount kit.

(E) **WINCH FRAME EXTENSION** – Used on winch models to extend frame for mounting of winch.

(F) **EXPANSIBLE VAN BODY** – Is designed with hardware and electrical attachments to facilitate electronic equipment operation or maintenance.

(G) **SPARE TIRE DAVIT** – Used on M934A1/A2 series vehicles for lifting and lowering spare tire. M934 models use a lifting eye located above spare tire.

(H) **EXTENDED LONG CARGO BODY** – Is a fixed side bed designed to carry large awkward loads which will not fit in standard beds.

1-12. LOCATION AND DESCRIPTION OF MAJOR COMPONENTS (Contd)

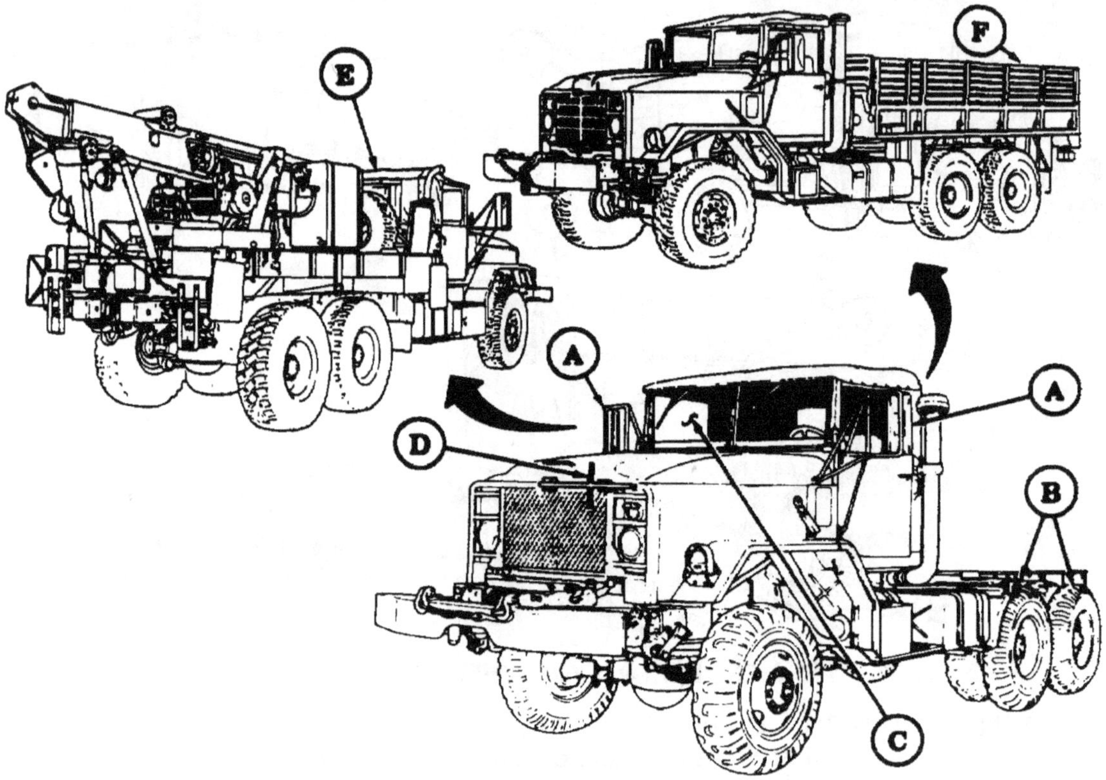
E
F
A
A
D
B
C

179 IN. (454.7 CM) WHEELBASE CHASSIS

(A) **REAR VIEW MIRROR** – Provides wide angle rear view of both right and left side, and rear sides of vehicle.

(B) **REAR BOGIE** – Consists of two axles on M939/A1/A2 series vehicles. M939 series vehicles utilize 11:00R20 tires (four per axle), and M939A1/A2 series vehicles utilize 14:00R20 tires (two per axle).

(C) **FOLDDOWN WINDSHIELD** – Allows for reduction in overall height of vehicle.

(D) **HOOD TILTING HANDLE AND LATCH** – Used to pull on top front of hood to tilt and latch it in a secure position.

(E) **CRANE AND BODY ASSEMBLY** – Used on M936/A1/A2 model vehicles. Used for recovering disabled and mired vehicles, and lifting operations.

(F) **DROPSIDE CARGO BODY** – Provides unobstructed access to side for loading with fork lift (M923/A1/A2 and M925/A1/A2 model vehicles).

1-12. LOCATION AND DESCRIPTION OF MAJOR COMPONENTS (Contd)

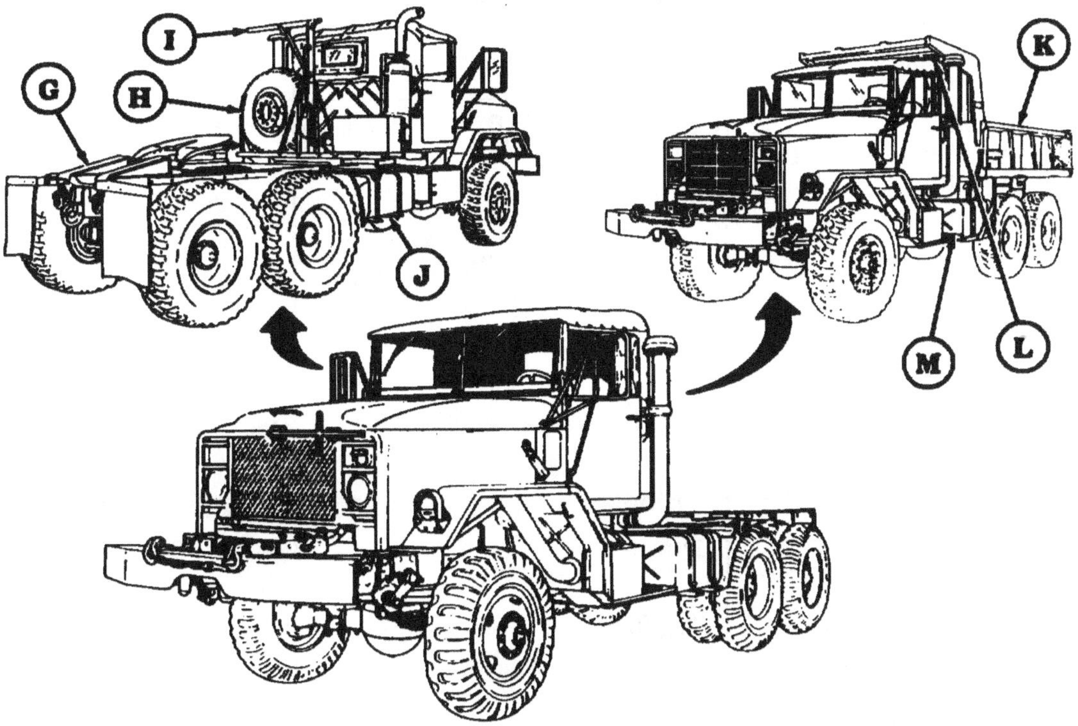

I
G
H
J
K
M
L

167 IN. (424.2 CM) WHEELBASE CHASSIS

(G) **FIFTH WHEEL APPROACH PLATE AND DECK PLATE** – Used on M931/A1/A2 and M932/A1/A2 model vehicles. Provides mechanical connection between semitractor and semitrailer.

(H) **SPARE TIRE AND MOUNTING BRACKET** – Provides storage location for spare tire.

(I) **SPARE TIRE DAVIT** – Used for lifting and lowering tire. Use on M923/A1/A2, M925/A1/A2, M927/A1/A2, M928/A1/A2, M931/A1/A2, and M932/A1/A2 model vehicles.

(J) **FUEL TANK** – Second tank for vehicles equipped with dual tank capacity.

(K) **DUMP BODY** – Used on M929/A1/A2 and M930/A1/A2 model vehicles for carrying various loads over different terrains.

(L) **SPARE TIRE LIFTING EYE** – Used for lifting and lowering spare tire on M929/A1/A2 and M930/A1/A2 model vehicles.

(M) **FUEL TANK** – Used on all vehicles for storage of fuel.

1-13. DIFFERENCE BETWEEN MODELS

Table 1-1. Differences Between Models.

Vehicle	Equipment/Function	Description (para/page)
M923A2,M925A2,M927A2, M928A2,M929A2,M930A2, M931A2,M932A2,M934A2, M936A2	**Central Tire Inflation System**	(para 1-20/page 1-57)
	Body Type	
M923,M923A1,M923A2, M925,M925A1,M925A2	Cargo Dropside	(para 2-23/page 2-124)
M927,M927A1,M927A2, M928,M928A1,M928A2	Cargo Fixed side (XLWB)	(para 2-23/page 2-124)
M936,M936A1,M936A2	Crane	(para 2-24/page 2-127)
M929,M929A1,M929A2, M930,M930A1,M930A2	Dump	(para 2-25/page 2-146)
M931,M931A1,M931A2, M932,M932A1,M932A2	Tractor	(para 2-26/page 2-152)
M934,M934A1,M934A2	Van	(para 2-27/page 2-158)
M936,M936A1,M936A2	**Floodlights**	(para 2-24/page 2-127)
	Fuel Tanks	
M929,M929A1,M929A2, M930,M930A1,M930A2, M931,M931A1,M931A2, M932,M932A1,M932A2	Dual Tanks (116 gal) (439.1 L)	(para 1-12/page 1-10)
M936,M936A1,M936A2	Dual Tanks (139 gal) (526.1 L)	(para 1-12/page 1-10)
M923,M923A1,M923A2, M925,M925A1,M925A2, M927,M927A1,M927A2, M928,M928A1,M928A2, M934,M934A1,M934A2	Single Tank (81 gal) (306.6 L)	(para 1-12/page 1-10)
M934,M934A1,M934A2	**Heat/Air Conditioned Body**	(para 2-27f/page 2-164) (para 2-27g/page 2-166)

1-13. DIFFERENCE BETWEEN MODELS (Contd)

Table 1-1. Differences Between Models (Contd).

Vehicle	Equipment/Function	Description (para/page)
	Operations	
M934,M934A1,M934A2	Communications/Electronic Repair	(para 2-27/page 2-158)
M929,M929A1,M929A2, M930,M930A1,M930A2	Dump	(para 2-25/page 2-146)
M931,M931A1,M931A2, M932,M932A1,M932A2	Fifth Wheel	(para 2-26/page 2-152)
M923,M923A1,M923A2, M925,M925A1,M925A2, M927,M927A1,M927A2, M928,M928A1,M928A2, M929,M929A1,M929A2, M930,M930A1,M930A2	Personnel/Cargo	(para 2-23/page 2-124)
M936,M936A1,M936A2	Wrecker	(para 2-41/page 2-127)
	Wheelbases	
M929,M929A1,M929A2, M930,M930A1,M930A2, M931,M931A1,M931A2, M932,M932A1,M932A2	167 in. (424.2 cm)	(para 1-12/page 1-13)
M923,M923A1,M923A2, M925,M925A1,M925A2, M936,M936A1,M936A2	179 in. (454.7 cm)	(para 1-12/page 1-12)
M927,M927A1,M927A2, M928,M928A1,M928A2, M934,M934A1,M934A2	215 in. (546.1 cm)	(para 1-12/page 1-11)
	Winch	
M925,M925A1,M925A2, M928,M928A1,M928A2, M930,M930A1,M930A2, M932,M932A1,M932A2, M936,M936A1,M936A2	Front	(para 2-22/page 2-116)
M936,M936A1,M936A2	Rear	(para 2-24c/page 2-127)

1-14. EQUIPMENT DATA

a. General. This paragraph organizes vehicle specifications, special equipment, and model differences in table form for easy reference by operators.

b. Specifications.

(1) **Winch and Crane Data.** See table 1-2.

(2) **Vehicle Dimensions.** See table 1-3.

(3) **Weights.** See table 1-4.

(4) **Chassis Dimensions.** See table 1-5.

(5) **Capacities for Normal Operating Conditions.** See table 1-6.

(6) **General Service Data.** See table 1-7.

(7) **Engine and Cooling System Data.** See table 1-8.

(8) **Automatic Transmission Data.** See table 1-9.

(9) **Tire Inflation Data.** See table 1-10.

(10) **Shipping Dimensions.** See table 1-11.

Table 1-2. Winch and Crane Data.

Vehicle	Description	Capacities Standard	Capacities Metric	Ref. Para
M925, M925A1, M925A2, M928, M928A1, M928A2, M930, M930A1, M930A2, M932, M932A1, M932A2	Front Winch: Max. Load Cable Length	 20,000 lb 200 ft	 9,080 kg 61 m	(2-22)
M936, M936A1, M936A2	Front Winch: Max. Load Cable Length	 20,000 lb 280 ft	 9,080 kg 85.4 m	(2-24b)
M936, M936A1, M936A2	Rear Winch: Max. Load Cable Length	 45,000 lb 350 ft	 20,430 kg 106.8 m	(2-24c)
M936, M936A1, M936A2	Crane: Max. Load (w/boom jacks) Rotation Retracted Length: Extended Length: Cable Length:	 20,000 lb 360° 10 ft 18 ft 95 ft 5 in.	 9,080 kg 3.05 m 5.5 m 29.1 m	(2-24d)

Table 1-3. Vehicle Dimensions.

Vehicle	Length		Height		Width		Reducible Height	
	in.	cm	in.	cm	in.	cm	in.	cm
M923	307.2	780.3	118.3	300.5	97.5	247.7	91.2	231.6
M923A1	310.5	788.7	121.0	307.3	97.4	247.4	93.9	238.5
M923A2	310.5	788.7	121.0	307.3	97.4	247.4	93.9	238.5
M925	328.7	834.9	118.3	300.5	97.5	247.7	91.2	231.6
M925A1	332.0	843.3	121.0	307.3	97.4	247.4	93.9	238.5
M925A2	332.0	843.3	121.0	307.3	97.4	247.4	93.9	238.5
M927	383.2	973.3	118.1	300.0	97.5	247.7	91.0	231.1
M927A1	385.5	979.2	120.6	306.3	97.4	247.4	93.5	237.5
M927A2	385.5	979.2	120.6	306.3	97.4	247.4	93.5	237.5
M928	404.7	1027.9	118.1	300.0	97.5	247.7	91.0	231.1
M928A1	408.0	1036.3	120.6	306.3	97.4	247.4	93.5	237.5
M928A2	408.0	1036.3	120.6	306.3	97.4	247.4	93.5	237.5
M929	273.0	693.4	120.8	306.8	97.5	247.7	90.3	229.4
M929A1	273.0	693.4	125.0	317.5	97.4	247.4	93.5	237.5
M929A2	273.0	693.4	125.0	317.5	97.4	247.4	93.5	237.5
M930	294.5	748.0	120.8	306.8	97.5	247.7	90.3	229.4
M930A1	294.5	748.0	125.0	317.5	97.4	247.4	93.5	237.5
M930A2	294.5	748.0	125.0	317.5	97.4	247.4	93.5	237.5
M931	264.5	671.8	118.5	301.0	97.5	247.7	91.4	232.2
M931A1	264.5	671.8	121.2	307.8	97.4	247.4	94.1	239.0
M931A2	264.5	671.8	121.2	307.8	97.4	247.4	94.1	239.0
M932	286.0	726.4	118.5	301.0	97.5	247.7	91.4	232.2
M932A1	286.0	726.4	121.2	307.8	97.4	247.4	94.1	239.0
M932A2	286.0	726.4	121.2	307.8	97.4	247.4	94.1	239.0
M934	362.6	921.0	138.0	350.5	98.0	248.9	138.0	350.5
M934A1	362.6	921.0	142.3	361.4	98.0	248.9	142.3	361.4
M934A2	362.6	921.0	142.3	361.4	98.0	248.9	142.3	361.4
M936	362.0	919.5	117.6	298.7	97.5	247.7	114.7	291.3
M936A1	362.0	919.5	120.0	304.8	97.4	247.4	108.5	275.6
M936A2	362.0	919.5	120.0	304.8	97.4	247.4	108.5	275.6

Model	Ground Clearance			
	Under Axle		Under Chassis	
	in.	cm	in.	cm
M939	11.5	29.2	10.5	26.7
M939A1	13.9	35.3	13.1	33.3
M939A2	13.9	35.3	13.1	33.3

Table 1-4. Weights.

Vehicle	Empty		Payload		Towed Load (Pintle)	
	lbs	kg	lbs	kg	lbs	kg
M923	21,600	9,806	10,000	4,540	15,000	6,810
M923A1	22,175	10,067	10,000	4,540	15,000	6,810
M923A2	20,930	9,502	10,000	4,540	15,000	6,810
M925	22,360	10,151	10,000	4,540	15,000	6,810
M925A1	23,275	10,567	10,000	4,540	15,000	6,810
M925A2	22,030	10,002	10,000	4,540	15,000	6,810
M927	27,749	12,598	10,000	4,540	15,000	6,810
M927A1	25,035	11,366	10,000	4,540	15,000	6,810
M927A2	23,790	10,801	10,000	4,540	15,000	6,810
M928	27,811	12,626	10,000	4,540	15,000	6,810
M928A1	26,135	11,865	10,000	4,540	15,000	6,810
M928A2	24,890	11,300	10,000	4,540	15,000	6,810
M929	25,888	11,753	10,000	4,540	15,000	6,810
M929A1	25,065	11,380	10,000	4,540	15,000	6,810
M929A2	23,820	10,814	10,000	4,540	15,000	6,810
M930	26,624	12,087	10,000	4,540	15,000	6,810
M930A1	26,165	11,879	10,000	4,540	15,000	6,810
M930A2	24,920	11,314	10,000	4,540	15,000	6,810
M931*	22,089	10,028	15,000	6,810	15,000	6,810
M931A1*	21,140	9,598	15,000	6,810	15,000	6,810
M931A2*	19,895	9,032	15,000	6,810	15,000	6,810
M932*	22,841	10,370	15,000	6,810	15,000	6,810
M932A1*	22,242	10,098	15,000	6,810	15,000	6,810
M932A2*	20,995	9,532	15,000	6,810	15,000	6,810
M934	29,946	13,595	5,000	2,270	15,000	6,810
M934A1	29,280	13,293	5,000	2,270	15,000	6,810
M934A2	28,035	12,728	5,000	2,270	15,000	6,810
M936**	39,334	17,858	7,000	3,178	20,000	9,080
M936A1**	38,155	17,322	7,000	3,178	20,000	9,080
M936A2**	36,910	16,757	7,000	3,178	20,000	9,080

* Loaded trailer weight on fifth wheel is 15,000 lb (6,810 kg); total semitrailer weight with payload is 37,500 lb (17,025 kg).

** On crane w/boom shipper raised and secured.

Table 1-5. Chassis Dimensions.

Vehicle	Wheelbase		Chassis		Length		Turning Radius Angle (Degrees) of Approach Departure	
	in.	cm	in.	cm	ft	M		
M923	179	454.7	307.5	781.1	38.0	11.6	46	37
M923A1	179	454.7	307.5	781.1	40.8	12.4	46	38
M923A2	179	454.7	307.5	781.1	40.8	12.4	46	38
M925	179	454.7	326.5	829.3	39.0	11.9	31	37
M925A1	179	454.7	326.5	829.3	42.10	13.1	31	38
M925A2	179	454.7	326.5	829.3	42.10	13.1	31	38
M927	215	546.1	380.9	967.5	45.2	13.8	46	22.5
M927A1	215	546.1	380.9	967.5	47.2	14.4	46	21
M927A2	215	546.1	380.9	967.5	47.2	14.4	46	21
M928	215	546.1	402.3	1021.8	46.2	14.1	31	22.5
M928A1	215	546.1	402.3	1021.8	49.4	15.1	31	21
M928A2	215	546.1	402.3	1021.8	49.4	15.1	31	21
M929	167	424.2	256.9	652.5	35.6	10.8	46	60
M929A1	167	424.2	256.9	652.5	39.2	11.9	46	77
M929A2	167	424.2	256.9	652.5	39.2	11.9	46	77
M930	167	424.2	278.4	707.1	36.6	11.1	31	60
M930A1	167	424.2	278.4	707.1	41.4	12.6	31	77
M930A2	167	424.2	278.4	707.1	41.4	12.6	31	77
M931	167	424.2	265.0	673.1	35.6	10.8	46	68
M931A1	167	424.2	265.0	673.1	39.2	11.9	46	77
M931A2	167	424.2	265.0	673.1	39.2	11.9	46	77
M932	167	424.2	278.4	707.1	36.6	11.1	31	63
M932A1	167	424.2	278.4	707.1	41.4	12.6	31	77
M932A2	167	424.2	278.4	707.1	41.4	12.6	31	77
M934	215	546.1	360.0	914.4	45.2	13.8	46	24
M934A1	215	546.1	360.0	914.4	47.2	14.4	46	32
M934A2	215	546.1	360.0	914.4	47.2	14.4	46	32
M936	179	454.7	322.7	819.7	39.0	11.9	31	37
M936A1	179	454.7	322.7	819.7	42.10	13.1	31	35
M396A2	179	454.7	322.7	819.7	42.10	13.1	31	35

WARNING

Accidental or intentional introduction of liquid contaminants into the environment is in violation of state, federal, and military regulations. Refer to Lubrication Order (para. 3-1) for information concerning storage, use, and disposal of these liquids. Failure to do so may result in injury or death.

Table 1-6. Capacities for Normal Operating Conditions.

Vehicle	Description	Capacity		In Normal Operating Conditions +32°F to +90°F (0°C to +32°C)
		Standard	Metric	
All	Cooling System	47 qt	44.5 L	1/2 Ethylene Glycol, 1/2 Water
M939 & M939A1 series	Engine (crankcase only)	23 qt	21.8 L	OE/HDO 15/40
M939A2 series	Engine (crankcase only)	18 qt	17.0 L	OE/HDO 15/40
M939 & M939A1 series	Engine (crankcase with new filter)	27 qt	25.5 L	OE/HDO 15/40
M939A2 series	Engine (crankcase with new filter)	20 qt	18.9 L	OE/HDO 15/40
M923,M923A1,M923A2, M925,M925A1,M925A2, M927,M927A1,M927A2, M928,M928A1,M928A2, M934,M934A1,M934A2	Fuel Tank (single tank)	81 gal.	306.6 L	Diesel Fuel (grade DF1, DF2, DFA, or JP8)
M936,M936A1,M936A2	Fuel Tanks (dual tanks)	139 gal.	526.1 L	Diesel Fuel (grade DF1, DF2, DFA, or JP8)
M929,M929A1,M929A2, M930,M930A1,M930A2, M931,M931A1,M931A2, M932,M932A1,M932A2	Fuel Tanks (dual tanks)	116 gal.	439.1 L	Diesel Fuel (grade DF1, DF2, DFA, or JP8)
M925,M925A1,M925A2, M928,M928A1,M928A2, M932,M932A1,M932A2	Hydraulic System	8 gal.	30.3 L	OE/HDO 10
M929,M929A1,M929A2	Hydraulic System	5 gal.	18.9 L	OE/HDO 10
M930,M930A1,M930A2	Hydraulic System	6.25 gal.	23.7 L	OE/HDO 10
M936,M936A1,M936A2	Hydraulic System	100 gal.	378.5 L	OE/HDO 10

Table 1-6 (Cont'd). Capacities for Normal Operating Conditions.

Vehicle	Description	Capacity		In Normal Operating Conditions +32°F to +90°F (0°C to +32°C)
		Standard	Metric	
All	Differentials (each)	12 qt	11.3 L	GO 80/90
M939/A1 Series	Steering System	5 qt	4.7 L	OE/HDO 10
M939/A2 Series	Steering System	3 qt	2.83 L	OE/HDO 10
All	Transmission (drain and refill)	17 qt	16.1 L	OE/HDO 15/40
	(W/PTO)	17 qt	16.1 L	OE/HDO 15/40
All	Transmission (dry)	23 qt	22.1 L	OE/HDO 15/40
	(W/PTO)	25 qt	23.7 L	OE/HDO 15/40
All	Transfer Case	6.5 qt	6.1 L	GO 80/90
All W/Front Winch	Winch Gearcase (front winch)	2.6 pt	1.2 L	GO 80/90
M936,M936A1, M936A2	Winch Gearcase (rear winch)	7 pt	3.3 L	GO 80/90

Table 1-7. General Service Data.

Vehicle	Description	Above +15°F (Above -9°C)	+40° to -15°F (+4° to -26°C)	+40° to -65°F (+4° to -54°C)	Arctic Conditions
All	Cooling System	1/4 Ethylene Glycol, 3/4 Water	2/5 Ethylene Glycol, 3/5 Water	3/5 Ethylene Glycol, 2/5 Water	Refer to FM 9-207
All	Engine	OE/HDO 15/40	OE/HDO 15/40	OEA	
All	Fuel Tank(s)	DF1, DF2, DFA, or JP8	DF1, DFA, or JP8	DFA	
All	Hydraulic Systems	OE/HDO 10	OE/HDO 10	OEA	
All	Differentials	GO 80/90	GO 80/90	GO 75	
All	Steering System	OE/HDO 10	OE/HDO 10	OEA	
All	Transmission	(See Table 1-9)	(See Table 1-9)	OEA	
All	Transfer Case	GO 80/90	GO 80/90	GO 75	
All W/Winch	Winch Gearcase	GO 80/90	GO 80/90	GO 75	
All	Windshield Washer	1/3 Cleaning Compound, 2/3 Water	1/2 Cleaning Compound, 1/2 Water	2/3 Cleaning Compound, 1/3 Water	

Table 1-8. Engine and Cooling System Data.

ENGINE CUMMINGS NHC 250 (M939 AND M939A1 SERIES)	
Type	Diesel, naturally-aspirated, liquid cooled
Cylinders	6 (in-line)
Brake Horsepower	250 horsepower @ 2100 rpm
Idle Speed (engine rpm)	600-650 rpm
Operating Speed (engine rpm)	1500-2100 rpm
Oil Pressure at Idle (minimum)	15 psi (103 kPa)
Coolant (normal operating temperature)	175°-195°F (79°-91°C)
Fuel Consumption (approximate)	3-4 mpg (1.3-1.7 km/l)
COOLING SYSTEM	
Surge Tank Cap Pressure	14 psi (97 kPa)
Thermostat:	
Starts to Open	175°F (79°C)
Fully Open	185°F (85°C)
Radiator	Crossflow Type
Fan	26 in. (660 mm), 6-blade
ENGINE CUMMINS 6CTA8.3 (M939A2 SERIES)	
Type	Diesel, turbocharged, aftercooled
Cylinders	6 (in-line)
Brake Horsepower	240 horsepower @ 2100 rpm
Idle Speed (engine rpm)	565-635 rpm
Operating Speed (engine rpm)	2100 rpm
Oil Pressure at Idle (minimum)	10 psi (69 kPa)
Coolant (normal operating temperature)	190°-200°F (88°-93°C)
Fuel Consumption (approximate)	5.5-6.0 mpg (2.3-2.6 km/l)
COOLING SYSTEM	
Surge Tank Cap Pressure	14 psi (97 kPa)
Thermostat:	
Starts to Open	181°F (83°C)
Fully Open	203°F (95°C)
Radiator	Crossflow type
Fan	7 blade

Table 1-9. Automatic Transmission Data.

TRANSMISSION	
Oil Type:	
OE/HDO 10	-4°F to +55°F (-40°C to +13°C)
OE/HDO 15/40	+10°F to +110°F (-12°C to +43°C)
Dexron III	-15°F to +75°F(-26°C to +24°C)
OEA	-65°F to +40°F(-54°C to +4°C)
Oil Capacity:	
WO/PTO (drain and refill)	4.25 gal. (16.1 L)
W/PTO (drain and refill)	4.25 gal. (16.1 L)
WO/PTO (dry)	5.75 gal. (21.8 L)
W/PTO (dry)	6.25 gal. (23.7 L)
Oil Temperature:	
Maximum	300°F (149°C)
Normal Operating Temperature	120°-220°F (49°-104°C)
Power Takeoff	Converter driven

Table 1-9. Automatic Transmission Data (Contd).

TRANSMISSION DRIVING RANGE SELECTION					
Range Selection	Condition	Maximum Operating Speeds W/Transfer Case			
		M939 Series		M939A1 and M939A2 Series	
		In High	In Low	In High	In Low
R (reverse)	Easy grades clear of traffic with ground guide	5 mph (8 km/h)		5 mph (8 km/h)	
N (neutral)		—	—	—	—
1-5 (drive)	Good roads, grades, traffic conditions	55 mph (80 km/h)	22 mph (35 km/h)	55 mph (80 km/h)	26 mph (42 km/h)
1-4 (fourth)	Moderate grades, traffic restricted speed limits	43 mph (69 km/h)	17 mph (27 km/h)	50 mph (80 km/h)	20 mph (32 km/h)
1-3 (third)	Moderate grades, heavy traffic, restricted speed limits	33 mph (53 km/h)	13 mph (21 km/h)	38 mph (61 km/h)	16 mph (26 km/h)
1-2 (second)	Steep grades, heavy traffic, rough terrain	25 mph (40 km/h)	10 mph (16 km/h)	29 mph (47 km/h)	12 mph (19 km/h)
1 (first)	Starting heavy loads, extreme grades, rough terrain	12 mph (19 km/h)	5 mph (8 km/h)	15 mph (24 km/h)	6 mph (10 km/h)

Table 1-9A. Maximum Safe Operating Speeds.

Terrain	W/ABS	W/O ABS
Highway/secondary roads	55 mph (88 km/h)	40 mph (64 km/h)
Cross country/off-road	40 mph (64 km/h)	35 mph (56 km/h)
Sand/snow	25 mph (40 km/h)	25 mph (40 km/h)
Ice/road emergencies	5-12 mph (8-19 km/h)	5-12 mph (8-19 km/h)

Table 1-10. Tire Inflation Data.

	M923	M925	M927	M928	M929	M930	M931	M932	M934	M936
M939 SERIES (11:00 X 20 TIRE) (11:00 X R20 TIRE)										
HIGHWAY										
FRONT										
Standard (psi)	90	90	90	90	90	90	90	90	90	90
Metric (kPa)	621	621	621	621	621	621	621	621	621	621
REAR										
Standard (psi)	70	70	70	70	70	70	70	70	70	90
Metric (kPa)	483	483	483	483	483	483	483	483	483	483
CROSS-COUNTRY										
FRONT										
Standard (psi)	60	60	60	60	60	60	60	60	60	60
Metric (kPa)	414	414	414	414	414	414	414	414	414	414
REAR										
Standard (psi)	30	30	30	30	30	30	30	30	30	60
Metric (kPa)	207	207	207	207	207	207	207	207	207	414
MUD, SAND, AND SNOW										
FRONT and REAR										
Standard (psi)	25	25	25	25	25	25	25	25	25	25
Metric (kPa)	172	172	172	172	172	172	172	172	172	172

Table 1-10. Tire Inflation Data (Contd).

	M923	M925	M297	M928	M929	M930	M931	M932	M934	M936
NOTE For M939A2 vehicles, highway inflation levels pertain to extended highway use. Normal CTIS pressure can be used for other than extended highway use. Refer to PMCS Table 2-3 CTIS Tire Pressures Chart.										
M939A1/A2 SERIES (14:00 X R20)										
HIGHWAY										
FRONT										
Standard (psi)	70	70	70	70	70	70	70	70	70	90
Metric (kPa)	483	483	483	483	483	483	483	483	483	551
REAR										
Standard (psi)	70	70	70	70	70	70	70	70	70	90
Metric (kPa)	483	483	483	483	483	483	483	483	483	621
CROSS COUNTRY										
FRONT AND REAR										
Standard (psi)	35	35	35	35	35	35	35	35	35	35
Metric (kPa)	241	241	241	241	241	241	241	241	241	241
MUD, SAND, AND SNOW										
FRONT AND REAR										
Standard (psi)	25	25	25	25	25	25	25	25	25	25
Metric (kPa)	172	172	172	172	172	172	172	172	172	172
EMERGENCY										
FRONT AND REAR										
Standard (psi)	12	12	12	12	12	12	12	12	12	12
Metric (kPa)	83	83	83	83	83	83	83	83	83	83
ALL MODELS:										
SPARE										
inflate to maximum highway pressure	—	—	—	—	—	—	—	—	—	—

Table 1-11. Shipping Dimensions.

Vehicle	Shipping Height in.	Shipping Height cm	Shipping Weight lb	Shipping Weight kg	Shipping Cubage cu. ft.	Shipping Cubage cu. m.	Shipping Tonnage Tons	Shipping Tonnage Metric Ton
M923	91.2	231.6	21,600	9,806	1,581.0	447.4	10.8	9.8
M923A1	93.9	238.5	22,175	10,067	1,644.0	465.3	11.1	10.1
M923A2	93.9	238.5	20,930	9,502	1,644.0	465.3	10.5	9.5
M925	91.2	231.6	22,360	10,151	1,692.0	478.8	11.2	10.1
M925A1	93.9	238.5	23,275	10,567	1,758.0	497.5	11.6	10.6
M925A2	93.9	238.5	22,030	10,002	1,758.0	497.5	11.0	10.0
M927	91.0	231.1	27,749	12,598	1,970.0	557.5	13.9	12.6
M927A1	93.5	237.5	25,035	11,366	2,032.0	575.1	12.5	11.4
M927A2	93.5	237.5	23,790	10,801	2,032.0	575.1	11.9	10.8
M928	91.0	231.1	27,811	12,626	2,078.0	588.1	13.9	12.6
M928A1	93.5	237.5	26,135	11,865	2,151.0	608.7	13.1	11.9
M928A2	93.5	237.5	24,890	11,300	2,151.0	608.7	12.4	11.3
M929	90.3	229.4	25,888	11,753	1,391.0	393.7	12.9	11.7
M929A1	93.5	237.5	25,065	11,380	1,441.0	407.8	12.5	11.4
M929A2	93.5	237.5	23,820	10,814	1,441.0	407.8	11.9	10.8
M930	90.3	229.4	26,624	12,087	1,501.0	424.8	13.3	12.1
M930A1	93.5	237.5	26,165	11,879	1,552.0	439.2	13.1	11.9
M930A2	93.5	237.5	24,920	11,314	1,552.0	439.2	12.5	11.3
M931	91.4	232.2	22,089	10,028	1,364.0	386.0	11.0	10.0
M931A1	94.1	239.0	21,140	9,598	1,403.0	397.0	10.6	9.6
M931A2	94.1	239.0	19,895	9,032	1,403.0	397.0	9.9	9.0
M932	91.4	232.2	22,841	10,370	1,475.0	417.4	11.4	10.4
M932A1	94.1	239.0	22,242	10,098	1,517.0	429.3	11.1	10.1
M932A2	94.1	239.0	20,995	9,532	1,517.0	429.3	10.5	9.5
M934	138.0	350.5	29,946	13,595	2,838.0	803.2	15.0	13.6
M934A1	142.3	361.4	29,280	13,293	2,925.0	827.8	14.6	13.3
M934A2	142.3	361.4	28,035	12,728	2,925.0	827.8	14.0	12.7
M936	114.7	291.3	39,334	17,858	2,174.0	615.2	19.7	17.8
M936A1	108.5	275.6	38,155	17,322	2,203.0	623.4	19.1	17.3
M936A2	108.5	275.6	36,910	16,757	2,203.0	623.4	18.5	16.7

1-15. GENERAL

This section explains how components of the 5-ton M939/A1/A2 series vehicles work together. A functional description of these components and their related parts will be covered in the following paragraphs. To find the operation of a specific system or component, see the principles of operation reference index below.

1-16. PRINCIPLES OF OPERATION REFERENCE INDEX

1-17. CONTROL SYSTEM OPERATION (Contd)

The control system includes those controls and their related parts that are essential to the operation of the vehicle. These controls are common to all vehicles with the exception of the transmission and transfer case power takeoff controls. All originate from the cab. Each of these controls and their related parts will be described as part of the following systems:

a. Starting and Ether Starting System Operation (page 1-28).

b. Accelerator Controls System Operation (page 1-30).

c. Parking Brake System Operation (page 1-31).

d. Steering System Operation (page 1-32).

e. Transmission Control System Operation (page 1-33).

f. Transfer Case Control System Operation (page 1-34).

1-17. CONTROL SYSTEM OPERATION (Contd)

a. Starting and Ether Starting System Operation.

The starting system is identical on all models covered in this manual. It will start the engine in all types of weather and has built-in protection that prevents starting components from reengaging once the engine has been started. Major components of the starting and ether starting system are:

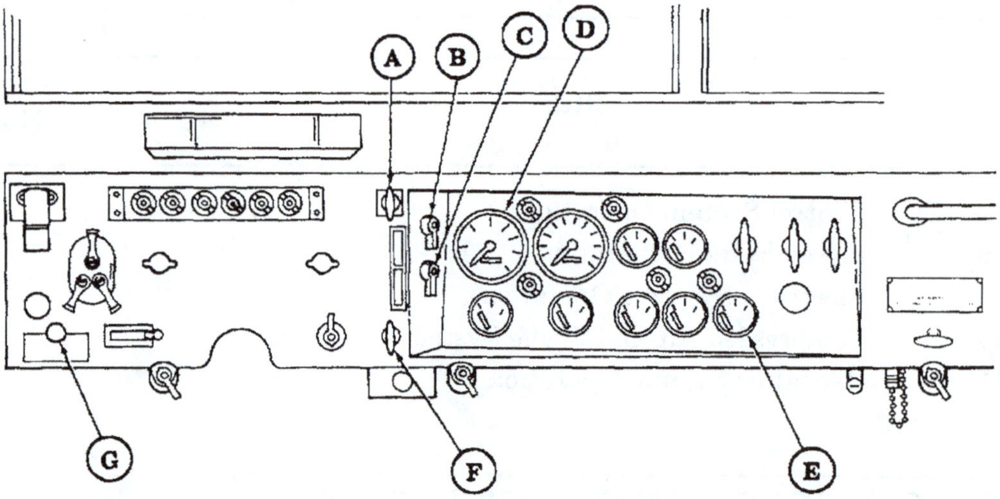
A
B
C
D
E
F
G

(A) **HAND THROTTLE CONTROL** – Used to set engine speed without applying pressure to the accelerator (rotated to lock).

(B) **BATTERY SWITCH** – Activates all electrical circuits except arctic heaters.

(C) **IGNITION SWITCH** – Has OFF, RUN, and START positions. Switch automatically returns from START to RUN when hand pressure is released.

(D) **TACHOMETER** – Used to indicate speed of engine.

(E) **VOLTMETER** – Indicates charging condition of the battery.

(F) **EMERGENCY ENGINE STOP** – Used to shut down engine during emergencies (M934/A1 series vehicles must be reset by unit maintenance).

(G) **ETHER START SWITCH** – Injects ether into engine for cold weather starting.

1-17. CONTROL SYSTEM OPERATION (Contd)

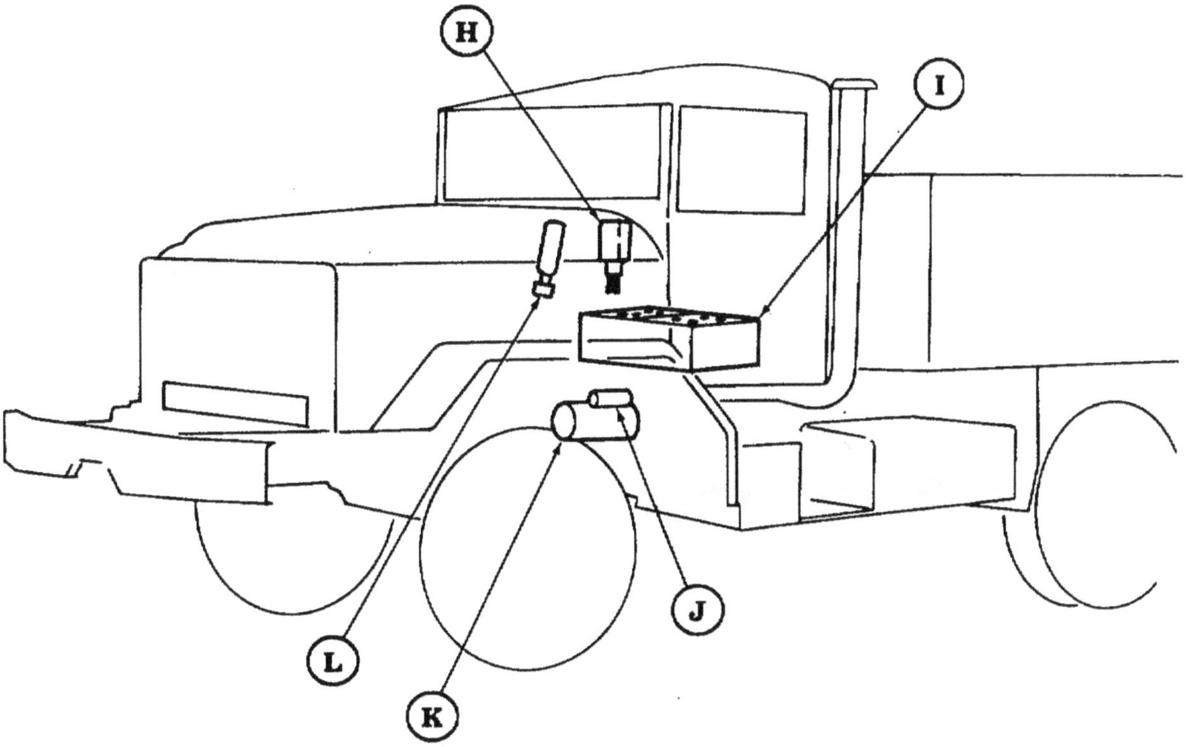
H
I
J
K
L

(H) **PROTECTIVE CONTROL BOX** – Prevents reengagement of starter motor once engine is running.

(I) **BATTERIES** – Provide 24-volt electrical current for energizing electrical circuits.

(J) **STARTER SOLENOID** – Relays 24-volt battery power to energize starter motor.

(K) **STARTER MOTOR** – When energized, it converts electrical energy to mechanical power as it engages the flywheel to crank engine.

(L) **ETHER START CYLINDER** – Stores ether used for cold weather starting.

1-17. CONTROL SYSTEM OPERATION (Contd)

b. Accelerator Controls System Operation.

The accelerator controls system permits the operator to control vehicle speed and engine power. It is identical on all models covered in this manual. Major components of the accelerator control system are:

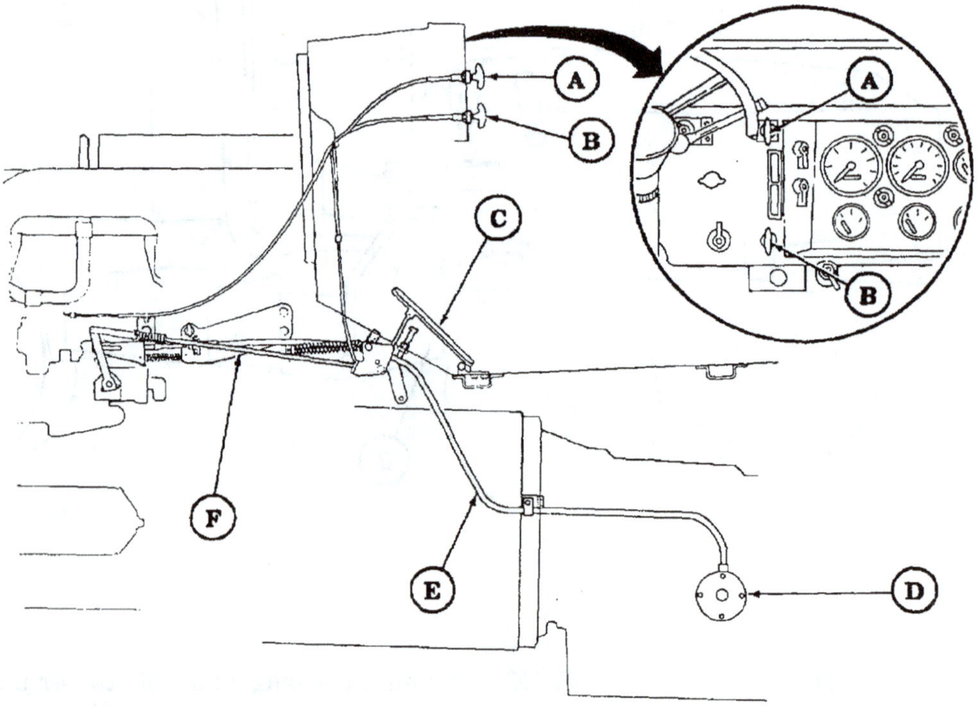
A
B
C
A
B
F
E
D

(A) **HAND THROTTLE CONTROL** – Sets engine speed at desired rpm without maintaining pressure on accelerator pedal.

(B) **EMERGENCY ENGINE STOP CONTROL** – Is pulled out to cut off fuel to engine. It is used only in an emergency.

(C) **ACCELERATOR PEDAL** – Controls engine speed.

(D) **MODULATOR** – With transmission selector lever in drive, modulator controls transmission upshifting and downshifting as engine rpm changes.

(E) **CABLE** – Connects modulator to fuel pump.

(F) **ACCELERATOR LINKAGE** – Links accelerator pedal and throttle control to fuel pump.

1-17. CONTROL SYSTEM OPERATION (Contd)

c. Parking Brake System Operation.

A mechanical and air-actuated brake system performs the following for all vehicles covered in this manual:

(1) Keeps vehicle from rolling once it has stopped.

(2) Slows down or stops vehicle movement.

(3) Provides emergency stopping if there is a complete air system failure.

The mechanical brake system is covered below. The compressed air function of the brake system will be covered in a following paragraph. Major components of the parking brake system are:

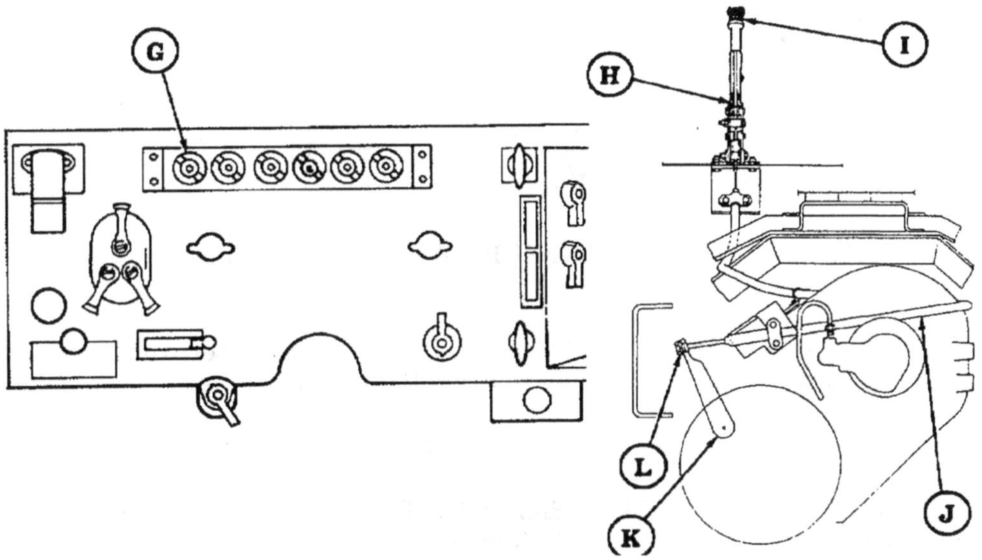
G
H
I
L
K
J

(G) **PARKING BRAKE WARNING LIGHT** – Illuminates when parking brake is engaged.

(H) **PARKING BRAKE CONTROL LEVER** – Is positioned up to engage parking brake and down to disengage parking brake.

(I) **PARKING BRAKE CONTROL LEVER ADJUSTING KNOB** – Permits operator to make minor tension adjustment of parking brake.

(J) **PARKING BRAKE CABLE** – Links parking brake lever to brakeshoe lever.

(K) **BRAKESHOE LEVER** – Lever turns cam which pushes brakeshoes against drum.

(L) **PARKING BRAKE ADJUSTING NUT** – Permits major tension adjustment between parking brake lever and brakeshoes.

1-17. CONTROL SYSTEM OPERATION (Contd)

d. Steering System Operation.

The steering system is identical for all models covered in this manual. It is a hydraulically assisted system that provides ease of turning and control for the operator. Major components of the steering system are:

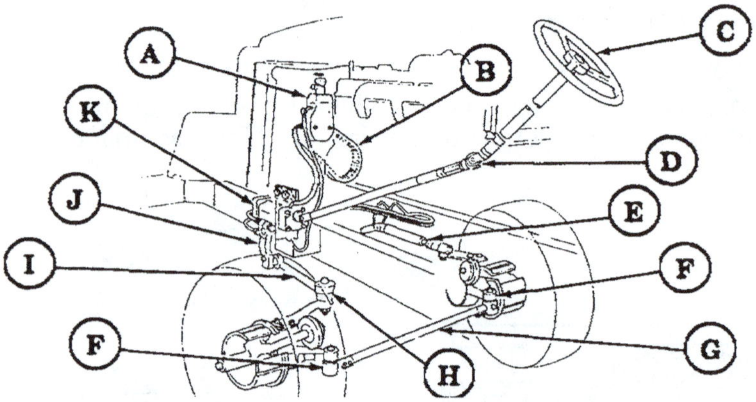
A
B
C
D
E
F
G
H
I
J
K
F

(A) **OIL RESERVOIR AND STEERING PUMP** – Combined in one unit, the reservoir serves as an oil fill point and the pump creates pressure.

(B) **ACCESSORY DRIVE BELTS** – Transmits mechanical power from accessory drive pulley to steering pump pulley to drive the steering pump.

(C) **STEERING WHEEL** – Serves 'as manual steering control for the operator.

(D) **STEERING COLUMN UNIVERSAL JOINT** – Connects, at an angle, the steering wheel column and input shaft of power steering gear.

(E) **POWER STEERING ASSIST CYLINDER** – Receives hydraulic pressure from the steering gear to assist in turning the front wheels.

(F) **STEERING KNUCKLE** – Serves as the pivot point and link for the front wheels from the tie rod assembly.

(G) **TIE ROD ASSEMBLY** – Connects steering knuckles so both wheels will turn at the same time.

(H) **STEERING ARM** – Connects drag link to steering knuckle.

(I) **DRAG LINK** – Transmits movement from steering arm to pitman arm.

(J) **PITMAN ARM** – Transfers torque from power steering gear to drag link.

(K) **STEERING GEAR** – Converts hydraulic pressure from steering pump to mechanical power at pitman arm.

1-17. CONTROL SYSTEM OPERATION (Contd)

e. Transmission Control System Operation.

The transmission control system permits shifting of transmission, prevents starting of engine with transmission in gear, and prevents shifting of transfer case unless transmission is in neutral. This system also permits engagement of the Transmission Power Takeoff (PTO) to provide hydraulic power for auxiliary equipment on M925/A1/A2, M928/A1/A2, M929/A1/A2, M930/A1/A2, M932/A1/A2 and M936/A1/A2 vehicles. Major components of the transmission control system are:

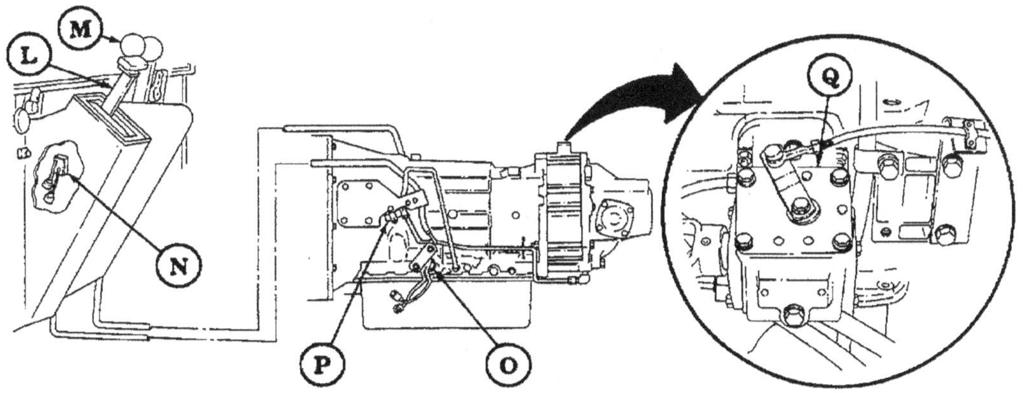
M
L
N
P
O
Q

(L) **TRANSMISSION SELECTOR LEVER** – Is used to select vehicle driving gear range.

(M) **POWER TAKEOFF CONTROL LEVER** – Engages transmission power takeoff to provide power for auxiliary equipment.

(N) **TRANSMISSION CONTROL SWITCH** – Actuates transmission lock-up solenoid valve when transmission selector lever is placed in neutral and transfer case shift lever lock-out switch is pressed.

(O) **TRANSMISSION sNEUTRAL START SWITCH** – The neutral start switch, wired to the starter switch, prevents the engine from being started with transmission in gear.

(P) **TRANSMISSION 5TH-GEAR LOCKUP SOLENOID VALVE** – Activated by transmission control switch and transfer case switch, the 5th-gear lockup solenoid valve directs main oil pressure of transmission to the transmission governor system. This puts the transmission in 5th-gear, creating less drag on the transfer case synchronizer which permits smoother shifting from one transfer case drive range to another. Refer to paragraph 1-17f, Transfer Case Control System Operation, for further details.

(Q) **TRANSMISSION POWER TAKEOFF (PTO)** – Driven by the transmission, the PTO drives the hydraulic pump which provides hydraulic pressure to power the front winch on M925/A1/A2, M928/A1/A2, M930/A1/A2, M932/A1/A2, and M936/A1/A2 vehicles, and to power the dump body on M929/AVA2 and M930/A1/A2 vehicles. The PTO is mounted on the right front side of the transmission.

1-17. CONTROL SYSTEM OPERATION (Contd)

f. Transfer Case Control System Operation.

The transfer case control system converts four-wheel driving power into six-wheel driving power, provides smooth shifting of transfer case into high or low driving rangs while vehicle is in motion, prevents transfer case from being shifted with transmission in gear, and provides hydraulic power for auxiliary equipment through a Power Takeoff (PTO).

(1) Six-wheel drive is achieved two different ways depending on the drive range (high or low) desired. In low range, the transfer case shift linkage automatically moves a cam-actuated valve which dumps air into the front drive cylinder. This forces a piston against the transfer case clutch to engage front-wheel drive. In high range, front-wheel drive is engaged in the same manner except that the front-wheel drive valve is manually actuated by the front-wheel drive lock-in switch on the instrument panel.

(2) In order to shift the transfer case from one driving range to another, an interlock system working in conjunction with the 5th-gear lock up solenoid is used. This system prevents the transfer case form being shifted unless the transmission is in neutral.

(3) With the automatic transmission, several actions must occur in order to shift the transfer case from one driving range to another. Because of the interlock system, the transmission must be placed in neutral. The transfer case shift lever switch must also be depressed.

(4) The transfer case control system, through the use of a PTO driven by the transfer case, also provides hydraulic power to operate the crane and rear winch on the M936/A1/A2 wreckers.

(5) Major components of this system are:

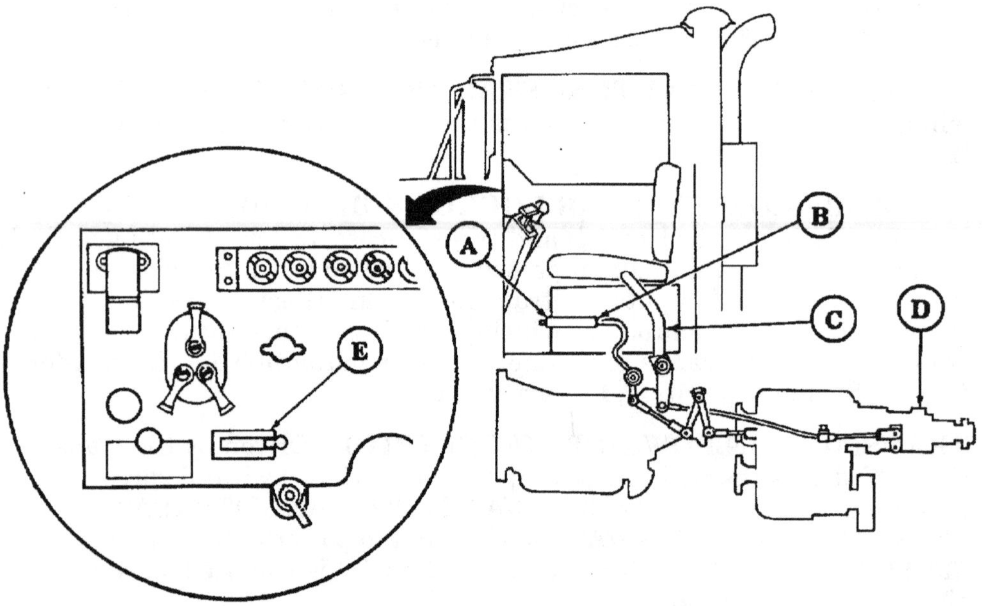
A
B
C
D
E

1-17. CONTROL SYSTEM OPERATION (Contd)

(A) **TRANSFER CASE SHIFT LEVER SWITCH** – When depressed with transmission in neutral, signals interlock solenoid valve to exhaust air pressure from interlock air cylinder and actuates lock-up solenoid.

(B) **TRANSFER CASE SHIFT LEVER** – Is pushed down to high for light load operations, and up to low for heavy load operations. Six-wheel drive is achieved automatically when transfer case shift lever is placed in low.

(C) **TRANSFER CASE POWER TAKEOFF CONTROL LEVER** – Manual control for engaging power takeoff.

(D) **TRANSFER CASE POWER TAKEOFF** – Mounted and mechanically driven at rear of transfer case, the PTO drives a pump to supply hydraulic pressure to power the rear winch and crane on the M936/A1/A2 wreckers.

(E) **FRONT-WHEEL DRIVE LOCK-IN SWITCH** – Manual control for activating front-wheel drive valve to provide front-wheel drive with transfer case in high drive range.

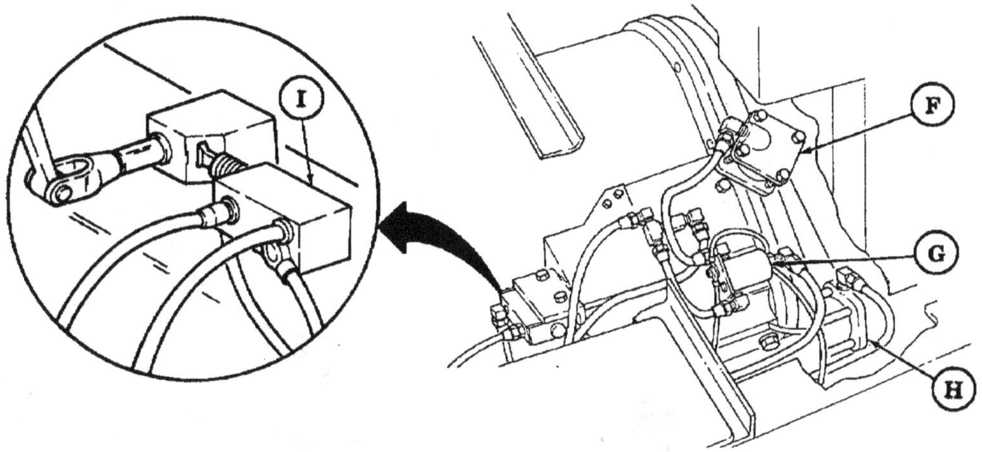
I
F
G
H

(F) **INTERLOCK AIR CYLINDER** – Under air pressure, a piston in the interlock air cylinder forces a shaft against one of three grooves in transfer case shift lever. This prevents transfer case from being shifted with transmission in gear.

(G) **INTERLOCK SOLENOID VALVE** – Releases air from interlock air cylinder when transmission is in neutral and transfer case shift lever switch is depressed. This permits the transfer case high/low shift shaft to move.

(H) **FRONT-WHEEL DRIVE AIR CYLINDER** – When under pressure, it moves transfer case clutch forward to engage front-wheel drive.

(I) **FRONT-WHEEL DRIVE VALVE** – When tripped by transfer case shift shaft, the front wheel drive valve routes air to front-wheel drive air cylinder.

1-18. POWER SYSTEM OPERATION

The power system includes those components' that give all vehicles covered in this manual the power to move. Each of these components will be described as part of the following subsystems:

a. Air Intake System Operation (page 1-36).
b. Fuel System (Dual Tank) Operation (page 1-38).
c. Fuel System (Single Tank) Operation (page 1-40).
d. Exhaust System Operation (page 1-41).
e. Cooling System Operation (page 1-42).
f. Engine Oil System Operation (page 1-44).
g. Powertrain System Operation (page 1-46).

a. Air Intake System Operation.

The air intake system channels and cleans air going to the combustion chamber where it mixes with fuel from the injectors to provide power for the engine. This system is identical on all models, except where indicated. Major components of the air intake system are:

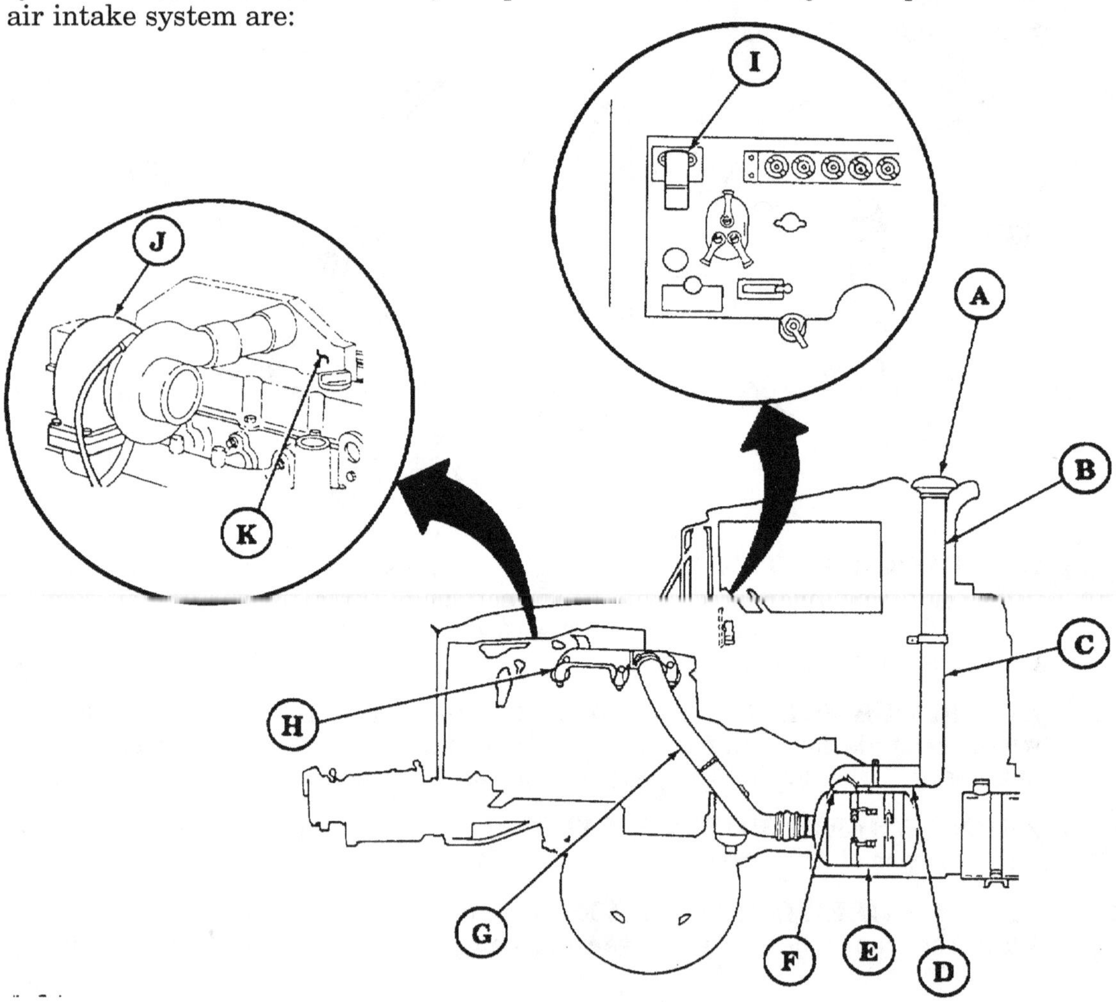
I
J
A
B
K
C
H
G
F
E
D

1-18. POWER SYSTEM OPERATION (Contd)

(A) **RAIN CAP** – Prevents rain and large objects from entering air intake system.

(B) **AIR INTAKE EXTENSION TUBE** – Routes air to air intake system. Can be removed for shipping.

(C) **STACK-TO-AIR INTAKE EXTENSION TUBE** – Routes air to air cleaner and is high enough to keep intake opening above fording level.

(D) **STACK-TO-AIR CLEANER ELBOW** – Flexible connection between air stack and air cleaner.

(E) **AIR CLEANER** – Filters dirt and dust from air.

(F) **HUMP HOSE** – Flexible connection between air cleaner and air cleaner outlet tube.

(G) **AIR CLEANER OUTLET TUBE** – Routes air from air cleaner to intake manifold.

(H) **INTAKE MANIFOLD** – Distributes air to combustion chambers in each cylinder head (M939/A1 series only).

(I) **AIR CLEANER INDICATOR** – Shows red when engine air filter needs servicing.

(J) **TURBOCHARGER** – Mounts on exhaust manifold and uses spent exhaust gases to dive impeller and pressurize air entering aftercooler (M939A2 series only).

(K) **AFTERCOOLER** – Distributes compressed air from turbocharger to combustion chambers while cooling air intake out of the turbocharger (M939A2 series only).

1-18. POWER SYSTEM OPERATION (Contd)

b. Fuel System (Dual Tank) Operation.

(1) The fuel system stores, cleans, and supplies fuel to the fuel injectors where it is mixed with air to initiate engine combustion.

(2) The fuel system is not identical for all models. Vehicles covered in this manual have either one or two tanks. These tanks can also differ in capacity. See table 1-6, Capacities for Normal Operating Conditions, for these differences.

(3) A typical two-tank fuel system is described below. A single tank is described later in paragraph 1-18c. Both systems include fuel supply, return, and vent lines to provide fuel flow and release the fumes throughout the system. Major components of fuel system (dual tank) are:

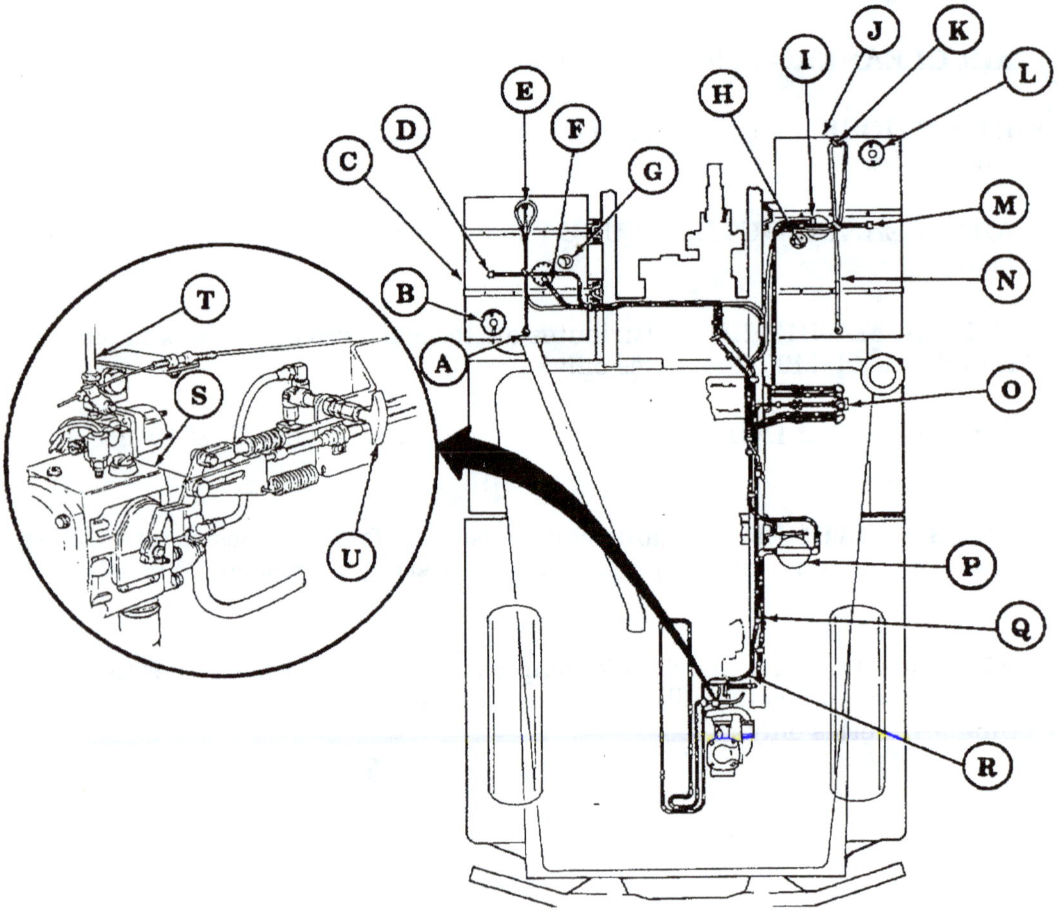
A
B
C
D
E
F
G
H
I
J
K
L
M
N
O
P
Q
R
S
T
U

(A) **RIGHT TANK (FRONT) VENT LINE** – Vents vapors from fuel tank to vent hole in air intake stack.

(B) **RIGHT TANK FILLER CAP** – Covers fuel filler opening on right fuel tank.

1-18. POWER SYSTEM OPERATION (Contd)

(C) **RIGHT FUEL TANK** – Stores fuel for vehicle use.

(D) **RIGHT TANK FUEL RETURN LINE** – Returns unused fuel back to fuel tank.

(E) **RIGHT TANK (REAR) VENT LINE** – Vents vapors from fuel tank to vent hole in air intake stack.

(F) **RIGHT TANK FUEL SUPPLY LINE** – Directs fuel from tank to fuel filter.

(G) **RIGHT TANK FUEL LEVEL SENDING UNIT** – Electrical signal registers fuel level in right tank at gauge on instrument panel.

(H) **LEFT TANK FUEL LEVEL SENDING UNIT** – Electrical signal registers fuel level in left tank at gauge on instrument panel.

(I) **LEFT TANK FUEL SUPPLY LINE** – Directs fuel from tank to fuel filter.

(J) **LEFT FUEL TANK** – Stores fuel for vehicle use.

(K) **LEFT TANK (REAR) VENT LINE** – Vents vapors from fuel tank to vent hole in air intake stack.

(L) **LEFT TANK FILLER CAP** – Covers fuel filler opening on left fuel tank.

(M) **LEFT TANK FUEL RETURN LINE** – Returns unused fuel back to fuel tank.

(N) **LEFT TANK (FRONT) VENT LINE** – Vents vapors from fuel tank to vent hole in air intake stack.

(O) **FUEL SELECTOR VALVE** – Manual control valve that opens fuel flow to engine from left or right fuel tank.

(P) **FUEL FILTER/WATER SEPARATOR** – Filters water and dirt from fuel.

(Q) **FUEL FILTER-TO-PUMP SUPPLY LINE** – Directs fuel from fuel filter to fuel pump.

(R) **FUEL RETURN LINE** – Returns unused fuel back to fuel tanks.

(S) **FUEL PUMP** – Draws fuel from tank(s) and pumps it through supply line to fuel injectors.

(T) **FUEL SUPPLY LINE** – Directs fuel from fuel pump to fuel injectors.

(U) **FUEL PRIMER PUMP** – Purges air from fuel system.

1-18. POWER SYSTEM OPERATION (Contd)

c. Fuel System (Single Tank) Operation.

Major components of the fuel system (single tank) are:

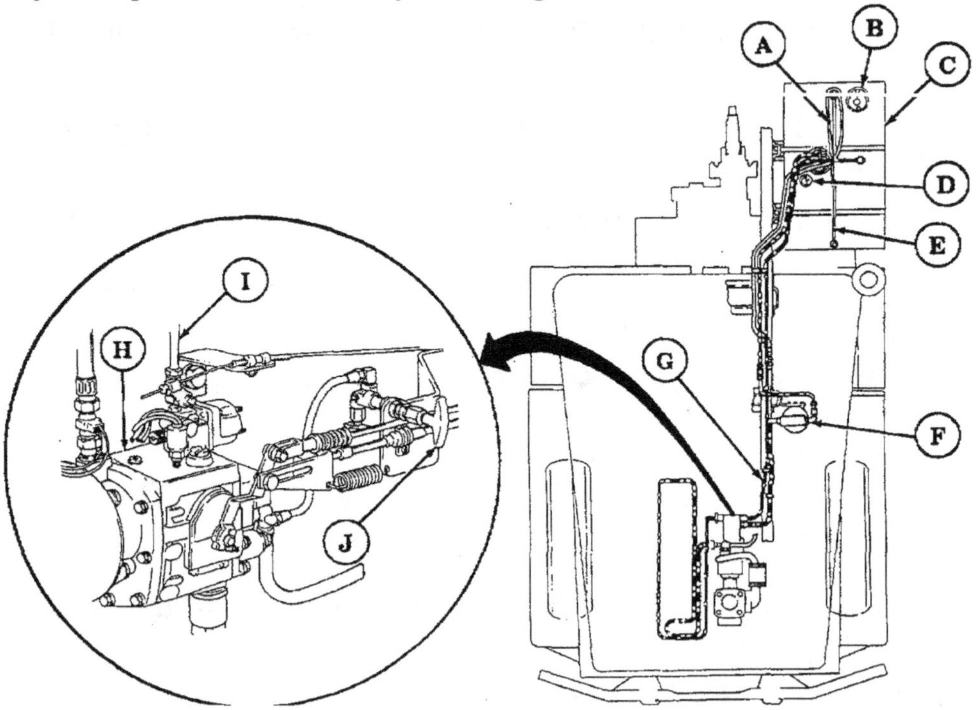
A
B
C
D
E
F
G
H
I
J

(A) **TANK (REAR) VENT LINE** – Vents vapors from fuel tank to vent hole in air intake stack.

(B) **TANK FILLER CAP** – Covers fuel fill opening.

(C) **FUEL TANK** – Stores fuel for vehicle use.

(D) **FUEL TANK LEVEL SENDING UNIT** – Electrical signal registers fuel level in tank at gauge on instrument panel.

(E) **TANK (FRONT) VENT LINE** – Vents vapors from fuel tank to vent hole in air intake stack.

(F) **FUEL FILTER/WATER SEPARATOR** – Filters water and dirt from fuel.

(G) **FUEL RETURN LINE** – Returns unused fuel back to fuel tank.

(H) **FUEL PUMP** – Draws fuel from tank and pumps it through supply line to fuel injectors.

(I) **FUEL SUPPLY LINE** – Directs fuel from fuel pump to fuel injectors.

(J) **FUEL PRIMER PUMP** – Purges air from fuel system.

1-18. POWER SYSTEM OPERATION (Contd)

d. Exhaust System Operation.

The exhaust system directs exhaust gases away from the vehicle for all models covered in this manual. Major components of the exhaust system are:

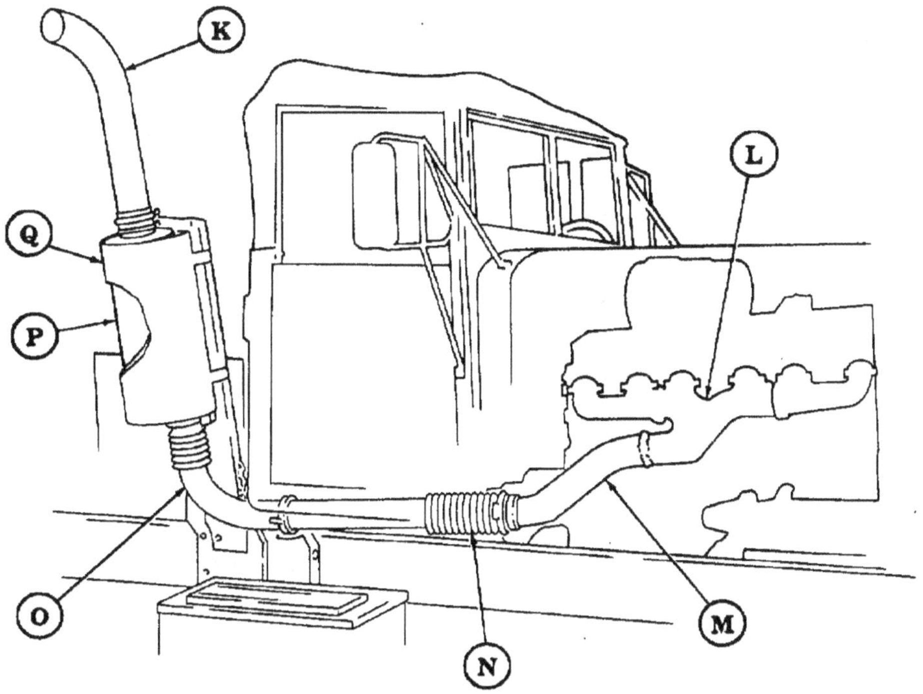
K
L
Q
P
O
N
M

(K) **EXHAUST STACK** – Directs exhaust from muffler away from vehicle.

(L) **EXHAUST MANIFOLD** – Collects exhaust from cylinder head ports and directs it to front exhaust pipe.

(M) **FRONT EXHAUST PIPE** – Directs exhaust to rear exhaust pipe.

(N) **FLEX PIPE** – Part of rear exhaust pipe and allows flexibility for vibration and expansion in system.

(O) **REAR EXHAUST PIPE** – Directs exhaust to muffler.

(P) **MUFFLER** – Quiets exhaust noises.

(Q) **MUFFLER SHIELD** – Protects personnel from muffler heat.

1-18. POWER SYSTEM OPERATION (Contd)

e. Cooling System Operation.

This system provides cooling of the engine and transmission. It differs slightly between the M939/A1 series and M939A2 series vehicles because different engines are used. Major components of the cooling system are:

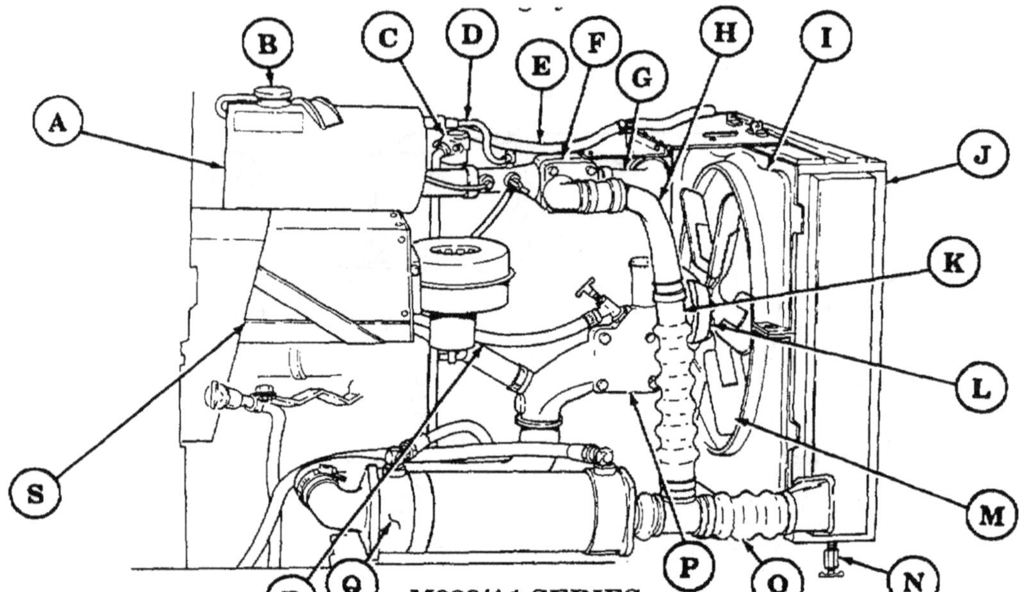
A
B
C
D
E
F
G
H
I
J
K
L
M
N
O
P
Q
R
S

M939/A1 SERIES

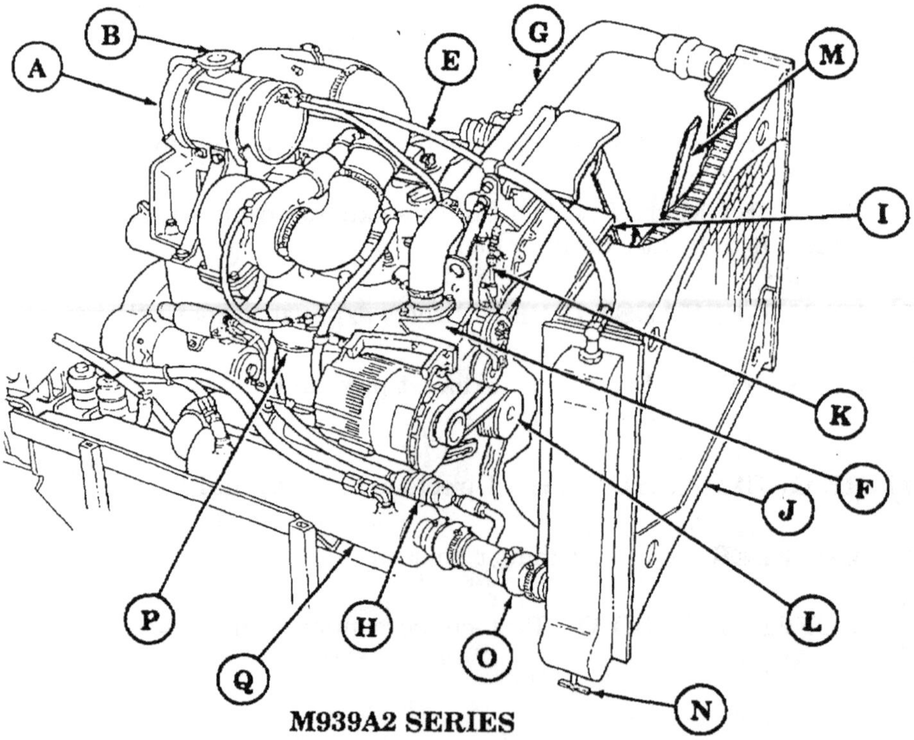
A
B
E
G
M
I
K
F
J
L
P
H
O
Q
N

M939A2 SERIES

1-18. POWER SYSTEM OPERATION (Contd)

(A) **SURGE TANK** – Filling point for cooling system. On M939A2 vehicles, a float sensor monitors water level and illuminates a light on instrument panel.

(B) **COOLANT PRESSURE CAP** – Designed to depressurize cooling system and to access cooling system for filling.

(C) **WATER MANIFOLD** – Collects coolant from cylinder heads and directs it to the thermostat housing (M939/A1 series only).

(D) **SURGE TANK-TO-WATER MANIFOLD VENT** – Vents air trapped in water manifold (M939/A1 series only).

(E) **SURGE TANK-TO-RADIATOR VENT** – Vents air in cooling system.

(F) **THERMOSTAT** – Shuts off coolant flow to radiator until temperature reaches 175°F (79°C) on M939/A1 series vehicles, and 181°F (83°C) on M939A2 series vehicles. Coolant is then directed to the radiator through the radiator inlet hose.

(G) **RADIATOR INLET HOSE** – Directs coolant from water manifold to radiator after thermostat has opened.

(H) **BYPASS TUBE** – Directs coolant back to transmission oil cooler where it is then recirculated through the engine block until the thermostat opens.

(I) **RADIATOR SHROUD** – Concentrates air flow through the radiator.

(J) **RADIATOR** – Directs coolant through a series of fins or baffles so outside air can remove excessive heat from coolant.

(K) **FAN CLUTCH** – Regulates use of fan to control engine temperature fan to belt driven pulley when conditions require additional cooling.

(L) **WATER PUMP** – Provides force to move coolant through engine.

(M) **FAN** – Provides force to pull air through radiator.

(N) **RADIATOR DRAINVALVE** – Permits coolant to be drained from radiator.

(O) **TRANSMISSION OIL COOLER HOSE** – Directs coolant to transmission oil cooler.

(P) **ENGINE OIL COOLER** – Reduces heat of engine oil (M939/A1 series only).

(Q) **TRANSMISSION OIL COOLER** – Reduces heat of transmission oil.

(R) **ENGINE OIL COOLER TO HEATER HOSE** – Directs coolant to personnel water heater when shutoff valve is open (M939/A1 series only).

(S) **PERSONNEL WATER HEATER** – Provides heat for cab and personnel.

1-18. POWER SYSTEM OPERATION (Contd)

f. Engine Oil System Operation.

The engine oil system provides lubricating oil for internal moving parts. Major components of the engine oil system are:

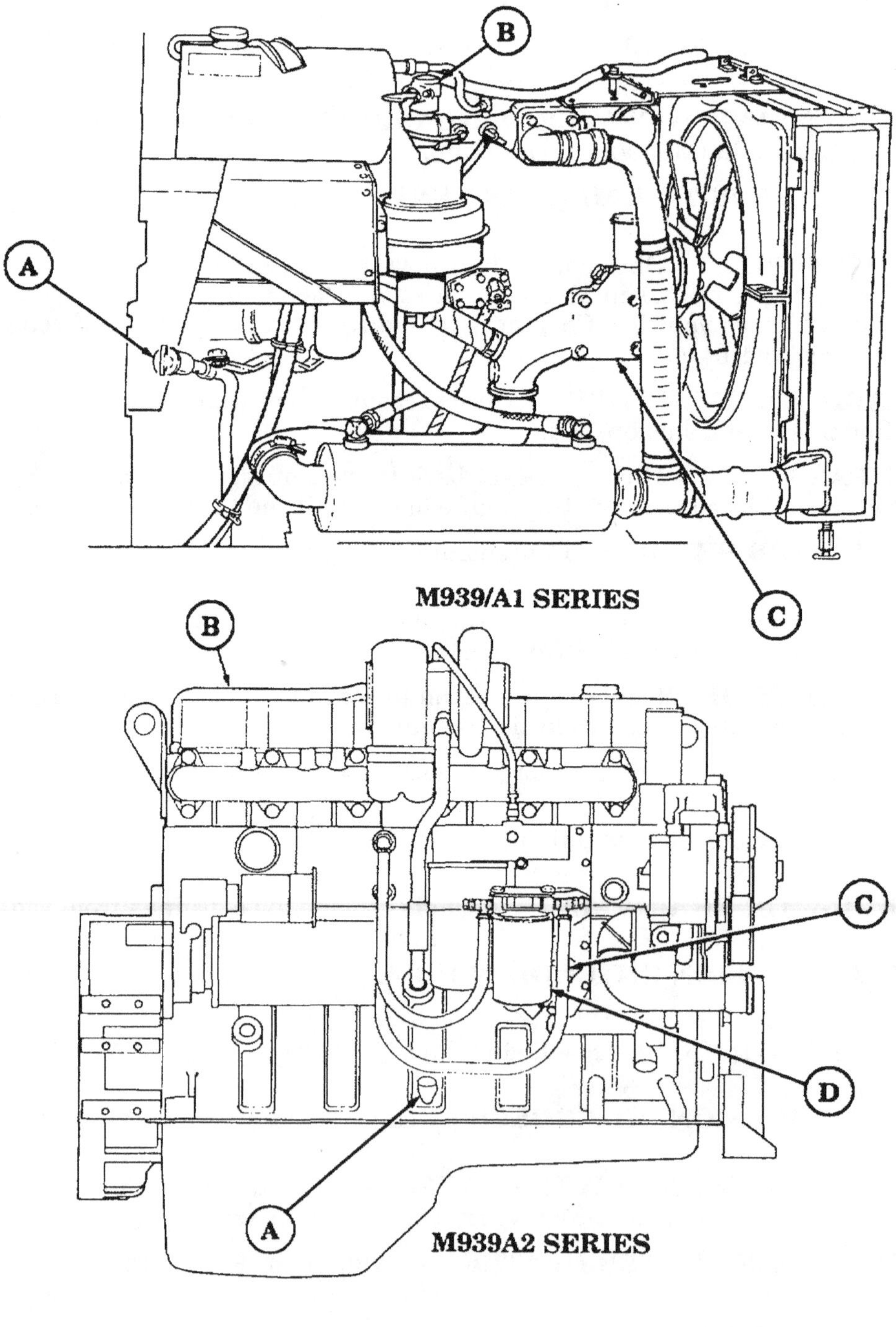
B
A
C
M939/A1 SERIES
B
C
D
A
M939A2 SERIES

1-18. POWER SYSTEM OPERATION (Contd)

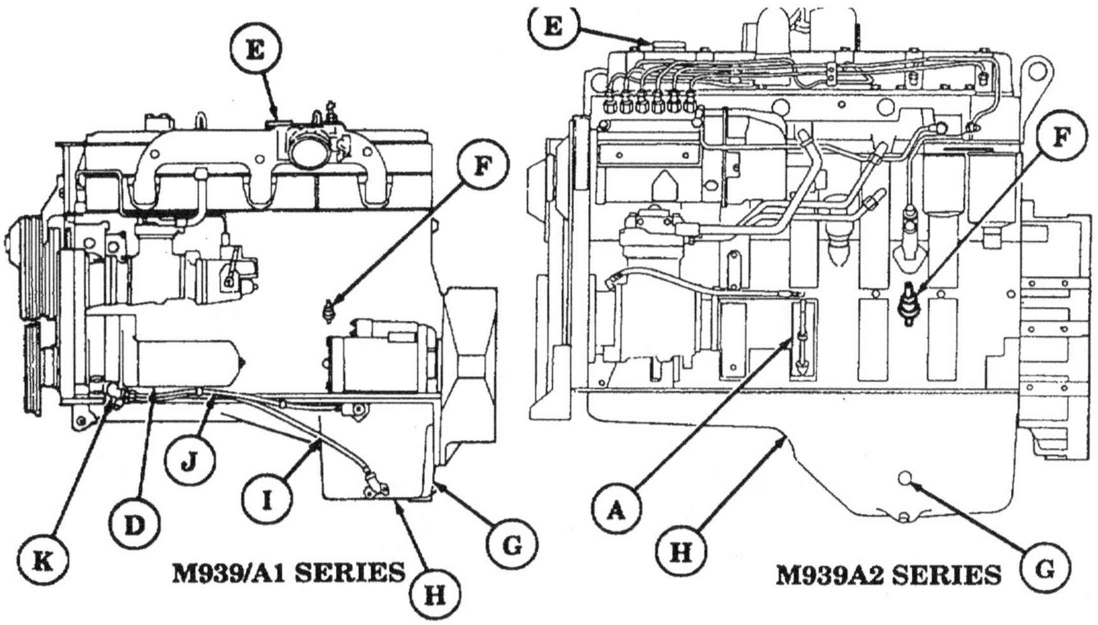
E
F
J
I
D
K
G
H
M939/A1 SERIES
E
F
A
H
G
M939A2 SERIES

(A) **OIL DIPSTICK** – Indicates engine oil level.

(B) **CRANKCASE BREATHER** – Vents hot engine oil fumes from engine and allows fresh air to enter.

(C) **ENGINE OIL COOLER** – Removes heat from engine oil as coolant circulates through internal tubes of oil cooler.

(D) **OIL FILTER** – Filters out foreign particals suspended in oil.

(E) **OIL FILLER CAP** – Located on rocker lever cover, cap covers engine oil fill opening.

(F) **OIL PRESSURE TRANSMITTER** – Sends an electrical signal that indicates engine oil pressure to gauge on instrument panel.

(G) **OIL PAN DRAINPLUG** – Plugs engine oil drain opening.

(H) **OIL PAN** – Reservoir for engine oil.

(I) **OIL SUPPLY LINE** – Carries oil from oil pan to the oil pump.

(J) **OIL BYPASS RETURN LINE** – Returns oil from oil pump to the oil pan.

(K) **OIL PUMP** – Provides mechanical pressurization of oil to circulate it through the oil system.

1-18. POWER SYSTEM OPERATION (Contd)

g. Powertrain System Operation.

The powertrain system is the same on all models in this manual except the extra-long wheelbase models which have an additional propeller shaft and center bearing. This system transmits engine power to the axles to put the vehicle in motion. Major components of the powertrain system are:

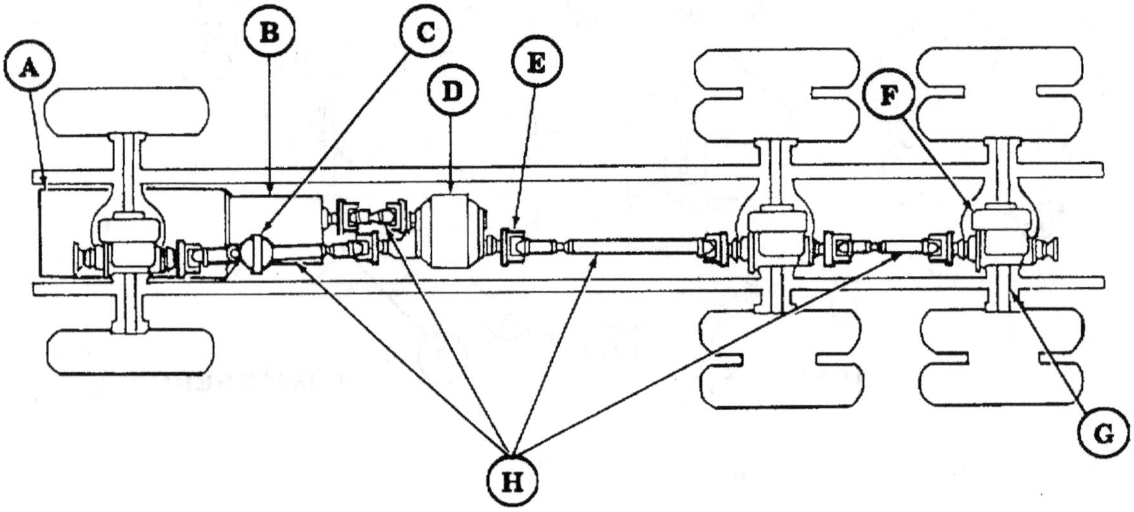
A
B
C
D
E
F
G
H

(A) **ENGINE** – Provides power needed for powertrain component operation.

(B) **TRANSMISSION** – Adapts engine power to meet different driving conditions.

(C) **CENTER BEARING** – Provides support for propeller shaft to decrease vibration and wear on universal joints (M927/A1/A2, M928/A1/A2, and M934/A1/A2 series only).

(D) **TRANSFER CASE** – Distributes power evenly to front and rear axles.

(E) **UNIVERSAL JOINTS** – Connections between two propeller shafts that permit one to drive the other even though they may be at different angles.

(F) **DIFFERENTIALS** – Distribute power to left and right axle shafts.

(G) **AXLES** – Transmit power from differentials to rotate wheels.

(H) **PROPELLER SHAFTS** – Serve as driving shafts that connect the transmission to the transfer case and the transfer case to the differentials.

1-19. ELECTRICAL SYSTEMS OPERATION

Nearly every component of the models covered in this manual is affected by the electrical system. These components and their electrical connections are described as part of the following electrical subsystems:

a. Battery System Operation (page 1-48).

b. Starting System Operation (page 1-49).

c. Ether Starting System Operation (page 1-50).

d. Generating System Operation (page 1-51).

e. Directional Signal System Operation (page 1-52).

f. Heating System Operation (page 1-53).

g. Indicator, Gauge, and Warning System Operation (page 1-54).

h. Trailer and Semitrailer Connection System Operation (page 1-56).

Electrical Terms and Definitions.

The following electrical terms and definitions will be frequently referred to throughout this section and should be understood before proceeding:

Alternating Current (AC signal) – Current in a circuit that flows, in one direction first, then in the other direction.

Circuit – A complete path for electric current flow between components.

Circuit Breaker – An automatic switch that interrupts current flow in a circuit when the current limit is exceeded.

Direct Current (DC signal) – Current in a circuit that flows in one direction.

Female Connector – One-half of a connector which fits over the other half.

Ground – A common return to complete a path for current flow in a circuit.

Harness – A group of wires connected between devices that are bundled and routed together to prevent damage and make repair and replacement easier.

Male Connector – One-half of a connector which fits the other half.

Polarity – The direction current flows in a circuit (usually positive to negative).

Relay – An electromagnetic device that operates like an automatic switch to control the flow of current in the same or different circuit.

Reverse Polarity – The condition that exists when circuit polarity is connected opposite of that which was intended.

Sending Unit – A device that produces an electrical signal and sends this signal to the device which will make use of it.

Sensor – An electrical sensor takes a physical condition (temperature, oil presence or absence) and converts this into an electrical signal.

Splice – A permanent physical connection of two or more wires.

Terminal – Fastener at end of wire used to connect the wire to an electrically-powered device.

1-19. ELECTRICAL SYSTEMS OPERATION (Contd)

a. Battery System Operation.

The battery system is identical for all models covered in this manual and consists of the following major components and circuits:

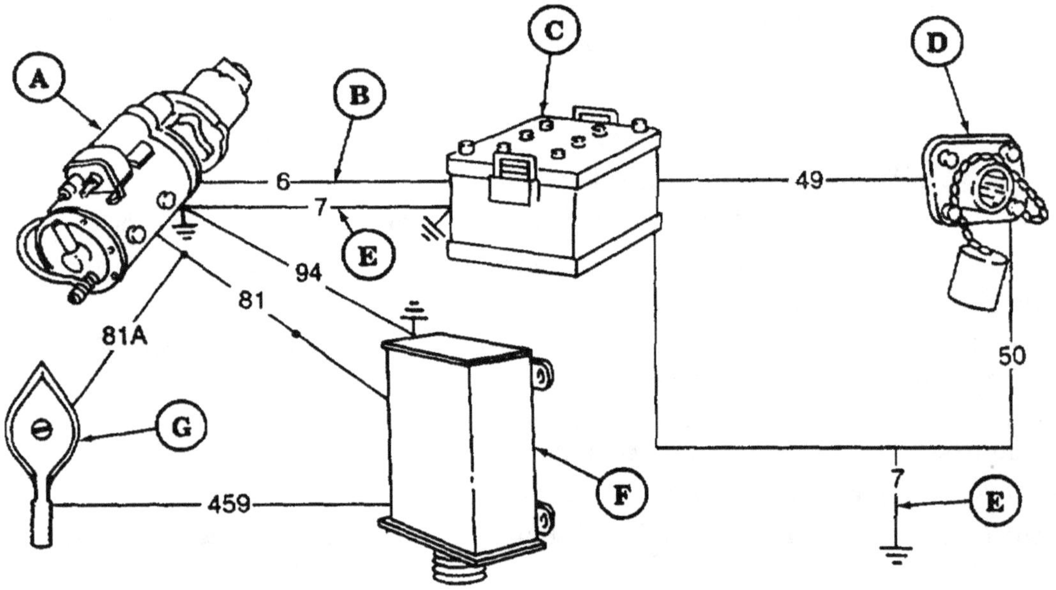

A
B
C
D
E
F
G
6
7
49
94
81
81A
50
459
7
E

(A) **STARTER SOLENOID** – Junction point for battery positive lead (circuit 6) and vehicle electrical feed wire (circuit 81).

(B) **CIRCUIT 6** – Connects the batteries to the starting motor and to the protective control box through circuit 81.

(C) **BATTERIES** – Four 6TN batteries are connected in series parallel to provide 24-volts DC for the electrical starter system and 12-volts DC for the heater fan low speed.

(D) **SLAVE RECEPTACLE** – Links an external power source directly to the slaved vehicle's batteries to assist in cranking the engine when the batteries are not sufficiently charged.

(E) **CIRCUIT 7** – Provides a ground between starter, battery, and chassis.

(F) **PROTECTIVE CONTROL BOX** – Protects the vehicle electrical system in the event the battery system polarity is reversed. Connects battery power to vehicle electrical lead through circuit 81 and circuit 5. Connects positive ground through circuit 94 to the starter.

(G) **BATTERY SWITCH** – Controls a relay in the protective control box through circuit 459 that connects the batteries to the vehicle electrical load.

1-19. ELECTRICAL SYSTEMS OPERATION (Contd)

b. Starting System Operation.

The starting system is identical for all models covered in this manual and consists of the following major components and circuitry:

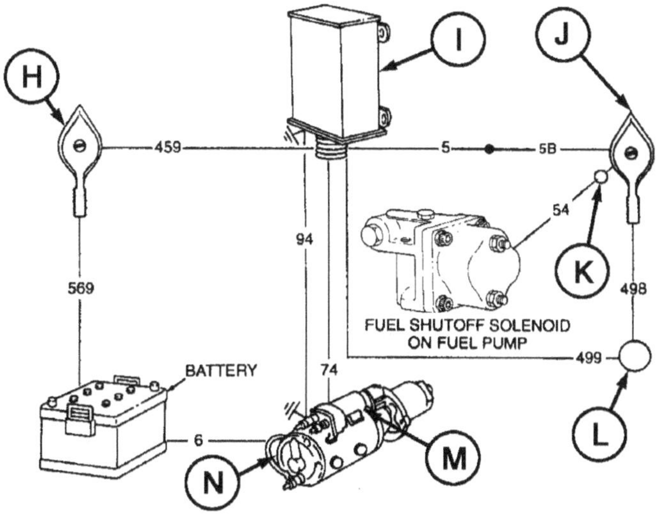

H
I
J
459
5
5B
54
94
K
569
498
FUEL SHUTOFF SOLENOID
ON FUEL PUMP
499
BATTERY
74
6
L
M
N

(H) **BATTERY SWITCH** – Completes circuit 459, closing a relay in the protective control box to supply power to the ignition switch through circuits 5 and 5B.

(I) **PROTECTIVE CONTROL BOX** – Locks out starter circuit, which prevents starter from reengaging while engine is running.

(J) **IGNITION SWITCH** – Provides battery power through circuit 54 and antilock brake system (ABS) to the fuel solenoid and through circuit 498 to the neutral start safety switch.

(K) **(ABS) WARNING LAMP** – Intended to give the operator visual signal that the antilock brake system has a malfunction if warning lamp is lit for more then three seconds after ignition switch is place in ON position.

(L) **NEUTRAL START SAFETY SWITCH** – Prevents starter from energizing when vehicle is not in neutral, by deenergizing circuit 499 and a relay in the protective control box, which disconnects power from circuit 74 and the starter solenoid.

(M) **STARTER SOLENOID** – A magnetic relay that is powered by circuit 74 to transmit 24-volt battery power to the starter motor through circuit 6.

(N) **STARTER MOTOR** – Cranks the engine for starting. Supplied with 24-volt battery power through circuit 6.

1-19. ELECTRICAL SYSTEMS OPERATION (Contd)

c. Ether Starting System Operation.

The ether starting system is identical for all models covered in this manual and consists of the following major components and circuitry:

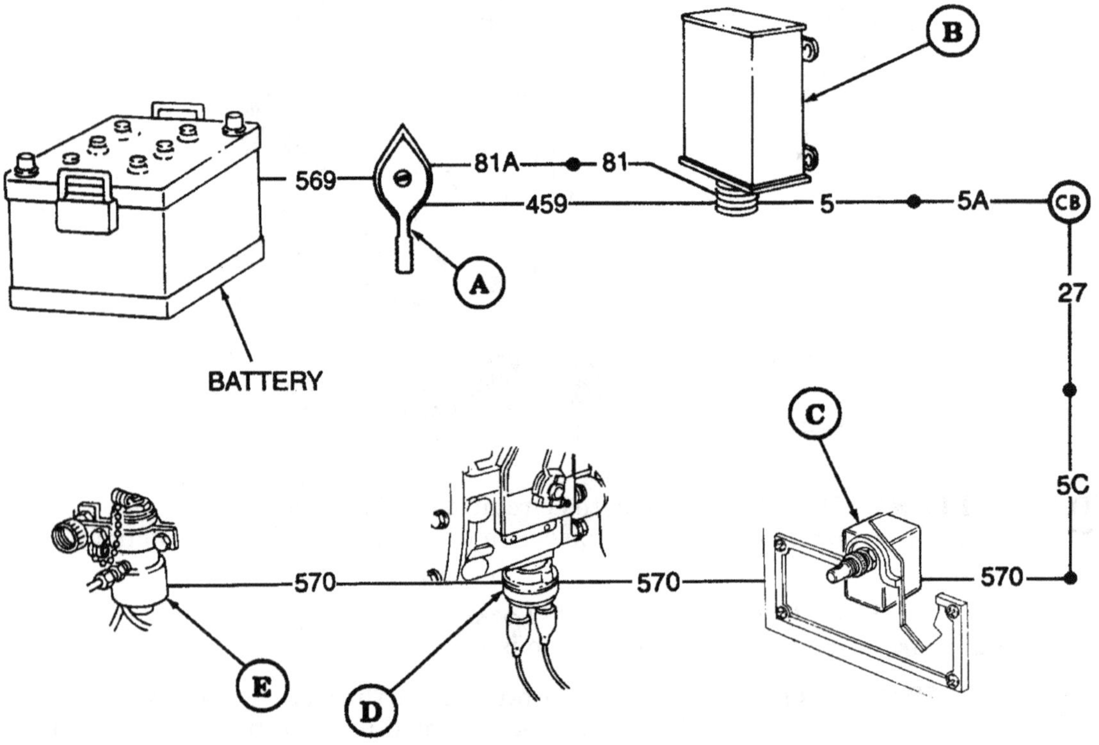

B
569
81A
81
459
5
5A
CB
A
27
BATTERY
C
5C
570
570
570
E
D

(A) **BATTERY SWITCH** – Provides 24-volt battery power to the protective control box through circuits 459, 81A, and 81.

(B) **PROTECTIVE CONTROL BOX** – Energizes the ether feed switch through circuits 5, 5A, 27, 5C, and 570.

(C) **ETHER FEED SWITCH** – Controls 24-volt power to the ether pressure switch through circuit 570.

(D) **ETHER PRESSURE SWITCH** – Connects the ether feed switch to the ether tank valve through circuit 570.

(E) **ETHER TANK VALVE** – Is activated through circuit 570 when the ether pressure switch is closed and the ether feed switch is pressed.

1-19. ELECTRICAL SYSTEMS OPERATION (Contd)

d. Generating System Operation.

The generating system is identical for all vehicles covered in this manual and consists of the following major components and circuitry:

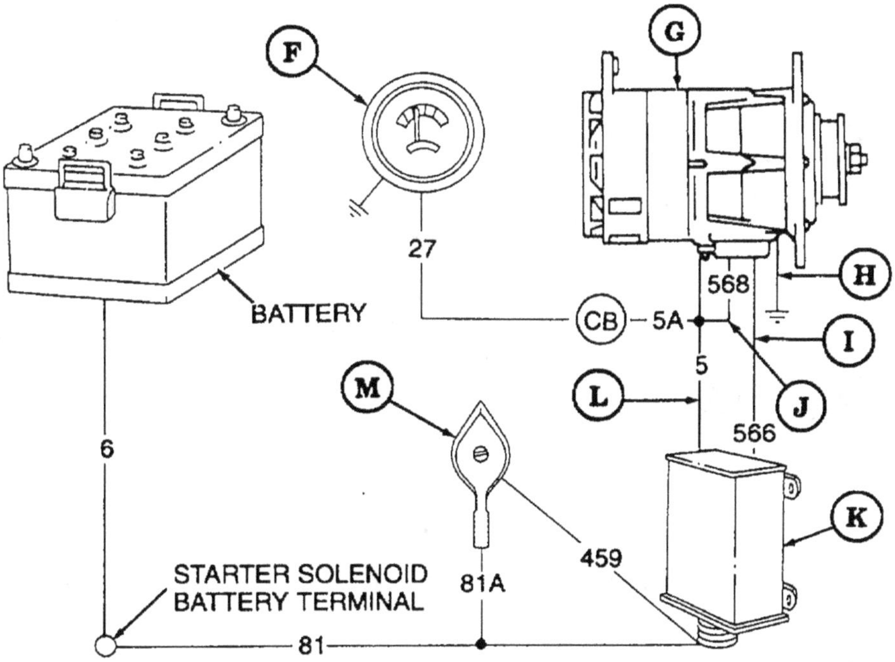
F
G
BATTERY
27
568
H
CB
5A
I
5
M
L
J
566
6
K
459
STARTER SOLENOID
BATTERY TERMINAL
81A
81

(F) **VOLTMETER** – Indicates electrical system voltage. It is connected to the electrical system through circuit 27.

(G) **ALTERNATOR** – Rated at 26-30 volts, 60 amperes, the alternator assists and recharges the batteries during operation. A 100 ampere model is available as a kit.

(H) **CIRCUIT 3** – Provides a ground circuit to the alternator.

(I) **CIRCUIT 566** – Controls a relay in the protective control box that prevents the starter from reactivating while the engine is running.

(J) **CIRCUIT 568** – Senses system voltage and excites the alternator field.

(K) **PROTECTIVE CONTROL BOX** – Connects circuit 5 to 81 to power the electrical system and charge the batteries.

(L) **CIRCUIT 5** – Conducts alternator output to charge the batteries and maintain vehicle voltage.

(M) **BATTERY SWITCH** – Closes the relay in the protective control box that connects battery circuits.

1-19. ELECTRICAL SYSTEMS OPERATION (Contd)

e. Directional Signal System Operation.

The directional signal system is identical on all models covered in this manual and consists of the following major components and circuitry:

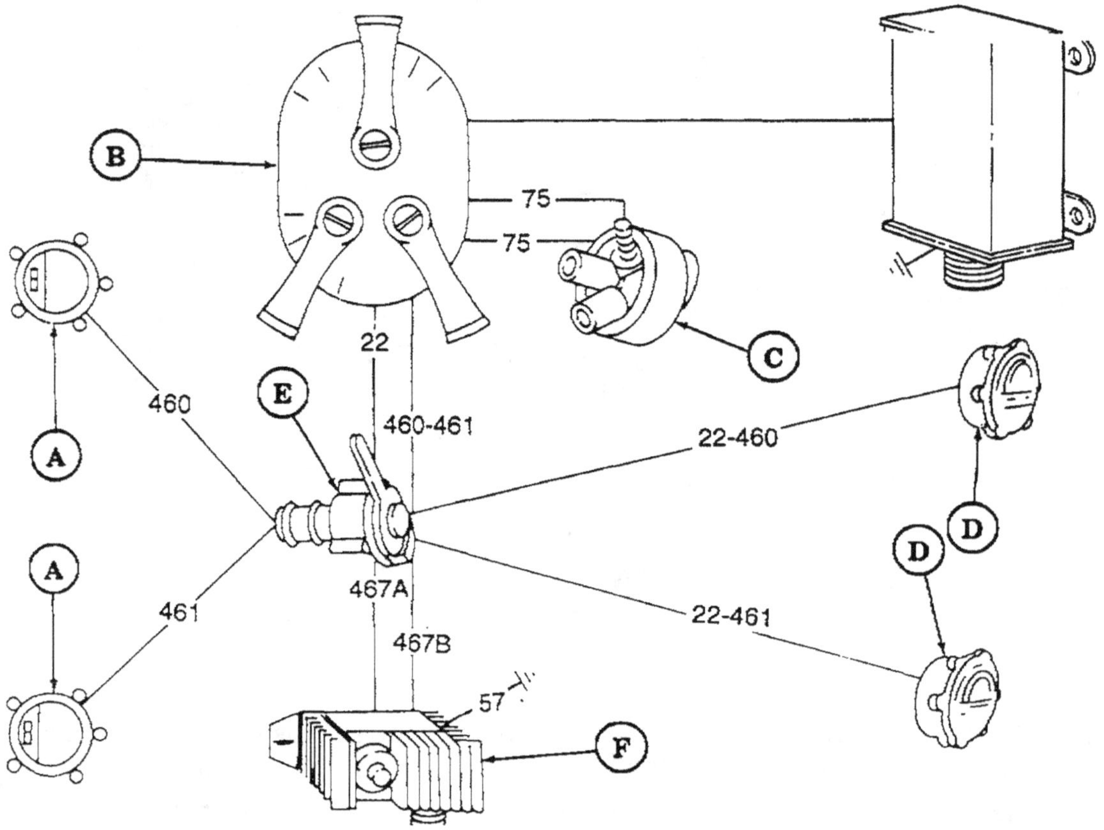
B
75
75
C
22
460
E
460-461
22-460
A
D
D
A
461
467A
467B
22-461
57
F

(A) **FRONT COMPOSITE LAMP** – Receives power from turn signal control through circuits 460 and 461 to indicate turning direction.

(B) **LIGHT SWITCH** – Provides battery power to the directional signal switch through circuits 460 and 461, and to the stoplight switch through circuit 75.

(C) **STOPLIGHT SWITCH** – Closing this switch allows power to flow from the light switch through circuit 75 to circuit 22 to the directional signal switch.

(D) **REAR COMPOSITE LAMP** – Receives power from turn signal control through circuit 22-460 and 22-461 to indicate turning direction.

(E) **DIRECTIONAL SIGNAL SWITCH** – A four-position switch that directs power to the composite and signal lamps through circuits 460, 461, 22-460, and 22-461 to indicate direction of turn.

(F) **TURN SIGNAL FLASHER** – Receives power through circuit 467A and sends intermittent current to the signal lamp through circuit 467B.

1-19. ELECTRICAL SYSTEMS OPERATION (Contd)

f. Heating System Operation.

The electrical portion of the heating system is identical for all models covered in the manual and consists of the following major components and circuitry:

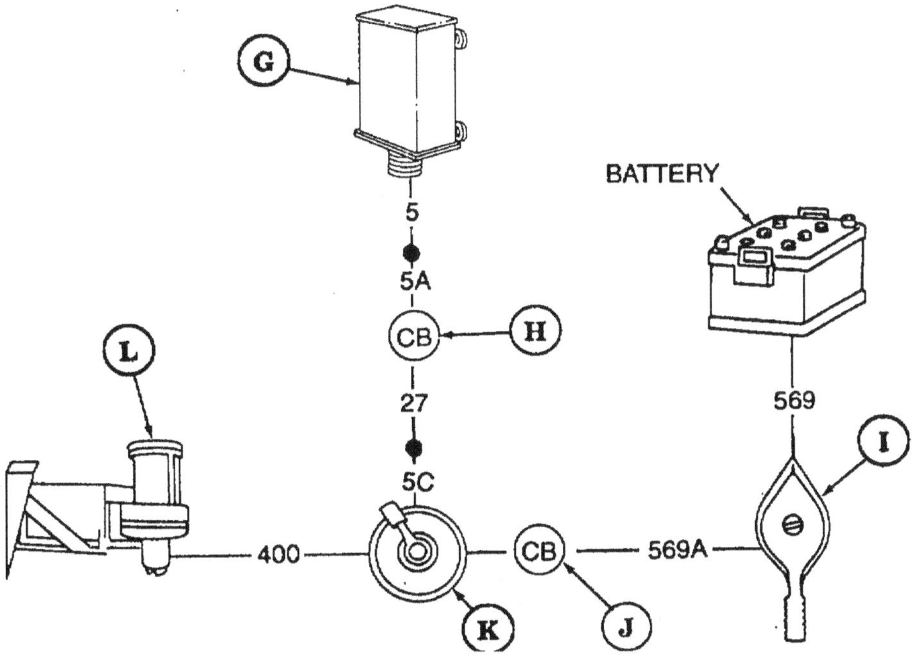

G
BATTERY
5
5A
CB
H
L
27
569
I
5C
400
CB
569A
K
J

(G) **PROTECTIVE CONTROL BOX** – Provides 24-volt power to circuit breaker through circuits 5 and 5a and to the heater switch through circuits 27 and 5c.

(H) **CIRCUIT BREAKER** – Provides overload protection for 24-volt circuits 5, 5a, 27, and 5c leading to the heater switch.

(I) **BATTERY SWITCH** – Provides 12-volt battery power from circuit 569 through 569A to the heater.

(J) **CIRCUIT BREAKER** – Provides overload protection for 12-volt circuit 569A leading to heater switch.

(K) **HEATER SWITCH** – Controls low and high blower motor speed and has two sources of power; 12-volt power is supplied through circuit 569A from the battery switch and is used to provide low speed; 24-volt power is supplied through circuit 5C from the protective control box and is used to provide high speed.

(L) **HEATER BLOWER MOTOR** – A direct current motor controlled by the heater switch through circuit 400.

1-19. ELECTRICAL SYSTEMS OPERATION (Contd)

g. Indicator, Gauge, and Warning System Operation.

The indicator, gauge, and warning system is comprised of several subsystems:

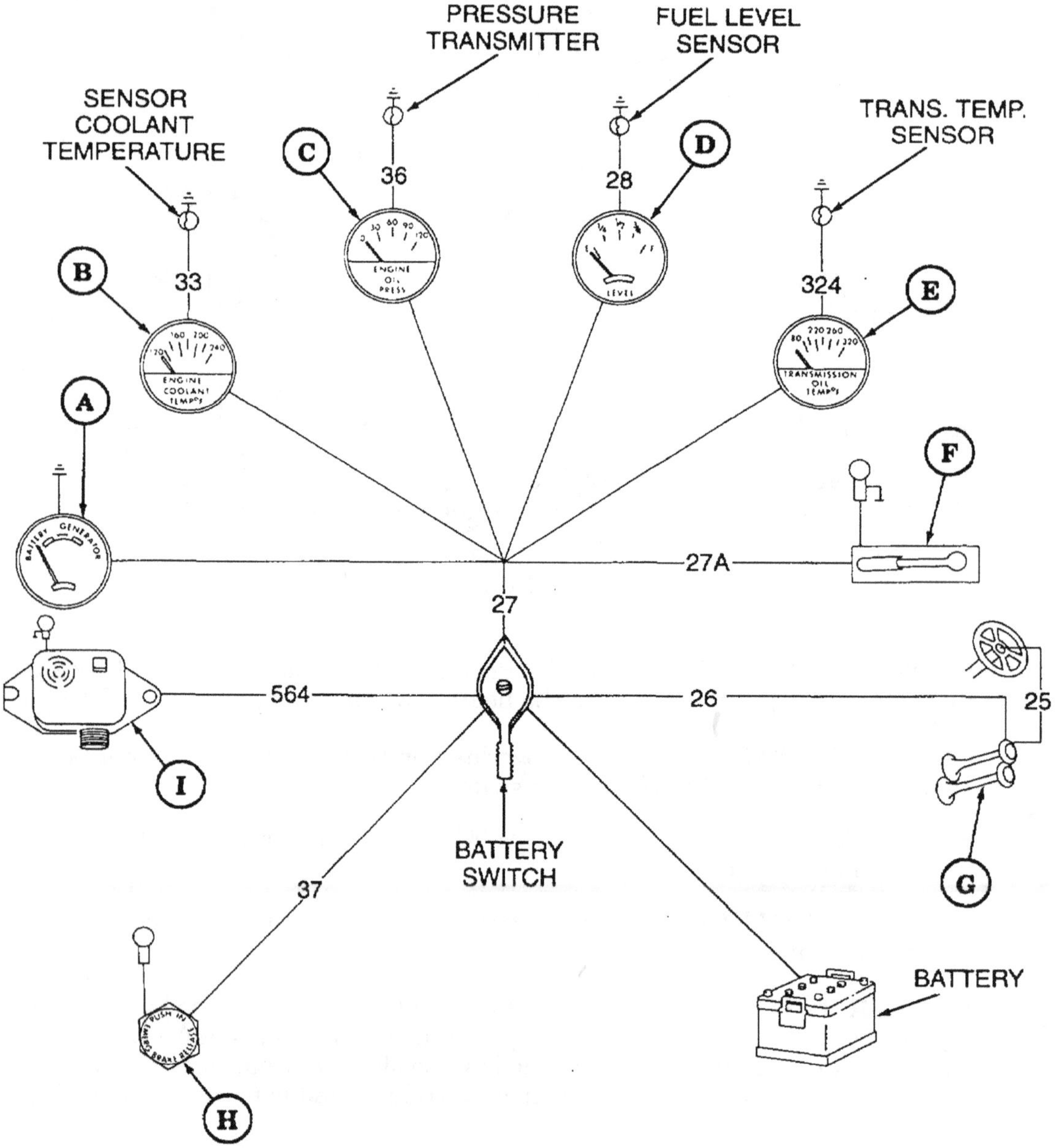

PRESSURE TRANSMITTER
FUEL LEVEL SENSOR
SENSOR COOLANT TEMPERATURE
TRANS. TEMP. SENSOR
A
B
C
D
E
F
G
H
I
36
28
33
324
27A
27
564
26
25
37
BATTERY SWITCH
BATTERY

1-19. ELECTRICAL SYSTEMS OPERATION (Contd)

(A) **VOLTMETER** – Indicates system voltage and is connected to the batteries through circuit 27 and to chassis ground through instrument panel.

(B) **ENGINE COOLANT TEMPERATURE INDICATOR** – Indicates engine coolant temperature and receives battery power through circuit 27. Circuit 33 completes the circuit to ground through a coolant temperature sensor that reacts to changes in engine coolant temperature by increasing or decreasing the resistance in the ground circuit.

(C) **ENGINE OIL PRESSURE INDICATOR** – Indicates engine oil pressure and receives battery power through circuit 27. Circuit 36 completes the circuit to ground through the oil pressure transmitter located on the engine block.

(D) **FUEL INDICATOR** – Indicates fuel level. Receives battery power through circuit 27. Circuit 28 or 29, depending upon which position the fuel selector switch is in, completes the circuit to ground through the fuel level sensor.

(E) **TRANSMISSION OIL TEMPERATURE INDICATOR** – Indicates transmission oil temperature and receives battery power through circuit 27. Circuit 324 completes the circuit to ground through a temperature sensor located in the transmission.

(F) **FRONT-WHEEL DRIVE ENGAGEMENT LIGHT** – Informs the operator that the front-wheel drive is engaged. The system consists of a normally open pressure switch, which is powered through circuit 27A and an indicator lamp powered through circuit 27A.

(G) **HORN SYSTEM** – The horn system consists of an air-operated horn that is controlled by an electric solenoid. The solenoid is powered through circuit 26 and controlled by the horn switch through circuit 25.

(H) **SPRING BRAKE WARNING SYSTEM** – Warns the operator that the spring brakes are applied. The system consists of normally open pressure switch powered through circuit 37 and an indicator lamp which is powered through circuit 37.

(I) **FAILSAFE WARNING SYSTEM** – Intended to give the operator an audible as well as visual signal of a malfunction in one of the primary systems. Power for the system is supplied from the ignition switch through circuit 564. The failsafe module causes an indicator lamp to illuminate and an alarm to sound when the air pressure falls below 60 psi (414 kPa) or when the parking brake is set.

1-19. ELECTRICAL SYSTEMS OPERATION (Contd)

h. Trailer and Semitrailer Connection System Operation.

The trailer receptacle is identical on all models covered in this manual. The semitrailer receptacle is on the tractor body only.

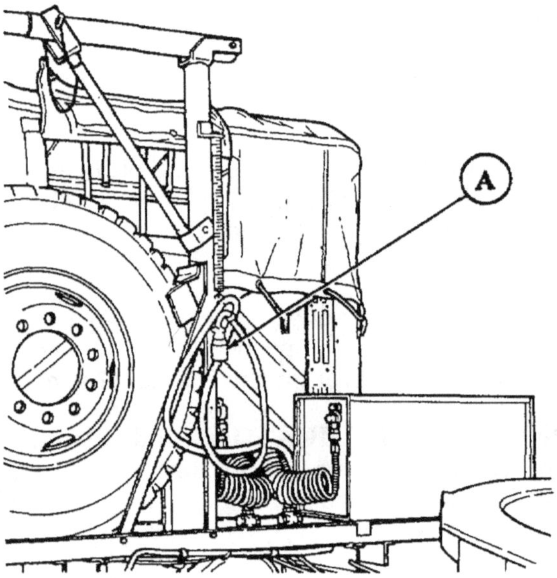
A

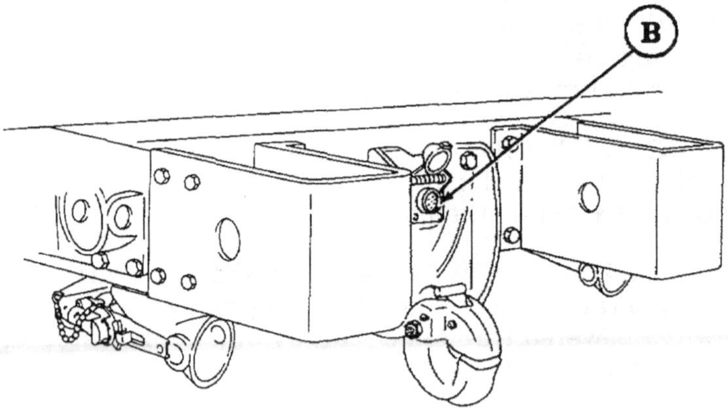
B

(A) **TRAILER RECEPTACLE** – Provides vehicle lighting, auxiliary power, and a ground circuit for trailers.

(B) **SEMITRAILER RECEPTACLE** – M931/A1/A2 and M932/A1/A2 vehicles equipped with a fifth wheel are provided with a semitrailer receptacle. This receptacle provides vehicle lighting, auxiliary power, and a ground circuit for semitrailers.

1-20. COMPRESSED AIR AND BRAKE SYSTEM OPERATION

NOTE

When vehicles are equipped with antilock brake system (ABS), they will have different front and rear relay valves, double check valves, LQ-2 valve replaces limiting valve, new air dryer with heater, and inversion valve installed.

The compressed air and brake system takes filtered air, compresses it, and supplies it to various components that enable the operator to slow down or stop the vehicle. This system also supplies compressed air to air-actuated accessories throughout the vehicle. These components and accessories will be described as part of the following systems:

a. Medium Wrecker Automatic Brake Lock System Operation (page 1-57).

b. Air Pressure Supply System Operation (page 1-58).

c. Secondary Service Airbrake System Operation (page 1-62).

d. Spring Airbrake System Operation (page 1-64).

e. Primary Service Airbrake System Operation (page 1-65).

f. Auxiliary Air-Powered System Operation (page 1-68).

g. Air Venting System Operation (page 1-70).

h. Central Tire Inflation System (CTIS) (M939A2 series vehicles) (page 1-72).

a. Medium Wrecker Automatic Brake Lock System Operation.

The M936/A1/A2 Medium Wrecker Automatic Brake Lock System locks the service airbrakes when the transfer case power takeoff lever is engaged. Major components of the automatic brake lock system are:

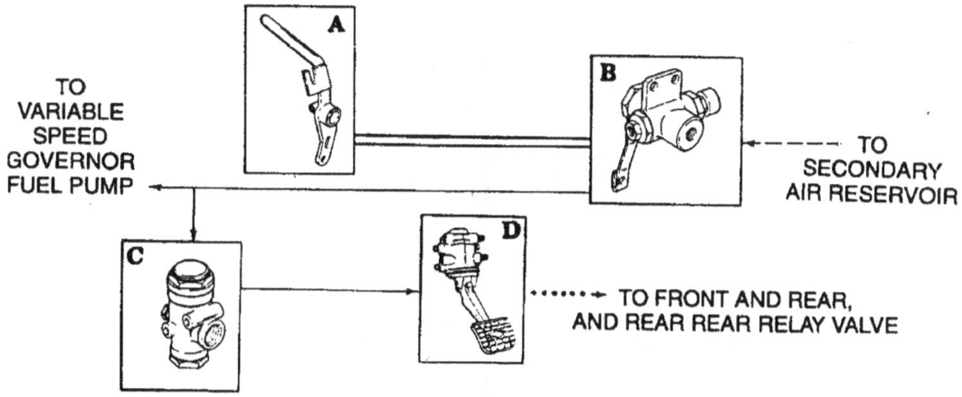
A
B
C
D
TO
VARIABLE
SPEED
GOVERNOR
FUEL PUMP
TO
SECONDARY
AIR RESERVOIR
TO FRONT AND REAR,
AND REAR REAR RELAY VALVE

(A) **TRANSFER CASE POWER TAKEOFF LEVER** – Opens the brake lock control valve through mechanical linkage when engaged.

(B) **BRAKE LOCK CONTROL** – Allows air pressure to flow from secondary air reservoir to pressure regulator and activate variable speed governor.

(C) **PRESSURE REGULATOR** – Reduces and regulates system air pressure to 70 psi (483 kPa) for automatic brake lock application.

(D) **TREADLE VALVE** – Connects pressure regulator and service airbrakes.

1-20. COMPRESSED AIR AND BRAKE SYSTEM OPERATION (Contd)

b. Air Pressure Supply System Operation.

(1) A constant air pressure supply is developed by the compressor which is regulated by the governor to maintain 90 to 120 psi (621 to 827 kPa) for the airbrake system. Moisture within the system is controlled through the use of either the alcohol evaporator or air dryer. The major components of the system are:

NOTE

- Vehicles equipped with antilock brake system (ABS) will have a new air dryer installed in position A shown below and items F and H will be removed from vehicle.
- Vehicles equipped with ABS have an added 250 psi (1724 kPa) safety valve installed in air supply line from compressor to air dryer.

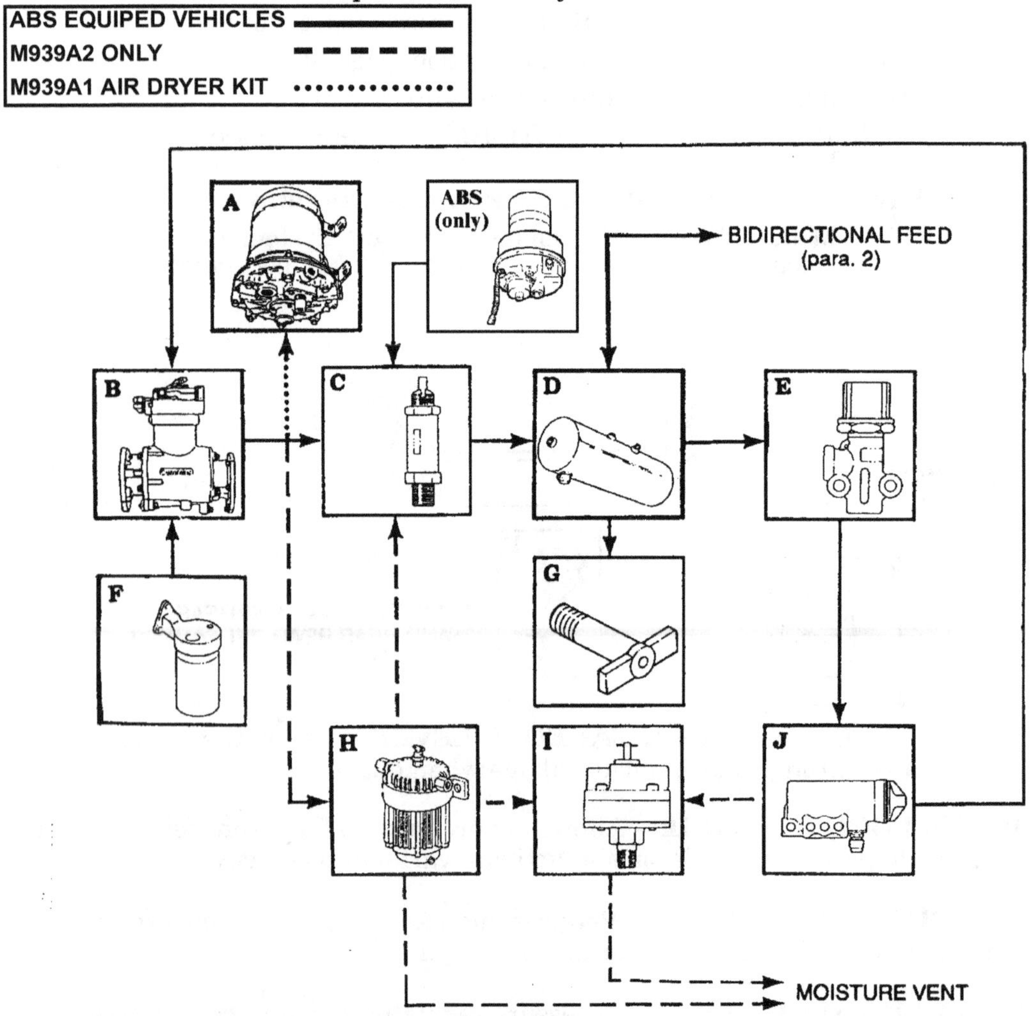
ABS EQUIPED VEHICLES
M939A2 ONLY
M939A1 AIR DRYER KIT
A
ABS (only)
BIDIRECTIONAL FEED (para. 2)
B
C
D
E
F
G
H
I
J
MOISTURE VENT

1-20. COMPRESSED AIR AND BRAKE SYSTEM OPERATION (Contd)

(A) **AIR DRYER** – Installed in supply line to wet tank and removes moisture from inlet air to wet tank (M939/A1 air dryer kit installed only), or (replaced with new type for vehicles with ABS kit installed).

(B) **AIR COMPRESSOR** – Draws in air from the intake manifold and forces it into the brake system and wet tank reservoir.

(C) **SAFETY VALVE** – Located at the inlet side of the wet reservoir, it prevents pressure build-up by releasing air pressure exceeding 150 psi (1034 kPa) when the governor fails to regulate air supplied by the compressor.

(D) **WET TANK RESERVOIR** – Performs two functions:
- Traps water in air reservoir to protect other air systems from freezing or corroding.
- Stores reserve air supply enabling operator to make normal stops when engine stalls or compressor fails.

(E) **PRESSURE PROTECTION VALVE** – Performs two functions:
- Allows air pressure to build to 60-65 psi (414-448 kPa) before supplying air to auxiliary air-powered equipment.
- Closes off auxiliary air system from other systems if an accessory fails and prevents loss of air from secondary reservoir.

(F) **ALCOHOL EVAPORATOR (M939/A1 series)** – Helps protect air lines from freezing (not used if ABS kit is installed).

(G) **WET TANK RESERVOIR DRAINVALVE** – Provides a drain for moisture and air from reservoir.

(H) **AIR DRYER** – Removes moisture from inlet air to wet tank (M939A2 only) or (not used if ABS kit is installed).

(I) **EXPELLO VALVE** – Augments air dryer condensation blowdown by venting moisture when compressor cycles (not used if ABS kit is installed).

(J) **GOVERNOR** – Trips valve inside compressor to regulate flow of air to the system.When pressure builds to 120-127 psi (827-876 kPa), the governor will close valve.

1-20. COMPRESSED AIR AND BRAKE SYSTEM OPERATION (Contd)

(2) The constant air pressure supply is distributed to the primary service airbrake system (para e.) and secondary airbrake system (para c.) through a shutoff and check valve. Air pressure can either be fed from or supplied to another vehicle through the emergency couplings.

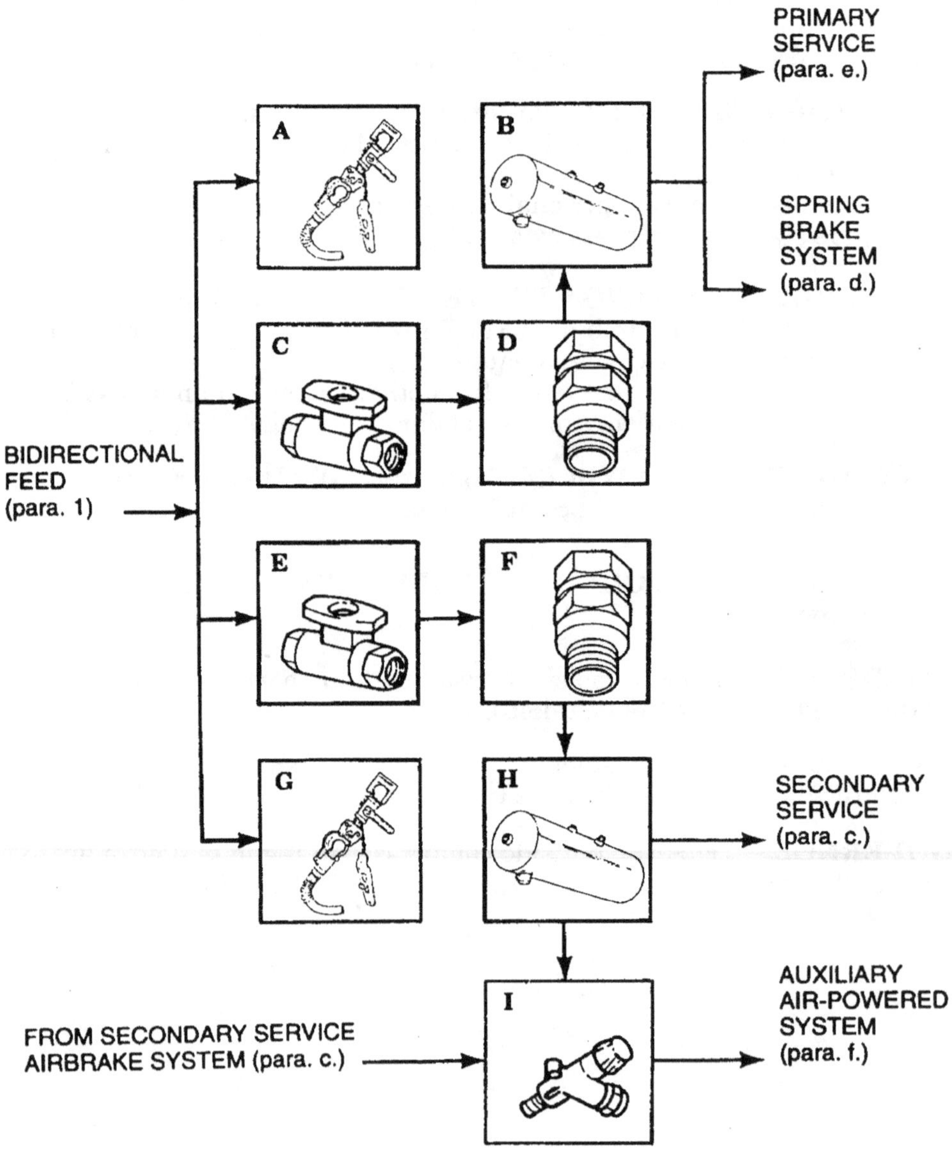
PRIMARY
SERVICE
(para. e.)
A
B
SPRING
BRAKE
SYSTEM
(para. d.)
C
D
BIDIRECTIONAL
FEED
(para. 1)
E
F
G
H
SECONDARY
SERVICE
(para. c.)
I
AUXILIARY
AIR-POWERED
SYSTEM
(para. f.)
FROM SECONDARY SERVICE
AIRBRAKE SYSTEM (para. c.)

1-20. COMPRESSED AIR AND BRAKE SYSTEM OPERATION (Contd)

(A) **FRONT EMERGENCY COUPLING** – When vehicle is being towed, coupling receives compressed air from towing vehicle's brake system to charge its own brake system.

(B) **PRIMARY AIR RESERVOIR** – Stores sufficient air pressure to allow operator to make normal brake applications, should system pressure fail or engine stall.

(C) **PRIMARY FEED CUTOFF** – Manually-operated valve used to isolate pressure leaks in primary air system from draining wet tank (para. 3-12).

(D) **PRIMARY AIR RESERVOIR CHECKVALVE** – Presents backflow of air from primary tank if wet system develops a leak.

(E) **SECONDARY FEED CUTOFF** – Manually-operated valve used to isolate pressure leaks in secondary air system from draining wet tank (para. 3-12).

(F) **SECONDARY AIR RESERVOIR CHECKVALVE** – Prevents backflow of air from secondary tank if wet system develops a leak.

(G) **REAR EMERGENCY COUPLING** – When towing another vehicle, coupling allows pressurized air from wet tank to charge towed vehicle's wet tank.

(H) **SECONDARY AIR RESERVOIR** – Stores enough air pressure in case constant pressure system fails or engine stalls. The operator can make normal brake application before running out of air.

(I) **PRESSURE PROTECTION VALVE** – Performs two functions:

- Allows air pressure to build to 60-65 psi (414-448 kPa) before supplying air to auxiliary air-powered equipment.
- Closes off auxiliary air system from other systems if an accessory fails and prevents loss of air from secondary reservoir.

1-20. COMPRESSED AIR AND BRAKE SYSTEM OPERATION (Contd)

c. Secondary Service Airbrake System Operation.

(1) The secondary service airbrake system is made up of two subsystems:

(a) Secondary constant pressure system provides continuous air pressure to:

- Pedal valve.

NOTE

When vehicles are equipped with ABS, they have a new type of rear relay valve that replaces item I below.

- Rear relay valve.
- Spring brake air reservoir.
- Spring parking brake valve.

(b) Secondary signal system serves three functions:

- Contains air pressure only when operator steps on brake pedal.
- Is regulated by various valves to control amount of braking.

NOTE

On vehicles equipped with ABS, the rear relay valve operates the rear-rear right hand brake chamber and itermediate rear right hand brake chamber. The front relay valve operates the rear-rear left hand brake chamber and intermediate rear left hand breake chamber.

- Provides pressure to apply the rear two service brakes and the intermediate and rear axles stamped with B. Service brakes on the rear axle are piggybacked to spring brakes but operate independently of them.

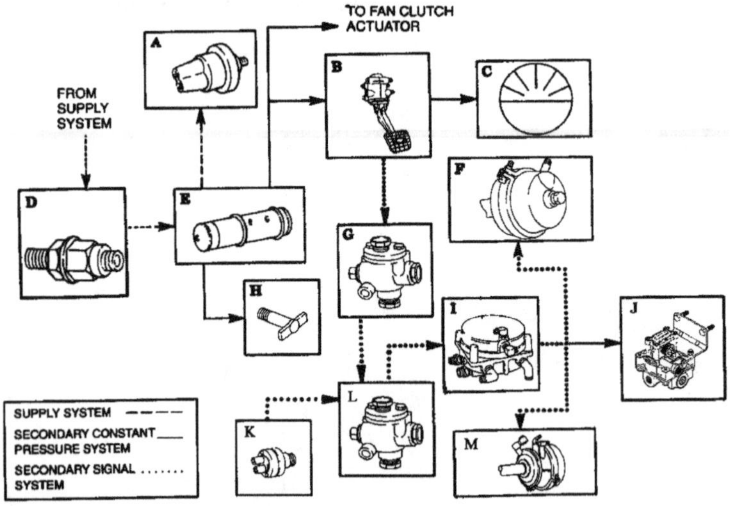
TO FAN CLUTCH ACTUATOR
FROM SUPPLY SYSTEM
A
B
C
D
E
F
G
H
I
J
K
L
M
SUPPLY SYSTEM
SECONDARY CONSTANT PRESSURE SYSTEM
SECONDARY SIGNAL SYSTEM

1-20. COMPRESSED AIR AND BRAKE SYSTEM OPERATION (Contd)

(2) The secondary constant pressure system is made up of the following components:

(A) **LOW AIR PRESSURE SWITCH** – Activates warning buzzer and warning lights when air pressure goes below 60 psi (414 kPa).

(B) **PEDAL VALVE** – Allows air pressure from secondary constant pressure system to flow into secondary signal system when operator depresses brake pedal.

(C) **SECONDARY AIR PRESSURE GAUGE** – Indicates amount of air pressure in secondary system.

(D) **ONE-WAY CHECK VALVE** – Allows air pressure to flow into secondary reservoir but prevents it from coming out if constant pressure system fails or engine stalls.

(E) **SECONDARY AIR RESERVOIR** – Stores enough air pressure so the operator can make five normal brake applications before running out of air if constant pressure fails or engine stalls.

(F) **INTERMEDIATE REAR BRAKE CHAMBERS** – Converts air pressure to mechanical force which applies intermediate rear service brake.

(G) **DOUBLECHECK VALVE #1** – Serves two functions:
- Allows system to receive signal pressure from either pedal valve or, when towed, from brake system of towing vehicle.
- Serves as a tee between front and rear primary signal lines.
- Removed when ABS kit is installed and replaced with a cross tee.

(H) **SECONDARY AIR RESERVOIR DRAINVALVE** – Provides a drain for moisture and air from secondary air reservoir.

(I) **FRONT RELAY VALVE** – Boosts signal air to rear brake chambers; regulates air pressure to rear brake chambers so operator has control over amount of braking; and releases air pressure to rear brake chambers directly to vent when brake pedal is released (not used if ABS kit is installed).

(J) **(ABS) FRONT RELAY VALVE WITH ELECTRONIC CONTROL UNIT (ECU)** – Boosts signal air to left rear brake chamber and left intermediate bake chamber controling brake lock up by venting air to chambers on left side of vehicle if left wheels lock up when braking.

(K) **STOPLIGHT SWITCH** – As the brake pedal is depressed, switch receives an air pressure signal at electrical contacts which close to activate circuits to taillights.

(L) **DOUBLECHECK VALVE #2** – Allows either primary or secondary signal air pressure to activate stoplight switch while keeping the two systems separate.

(M) **REAR-REAR BRAKE CHAMBERS** – Converts air pressure to mechanical force which applies rear-rear brakes.

1-20. COMPRESSED AIR AND BRAKE SYSTEM OPERATION (Contd)

d. Spring Airbrake System Operation.

The spring airbrake system applies rear brakes when vehicle parking brake is applied or in event of a major brake failure. The spring brake is located on one of the two service brake chambers at each rear wheel. Major components of the spring airbrake system are:

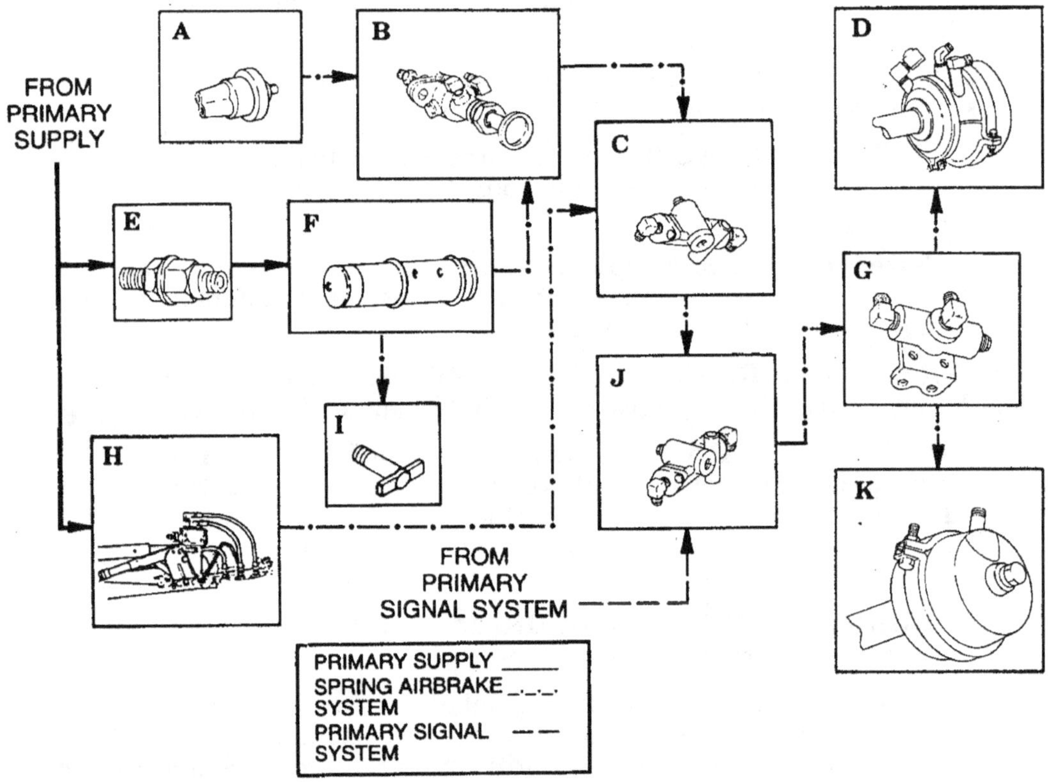
A
B
C
D
E
F
G
H
I
J
K
FROM PRIMARY SUPPLY
FROM PRIMARY SIGNAL SYSTEM
PRIMARY SUPPLY
SPRING AIRBRAKE SYSTEM
PRIMARY SIGNAL SYSTEM

(A) **SPRING BRAKE WARNING LIGHT SWITCH** – Activates warning light when spring brakes are engaged.

(B) **SPRING BRAKE RELEASE CONTROL VALVE** – Pushed in to release spring brakes independently of mechanical parking brake. Control is also used to release spring brakes in order to test and adjust mechanical brake.

(C) **DOUBLECHECK VALVE #4** – Allows spring brake air pressure to come from either release control valve or spring parking brake valve directly to doublecheck valve #3.

1-20. COMPRESSED AIR AND BRAKE SYSTEM OPERATION (Contd)

(D) **INTERMEDIATE FRONT SPRING BRAKE CHAMBER** – Contains a large spring which applies rear brakes when spring brake air pressure is released.

(E) **ONE-WAY CHECK VALVE** – Allows air pressure to flow into spring brake reservoir but prevents it from coming out if constant pressure system or primary system fails.

(F) **SPRING BRAKE AIR RESERVOIR** – Stores enough air pressure to release spring brakes for emergency operation in event of primary or secondary air system failure.

(G) **QUICK-RELEASE VALVE** – Releases spring brake air pressure directly to vent if parking brake has been set or brake system fails.

(H) **SPRING BRAKE VALVE** – Automatically sets spring brakes when parking brake is set. Valve can be released independently of parking brake when spring brake control valve is pushed in.

(I) **SPRING BRAKE RESERVOIR DRAINVALVE** – Provides a drain for moisture and air from spring brake reservoir.

(J) **DOUBLECHECK VALVE #3** – Allows spring brake air pressure to come from either release control valve or spring parking brake valve directly to doublecheck valve #4.

(K) **REAR-REAR SPRING BRAKE CHAMBER** – Contains a large spring which applies rear brakes when spring brake air pressure is released.

e. Primary Service Airbrake System Operation.

(1) The primary service airbrake system is made up of two subsystems:

(a) Primary constant pressure system provides continuous air pressure to:

- Pedal valve.
- Rear relay valve.
- Spring brake air reservoir.
- Spring parking brake valve.

(b) Primary signal system serves three functions:

- Contains pressure only when operator steps on brake pedal.
- Is regulated by various valves to give operator control over amount of braking.
- Provides pressure to apply front service brakes and the front two service brakes on the intermediate and rear axles stamped with an A. Service brakes on the intermediate axle are piggybacked to spring brakes but operate independently of them.

1-20. COMPRESSED AIR AND BRAKE SYSTEM OPERATION (Contd)

(2) The primary constant pressure system is made up of the following components:

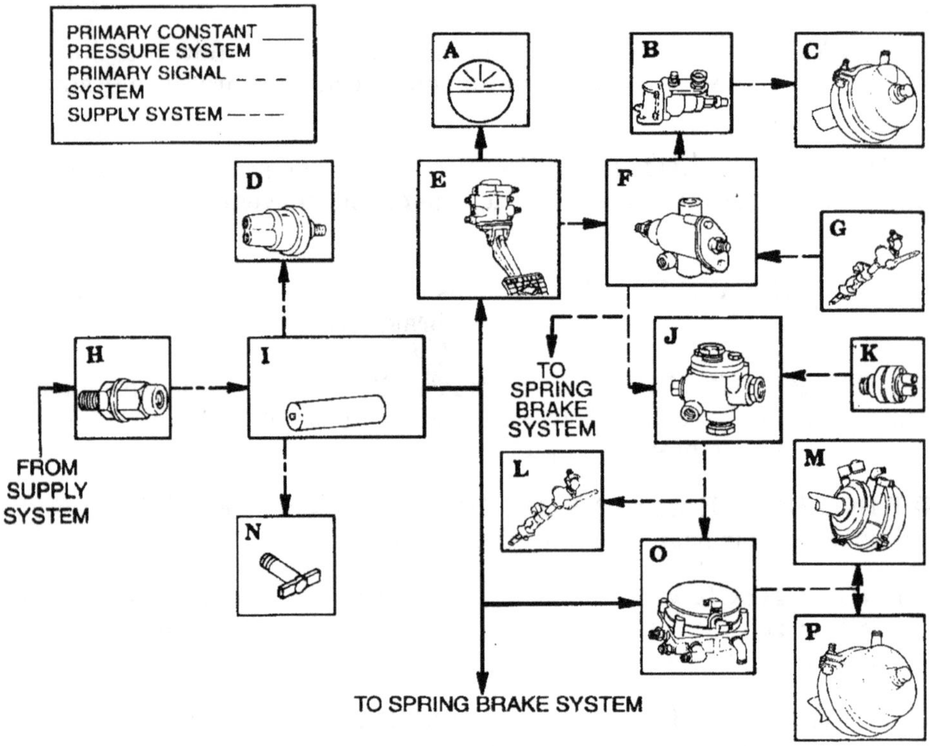
PRIMARY CONSTANT PRESSURE SYSTEM
PRIMARY SIGNAL SYSTEM
SUPPLY SYSTEM
A
B
C
D
E
F
G
H
I
J
K
L
M
N
O
P
TO SPRING BRAKE SYSTEM
FROM SUPPLY SYSTEM
TO SPRING BRAKE SYSTEM

(A) **PRIMARY AIR PRESSURE GAUGE** – Indicates amount of air pressure in primary system.

NOTE

Vehicles equipped with ABS will have a LQ-2 valve inplace of limiting valve or front brake valve to control air pressure to front brake chambers.

(B) **LIMITING VALVE** – Serves three functions:

- Regulates signal air pressure going to front brake chambers so rear brakes are applied first.
- Regulates signal air pressure to front brake chambers so operator has control over amount of braking.
- Releases air pressure in front brake chambers directly to vent in the valve when brake pedal is released.

(C) **FRONT BRAKE CHAMBERS** – Converts air pressure to mechanical force which applies front service brakes.

(D) **PRIMARY RESERVOIR LOW AIR PRESSURE SWITCH** – Activates warning buzzer and warning light when air pressure goes below 60 psi (414 kPa).

1-20. COMPRESSED AIR AND BRAKE SYSTEM OPERATION (Contd)

(E) **PEDAL INTO VALVE** – Allows air pressure from primary constant pressure system to flow into primary signal system when operator depresses brake pedal.

(F) **DOUBLECHECK VALVE #1** – Serves two functions:

- Allows system to receive signal pressure from either pedal valve or, when towed, from brake system of towing vehicle.
- Serves as a tee between front and rear primary signal lines.
- Removed when ABS kit is installed and replaced with cross tee.

(G) **FRONT SERVICE COUPLING** – When vehicle is being towed, coupling is connected to towing vehicle so that the brake systems of the two vehicles work together.

(H) **ONE-WAY CHECK VALVE** – Allows air pressure to flow into primary reservoir but prevents it from coming out if constant pressure system fails or engine stalls.

(I) **PRIMARY AIR RESERVOIR** – Stores enough air pressure so the operator can make five normal brake applications before running out of air if constant pressure fails or engine stalls.

(J) **DOUBLECHECK VALVE #2** – Allows either primary or secondary signal air pressure to activate stoplight switch while keeping the two systems separate.

(K) **STOPLIGHT SWITCH** – As the brake pedal is depressed, switch receives an air pressure signal which closes electric contacts turning on stoplight.

(L) **REAR SERVICE COUPLING** – When towing another vehicle, coupling is connected to towed vehicle so that the brake system of the two vehicles work together.

(M) **INTERMEDIATE FRONT BRAKE CHAMBERS** – Converts air pressure to mechanical force which applies intermediate rear service brake.

(N) **PRIMARY RESERVOIR DRAINVALVE** – Provides a drain for moisture and air from primary air reservoir.

NOTE

On vehicles equipped with ABS, the rear relay valve operates the rear-rear right hand brake chamber and intermediate rear right hand brake chamber.

(O) **REAR RELAY VALVE** – Serves three functions:

- Boosts signal air pressure to rear brake chambers. Air signal from brake pedal opens valve to route constant air pressure to rear brake chambers.
- Regulates signal air pressure from brake pedal to rear brake chambers so operator has control over amount of braking. Regulates amount of constant air pressure going to brake chambers as the operator depresses the brake pedal.
- Releases air pressure in rear brake chamber directly to vent system when brake pedal is released.

(P) **REAR FRONT BRAKE CHAMBERS** – Converts air pressure to mechanical force which applies rear service brakes.

1-20. COMPRESSED AIR AND BRAKE SYSTEM OPERATION (Contd)

f. Auxiliary Air-Powered System Operation.

The auxiliary air-powered system consists of air-actuated vehicle accessories. All of these accessories receive air pressure through the accessory manifold and off the pressure protection valve with the exception of the horns. Components of the auxiliary air-powered system are:

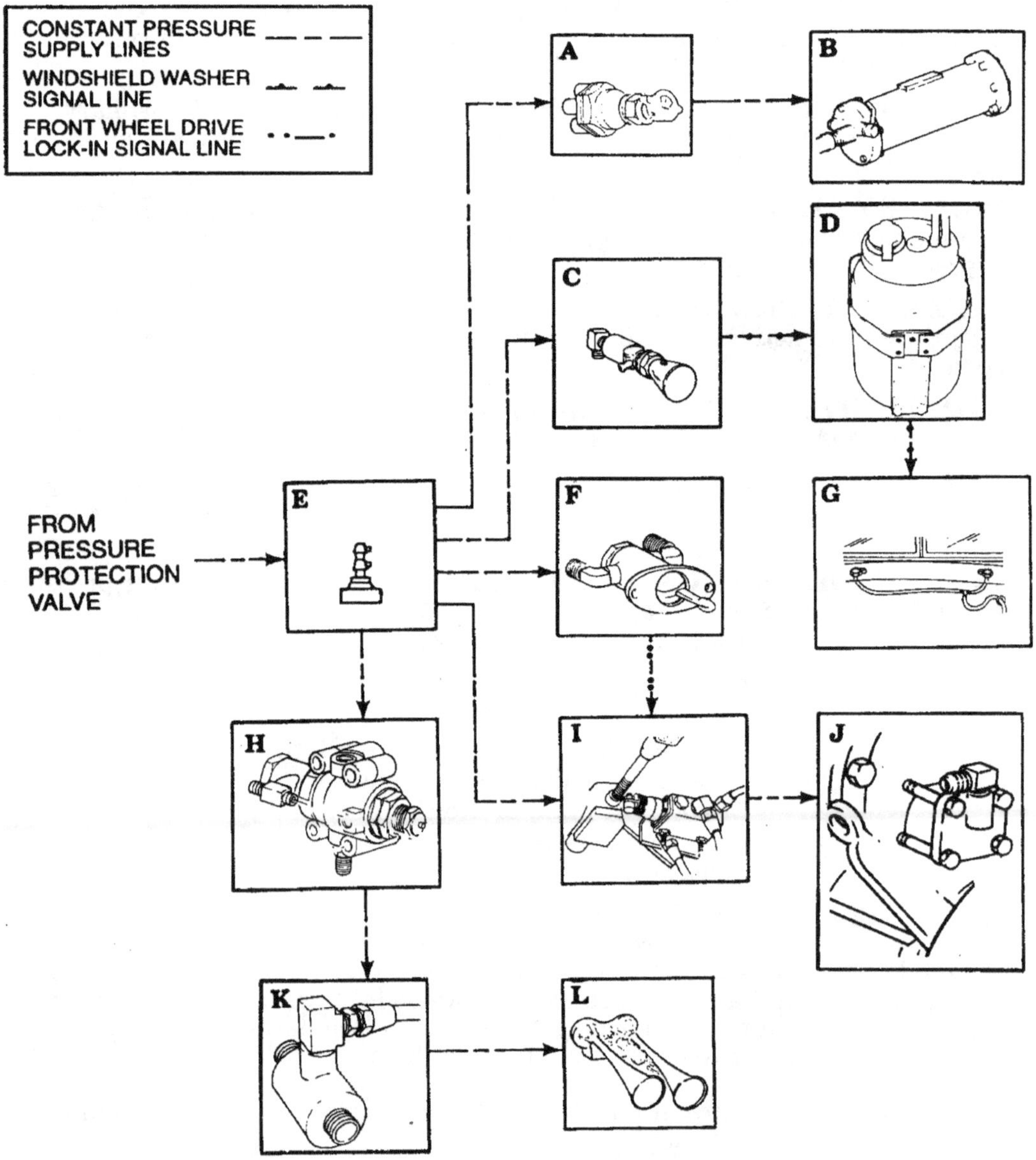

CONSTANT PRESSURE SUPPLY LINES
WINDSHIELD WASHER SIGNAL LINE
FRONT WHEEL DRIVE LOCK-IN SIGNAL LINE
FROM PRESSURE PROTECTION VALVE
A
B
C
D
E
F
G
H
I
J
K
L

1-20. COMPRESSED AIR AND BRAKE SYSTEM OPERATION (Contd)

(A) **WINDSHIELD WIPER CONTROL SWITCH** – Opens air pressure valve in wiper motor to operate wipers.

(B) **WINDSHIELD WIPER MOTOR** – Air-actuated motor powers windshield wipers.

(C) **WINDSHIELD WASHER CONTROL** – Spring-loaded valve that allows air pressure to force washer fluid from washer reservoir to windshield.

(D) **WINDSHIELD WASHER RESERVOIR** – Container for windshield washer fluid.

(E) **ACCESSORY MANIFOLD** – Receives air pressure from the pressure protection valve and distributes it to the various accessories.

(F) **FRONT-WHEEL DRIVE LOCK-IN SWITCH** – Air-actuated switch that engages front-wheel drive when transfer case is in HIGH.

(G) **WINDSHIELD WASHER NOZZLES** – Direct washer fluid on windshield.

(H) **GOVERNOR** – Serves as a tee between accessory manifold and horn relay valve. It also signals the air compressor to stop compressing air for the supply system when operating pressure has been reached.

(I) **TRANSFER CASE AIR SHIFT CYLINDER** – Engages front-wheel drive when it receives air pressure from lock-in switch or engagement control valve.

(J) **FRONT AXLE ENGAGEMENT CONTROL VALVE** – Operates off cam on transfer case shift linkage so front-wheel drive engages automatically when transfer case is put into LOW.

(K) **HORN RELAY VALVE** – Electrical signal from horn button on steering wheel opens valve in horn relay, allowing air pressure to sound horns.

(L) **HORNS** – Receive air pressure from horn relay valve to sound off.

1-20. COMPRESSED AIR AND BRAKE SYSTEM OPERATION (Contd)

g. Air Venting System Operation.

The air venting system vents air from brake system and power train, and fuel vapors from fuel system into air intake stack where it is released into the atmosphere. The components of the air venting system are:

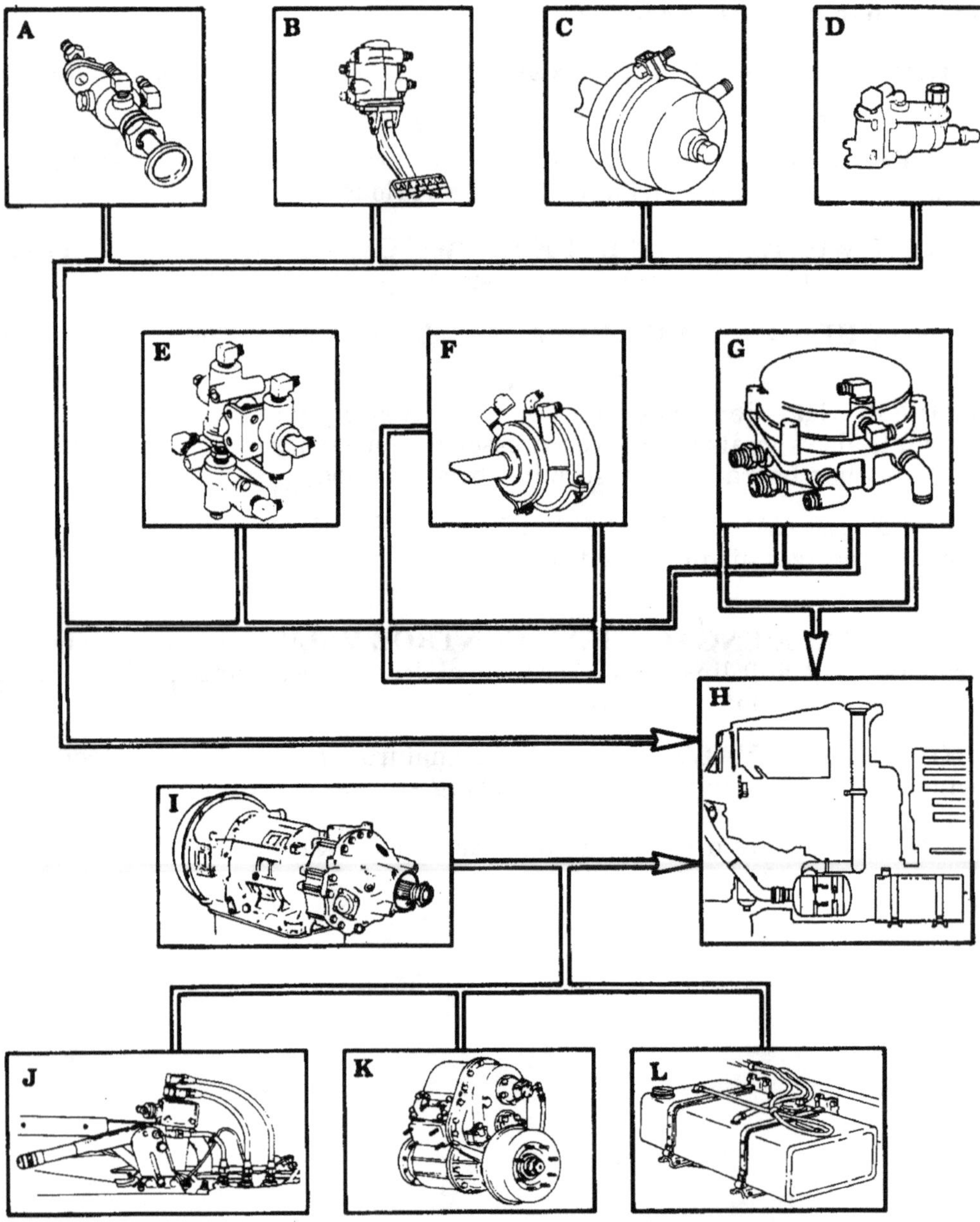
A
B
C
D
E
F
G
H
I
J
K
L

1-20. COMPRESSED AIR AND BRAKE SYSTEM OPERATION (Contd)

(A) **SPRING BRAKE RELEASE CONTROL VALVE** – This valve functions as an override when a failure in the air supply system (causing spring brakes to engage) occurs. When valve is manually pushed in, emergency air is supplied to the spring brake chambers. This releases the spring brakes, allowing vehicle movement.

(B) **PEDAL VALVE** – Vents primary or secondary signal air pressure when pedal is released.

(C) **FRONT BRAKE CHAMBER VENT** – Vents air pressure inside chambers when pedal valve is released.

(D) **LIMITING VALVE** – Vents signal air pressure going to front brake chambers so rear brakes apply first.

(E) **STEP BOX QUICK-RELEASE VALVE** – Vents air pressure from spring brake chambers when parking brake valve has been actuated.

(F) **REAR BRAKE CHAMBERS** – Vents ports on chambers to prevent air pressure build-up.

(G) **RELAY VALVES** – Vents air pressure in rear brake chambers directly to intake tube when brake pedal is released. Vents signal air pressure through upper port in valve.

(H) **AIR INTAKE STACK** – Venting point for the vent system.

(I) **TRANSMISSION VENT** – Vents internal air pressure build-up due to internal heat.

(J) **SPRING PARKING BRAKE VALVE** – Vents air pressure from air and doublecheck valves #3 and #4.

(K) **TRANSFER CASE VENT** – Vents internal air pressure build-up due to internal heat.

(L) **FUEL TANK VENTS** – Vent fuel vapors to prevent partial vacuum from stopping fuel flow.

1-20. COMPRESSED AIR AND BRAKE SYSTEM OPERATION (Contd)

h. Central Tire Inflation System (CTIS).

NOTE

Vehicles equipped with antilock brake system (ABS) will have a new air dryer installed in position D shown below.

The CTIS is common to all M939A2 series vehicles. This system maintains tire air pressure depending on which road type is selected. If this setting is changed, tires will automatically inflate or deflate to the new setting.

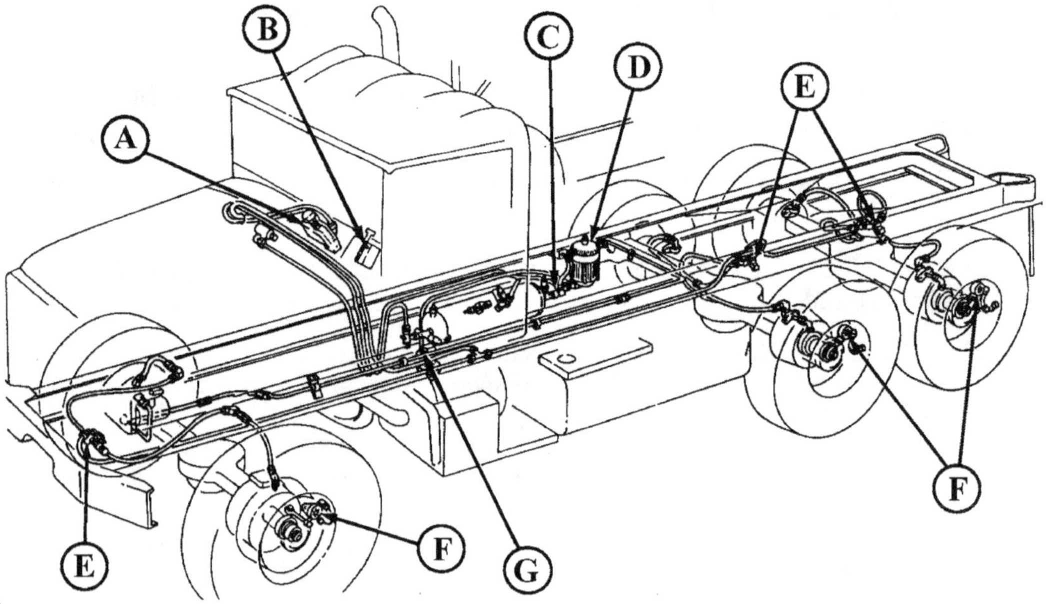
B
C
D
E
A
F
E
F
G

(A) **PNEUMATIC CONTROLLER** – Directs air pressure according to ECU commands.

(B) **ELECTRONIC CONTROL UNIT (ECU)** – Contains CTIS selector panel so that operator can change tire inflation during vehicle operation.

(C) **AIR PRESSURE SWITCH** – Protects air brake system for a minimum supply of 85 psi (586 kPa) of air.

(D) **AIR DRYER AND FILTER** – Separates moisture from compressed air system and filters impurities from compressed air before they enter CTIS.

(E) **EXHAUST VALVES** – Exhausts air from tires during deflation.

(F) **WHEEL VALVES** – Isolates air pressure in the tires during normal operation and for tire removal.

(G) **SPEED SIGNAL GENERATOR** – Signals ECU to automatically inflate CTIS when vehicle speed exceeds the top speed setting for the selected mode by 10 mph (16 km/h).

1-21. CONTROL SYSTEM OPERATION

Oil pressure (hydraulics) is used to provide operating power for the auxiliary equipment on the vehicles covered in this manual. The components that provide hydraulic power are discussed in the following order:

a. Front Winch Hydraulic System Operation (page 1-73).

b. Rear Winch Hydraulic System Operation (page 1-74).

c. Body Hydraulic System Operation (page 1-76).

d. Medium Wrecker Crane Hydraulic System Operation (page 1-78).

a. Front Winch Hydraulic System Operation.

A front winch is installed on M925/A1/A2, M928/A1/A2, M930/A1/A2, M932/A1/A2, and M936/A1/A2 series vehicles. The front winch hydraulic system converts mechanical power at the winch drive motor. The basic operating principles are the same for each model. Major components of this system are:

(A) **TRANSMISSION POWER TAKEOFF (PTO) CONTROL** - A manually operated control lever located inside the cab that permits engagement or disengagement of the transmission Power Takeoff (PTO).

(B) **WINCH CONTROL LEVER** - An operator control that determines the hydraulic oil pressure flow from the control valve to the winch motor. The flow of this oil determines the direction the winch drum will turn.

(C) **TRANSMISSION POWER TAKEOFF (PTO)** - Uses driving power of the transmission to provide mechanical driving power for the hydraulic pump.

(D) **POWER TAKEOFF (PTO) DRIVE SHAFT** - Transmits mechanical power from the PTO to the hydraulic pump.

(E) **HYDRAULIC PUMP** - Driven by the PTO drive shaft, it draws oil from the oil reservoir through hydraulic hoses, then pressurizes and directs this oil to the control valve.

1-21. CONTROL SYSTEM OPERATION (Contd)

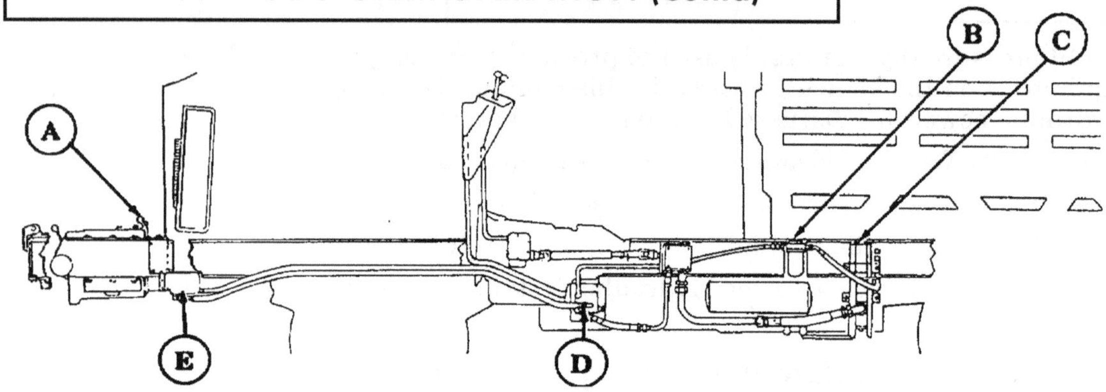
A
B
C
D
E

(A) CLUTCH LEVER - Manual control that engages the winch drum gear to the drive gear of the winch motor.

(B) OIL FILTER - Filters used or bypassed oil from the control valve before it returns to the hydraulic oil reservoir.

(C) HYDRAULIC OIL RESERVOIR - Storage tank for hydraulic oil.

(D) CONTROL VALVE - Four-port valve accepts pressurized oil from the hydraulic pump and directs this oil to the winch motor. It also directs oil returning from the winch back to the oil reservoir. The flow of this oil from the valve determines the directional drive of the winch motor.

(E) WINCH MOTOR - Converts hydraulic power into mechanical power as hydraulic oil is forced through the winch motor.

b. Rear Winch Hydraulic System Operation.

A rear winch is installed only on the M936/A1/A2 medium wrecker. It is primarily to rescue vehicles that have become deeply mired. The rear winch hydraulic system converts mechanical power of the engine into fluid power through use of the hydraulic pump and back into mechanical power at the winch drive motor. The major components of the rear winch hydraulic system are:

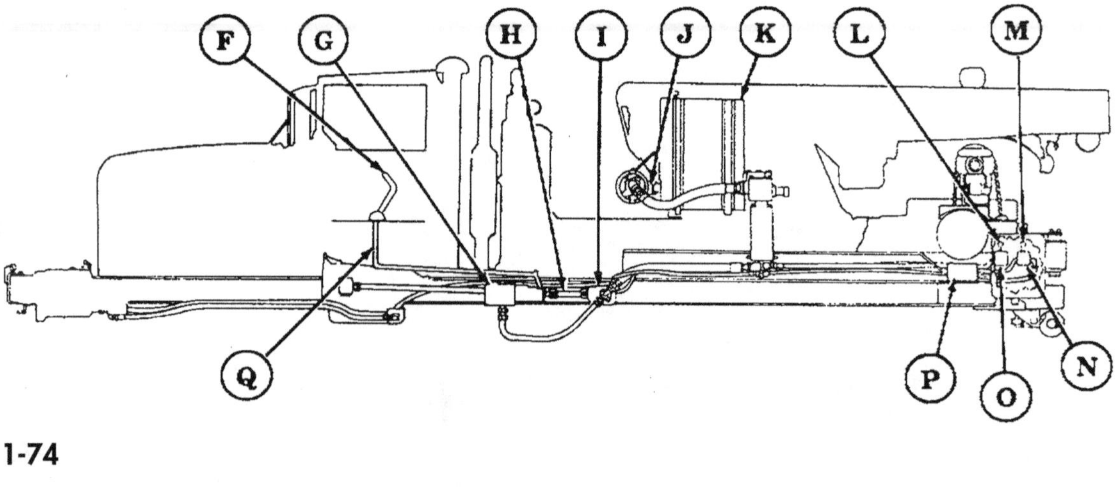
F
G
H
I
J
K
L
M
N
O
P
Q

1-21. CONTROL SYSTEM OPERATION (Contd)

(F) **TRANSFER CASE POWER TAKEOFF (PTO) CONTROL** - A manually operated control lever located inside the cab that permits engagement or disengagement of the Power Takeoff (PTO).

(G) **TRANSFER CASE POWER TAKEOFF (PTO)** - Uses driving power of the transfer case to provide mechanical driving power for the hydraulic pump.

(H) **POWER TAKEOFF (PTO) DRIVE SHAFT** - Transmits mechanical driving power from PTO to the hydraulic pump.

(I) **HYDRAULIC PUMP** - Draws oil from hydraulic oil reservoir and directs it to the rear winch control valve and winch drive motor.

(J) **OIL FILTER** - Filters used or bypassed oil from the control valve before it returns to the hydraulic oil reservoir.

(K) **HYDRAULIC OIL RESERVOIR** - Storage tank for hydraulic oil.

(L) **TORQUE CONTROL LEVER** - Controls the operating gear ratio of the winch drive motor. Lever is pulled outward to HIGH for heavy loads or pushed inward to LOW for light loads.

(M) **WINCH DIRECTIONAL CONTROL LEVER** - Manually-operated lever that controls the WIND and UNWIND direction of the rear winch drum. Lever does this by opening and closing the directional control valve to the winch motor, and reversing the direction of pressurized hydraulic fluid. Lever is pushed inward to wind and pulled outward to unwind winch cable.

(N) **DIRECTIONAL CONTROL VALVE** - Receives pressurized hydraulic oil from the hydraulic pump and directs it to the winch motor. The flow of the hydraulic oil to and from this control valve provides forward or reverse driving power to the winch motor. Valve also returns used oil back to the hydraulic oil reservoir from the winch.

(O) **TORQUE CONTROL VALVE** - Hydraulically controls the hydraulic oil pressure to engage rear winch drum clutch in HIGH or LOW gear range.

(P) **WINCH MOTOR** - Converts hydraulic power back into mechanical power needed to turn the rear winch drum.

(Q) **CONTROL LINKAGE** - Connects transfer case power takeoff control to transfer case power takeoff.

1-21. CONTROL SYSTEM OPERATION (Contd)

c. Dump Body Hydraulic System Operation.

The dump body is installed on M929/A1/A2 and M930/A1/A2 vehicles. These models are used to transport and deposit cargo. The dump body hydraulic system converts mechanical power from the engine into fluid power through use of the hydraulic pump. The pump draws fluid from the oil reservoir and then forces it into the control valve. This hydraulic pressure raises and lowers the dump body. Major components of the dump body hydraulic system are:

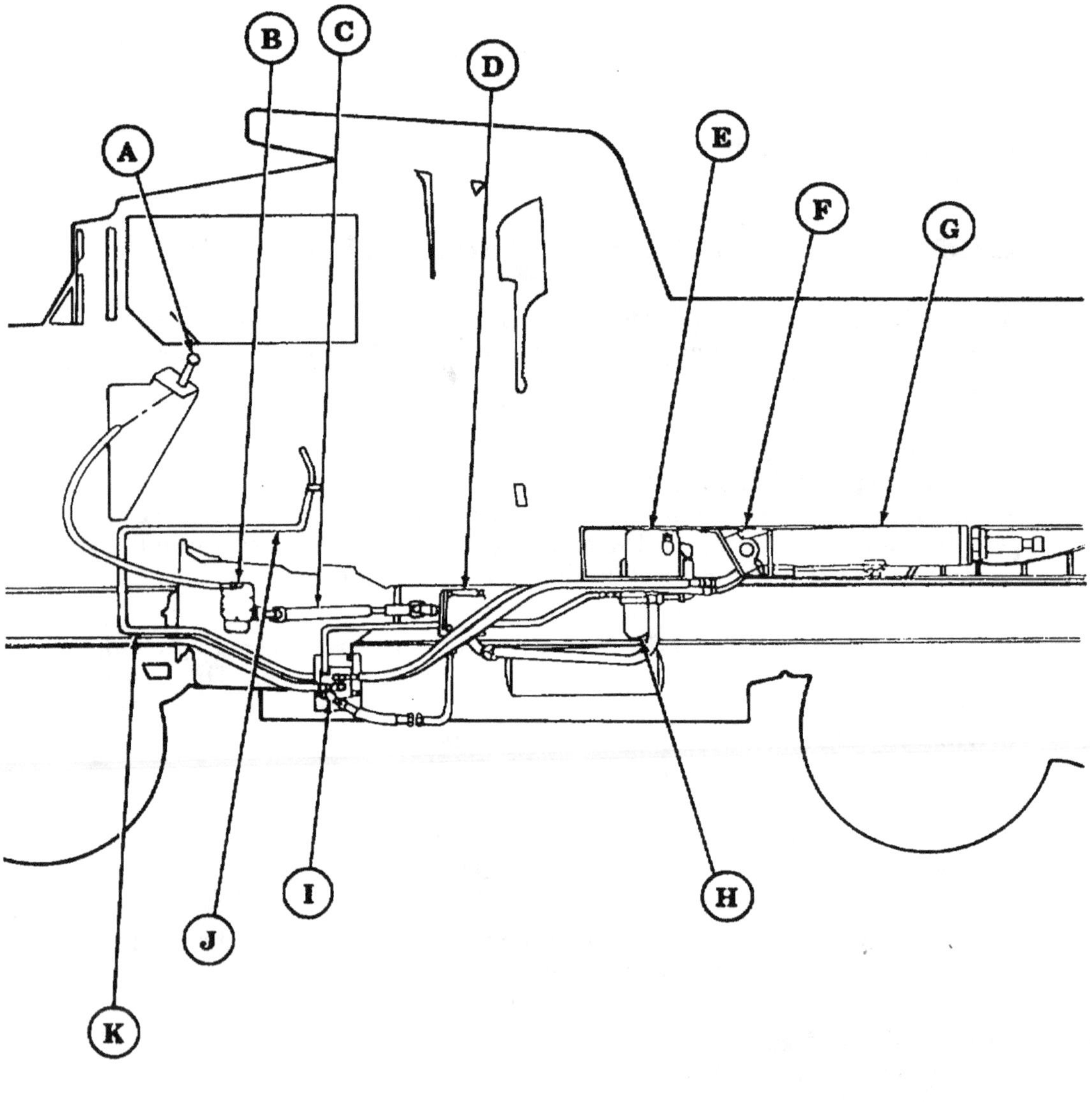

A
B
C
D
E
F
G
H
I
J
K

1-21. CONTROL SYSTEM OPERATION (Contd)

(A) **TRANSMISSION POWER TAKEOFF (PTO) CONTROL** - A manually operated control lever located inside the vehicle cab that permits engagement or disengagement of the transmission Power Takeoff (PTO).

(B) **TRANSMISSION POWER TAKEOFF (PTO)** - Uses driving power of the transmission to provide mechanical driving power for the hydraulic pump.

(C) **POWER TAKEOFF (PTO) DRIVE SHAFT** - Transmits mechanical driving power from the PTO to the hydraulic pump.

(D) **HYDRAULIC PUMP** - Driven by the PTO drive shaft, it draws oil from the oil reservoir through hydraulic hoses, then pressurizes and directs it to the control valve.

(E) **HYDRAULIC OIL RESERVOIR** - Storage tank for hydraulic oil.

(F) **DUMP BODY SAFETY LATCH** - Hydraulically-operated in conjunction with the dump body control lever, the safety latch locks the dump body in the lowered position and releases it when the control lever is pulled back to the raised position.

(G) **DUMP BODY CYLINDER ASSEMBLY** - Consists of two piston-type hydraulic cylinder hoists. Assembly raises and lowers dump body with hydraulic oil, forcing the cylinder upward or downward.

(H) **OIL FILTER** - Filters used or bypassed oil from the control valve before it returns to the hydraulic oil reservoir.

(I) **CONTROL VALVE** - Four-port valve accepts pressurized oil from the hydraulic pump and directs oil pressure flow from control valve to the hydraulic cylinders. It also directs oil returning from the hydraulic cylinders back to the hydraulic oil reservoir.

(J) **DUMP BODY CONTROL LEVER** - An operator control that determines the hydraulic oil pressure flow from control valve to the hydraulic cylinders. The route this oil takes will determine whether the dump will raise or lower.

(K) **CONTROL LINKAGE** - Connects dump body control lever inside cab to the control valve.

1-21. CONTROL SYSTEM OPERATION (Contd)

d. Medium Wrecker Crane Hydraulic System Operation.

The M936/A1/A2 medium wrecker is equipped with a hydraulically-operated crane that extends a maximum 18 feet (5 meters), elevates 45 degrees, and swings 360 degrees. It is capable of lifting loads up to 20, 000 lbs (9, 090 kg).

(1) The crane hydraulic system converts power of the engine into fluid power for use by the hydraulic pump. At this pump, oil pressure is supplied to different crane control valves: BOOM, HOIST, CROWD, and SWING. Each of these actions are dealt with separately. The major components for raising and lowering the wrecker boom are:

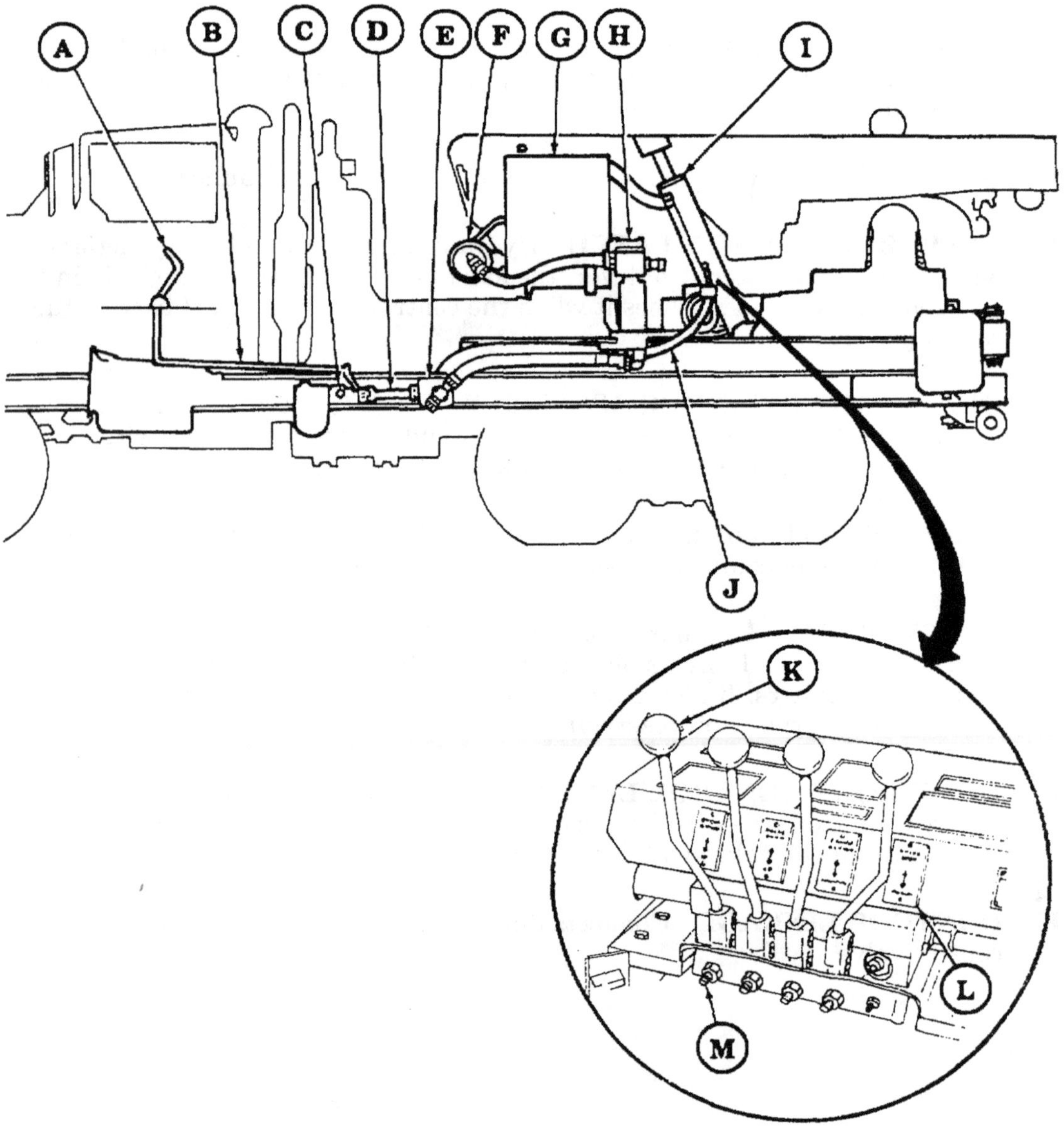
A
B
C
D
E
F
G
H
I
J
K
L
M

1-21. CONTROL SYSTEM OPERATION (Contd)

(A) **TRANSFER CASE POWER TAKEOFF (PTO) CONTROL** - A manually operated control lever located inside the cab that engages and disengages the transfer case power takeoff.

(B) **TRANSFER CASE POWER TAKEOFF (PTO) LINKAGE** - Connects transfer case power takeoff control to transfer case Power Takeoff (PTO).

(C) **TRANSFER CASE POWER TAKEOFF (PTO)** - Receives driving power from vehicle's engine through the transfer case to provide mechanical driving power for the hydraulic pump.

(D) **POWER TAKEOFF (PTO) DRIVE SHAFT** - Transmits mechanical driving power from the power takeoff to the hydraulic pump.

(E) **HYDRAULIC PUMP** - Draws oil from hydraulic oil reservoir and directs it to valves inside the crane control console.

(F) **OIL FILTER** - Filters used or bypassed oil from the control valve before it returns to the hydraulic oil reservoir.

(G) **HYDRAULIC OIL RESERVOIR** - Storage tank for hydraulic oil.

(H) **SWIVEL VALVE** - Permits oil to channel through pivot post while crane is swinging and eliminates twisting of the hydraulic lines connecting reservoir to the stationary pump.

(I) **BOOM LIFT CYLINDER** - A hydraulically-driven piston that extends upward when boom control lever is pulled back to UP position, raising the boom. A check valve located near hydraulic oil inlet hose prevents piston from lowering when control lever is in neutral. Oil returns through boom control valve back to hydraulic oil reservoir allowing piston to lower when control lever is pushed forward to DOWN position.

(J) **BOOM HYDRAULIC LINES** - Carry the hydraulic oil to and from boom lift cylinder. Oil pumped through the bottom lines pushes the lift cylinder piston upward. Oil pumped through the top lines pushes the lift cylinder piston downward. When this downward action occurs, the oil that originally pushes the cylinder upwards is returned to the hydraulic oil reservoir.

(K) **BOOM CONTROL LEVER** - Manual control attached to the control valve that determines hydraulic oil flow for raising and lowering action of the boom. Lever is pulled back to raise the boom and pushed forward to lower boom.

(L) **CRANE CONTROL CONSOLE** - Houses BOOM, HOIST, CROWD, and SWING levers and their control valves.

(M) **BOOM CONTROL VALVE** - Located directly below boom control lever. Valve directs hydraulic oil from the hydraulic pump to the boom lift cylinder for lifting, or out of the lift cylinder and back to the hydraulic oil reservoir for lowering.

1-21. CONTROL SYSTEM OPERATION (Contd)

(2) The major components for raising and lowering the crane cable and hook for the HOIST action are:

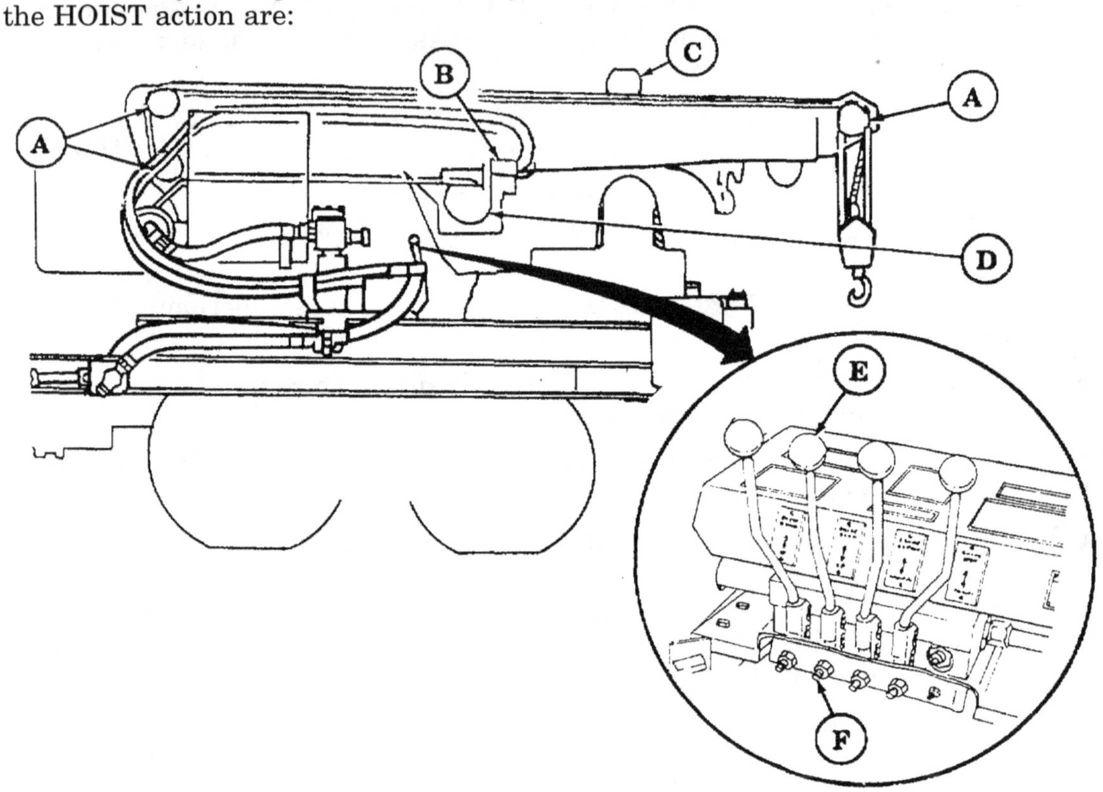

C
B
A
A
D
E
F

(A) **SHEAVES** - Grooved wheels that guide hoist cable through boom.

(B) **HOIST MOTOR ASSEMBLY** - Converts hydraulic power back into mechanical power needed to turn the hoist drum.

(C) **UPPER ROLLER ASSEMBLY** - Prevents cable from contacting inner boom during winding/unwinding.

(D) **CRANE HOIST CABLE DRUM** - Is turned by the worm gear in hoist motor assembly. Drum unwinds cable when turning toward front of vehicle. Drum winds cable when turning toward rear of vehicle.

(E) **HOIST CONTROL LEVER** - Manual control attached to the control valve that determines hydraulic oil flow for the raising and lowering action of the crane hoist cable and hook. Lever is pulled back to raise cable and hook and pushed forward to lower cable and hook.

(F) **HOIST CONTROL VALVE** - Two-way hydraulic valve located under the hoist control lever directs fluid from the hydraulic pump to the hoist motor assembly and back through the valve to the hydraulic oil reservoir.

1-21. CONTROL SYSTEM OPERATION (Contd)

(3) Major components for extending and retracting the boom for the CROWD action are:

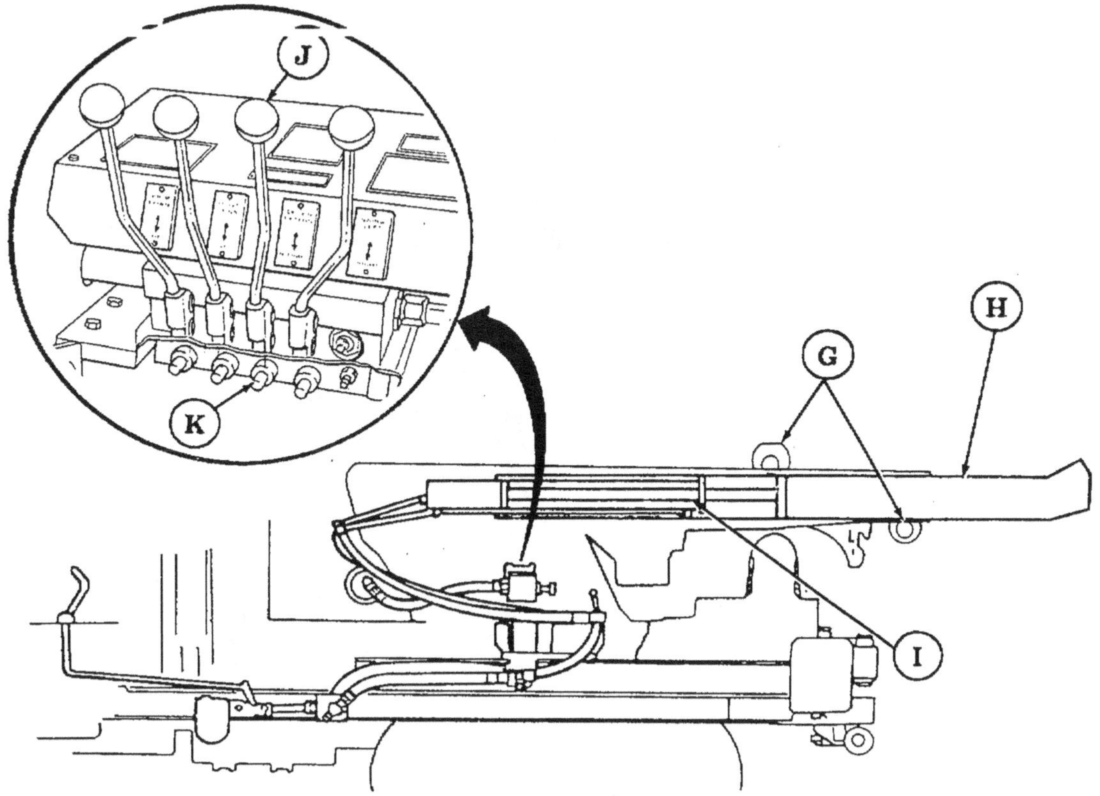
J
K
H
G
I

(G) **ROLLERS** - Guides inner boom assembly and permits smooth extension and retraction of boom.

(H) **INNER BOOM ASSEMBLY** - Extends when crowd control lever is pushed forward and retracts when control lever is pulled back.

(I) **CROWD CYLINDER** - A hydraulically-driven piston that extends outward when crowd control lever is pushed forward to EXTEND position. Piston is hydraulically driven back into the cylinder when crowd control lever is pulled back to RETRACT position. This cylinder is contained in the inner boom assembly.

(J) **CROWD CONTROL LEVER** - Manual control attached to the control valve that determines oil flow for extending and retracting the crane boom. Lever is pushed forward to extend the boom and pulled back to retract the boom.

(K) **CROWD CONTROL VALVE** - Two-way hydraulic valve located directly below crowd control lever. Valve directs hydraulic oil from the hydraulic pump to the crowd cylinder to extend and retract inner boom assembly.

1-21. CONTROL SYSTEM OPERATION (Contd)

(4) The major components for swinging the crane left and right for the SWING action are:

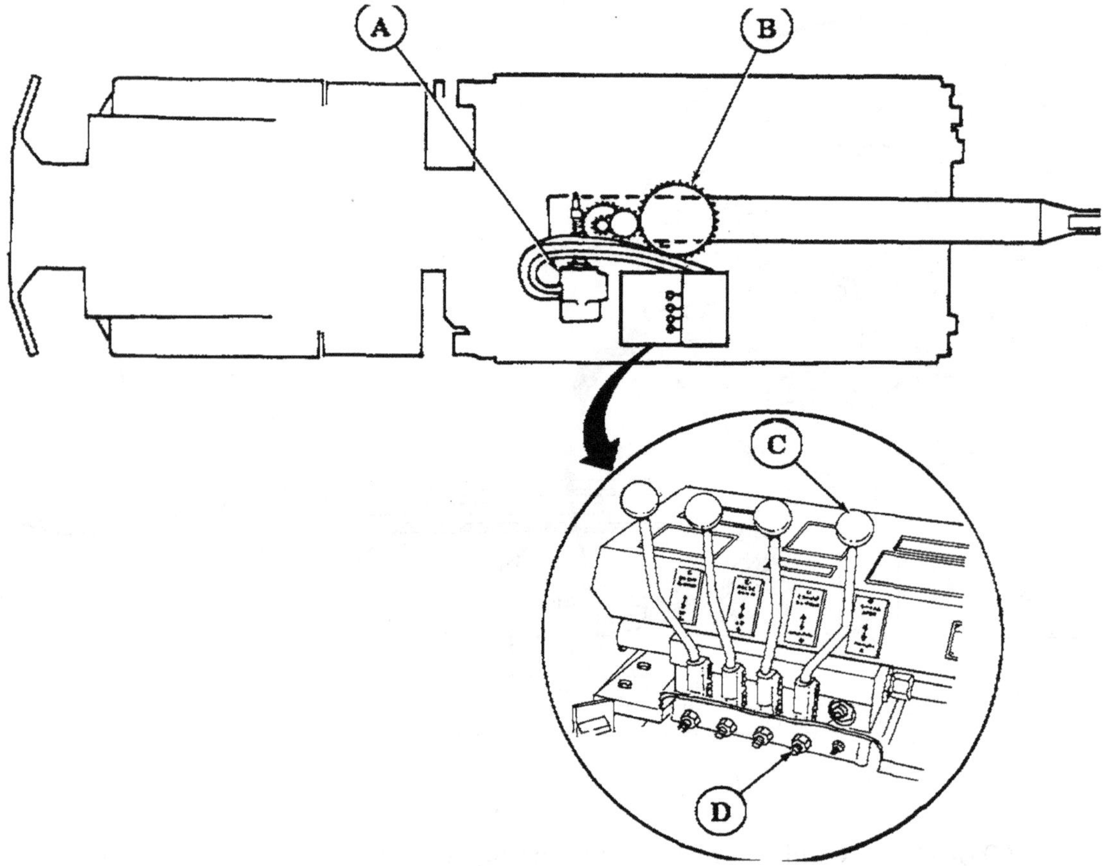
A
B
C
D

(A) **SWING MOTOR** - Converts hydraulic power back into mechanical power needed to turn the crane turntable when hydraulic fluid is forced through its worm gear. This gear turns a large gear at the base of the turntable to swing the crane.

(B) **TURNTABLE ASSEMBLY** - Driven by the swing motor through a ring gear at the base of the assembly, permits the crane to swing 360 degrees.

(C) **SWING CONTROL LEVER** - Manual control attached to the control valve that determines hydraulic oil flow for swinging wrecker boom to the left and to the right. Lever is pushed inward for left boom movement, and pulled outward for right boom movement.

(D) **SWING CONTROL VALVE** - Two-way hydraulic valve located directly below swing control lever. Valve directs hydraulic oil from the hydraulic pump to the swing motor assembly and back through the valve to the hydraulic oil reservoir.

CHAPTER 2
OPERATING INSTRUCTIONS

Section I. DESCRIPTION AND USE OF OPERATOR'S CONTROLS AND INDICATORS

2-1. KNOW YOUR CONTROLS AND INDICATORS

Before you attempt to operate your equipment, ensure you are familiar with the location and function of all controls and indicators. The location and function of controls and indicators for the M939, M939A1, and M939A2 series vehicles are described in this section as follows:

a. Chassis Controls and Indicators: paragraph 2-3.

b. Body Equipment Controls and Indicators: paragraph 2-4.

c. Special Kits Controls and Indicators: paragraph 2-5.

NOTE

- Except where specifically noted, the controls and indicators in this section are generally applicable to all vehicles covered in this manual.
- In this manual, the term left indicates the driver's side of the vehicle. The term right indicates the opposite, or crew's side of the vehicle.

2-2. PREPARATION FOR USE

When a vehicle is received by the using organization, it is the responsibility of the officer-in-charge to determine whether it has been properly prepared for service by the supplier. It is the responsibility of the officer-in-charge to ensure the vehicle is in condition to perform its functions. Unit maintenance will provide any additional service required to bring the vehicle to operating standards. Before operating the vehicle, the operator must become familiar with the vehicle controls and indicators as described in this chapter.

2-3. CHASSIS CONTROLS AND INDICATORS

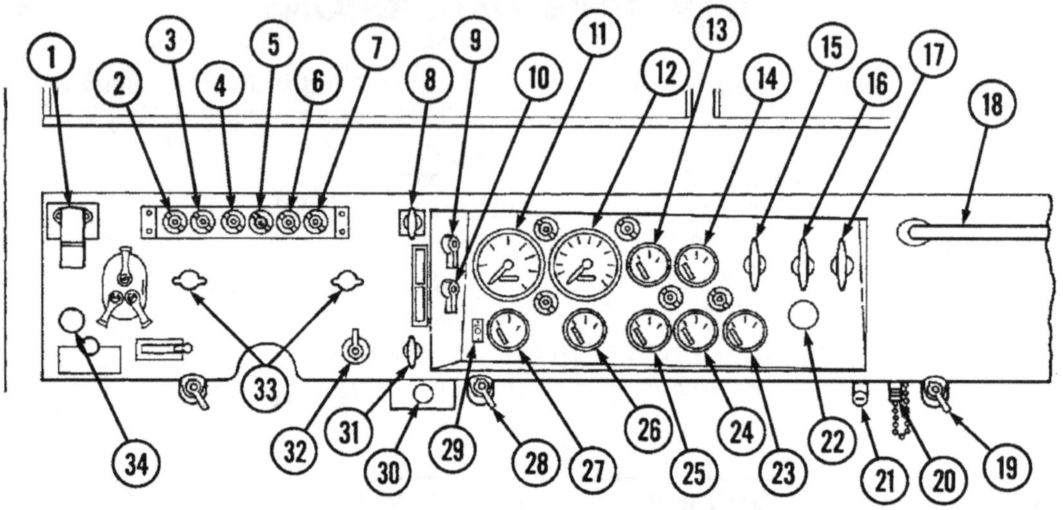
1
2
3
4
5
6
7
8
9
10
11
12
13
14
15
16
17
18
19
20
21
22
23
24
25
26
27
28
29
30
31
32
33
34

Key	Item and Function
1	AIR CLEANER INDICATOR shows red when engine air filter needs servicing.
2	PARKING BRAKE WARNING LIGHT lights when parking brake is on.
3	LOW AIR PRESSURE WARNING LIGHT illuminates when air brake system pressure drops below 50-60 psi (345-414 kPa).
4	SPRING BRAKE WARNING LIGHT lights when spring brakes are engaged.
5	LOW COOLANT LEVEL LIGHT (M939A2 series vehicles) lights when engine coolant level is low.
6	AXLE LOCK-IN LIGHT lights when front wheel drive lock-in switch is on.
7	HIGH BEAM INDICATOR lights when front headlights are on high beam.
8	HAND THROTTLE CONTROL sets engine speed at desired rpm without maintaining pressure on accelerator pedal. Throttle control locks in desired position when pulled out. Rotating control handle clockwise or counter—clockwise unlocks it.
9	BATTERY SWITCH activates and deactivates all electrical circuits on or off except arctic heater and lights.
10	IGNITION SWITCH has OFF, RUN, and START positions. Switch automatically returns from START to RUN when hand pressure is released.
11	TACHOMETER indicates engine speed in revolutions per minute (rpm) and operating hours in tenths.
12	SPEEDOMETER/ODOMETER indicates vehicle speed and total mileage.
13	ENGINE COOLANT TEMPERATURE GAUGE indicates engine coolant temperature. Normal engine coolant operating temperature for M939/A1 series vehicles is 175°-195°F (79°-91°C) and 190°-200°F (88°-93°C) for M939A2 series vehicles.

2-3. CHASSIS CONTROLS AND INDICATORS (Contd)

Key	Item and Function
14	PRIMARY AIR PRESSURE GAUGE indicates air pressure in the primary brake system. Normal pressure is 90-130 psi (621-896 kPa).
15	DEFROSTER CONTROL opens vents to direct heated air at the windshield.
16	HEAT VENT CONTROL controls the amount of heat blown into the cab by adjusting the opening of heat ventilation doors.
17	FRESH AIR VENT CONTROL pulls out to open ventilation doors, allowing outside air to circulate in the cab.
18	GRAB HANDLE aids crewmembers in entering and exiting vehicle cab. Handle is also a brace for crewmembers during travel.
19	FLOODLIGHT CONTROL SWITCH (M936/A1/A2 models) turns on floodlights installed on wrecker body for use in night crane operations.
20	ELECTRICAL RECEPTACLE OUTLET (M936/A1/A2 models) provides battery voltage for extension cord trouble light.
21	AUXILIARY OUTLET RECEPTACLE (M936/A1/A2 models) is plugged into electrical receptacle outlet. A trouble light is then plugged into the auxiliary outlet receptacle in order to be powered by vehicle batteries.
22	SPRING BRAKE RELEASE CONTROL is pushed in to release spring brakes independently of the mechanical parking brake. Control is used to release spring brakes in order to test and adjust mechanical brakes.
23	VOLTMETER indicates charging condition of the batteries.
24	SECONDARY AIR PRESSURE GAUGE indicates air pressure in the secondary brake system. Normal pressure is 90-130 psi (621-896 kPa).
25	TRANSMISSION OIL TEMPERATURE GAUGE indicates temperature of transmission oil. Normal operating temperature is 120°-220°F (49°-104°C).
26	ENGINE OIL PRESSURE GAUGE indicates oil pressure when engine is running. Normal operating pressure at idle is 15 psi (103 kPa).
27	FUEL GAUGE indicates fuel level in fuel tank(s).
28	AMBER WARNING LIGHT SWITCH (M936/A1/A2 models) controls operation of amber warning light used during crane operations or while towing disabled vehicle.
29	YELLOW ABS WARNING LAMP (all models with antilock brake system) illuminates for three seconds when ignition switch is placed in run position as ABS is performing self check and then goes out if ABS does not have any malfunctions.
30	TRAILER AIR SUPPLY VALVE (M931/A1/A2 and M932/A1/A2 models) is pushed in to supply air to the brake system of towed trailer or semitrailer.
31	EMERGENCY ENGINE STOP CONTROL is pulled out to cut off fuel to engine. It is used only in an emergency.
32	HEATER BLOWER MOTOR SWITCH activates heater blower.
33	WIPER MOTOR SWITCHES activate and controls speed of wiper.
34	WINDSHIELD WASHER CONTROL sprays cleaning solution on windshield when depressed.

2-3. CHASSIS CONTROLS AND INDICATORS (Contd)

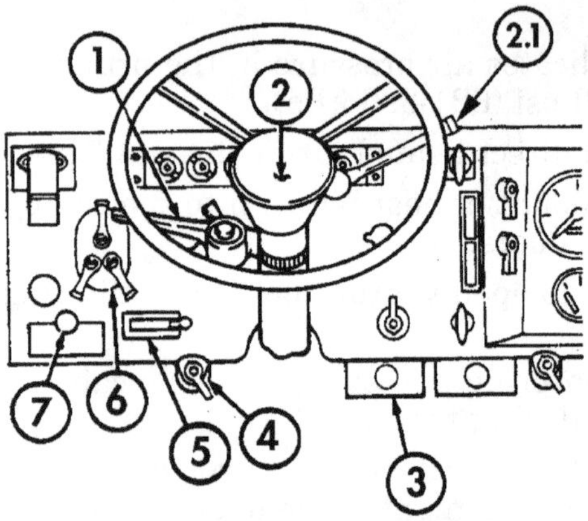
1
2
2.1
3
4
5
6
7

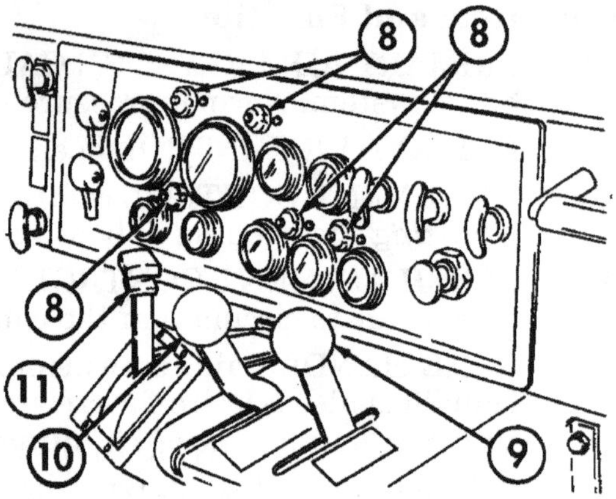
8
8
8
11
10
9

Key	Item and Function
1	TURN SIGNAL CONTROL LEVER is moved down to operate vehicle left turn signals and up to operate right turn signals. Lever must be returned to the center position to turn off signal. Turn signal control is equipped with hazard tab button.
2	HORN BUTTON is pressed to operate vehicle horn.
2.1	TRAILER BRAKE/JOHNNIE BAR (M931 and M932 series vehicles) is used ONLY to prevent vehicle movement while stopped on an incline or when coupling or uncoupling the tractor to or from trailer.
3	SPRING BRAKE OVERRIDE held in during self-recovery of M936/A1/A2 model vehicles with rear winch.
4	FUEL LEVEL GAUGE SWITCH (M929/A1/A2, M930/A1/A2, M931/A1/A2, M932/A1/A2, and M936/A1/A2 models) permits reading fuel level on the fuel gauge for each fuel tank when turned L (left) or R (right).
5	FRONT WHEEL DRIVE LOCK-IN SWITCH allows operator to engage front wheel drive and is used only when vehicle transfer case is in high range. In low range, the vehicle's front wheel drive engages automatically. Vehicle may be in motion or stopped to engage front wheel drive lock-in switch.
6	LIGHT SWITCH controls operation of vehicle lights.
7	ETHER START SWITCH injects ether into engine for cold weather starting.
8	INSTRUMENT PANEL LIGHTS illuminate instrument panel gauges.
9	TRANSMISSION POWER TAKEOFF CONTROL LEVER (M925/A1/A2, M928/A1/A2, M929/A1/A2, M930/A1/A2, M932/A1/A2, and M936/A1/A2 models) engages transmission power takeoff to provide power for auxiliary equipment.
10	FRONT WINCH CONTROL LEVER (M925/A1/A2, M928/A1/A2, M930/A1/A2, M932/A1/A2, and M936/A1/A2 models) is pulled back to wind front winch, and forward to unwind for lowering loads during A-frame operation.
11	AUTOMATIC TRANSMISSION SELECTOR LEVER is used to select vehicle driving gear.

2-3. CHASSIS CONTROLS AND INDICATORS (Contd)

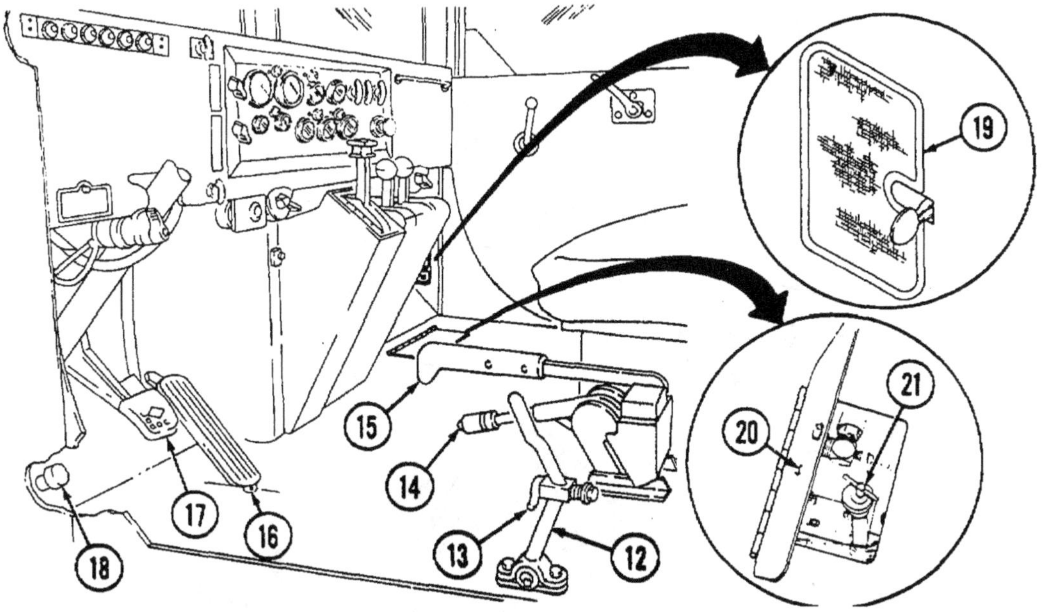
12
13
14
15
16
17
18
19
20
21

Key **Item and Function**

12 (a) DUMP BODY CONTROL LEVER (M929/A1/A2 and M930/A1/A2 models) is pushed back to raise dump body and pulled forward to lower dump body.

(b) TRANSFER CASE POWER TAKEOFF CONTROL LEVER (M936/A1/A2 models) is engaged to provide power to crane and rear winch.

13 SAFETY LATCH secures dump body control in neutral when not in use.

14 MECHANICAL PARKING BRAKE CONTROL LEVER is pulled up to engage parking brakes and down to disengage brakes. Knob on top of handle is turned clockwise to increase parking brake tension and counterclockwise to decrease parking brake tension. When parking brake lever is applied, it also trips a valve to release air pressure from spring brakes. This engages spring brakes.

15 TRANSFER CASE SHIFT LEVER is pushed down to high range for light load operations and up to low range for heavy load operations. Six-wheel drive is achieved automatically when transfer case shift lever is placed in low range.

16 ACCELERATOR PEDAL controls engine speed.

17 BRAKE PEDAL is depressed to stop vehicle.

18 DIMMER SWITCH is depressed to raise or lower headlight beam.

19 COWL VENTILATOR (one on each side of cab) is opened manually to provide fresh air ventilation.

20 ACCESS DOOR opens to provide access to transmission dipstick and oil fill (M939/A1 series vehicles).

21 TRANSMISSION DIPSTICK is turned counterclockwise to remove and check transmission oil level, (M939/A1 series vehicles).

2-3. CHASSIS CONTROLS AND INDICATORS (Contd)

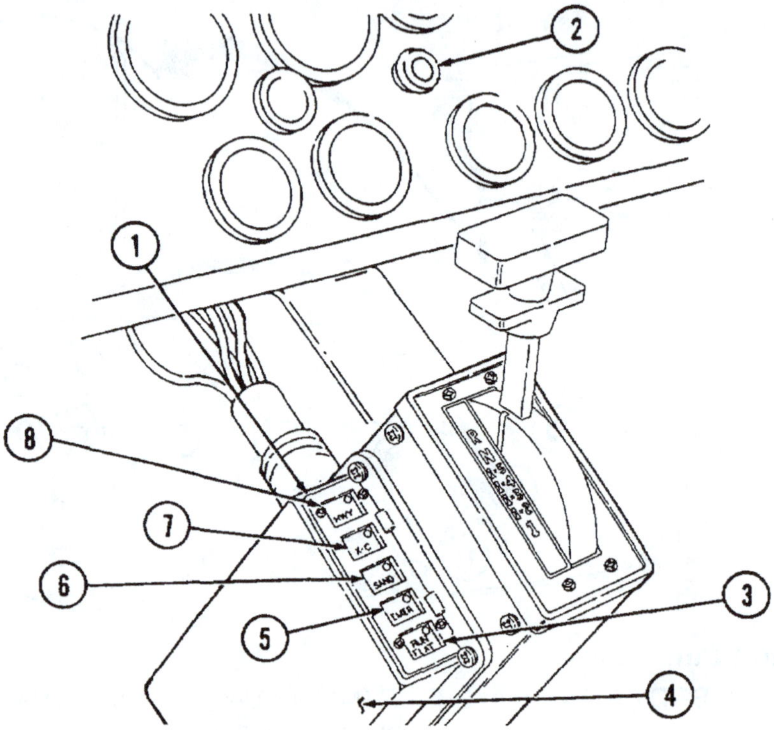
1
2
3
4
5
6
7
8
HWY
X-C
SAND
EMER
RUN FLAT

Key	Item and Function
1	SELECTOR PANEL (M939A2 series vehicles) contains selectors for the four preset tire pressure modes and a run-flat selector.
2	AMBER WARNING LIGHT (M939A2 series vehicles) flashes as an overspeed warning for cross-country and sand tire pressure modes. It will stay lit when the emergency tire pressure mode is selected.
3	RUN FLAT (M939A2 series vehicles) selector causes the CTIS to check tire pressures every fifteen seconds.
4	ELECTRONIC CONTROL UNIT (ECU) (M9393A2 series vehicles) is the microprocessor in the CTIS. It is contained in the selector panel housing.
5	EMERGENCY (EMER) (M939A2 series vehicles) tire pressure selector is used for operating the vehicle in extreme terrain conditions where maximum traction is required.
6	SAND (M939A2 series vehicles) tire pressure selector is used for operating the vehicle in sand, snow, and mud.
7	CROSS-COUNTRY (X-C) (M939A2 series vehicles) tire pressure selector is used for operating the vehicle on non-paved secondary roads and unimproved surfaces.
8	HIGHWAY (HWY) (M939A2 series vehicles) tire pressure selector is the normal operating modes of CTIS. The highway mode is automatically set each time the engine is started.

2-3. CHASSIS CONTROLS AND INDICATORS (Contd)

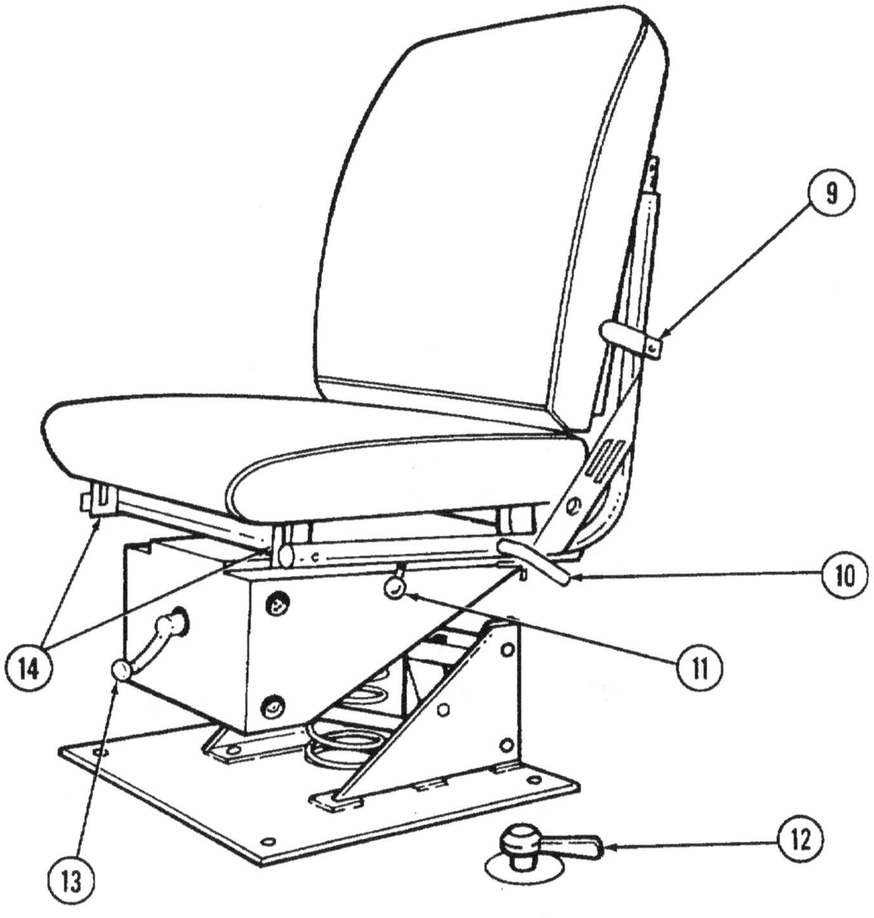
9
10
11
12
13
14

Key	Item and Function
9	BACKREST CONTROL adjusts angle of seat backrest.
10	SEAT CUSHION CONTROL adjusts height and angle of seat cushion.
11	SEAT HORIZONTAL CONTROL positions seat forward or backward.
12	FUEL TANK SELECTOR SWITCH (M929/A1/A2, M930/A1/A2, M931/A1/A2, M932/A1/A2, and M936/A1/A2 models) is turned L (left) or R (right) to select fuel supply source. Selector is located on the cab floor to the left side of the operator's seat. Occasionally switch fuel tanks to prevent the fuel from becoming dirty, moisture filled, and thick.
13	SPRING TENSION CONTROL increases seat spring tension when crank is turned clockwise.
14	SLOTTED BRACKETS at each corner permit front portion of seat to be raised or lowered.

2-3. CHASSIS CONTROLS AND INDICATORS (Contd)

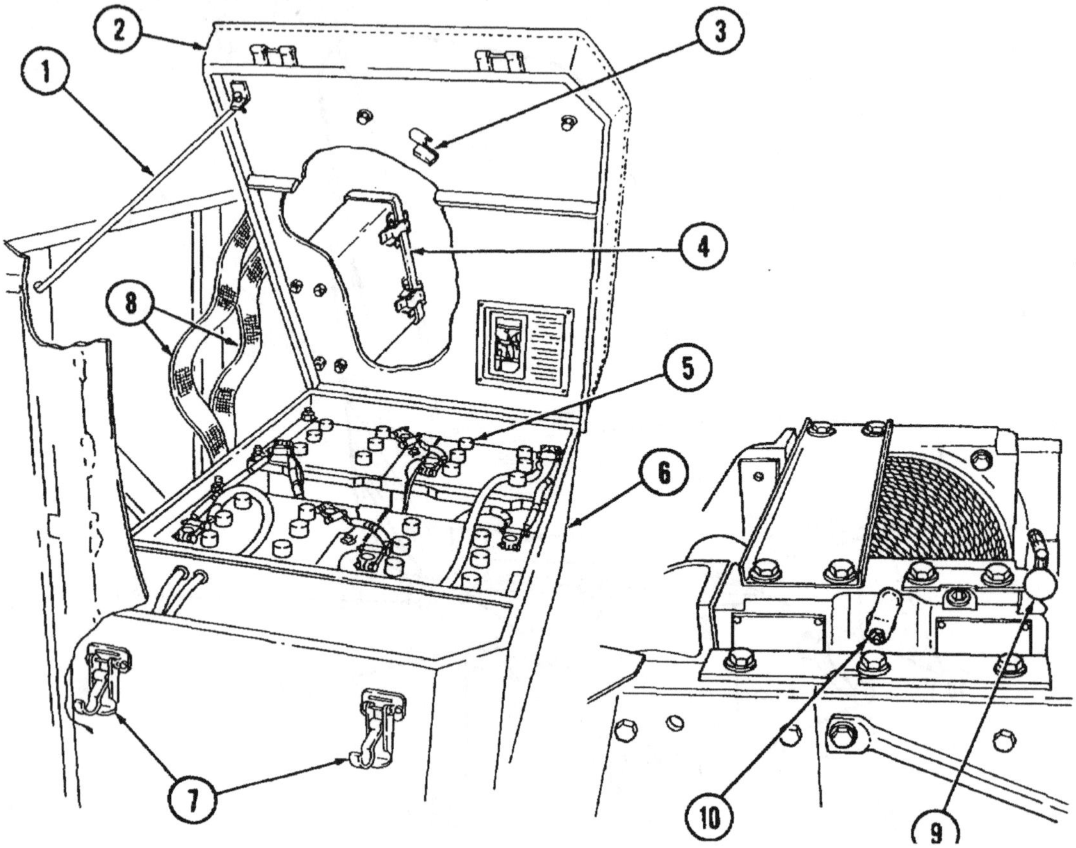
1
2
3
4
5
6
7
8
9
10

Key	Item and Function
1	RETAINING ROD holds crew seat in up position for inspection of batteries.
2	COMPANION AND BATTERY BOX COVER is lifted to provide access to batteries.
3	CLIP holds safety rod when companion seat is down.
4	MAP COMPARTMENT stores maps, manuals, forms, and papers.
5	BATTERY CAPS are removed to check fluid level.
6	BATTERY BOX provides compartment for four 12-volt batteries.
7	LATCHES lock companion seat down for travel.
8	COMPANION SEAT SEATBELTS provide two personnel restraints for crewmembers. (Ensure straps are not caught inside battery box when cover is closed.)
9	FRONT WINCH CONTROL (W/W models only) is pulled out to engage winch clutch and pushed in to disengage winch clutch.
10	FRONT WINCH DRUM LOCK KNOB locks drum when winch is not in use.

2-3. CHASSIS CONTROLS AND INDICATORS (Contd)

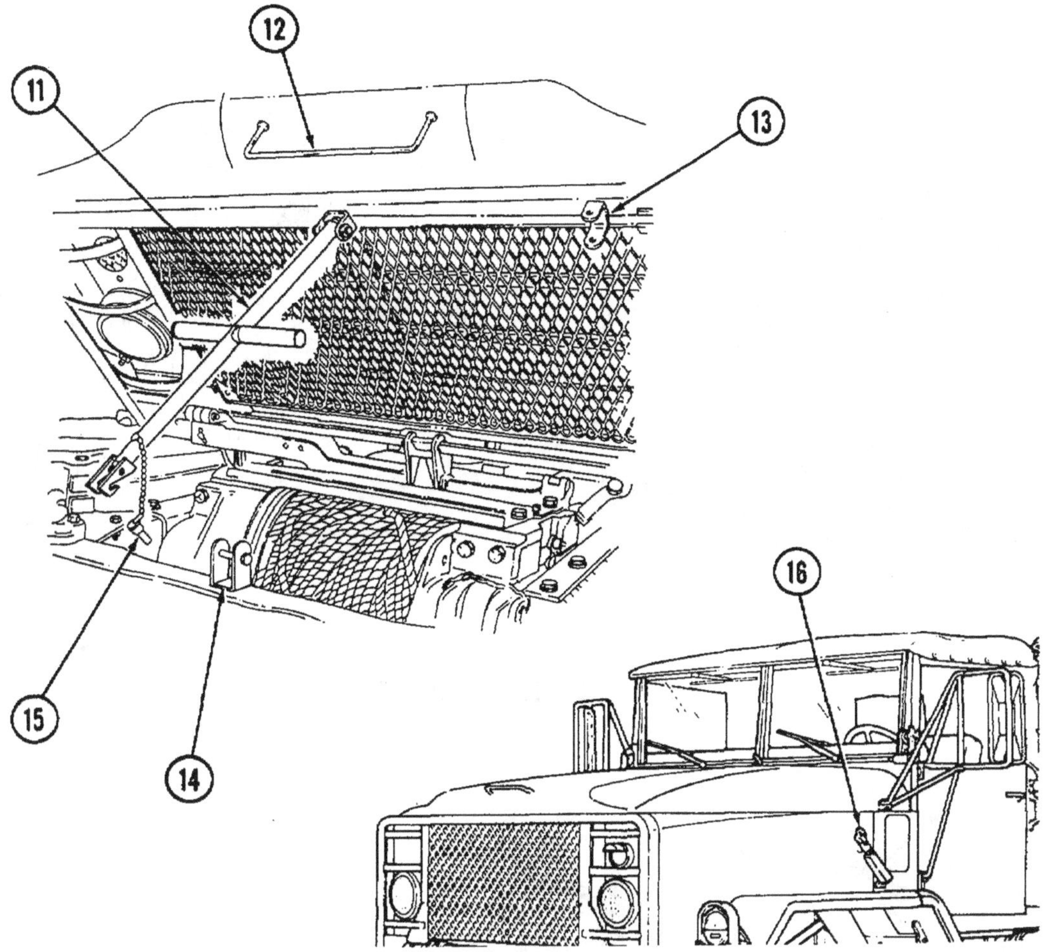
12
11
13
15
14
16

Key	Item and Function
11	HOOD RETAINING BAR is used to raise and lower hood. Bar secures raised hood to bumper while in up position. Bar is secured by hood retaining bar safety pin (15) to storage bracket (13) during travel.
12	HOOD HANDLE is used to assist in raising and lowering hood.
13	HOOD RETAINING BAR STORAGE BRACKET secures retaining bar (11) to hood.
14	BUMPER BRACKET secures hood retaining bar (11) to bumper when hood is in open position.
15	HOOD RETAINING BAR SAFETY PIN is attached to hood retaining bar (11). Pin secures retaining bar (11) to storage bracket (13) during travel and secures bar (11) to bumper bracket (14) during use.
16	HOOD LATCH (one on each side of vehicle) unhooks to release hood.

2-3. CHASSIS CONTROLS AND INDICATORS (Contd)

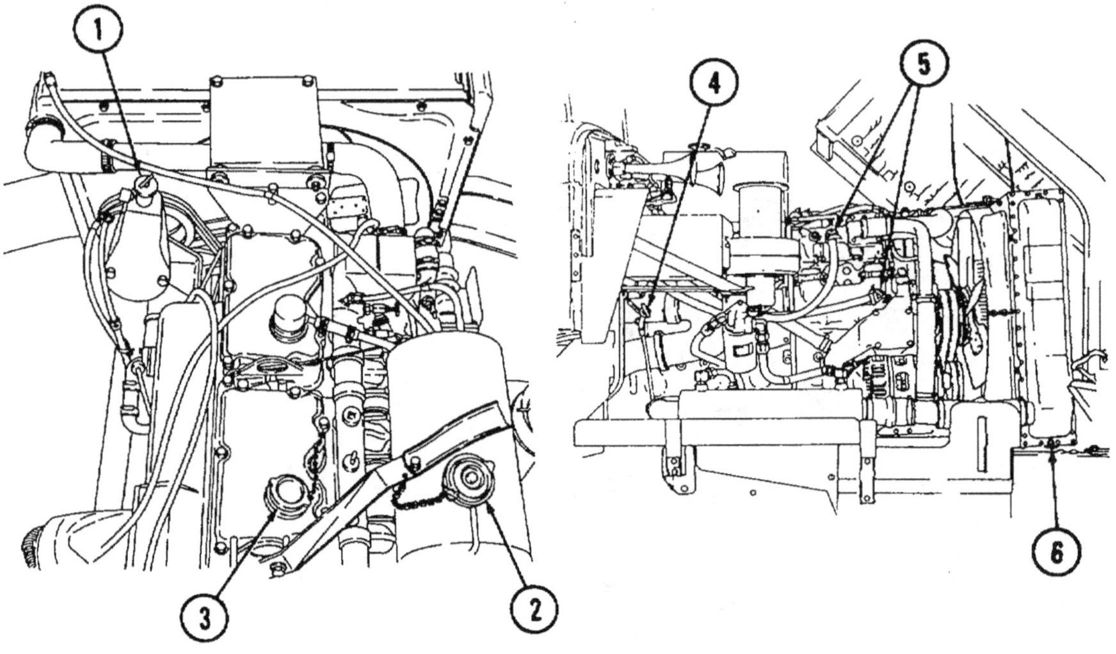
1
4
5
6
3
2

M939/A1 SERIES VEHICLES

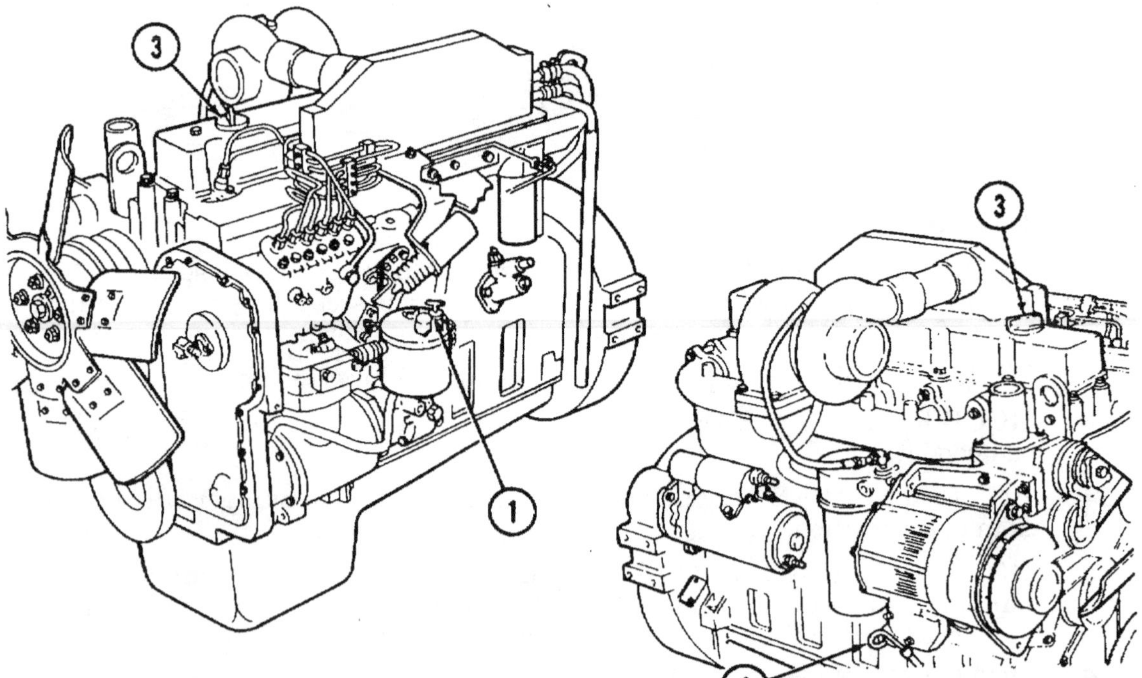
3
1
3
4

2-3. CHASSIS CONTROLS AND INDICATORS (Contd)

Key	Item and Function
1	POWER STEERING OIL RESERVOIR DIPSTICK is attached to reservoir fill cap. Dipstick is turned counterclockwise and removed to check power steering oil level.
2	COOLANT SURGE TANK CAP is turned counterclockwise and removed to add coolant. (Location common to all vehicles)
3	ENGINE OIL FILLER CAP is turned counterclockwise and removed to add oil.
4	OIL DIPSTICK is to check engine oil level. On M939/A1 series vehicles, turn counterclockwise to remove.
5	SHUTOFF VALVES are turned counterclockwise to circulate heated coolant through vehicle cab heating system.
6	RADIATOR DRAINVALVE is turned counterclockwise to drain coolant from radiator. (Location common to all vehicles)

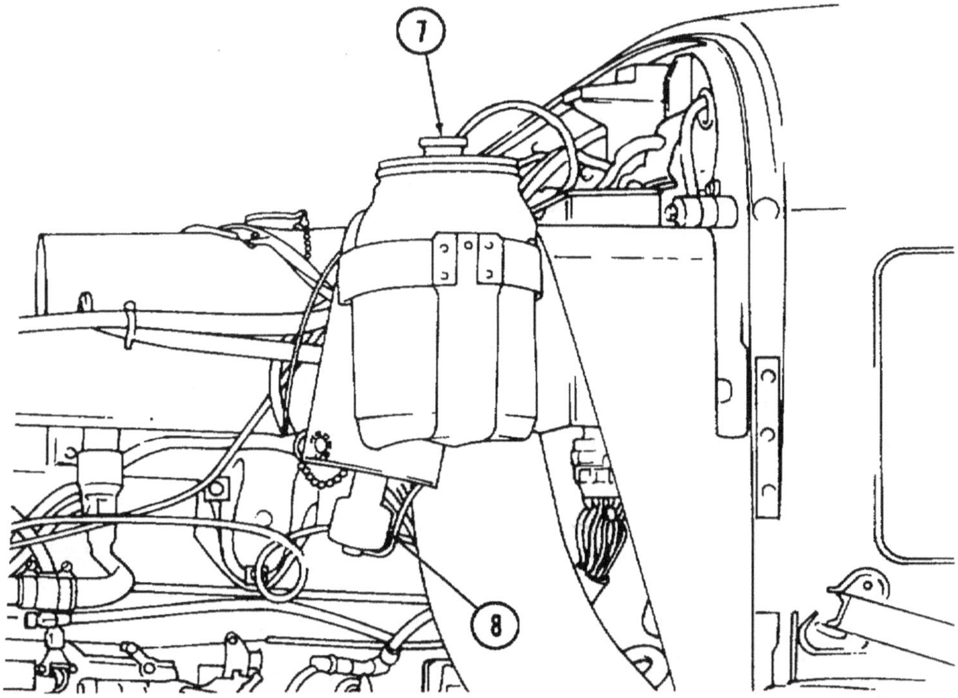
7
8

7	WINDSHIELD WASHER BOTTLE CAP is unsnapped to refill washer reservoir.
8	ETHER STORAGE CYLINDER stores ether used for starting during cold–weather operations.

2-3. CHASSIS CONTROLS AND INDICATORS (Contd)

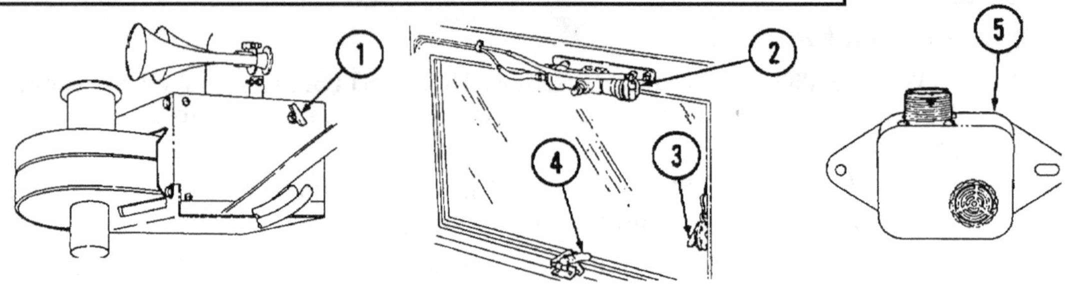
1
2
3
4
5

Key	Item and Function
1	PERSONNEL HEATER VENTVALVE, located in engine compartment, is turned counterclockwise to purge personnel hot water heater of air.
2	WINDSHIELD WIPER MOTOR, located on each front window, powers windshield wipers.
3	WING NUT, located on each side of window, is tightened to hold window in open position.
4	LATCHING HANDLE secures window in closed position.
5	WARNING ALARM BUZZER is located above left cowl vent. The alarm sounds when airbrake system pressure drops below 50-60 psi (345-414 kPa) or when parking brake is engaged.

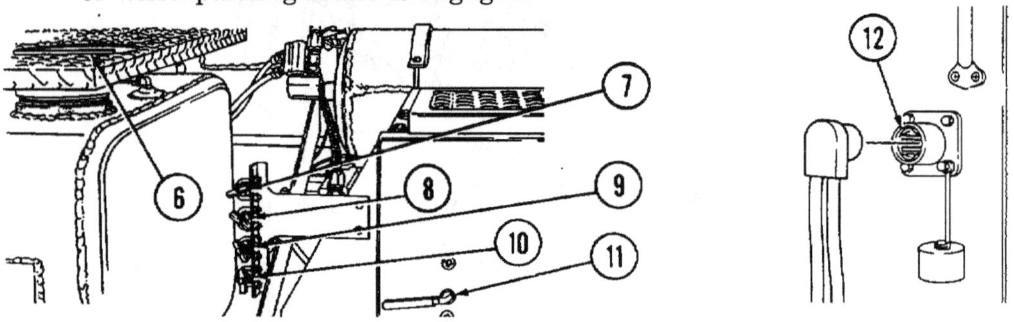
6
7
8
9
10
11
12

Key	Item and Function
6	FUEL TANK FILLER CAP is turned counterclockwise and removed for fuel servicing.
7	PRIMARY DRAINVALVE, located on right side of vehicle, is turned counterclockwise to drain water from primary brake system air reservoir.
8	SECONDARY DRAINVALVE, located on right side of vehicle, is turned counterclockwise to drain water from secondary brake system air reservoir.
9	WET TANK DRAINVALVE, located on right side of vehicle, is turned counterclockwise to drain water from brake system wet tank air reservoir.
10	SPRING BRAKE DRAINVALVE, located on right side of vehicle, is turned counterclockwise to drain water from spring brake system air reservoir.
11	TOOLBOX LATCH HANDLE is turned up to unlatch and open toolbox.
12	SLAVE RECEPTACLE, located on right rear side of cab, is plug-in point for an external power source required to slave start vehicle when batteries have become discharged.

2-3. CHASSIS CONTROLS AND INDICATORS (Contd)

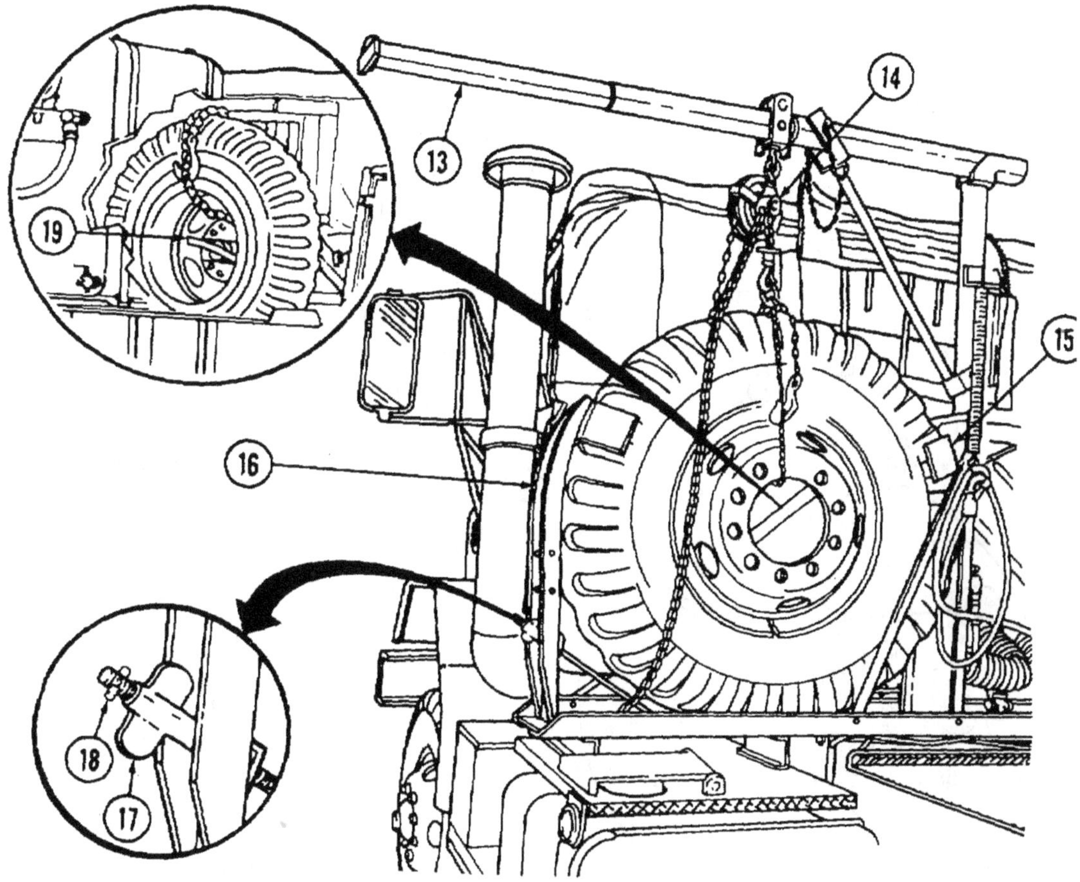
13
14
15
16
17
18
19

Key	Item and Function
13	SPARE TIRE DAVIT BOOM is extended to allow spare tire to be lifted and guided over side of vehicle. Davit boom is installed on all M939/A1/A2 series vehicles except M929/A1/A2, M930/A1/A2, and M934/A1/A2 models which are equipped with an eyebolt or special divot, and M936/A1/A2 models which use the vehicle boom to lift and lower spare tire.
14	CHAIN FALL lifts and lowers spare tire (all models except M936/A1/A2).
15	WHEEL BRACE holds tire in place once hinged wheel brace (16) is secured.
16	HINGED WHEEL BRACE (all models except M929/A1/A2, M930/A1/A2, and M936/A1/A2) is removed before spare tire removal.
17	WING LUG secures hinged wheel brace (16) to spare tire.
18	RETAINING PIN secures wing lug (17) to hinged wheel brace (16) for traveling.
19	SPARE TIRE WING LUG (M936/A1/A2 models) secures spare tire to wrecker.

2-3. CHASSIS CONTROLS AND INDICATORS (Contd)

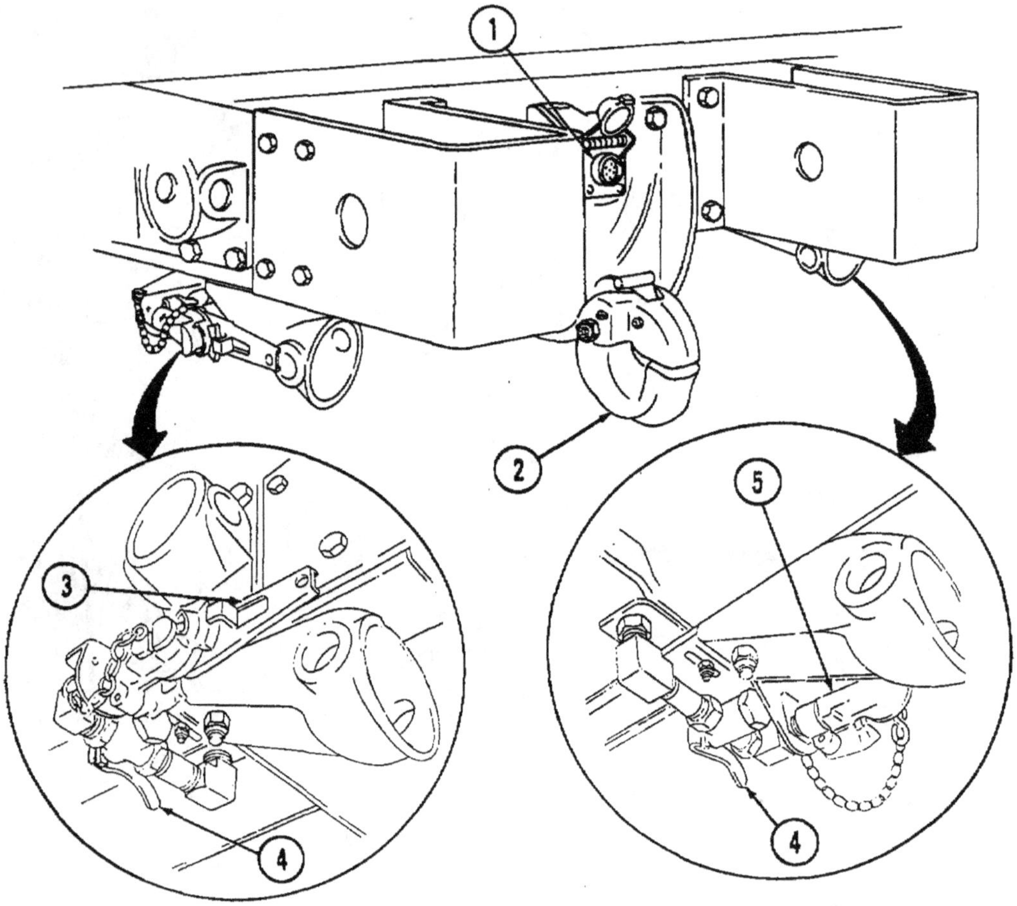
1
2
3
4
5
4

Key	Item and Function
1	TRAILER POWER OUTLET RECEPTACLE provides electric power for trailer.
2	TOWING PINTLE HOOK is opened to attach trailer towing bar.
3	TRAILER SERVICE AIR COUPLING is connected by an air coupling hose to the service coupling of a trailer or vehicle to be towed. This connection permits operator to engage brakes of the towed load when pressing brake pedal of the towing vehicle.
4	TRAILER AIR VALVE HANDLES are turned to release compressed air to trailer brake system.
5	EMERGENCY AIR COUPLING is connected by an air coupling hose to the emergency coupling of a trailer or vehicle to be towed. This connection permits towing vehicle to charge the brake system of a trailer or disabled vehicle with air.

2-4. BODY EQUIPMENT CONTROLS AND INDICATORS

a. Medium Wrecker.

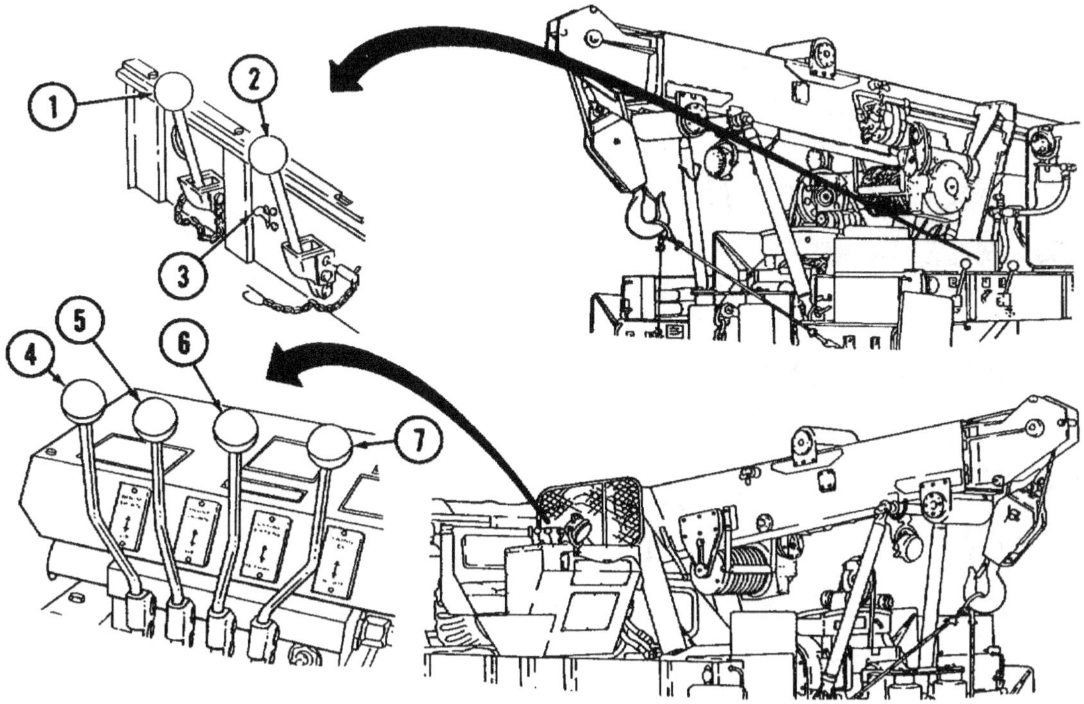
1
2
3
4
5
6
7

Key	Item and Function
1	DIRECTIONAL CONTROL LEVER controls rotation of rear winch drum. Control is pulled back to UNWIND and pushed forward to WIND. Center position is neutral.
2	TORQUE CONTROL LEVER is pushed forward to LOW for heavy winch loads and pulled back to HIGH for light winch loads. Lever must be engaged in HIGH or LOW before operation of directional control lever (1).
3	CABLE TENSIONER CONTROL VALVE controls tension on winch cable. Lever is positioned up to release tension and down to apply tension.
4	BOOM CONTROL LEVER raises boom when pulled toward operator and lowers boom when moved away from operator.
5	HOIST CONTROL LEVER raises boom hook when pulled toward operator and lowers hook when moved away from operator.
6	CROWD CONTROL LEVER retracts boom when pulled toward operator and extends boom when moved away from operator.
7	SWING CONTROL LEVER swings crane assembly right when pulled toward operator and swings crane assembly left when moved away from operator.

2-4. BODY EQUIPMENT CONTROLS AND INDICATORS (Contd)

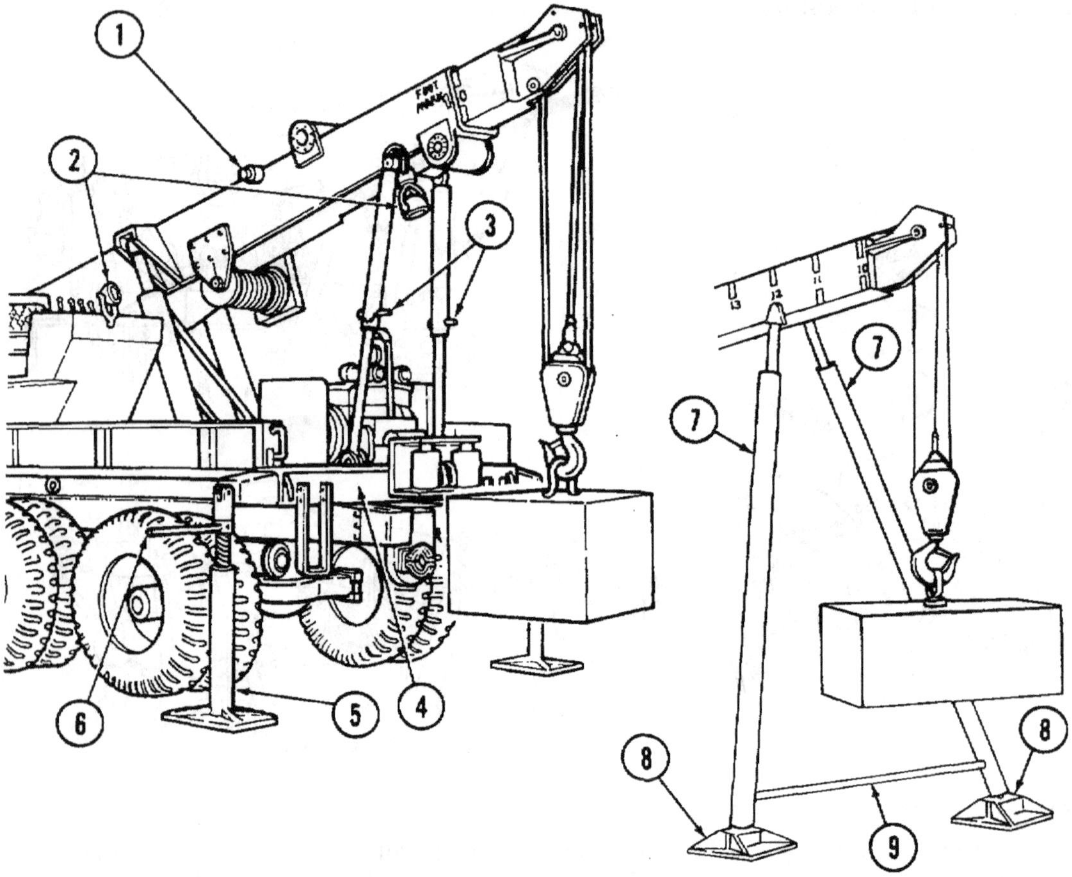
1
2
3
4
5
6
7
7
8
8
9

Key	Item and Function
1	SHIPPER BRACE RETAINING BRACKET holds shipper braces (3) when not in use.
2	FLOODLIGHTS are used for night crane operations and have individual on/off switches.
3	SHIPPER BRACE ASSEMBLIES support shipper and boom.
4	FRAME TUBE provides storage space for outriggers (5) when not in use.
5	OUTRIGGERS provide stabilization during crane operation.
6	HANDLES adjust length of outriggers (5).
7	BOOM JACKS provide stabilization as required for heavy lifting.
8	BOOM JACK BASE PLATES provide platforms for boom jacks (7) when in use.
9	TIE BAR provides stability for boom jacks (7) and is secured to boom with retainer pins.

2-4. BODY EQUIPMENT CONTROLS AND INDICATORS (Contd)

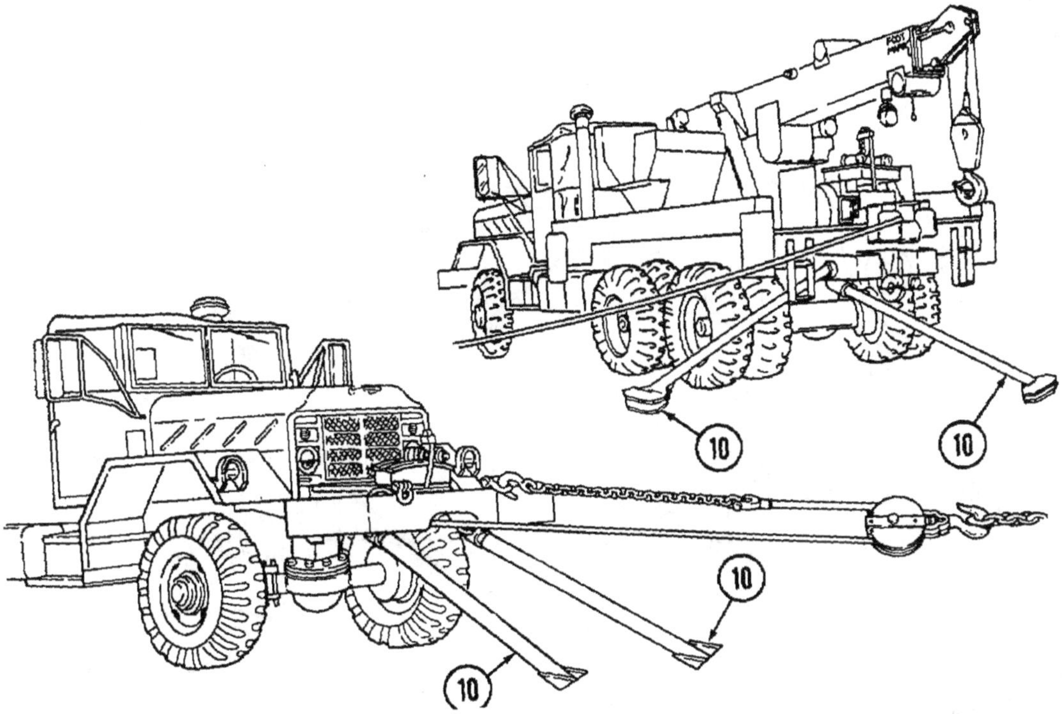
10
10
10
10

Key	Item and Function
10	FIELD CHOCKS anchor vehicle during winch operations.

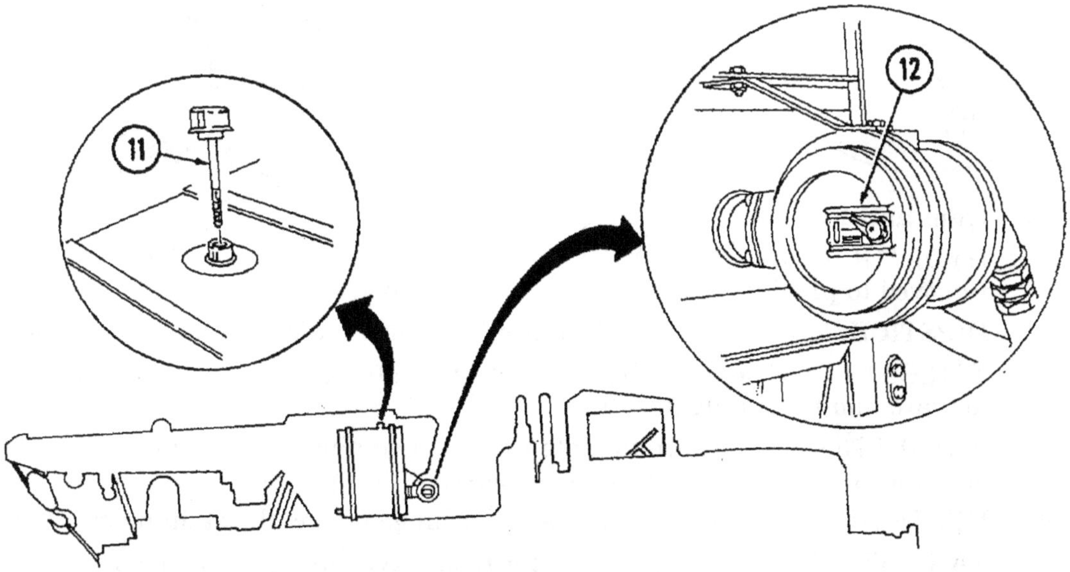
11
12

11	DIPSTICK on top of reservoir measures oil level in hydraulic oil reservoir.
12	FILTER INDICATOR on front of reservoir indicates whether crane oil filter is CLEAN or NEEDS CLEANING. Filter element must be changed whenever indicator is on NEED CLEANING.

2-4. BODY EQUIPMENT CONTROLS AND INDICATORS (Contd)

b. Dump Truck.

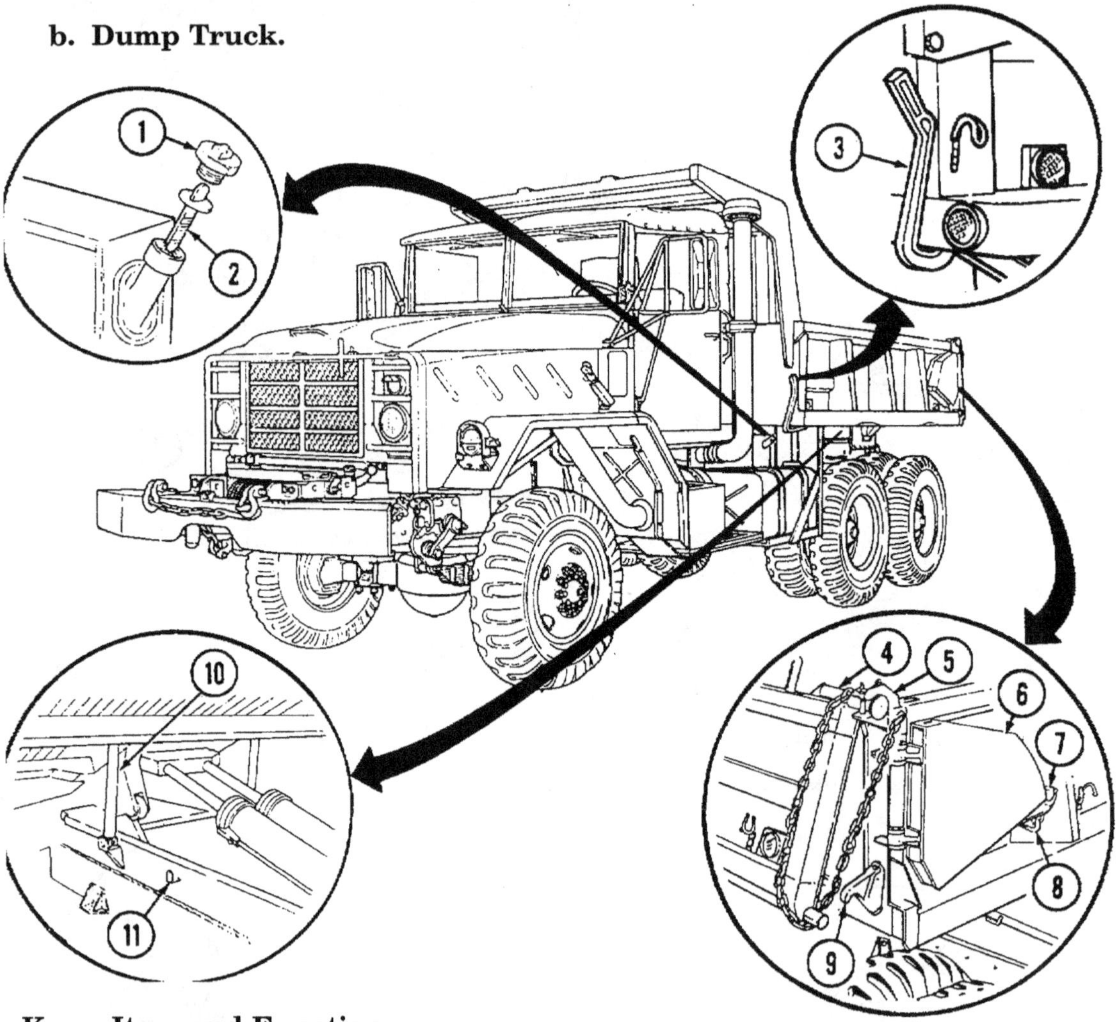
1
2
3
4
5
6
7
8
9
10
11

Key	Item and Function
1	HYDRAULIC OIL RESERVOIR CAP is turned counterclockwise and removed to provide access to hydraulic oil dipstick (2).
2	DIPSTICK, located inside hydraulic oil fill tube, indicates hydraulic oil level.
3	TAILGATE CONTROL LEVER unlocks tailgate latches (9) when pulled forward and locks tailgate when pushed back.
4	RETAINING PINS secure tailgate upper hinge pins during standard dump operations. Pins are removed for rocker-type dump operations.
5	UPPER HINGE BRACKETS holds upper hinge pins with retaining pins (4).
6	TAILGATE WINGS swing to rear for rocker-type dump operations.
7	TAILGATE WING BRACKETS used during rocker-type dump operations.
8	WING HARNESS HOOKS secure tailgate wings to dump body. Retaining pins (4) are removed for rocker-type dump operations.

2-4. BODY EQUIPMENT CONTROLS AND INDICATORS (Contd)

9 TAILGATE LATCHES unlock when tailgate control lever (3) is pulled forward. Latches lock tailgate when control lever (3) is pushed back.

10 DUMP BODY SUPPORT BRACES hold dump body in up position for safety during maintenance or cleaning of dump body underside.

11 BRACKETS hold dump body support braces (10) when not in use.

c. Tractor.

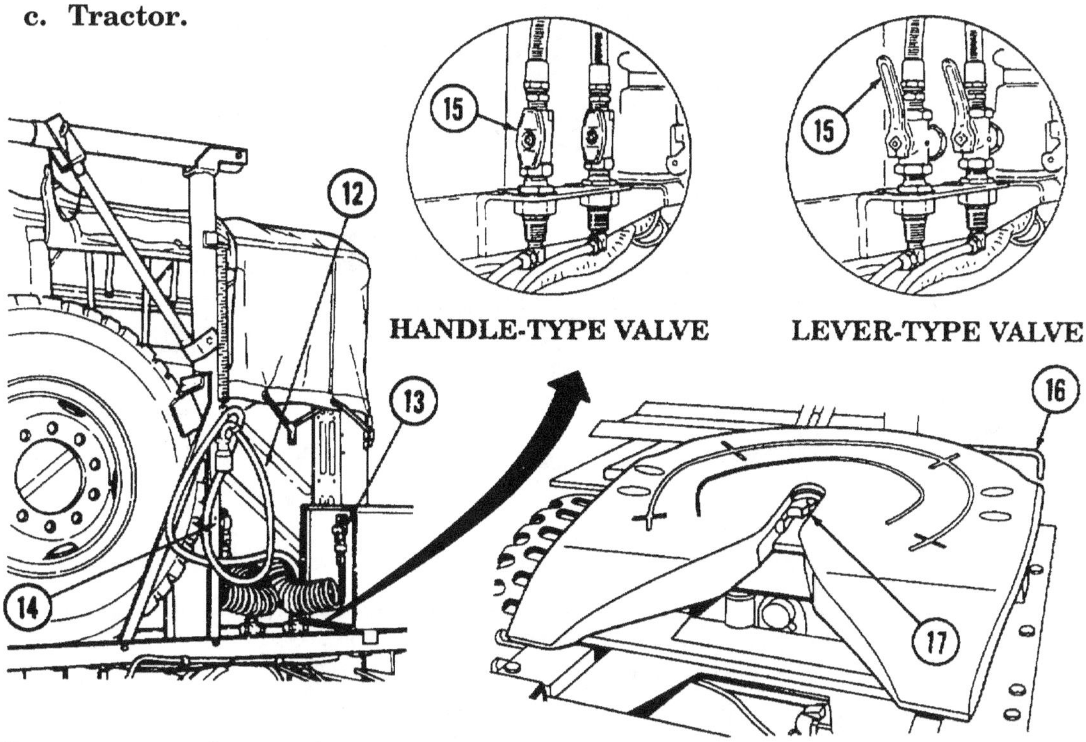
15
15
12
HANDLE-TYPE VALVE
LEVER-TYPE VALVE
16
13
14
17

Key Item and Function

12 ELECTRICAL CABLE AND CONNECTOR provides electrical power to semitrailer.

13 EMERGENCY AIRBRAKE HOSE AND COUPLING connect to the emergency airbrake coupling on the semitrailer. Semitrailer emergency brake system is activated when the service airbrakes fail.

14 SERVICE AIRBRAKE HOSE AND COUPLING connect to the service air coupling on the semitrailer.

15 AIRBRAKE HOSE COUPLING SHUTOFF VALVES are placed in the down (open) position to release compressed air to the semitrailer.

16 LOCKING PLUNGER HANDLE is pulled forward, then out to unlock fifth wheel coupling jaws (17).

17 FIFTH WHEEL COUPLING JAWS lock the semitrailer kingpin into the tractor fifth wheel.

2-4. BODY EQUIPMENT CONTROLS AND INDICATORS (Contd)

d. Cargo Body.

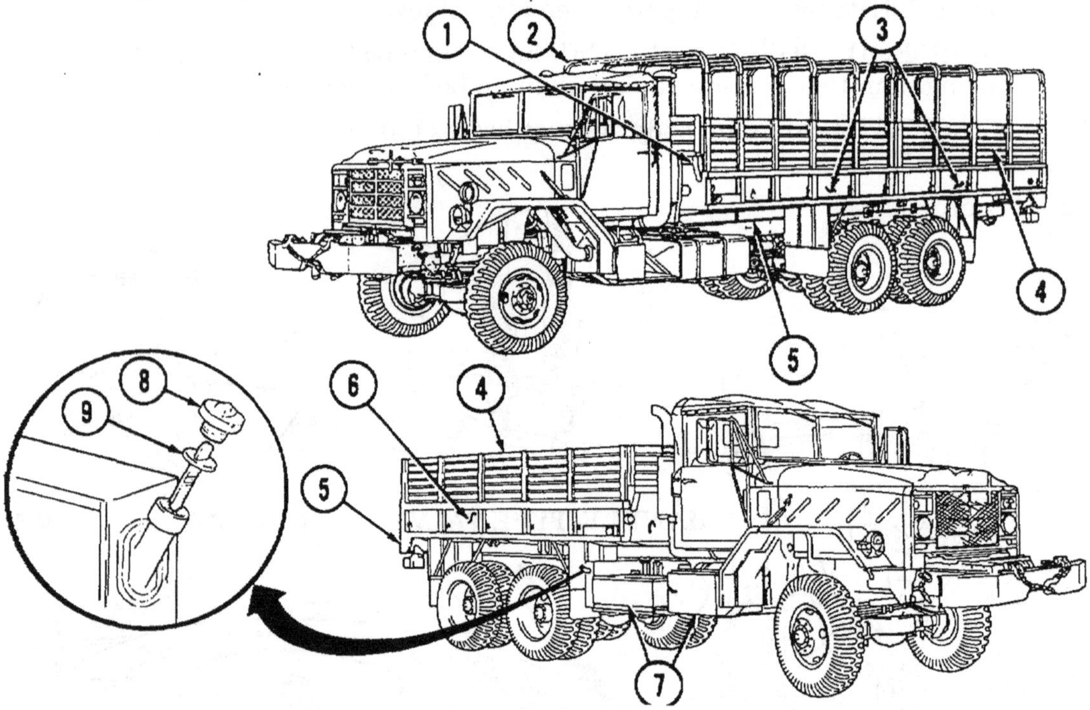
1
2
3
4
5
8
9
6
4
5
7

Key	Item and Function
1	BOW AND TARPAULIN STORAGE LOCATION provides fasteners and supports for storage of bow and tarpaulin kit (para. 2-42).
2	BOW AND TARPAULIN KIT provides a covering for the bed (5) and its contents to protect them from the weather (para. 2-42).
3	WELD SIDES provide greater stability to large shifting loads and less flexing of bed. (M927/A1/A2 and M928/A1/A2 models).
4	SIDE RACKS provides extension to sides of bed and troopseats (para. 2-23).
5	CARGO BED provides platform for moving troops and various types of cargo.
6	DROPSIDES are hinged sides which can be lowered to provide side loading of cargo and secured in upright position with quick locking handles (para. 2-23). (M923/A1/A2 and M925/A1/A2 models).
7	TOOL COMPARTMENT provides storage location for Basic Issue Items (BII) and Additional Authorized List (AAL) equipment.
8	HYDRAULIC OIL RESERVOIR CAP is turned counterclockwise and removed to provide access to hydraulic oil dipstick (9). (M925/A1/A2 and M928/A1/A2 models).
9	DIPSTICK, located inside hydraulic oil fill tube, indicates hydraulic oil level. (M925/A1/A2 and M928/A1/A2 models).

2-4. BODY EQUIPMENT CONTROLS AND INDICATORS (Contd)

e. Expansible Van.

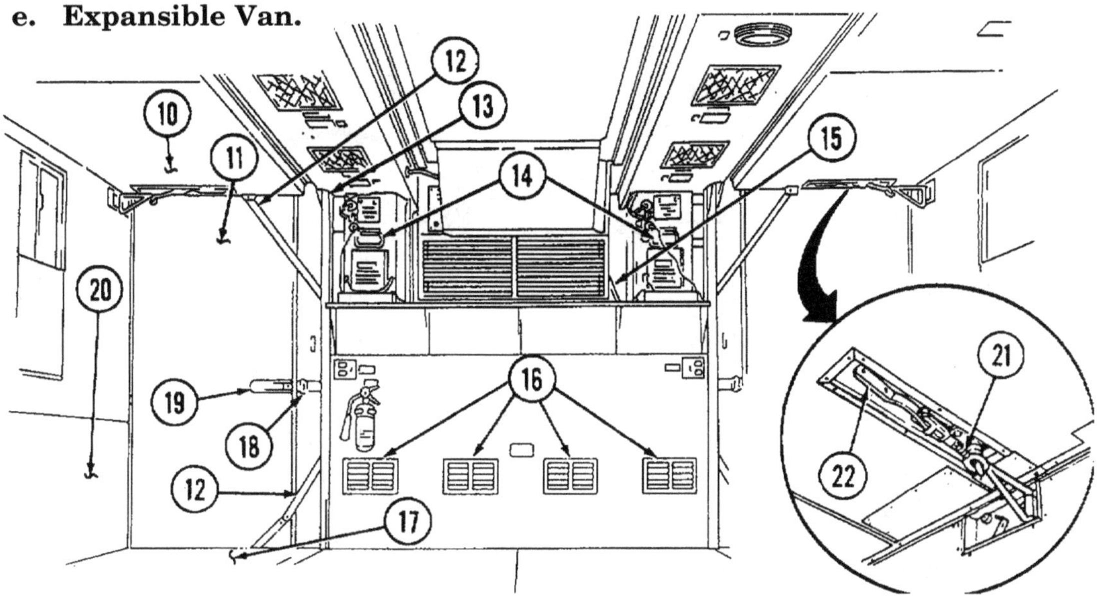

12
13
10
11
15
14
20
21
16
19
18
22
12
17

Key	Item and Function
10	HINGED ROOF PANEL is supported by swivel hooks (21) and toggle clamps (22).
11	END PANEL is hinged to van side (20) and secured to corner post (13) by sliding bolt (19).
12	BALANCE MECHANISM evenly controls lowering and raising of hinged floor (17) and roof panel (10).
13	CORNER POST provides brace for expanded van sides (20).
14	HEATERS in front of van provide heat.
15	BONNET DOOR CONTROL HANDLE is pushed forward to open bonnet door before operating air conditioner. Handle is pulled back to close bonnet door after air conditioner has been shut off.
16	HEAT REGISTERS are in use when van heaters (14) are operating. Registers must be closed when van heaters (14) are not in use.
17	FLOOR is hinged for up and down movement.
18	LATCH on van corner post (13) holds sliding bolt (19) in correct position.
19	SLIDING BOLT aligns end panel (11) with van corner post (13).
20	EXPANDED VAN SIDE is secured to hinged roof panel (10) by swivel hooks (21) and toggle clamps (22).
21	SWIVEL HOOKS are swung sideways to support hinged roof panel (10) when van sides (20) are expanded.
22	TOGGLE CLAMP locks with swivel hook (21) to secure hinged roof panel (10) to van side (20).

2-4. BODY EQUIPMENT CONTROLS AND INDICATORS (Contd)

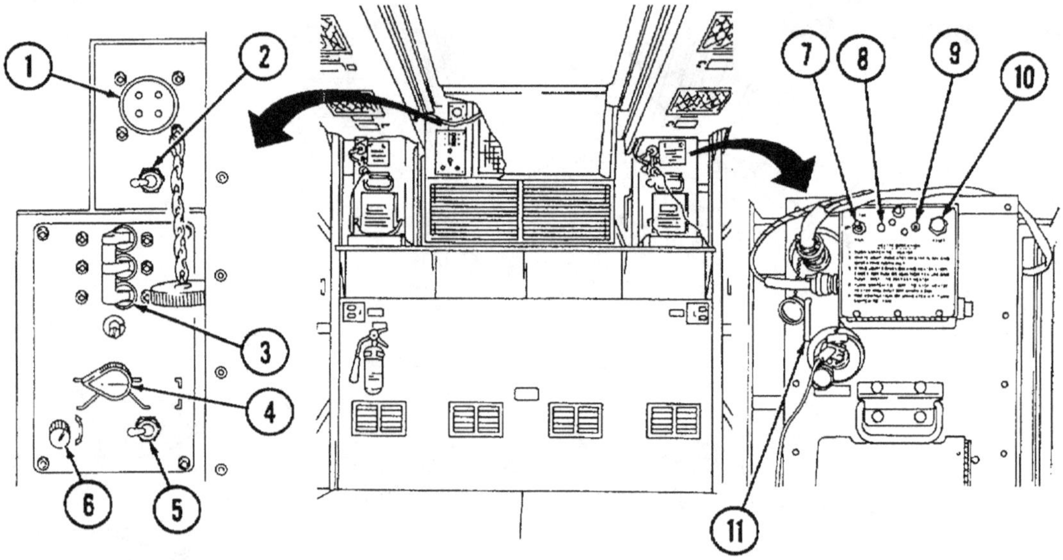
1
2
3
4
5
6
7
8
9
10
11

Key	Item and Function
1	POWER INPUT RECEPTACLE is connected by cable to outside power source to provide electric power for air conditioner.
2	CIRCUIT BREAKER must be in ON position before use of air conditioner.
3	COMPRESSOR CIRCUIT BREAKER shuts air conditioner off automatically if electrical or other malfunction develops in air conditioner. Circuit breaker (2) must be manually reset to ON position after malfunction has been corrected.
4	AIR CONDITIONER CONTROL regulates air circulation. COOLER position circulates cool air. VENT position circulates outside air into the van body.
5	FAN SPEED CONTROL provides high and low fan speed operation for air circulation.
6	TEMPERATURE SELECTOR CONTROL provides cool temperatures when in COOLER position. In WARMER position, air conditioner maintains 68°-72°F (20°-22°C) temperature. Air conditioner shuts off when temperature selector knob is in OFF position.
7	HEATER SWITCH provides heated air in HEATER position and unheated outside air in FAN position. Heater stops in OFF position.
8	WHITE INDICATOR LIGHT illuminates when heater is working properly.
9	RED INDICATOR LIGHT illuminates when heater stops because of fuel or ignition malfunction.
10	RESET BUTTON is pressed to restart heater when fuel or ignition malfunctions have been corrected.
11	HANDLE controls mixture of outside and inside air when heater is operating.

2-4. BODY EQUIPMENT CONTROLS AND INDICATORS (Contd)

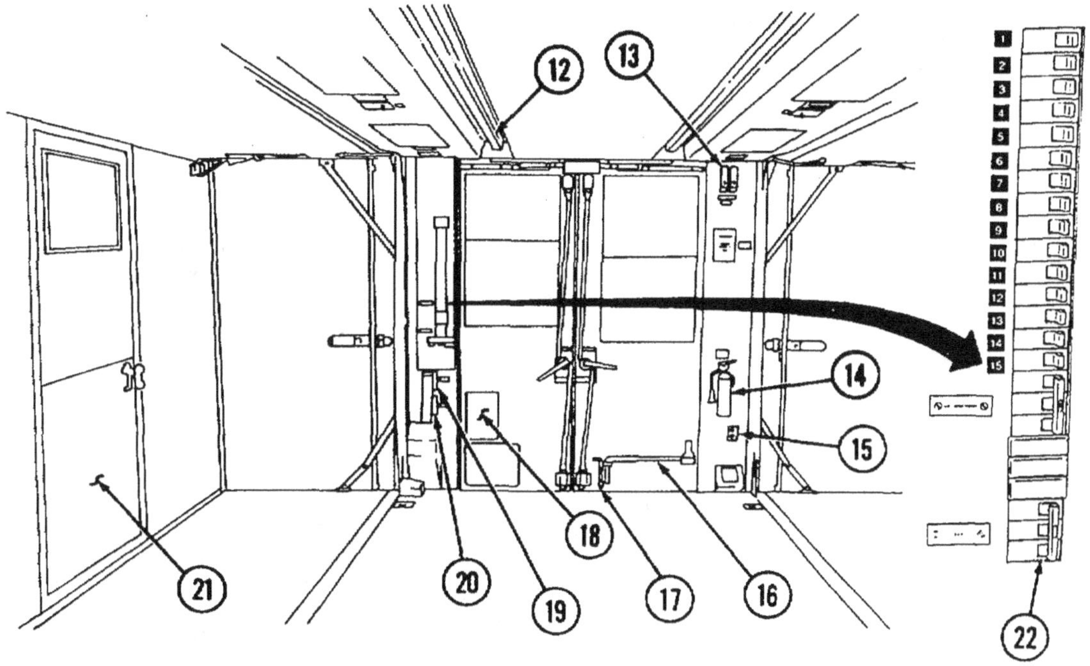
12
13
14
15
16
17
18
19
20
21
22

Key	Item and Function
12	AIR CONDITIONER VENT runs along entire length of ceiling and allows air to circulate.
13	HEATER THERMOSTAT regulates heater temperature.
14	FIRE EXTINGUISHERS are mounted on van front and rear walls.
15	TELEPHONE JACK connects van telephone to outside lines.
16	VAN BODY EXPANDING AND RETRACTING WRENCH is used to expand and retract van body.
17	SIDE PANEL LOCK WRENCH is used to lock outer edges of van roof, corner posts, and end panels when van is expanded.
18	PLATE contains instructions for operation of van.
19	BLACKOUT MAIN SWITCH is turned on for blackout operations.
20	EMERGENCY LIGHT SWITCH is turned on when normal service lights fail.
21	SIDE DOORS are used by personnel when van is expanded.

NOTE

An instruction plate near the circuit breaker panel lists circuits controlled by each switch.

22	CIRCUIT BREAKER PANEL controls electric power received from outside source.

2-4. BODY EQUIPMENT CONTROLS AND INDICATORS (Contd)

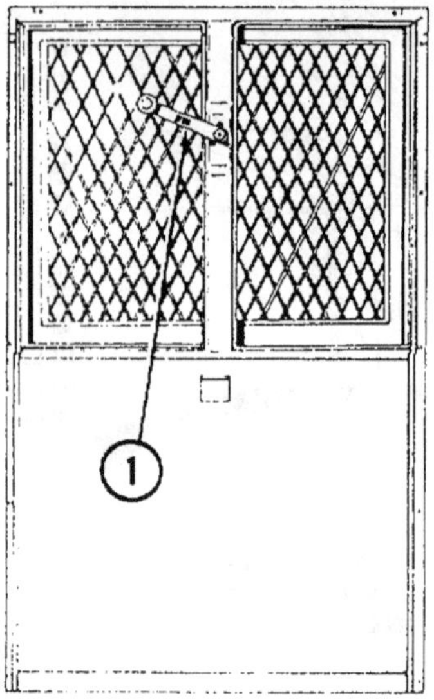
1

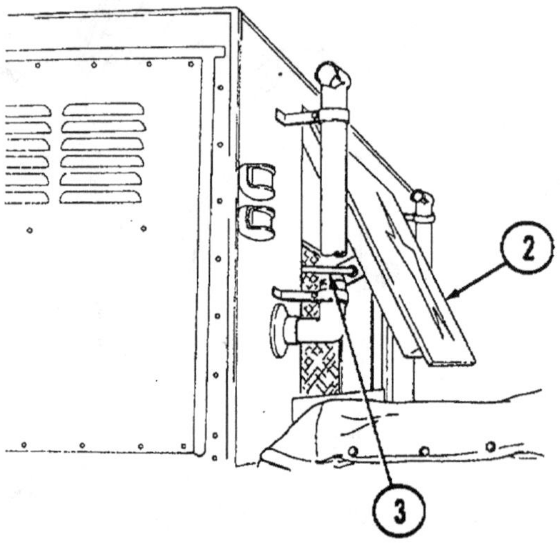
2
3

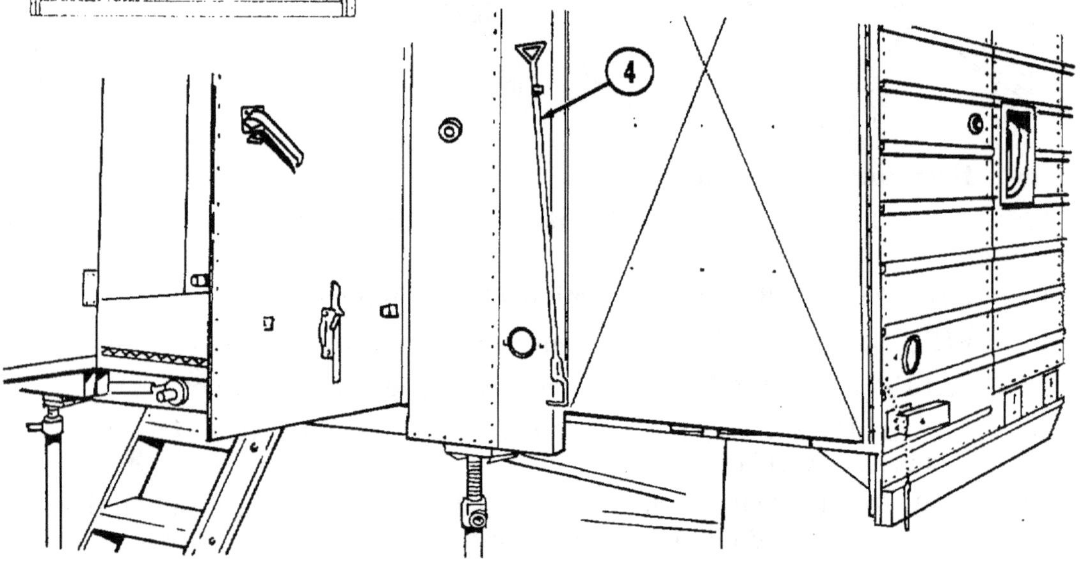
4

Key	Item and Function
1	WINDOW REGULATOR opens window when turned clockwise and closes window when turned counterclockwise.
2	BONNET DOOR allows fresh air to enter air conditioner.
3	ROD opens bonnet door (2) when bonnet door control handle inside van is pushed forward. Rod pulls bonnet door (2) closed when control handle is pulled back.
4	SIDE LOCKRODS stabilize expanded van sides when attached to lock handles.

2-4. BODY EQUIPMENT CONTROLS AND INDICATORS (Contd)

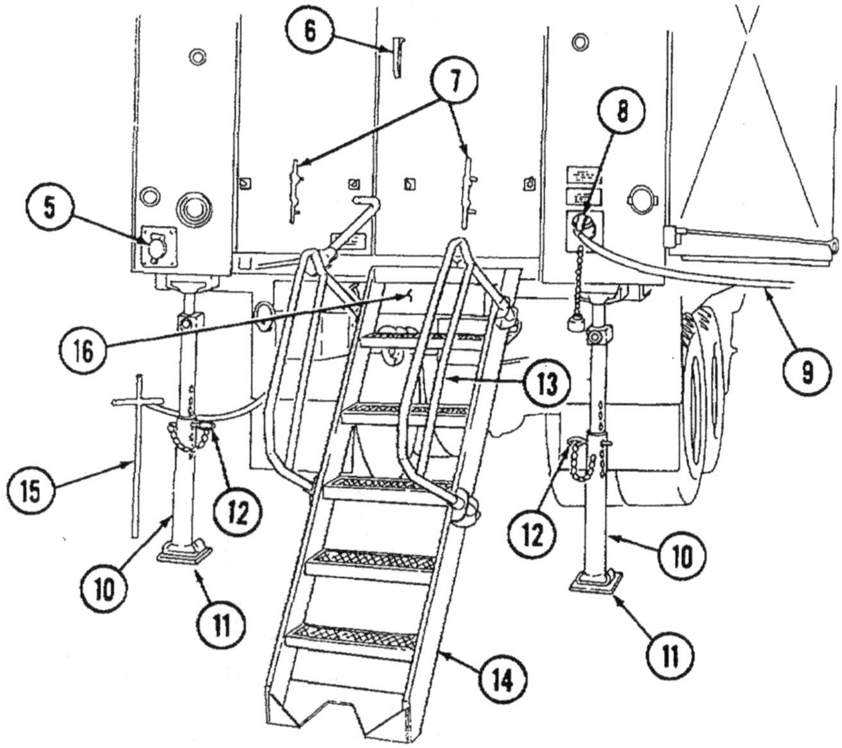
6
7
8
5
16
13
9
15
12
12
10
10
11
14
11

Key	Item and Function
5	PHONE JACK RECEPTACLE receives outside communication lines.
6	DOOR HANDLE is turned counterclockwise to open and clockwise to close rear van door.
7	LADDER CLAMPS (M934/A1/A2 models) secure lower ends of ladders.
8	POWER RECEPTACLE provides electrical power to van from outside source.
9	POWER CABLE connects to power receptacle (8) from outside power source.
10	STABILIZERS steady van when expanded.
11	STABILIZER FOOTPADS form base for stabilizer (10).
12	CHAINED PIN inserts into stabilizer (10).
13	VAN HAND RAIL MODIFICATION KIT is available for all models, to provide increased safety when using ladder (14).
14	LADDER to gain access to van from rear or sides of vehicle.
15	GROUND SPIKE provides electrical ground when external electric power is used.
16	STORAGE BOX stores ground spike (15), tools, cable reel canvas cover, and stabilizers (10).

2-4. BODY EQUIPMENT CONTROLS AND INDICATORS (Contd)

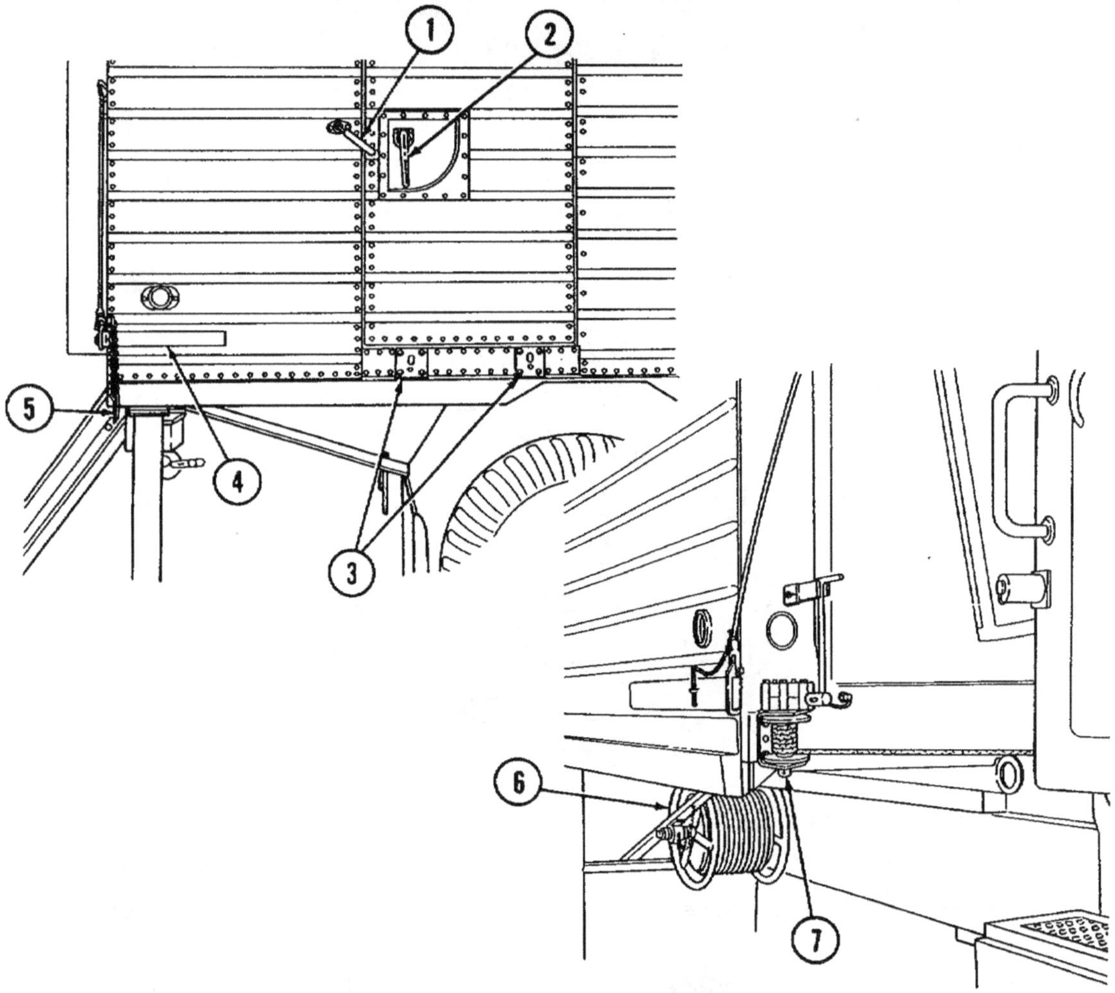
1
2
3
4
5
6
7

Key	Item and Function
1	LOCK WRENCH is turned counterclockwise to unlock expansible sides, hinged roof, and hinged floor, before expansion or retraction. Wrench is turned clockwise to lock these components after van is expanded or retracted.
2	DOOR LOCK HANDLE is turned up to unlock and down to lock van doors.
3	LADDER MOUNTING BRACKETS secure ladders when ladders are in use.
4	LOCK HANDLES engage side lockrods when van is expanded to secure expanded sides to frame.
5	PIN secures lock handle (4) in closed position.
6	POWER CABLE REEL stores power cable.
7	WINCH (M934A1/A2 models) raises and lowers spare tire on opposite side of vehicle when wire rope is attached to tire.

2-4. BODY EQUIPMENT CONTROLS AND INDICATORS (Contd)

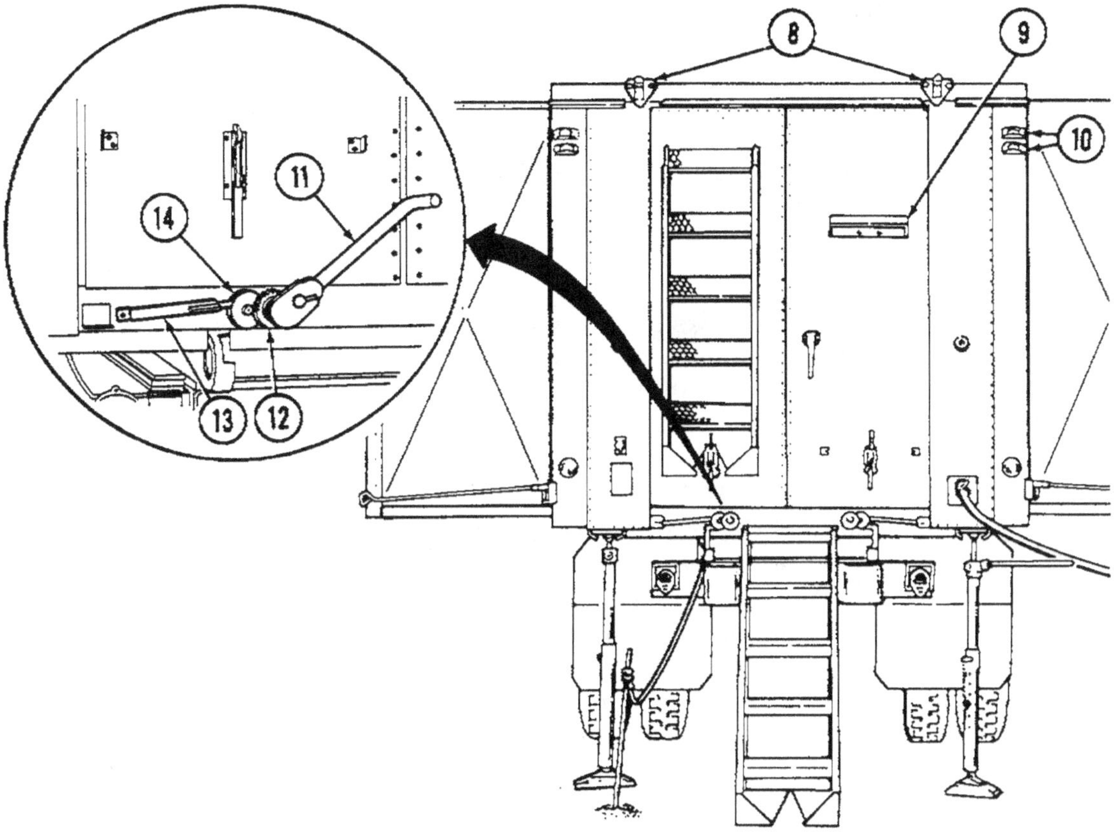
8
9
10
11
14
13
12

Key	Item and Function
8	LIFTING BRACKETS on rear and front of van body allow lifting van body from chassis.
9	LADDER RACK (M934/A1/A2 models) holds ladders.
10	CLEARANCE AND BLACKOUT LIGHTS are controlled from vehicle cab.
11	VAN BODY EXPANDING AND RETRACTING WRENCH (stowed in bracket on inner part of rear door) fits over ratchets (12). Wrench is turned counter-clockwise to expand left van side and clockwise to expand right van side.
12	RATCHET is turned by expanding and retracting wrench (11) to expand and retract van sides.
13	LOCKING PLUNGERS (located below left and right rear doors) are pushed downward to release ratchets (11) and pawls (14) before expanding or retracting sides. Plungers are pulled upward to lock van sides in expanded or retracted position.
14	PAWL attached to locking plunger (13) locks ratchet (12). Pawl releases ratchet when locking plunger (13) is pushed downward.

2-5. SPECIAL KITS CONTROLS AND INDICATORS

Controls and indicators which are added as part of a modification kit have been incorporated into paragraph 2-3 and 2-4. Controls and Indicators which are added as part of an optional kit can be found in section V.

Section II. PREVENTIVE MAINTENANCE CHECKS AND SERVICES (PMCS)

2-6. GENERAL

Preventive Maintenance Checks and Services (PMCS) means systematic caring, inspecting, and servicing of equipment to keep it in good condition and to prevent breakdowns. As the vehicle's operator, your mission is to:

a. Be sure to perform your PMCS each time you operate the vehicle. Always do your PMCS in the same order, so it gets to be a habit. Once you've had some practice, you'll quickly spot anything wrong.

b. Do your BEFORE checks and service just prior to the start of the mission to identify faults that will prevent performance of the mission. Pay attention to WARNINGS, CAUTIONS, and NOTES.

c. Do your DURING checks and services during the mission to identify faults in equipment performance. Pay attention to WARNINGS, CAUTIONS, and NOTES.

d. Do your AFTER checks immediately at the conclusion of the mission. Pay attention to WARNINGS, CAUTIONS, and NOTES.

NOTE

When a check and service procedure is required for both weekly and after intervals, it is not necessary to perform the weekly procedure during the same week in which the before procedure was done.

e. Do your WEEKLY checks once a week to identify faults that must be corrected to sustain equipment at fully mission capable standards until next unit maintenance service.

f. Do your MONTHLY checks for faults that do not need to be checked weekly, but must be checked more often than at next service by unit maintenance.

g. Use DA Form 2404 (Equipment Inspection and Maintenance Worksheet) to record any faults that you discover before, during, or after operation, unless you can fix them. You DO NOT need to record faults that you fix.

h. Be prepared to assist unit maintenance when they lubricate the vehicle. Perform any other services when required by unit maintenance.

i. A permanent record of the services, repairs, and modifications made to these vehicles must be recorded. See DA PAM 738-750 for a list of the forms and records required. Refer to chapter 3, for specific maintenance instructions.

2-7. PMCS PROCEDURES

a. Your PMCS, table 2-3, lists inspections and care required to keep your vehicle in good operating condition. It is set up so you can make your BEFORE (B) OPERATION checks as you walk around the vehicle.

b. The Item Number column of table 2-3 lists procedures in consecutive numerical order. The TM Number column on DA Form 2404, Equipment Inspection and Maintenance Worksheet, refers to these item numbers when recording PMCS results.

c. The INTERVAL column of table 2-3 tells you when to do a certain check or service.

d. The PROCEDURE column of table 2-3 tells you how to do required checks and services. Carefully follow these instructions. If you do not have tools, or if the procedure tells you to, notify unit maintenance.

e. The Not Fully Mission Capable If column of table 2-3 tells you why vehicle is not able to perform the described mission, and what equipment will be reported as not ready or unavailable. Refer to DA PAM 738-750.

f. If the truck does not perform as required, refer to Table 3-1, Troubleshooting Procedures.

g. If anything looks wrong and you can't fix it, write it on your DA Form 2404. IMMEDIATELY report it to unit maintenance.

WARNING

Drycleaning solvent is flammable and will not be used near an open flame. A fire extinguisher will be kept nearby when solvent is used. Use only in well-ventilated places. Failure to do this will result in injury to personnel and/or damage to equipment.

NOTE

Dirt, grease, oil, and debris may cover up a serious problem. Clean as you check. Use drycleaning solvent on all metal surfaces. Use soap and water on rubber or plastic material.

h. When you do your PMCS, you will always need a rag or two. Following are checks that are common to the entire vehicle.

(1) Check all bolts, nuts, and screws. If loose, bent, broken, or missing, either tighten or report condition(s) to unit maintenance.

(2) Look for loose or chipped paint and rust or gaps at welds. If a cracked or broken weld is found, report condition(s) to unit maintenance.

(3) Check electrical wires and connectors for cracked or broken insulation. Look for bare wires and loose or broken connections. Tighten loose connections. Report other problem(s) to unit maintenance.

(4) Check hoses and fluid lines for wear, damage, and leaks. Ensure clamps and fittings are tight. (Refer to para. 2-9 for information on leaks.)

(5) Check air lines for damage or leaks. Ensure clamps and fittings are tight. Tighten loose connections. If leaks or other problems still exist, report condition to unit maintenance.

2-7. PMCS PROCEDURES (Contd)

i. Correct Assembly or Stowage. Check each component or assembly for proper installation and ensure that there are no missing parts.

2-8. CLEANING INSTRUCTION AND PRECAUTIONS

WARNING

Accidental or intentional introduction of liquid contaminants into the environment is in violation of state, federal, and military regulations. Refer to Lubrication Order (para. 3-1) for information concerning storage, use, and disposal of these liquids. Failure to do so may result in injury or death.

Cleaning is an after-operation service performed by operator/crew to keep the vehicle in a state of readiness. Facilities and material available to operators for vehicle cleaning can vary greatly in differing operating conditions. However, vehicles must be kept as clean as available cleaning equipment, materials, and tactical situations permit.

a. General Cleaning Precautions.

(1) All cleaning procedures must be accomplished in well-ventilated areas.

(2) Protective gloves, clothing, and/or respiratory equipment must be worn whenever caustic, toxic, or flammable cleaning solutions are used.

(3) Diesel fuel or gasoline must never be used for cleaning.

(4) A fire extinguisher must be available and ready during all cleaning operations involving solvents.

b. Special Precautions.

(1) Do not allow cleaning compounds to come into contact with rubber, leather, vinyl, or canvas materials.

(2) Do not allow corrosion-removing cleaning compounds to contact painted surfaces.

(3) Do not use air in cleaning truck cab interiors or van body interiors.

(4) Do not steam–clean any part of vehicle that has been rustproofed.

(5) Mildew must be removed with a bristle brush before canvas tarpaulin can be properly cleaned and aired.

(6) The radiator is always cleaned first from behind in order to blow debris, insects, or other obstructions out and away from the radiator core. Low pressure water or air can be used in cleaning radiator core of obstructions.

c. Cleaning Materials. Detailed description of specific cleaning compounds, cleaning solvents, drycleaning solutions, and corrosion-removing compounds are found in TM 9-247.

2-8. CLEANING INSTRUCTION AND PRECAUTIONS (Contd)

d. General Guidelines. Table 2-1 provides a general guideline to cleaning materials used in removing contaminants from various vehicle surfaces.

Table 2-1. General Cleaning Instructions.

	Cleaning Materials Used to Remove		
Surface	**Oil/Grease**	**Salt/Mud/Dust/Debris**	**Surface Rust/Corrosion**
Body	Grease-cleaning compound, running water, and damp or dry rags.	High pressure water, soapy warm water, soft brush, and damp or dry rags.	Corrosion-removing compound, bristle brush, dry rags, and lubricating oil.*
Cab Interior (Metals)	Grease cleaning compound, and damp or dry rags.	Damp and dry rags.	Corrosion-removing compound, bristle brush, dry rags, and lubricating oil.*
Cab Interior/Cab Top (Material)	Saddle soap, warm water, soft brush, and dry rags.	Soft brush, soapy warm water, and damp or dry rags.	Not applicable.
Frame	Grease-cleaning compound, rinsed with running water, and rags.	High pressure water, soapy warm water, wire brush, and damp or dry rags.	Corrosion-removing compound, wire brush, dry rags, and lubricating oil.*
Engine/ Transmission	Mixed solution, 1 part grease-cleaning compound, 4 parts drycleaning solvent, running water, and rags.	High pressure water, soapy warm water, soft wire brush, and damp or dry rags.	Bristle brush, warm soapy water, and dry rags.
Glass	Glass cleaning solution and clean, dry rags.	Glass cleaning solution and clean, dry rags.	Not applicable.
Radiator	Not applicable.	Low pressure water, air, soapy warm water, and damp or dry rags.	Not applicable.

* After cleaning, apply light grade of lubricating oil to unprotected surfaces to prevent continued rust.

2-8. CLEANING INSTRUCTION AND PRECAUTIONS (Contd)

Table 2-1. General Cleaning Instructions (Contd).

	Cleaning Materials Used to Remove		
Surface	**Oil/Grease**	**Salt/Mud/Dust/Debris**	**Surface Rust/Corrosion**
Rubber Insulation	Damp or dry rags.	Damp or dry rags.	Not applicable.
Tires	Soapy water and bristle brush.	High pressure water and bristle brush.	Not applicable.
Wire Rope	Cleaning compound and wire brush.	Wire brush.	Wire brush and lubricating oil.**
Wood	Detergent, warm water, and damp or dry rags.	Low pressure water, soapy warm water, and damp or dry rags.	Not applicable.

** After cleaning, apply grease (MIL-B-18458).

2-9. CLASS LEAKAGE DEFINITIONS

WARNING

Accidental or intentional introduction of liquid contaminants into the environment is in violation of state, federal, and military regulations. Refer to Lubrication Order (para. 3-1) for information concerning storage, use, and disposal of these liquids. Failure to do so may result in injury or death.

Wetness around seals, gaskets, fittings, or connections indicates leakage. A stain also denotes leakage. If a fitting or connector is loose, tighten it. If broken or defective, report it. Use the following as a guide when referring to table 2-2, Fluid Leakage Criteria for Tactical Vehicles:

a. Class I. Leakage indicated by wetness or discoloration not great enough to form drops.

b. Class II. Leakage great enough to form drops, but not enough to cause drops to drip from item being checked/inspected.

c. Class III. Leakage great enough to form drops that fall from the item being checked/inspected.

2-9. CLASS LEAKAGE DEFINITIONS (Contd)

CAUTION

Operation is allowable with class I or II leakage. You must, of course, consider fluid capacity of the item/system. WHEN IN DOUBT, NOTIFY UNIT MAINTENANCE. When operating with class I or II leaks, check fluid levels more frequently. Class III leaks must be reported immediately to your supervisor or to unit maintenance.

Table 2-2. Fluid Leakage Criteria for Tactical Vehicles.

FLUID LEAKAGE CRITERIA FOR TACTICAL VEHICLES	DEADLINING LEAKAGE CRITERIA		
	I	II	III
ALL MODELS			
Air Compressor (Oil)			✓
Coolant System (Engine)			✓
Differentials			✓
Dump Hoist Cylinder			✓
Engine Lube Oil System			✓
Engine Surge Tank			✓
Front Winch (M925/A1/A2, M928/A1/A2, M930/A1/A2, M932/A1/A2, M936/A1/A2)			✓
Fuel Filter/Water Separator			✓
Fuel System			✓
Fuel Tank			✓
Hydraulic Hoist (except M923/A1/A2, M927/A1/A2, M931/A1/A2, and M934/A1/A2)			✓
Hydraulic Lines and Hoses (except M923/A1/A2, M927/A1/A2, M931/A1/A2, and M934/A1/A2)			✓
Hydraulic Pump (except M923/A1/A2, M927/A1/A2, M931/A1/A2, and M934/A1/A2)			✓
Hydraulic Tank			✓
Oil Filter			✓
Personnel Heater and Hoses			✓
Power Steering Assist Cylinder			✓
Power Steering Pump			✓
Power Train			
PTO (except M923/A1/A2, M927/A1/A2, M931/A1/A2, and M934/A1/A2)			✓
Radiator			✓
Transmission Cooler			✓
Transmission and Transmission Case			✓
Water Pump			✓
M936/A1/A2 MEDIUM WRECKER			
Crane in Operation			✓
Rear Winch and Crane			✓
Wrecker Crane			✓

2-10. PMCS TABLE

Walk-around inspections will begin at the front of the vehicle and proceed around the crewmember's side (right side), around the rear of the vehicle, and continue up the driver's side (left side). If inspection items are found in more than one location, cover the entire vehicle. Reinspection is required after a change in engine run-up condition, and is not complete until all areas have been inspected or reinspected.

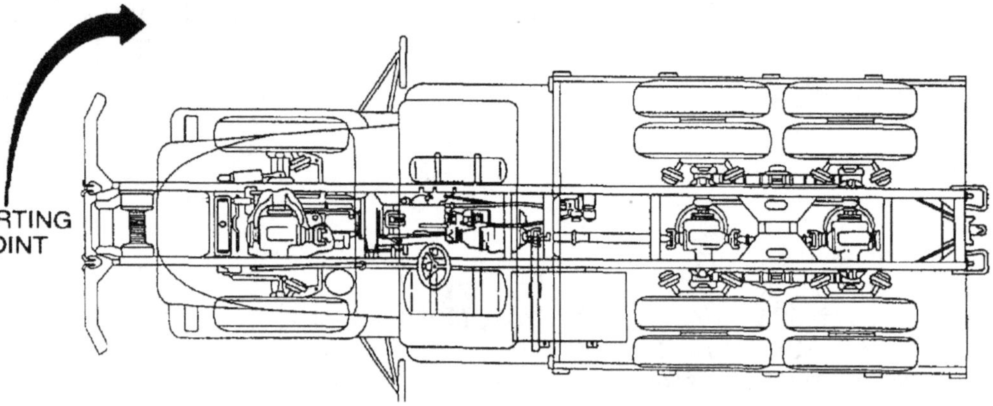
STARTING POINT

Table 2-3. Preventive Maintenance Checks and Services for Models M939/A1/A2.

ITEM NO.	INTERVAL	LOCATION ITEM TO CHECK/ SERVICE	CREWMEMBER PROCEDURE	NOT FULLY MISSION CAPABLE IF:
			WARNING Always remember the CAUTIONS, WARNINGS, and NOTES before operating this vehicle and prior to PMCS. Perform all before, during, after, and weekly checks if: **a.** You are the assigned driver but have not operated the vehicle since the last weekly inspection. **b.** You are operating the vehicle for the first time.	

Table 2-3. Preventive Maintenance Checks and Services for Models M939/A1/A2 (Contd).

ITEM NO.	INTERVAL	LOCATION ITEM TO CHECK/ SERVICE	CREWMEMBER PROCEDURE	NOT FULLY MISSION CAPABLE IF:
			<u>DRIVER</u> **NOTE** If leakage is detected, further investigation is needed to determine the location and cause of the leak (para. 2-9).	
1	Before	Front of Vehicle	**a.** Look under vehicle for obvious fluid leaks such as oil, fuel, and water (para. 2-9). **b.** Visually check for obvious damage that would prevent operation. **c.** Ensure service (6) and emergency (2) gladhand valves are closed. **d.** Ensure gladhand covers (5) are installed.	**a.** Any class III leak evident. **b.** Any damage will prevent operation.
			<u>DRIVER</u> **NOTE** Cracked or broken windshield may violate AR 385-55.	
2	Before	Wind-shield Wipers	**a.** Check windshield (1) for any cracks that would impair vision **b.** Check wiper arms (3) and blades (4) for damage.	

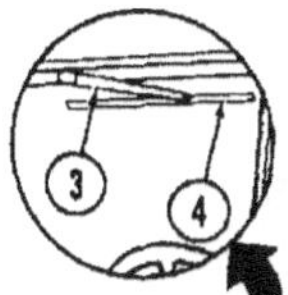
3
4

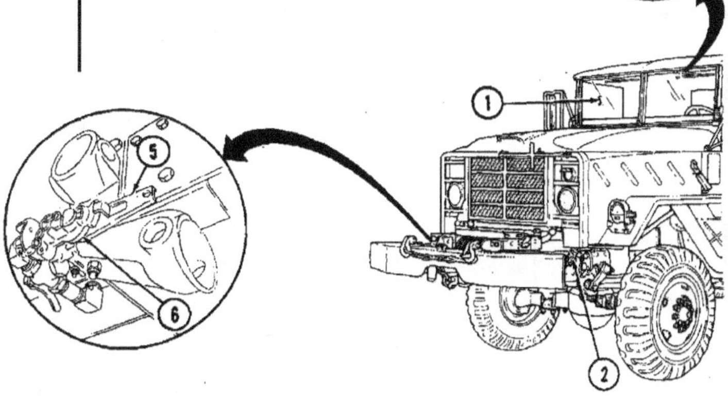
1
5
6
2

Table 2-3. Preventive Maintenance Checks and Services for Models M939/A1/A2 (Contd).

ITEM NO.	INTERVAL	LOCATION ITEM TO CHECK/ SERVICE	CREWMEMBER PROCEDURE	NOT FULLY MISSION CAPABLE IF:
3	Before	Right, Front, and Side Exterior	DRIVER **NOTE** If leakage is detected, further investigation is needed to determine the location and cause of the leak (para. 2-9). **a.** Visually check outside front, underneath, and right side of vehicle for fuel or oil leaks (para. 2-9). **b.** Visually check right side of vehicle for obvious damage that would impair operation.	**a.** Any class III leakage evident. **b.** Damage would prevent operation.
			DRIVER **NOTE** • On A2 models, CTIS will automatically set to previous inflation level when engine starts. Recheck inflation level after starting engine (para 3-11). • Quantities specified are for the entire vehicle. • On M931A2 and M932A2 models equipped with CTIS, the CTIS must be neutralized at unit maintenance if towing a 5,000-gallon semitrailer (M131 series, M967/A1, M969/A1/A2, and M970/A1). **CAUTION** Ensure all tires are properly inflated before operating vehicle (table 1-10).	Tire(s) do not hold required air pressure.

Table 2-3. Preventive Maintenance Checks and Services for Models M939/A1/A2 (Contd).

ITEM NO.	INTERVAL	LOCATION ITEM TO CHECK/ SERVICE	CREWMEMBER PROCEDURE	NOT FULLY MISSION CAPABLE IF:
4	Before	Right Side Tire(s)	**a.** Visually check right side tire(s) (2) for presence and under-inflation (para. 3-11).	**a.** M939A1/A2 series vehicles have any tires missing, or are unserviceable. M939 series vehicles have two or more tires missing or are unserviceable.
		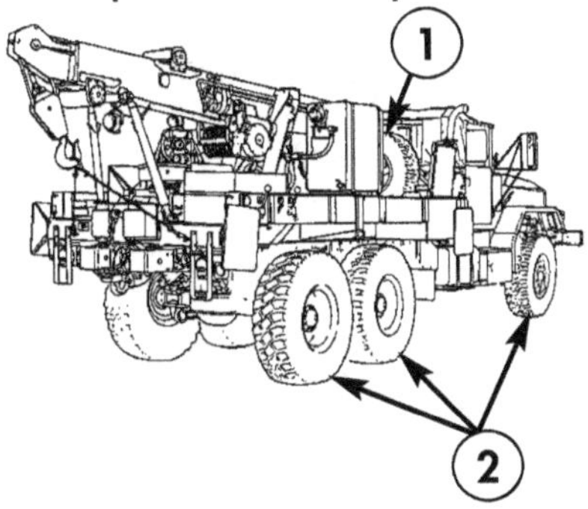 1 2	**b.** If a tire (2) is damaged, replace with spare (1) (para 3-11).	**b.** Tire(s) have cuts, gouges, cracks, or leaks that would cause tire failure.
			c. Check for cupping or worn tires (2), replace with spare (1) (para. 3-11).	**c.** Tires have cupping which cause erratic steering, or worn to within 4/32 in. (3 mm).
			DRIVER	
5	Before	Fifth Wheel	**NOTE** Item 5 applies to M931/A1/A2 and M932/A1/A2 models only.	
			a. Inspect fifth wheel (4), locking mechanism (5), and approach plates (3) for bends and damage.	**a.** Damage causes wheel to be inoperative.
			b. Check fifth wheel (4) subframe rivets (6).	**b.** Any loose or missing rivets.
			c. Inspect fifth wheel mounting brackets for broken welds, cracked or damaged components, or worn parts.	**c.** Fifth wheel mounting brackets have broken welds or damaged components.
		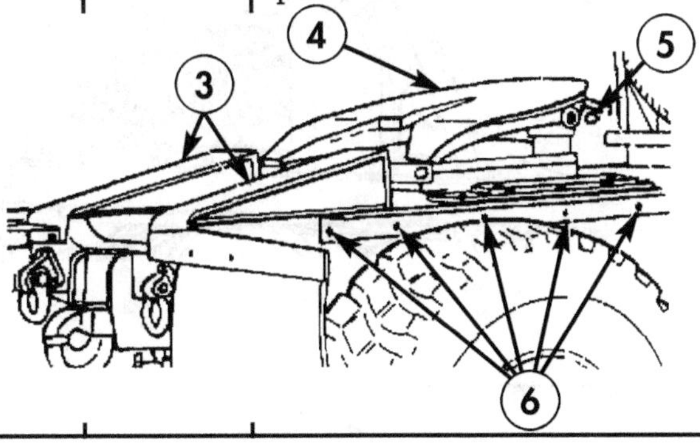 3 4 5 6		

Table 2-3. Preventive Maintenance Checks and Services for Models M939/A1/A2 (Contd).

ITEM NO.	INTERVAL	LOCATION ITEM TO CHECK/ SERVICE	CREWMEMBER PROCEDURE	NOT FULLY MISSION CAPABLE IF:
			DRIVER	
6	Before	Trailer Connecting Accessories	**a.** Inspect electrical cable (1) and connector (2) for cracks, breaks, and other damage.	**a.** Cable or connector is cracked, broken, missing, or unserviceable.
			b. Inspect emergency air/brake hose lines (3).	**b.** Air/brake hose lines are loose or missing.

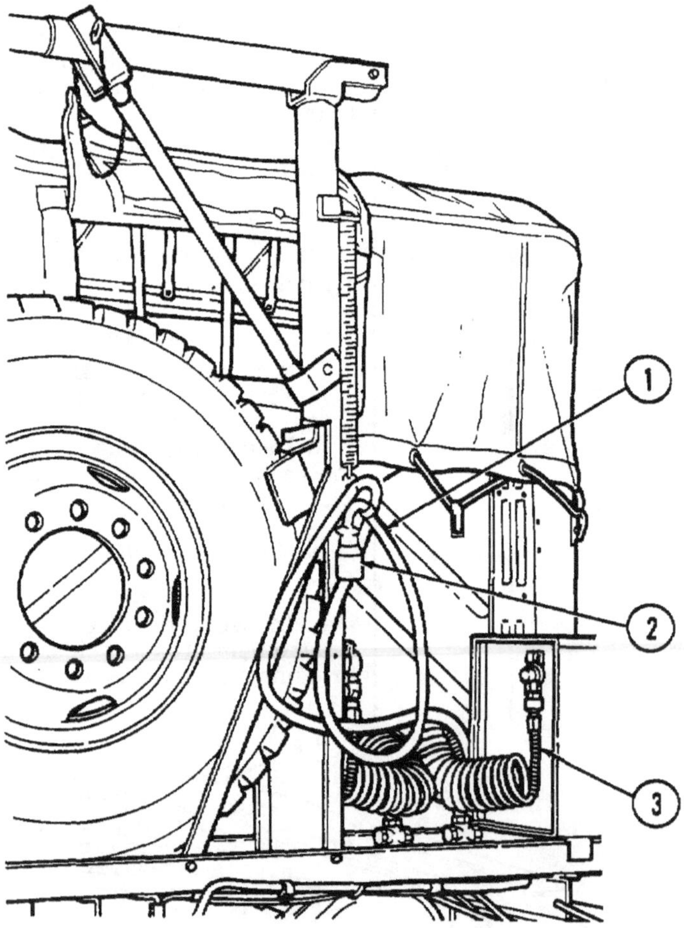
1
2
3

Table 2-3. Preventive Maintenance Checks and Services for Models M939/A1/A2 (Contd).

ITEM NO.	INTERVAL	LOCATION ITEM TO CHECK/ SERVICE	CREWMEMBER PROCEDURE	NOT FULLY MISSION CAPABLE IF:
			DRIVER **NOTE** If leakage is detected, further investigation is required to determine location and cause of leak (para. 2-9).	
7	Before	Rear of Vehicle and Under Rear of Vehicle	**a.** Look under vehicle for obvious fluid leaks such as oil and fuel. **b.** Visually check for obvious damage that would impair operation. **c.** Ensure service (6) and emergency (4) gladhand valves are closed. **d.** Ensure gladhand covers (5) are installed.	**a.** Any class III leak evident. **b.** Any damage that will prevent operation.

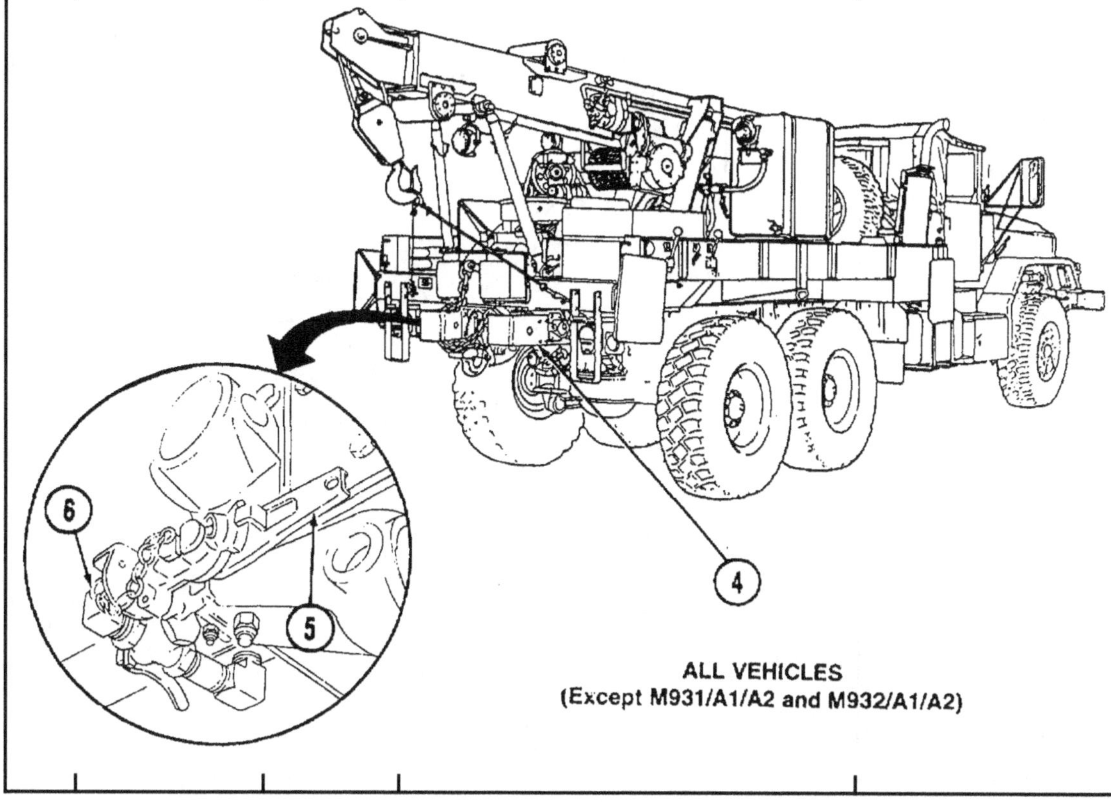
6
5
4

ALL VEHICLES
(Except M931/A1/A2 and M932/A1/A2)

Table 2-3. Preventive Maintenance Checks and Services for Models M939/A1/A2 (Contd).

ITEM NO.	INTERVAL	LOCATION ITEM TO CHECK/ SERVICE	CREWMEMBER PROCEDURE	NOT FULLY MISSION CAPABLE IF:

LUBRICANTS	EXPECTED TEMPERATURES Above +80°F (+27°C)	+80° TO + 30°F (+27° TO -1°C)	+30°F TO -30°F (-1°C TO -34°C)	-30°F TO -65°F (-34°C TO -54°C)
CW- Lubricating Oil (VV-L-751)	CW-11C	CW-11B	CW-11A	GO75

ITEM NO.	INTERVAL	ITEM TO CHECK/ SERVICE	CREWMEMBER PROCEDURE	NOT FULLY MISSION CAPABLE IF:
			DRIVER **WARNING** Wear hand protection when handling cable. Do not handle cable with bare hands. Broken wires will cause injury.	
8	Before	Rear Winch and Controls	**a.** Visually check winch hoses and lines for signs of deterioration and leakage.	**a.** Class III leakage is evident.
			b. Check cable (1) for kinks, frays, breaks, and missing.	**b.** Cable has kinks, frays, breaks, or missing.
			c. Clean and oil cable (1) after each operation. If used infrequently or in very damp or salty conditions, perform lubrication. Do not lubricate cable (1) in dry, dusty conditions.	

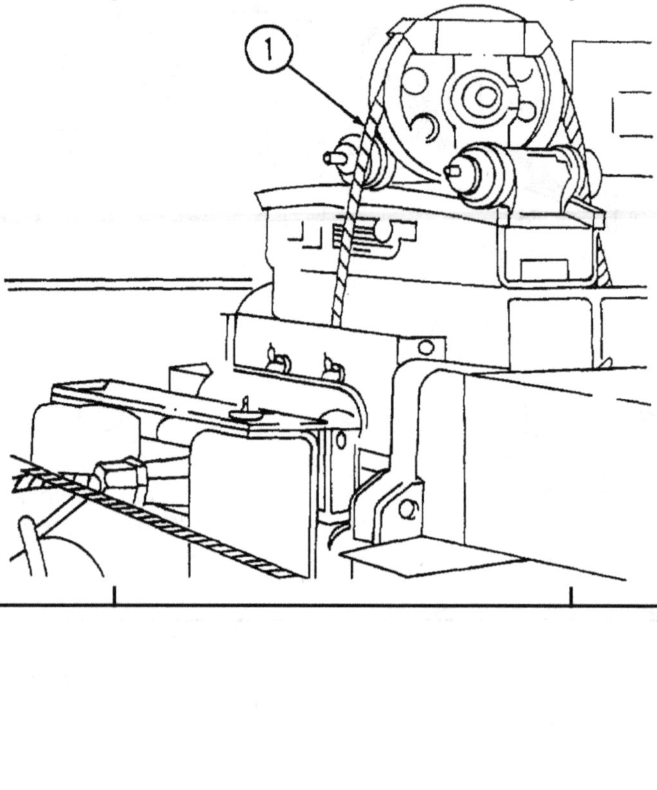
1

Table 2-3. Preventive Maintenance Checks and Services for Models M939/A1/A2 (Contd).

ITEM NO.	INTERVAL	LOCATION / ITEM TO CHECK/ SERVICE	CREWMEMBER PROCEDURE	NOT FULLY MISSION CAPABLE IF:
			DRIVER **NOTE** • On A2 models, CTIS will automatically set to previous inflation level when engine starts. Recheck inflation level after starting engine (para 3-11). • Quantities specified are for the entire vehicle. **CAUTION** Ensure all tires are properly inflated before operating vehicle (table 1).	Tire(s) do not hold required air pressure.
9	Before	Left Side Tires	**a.** Visually check left side tire(s) (3) for presence and under-inflation (para. 3-11).	**a.** M939A1/A2 series vehicle has any tire missing or is unserviceable. M939 series vehicle has two or more tires missing or unserviceable.
			b. If a tire (3) is damaged, replace with spare (2) (para. 3-11).	**b.** Tire(s) have cuts, gouges, cracks, or leaks that would cause tire failure.
			c. Check for cupping or worn tires (3) replace with spare (2) (para. 3-11).	**c.** Tire(s) have cupping which cause erratic steering, or worn to within 4/32 in. (3 mm).

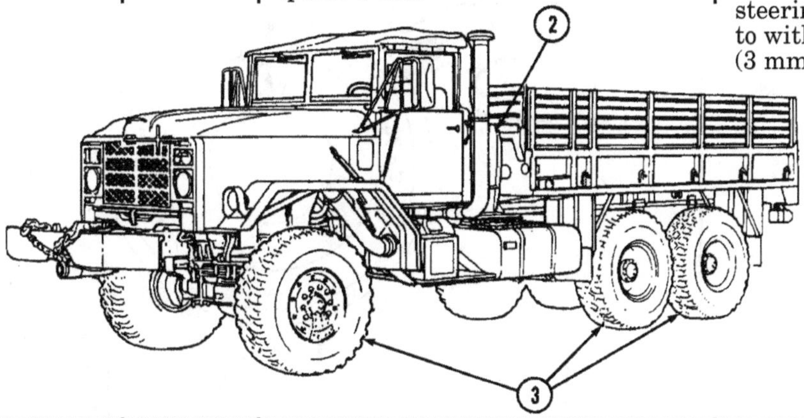
2
3

Table 2-3. Preventive Maintenance Checks and Services for Models M939/A1/A2 (Contd).

ITEM NO.	INTERVAL	LOCATION ITEM TO CHECK/ SERVICE	CREWMEMBER PROCEDURE	NOT FULLY MISSION CAPABLE IF:
			NOTE Two fire extinguishers are located inside van body on M934/A1/A2 series vehicles, and behind gondola cab on M936/A1/A2 series vehicles. Refer to Appendix E for exact locations. DRIVER	
10	Before	Fire Extin- guisher	**a.** Check for missing or damaged fire extinguisher (1).	**a.** Fire extinguisher missing or damaged.
			b. Check gauge (3) for proper pressure of about 150 psi (1034 kPa).	**b.** Pressure gauge needle in recharge area.
			c. Check for damaged or missing seal (2).	**c.** Seal broken or missing.

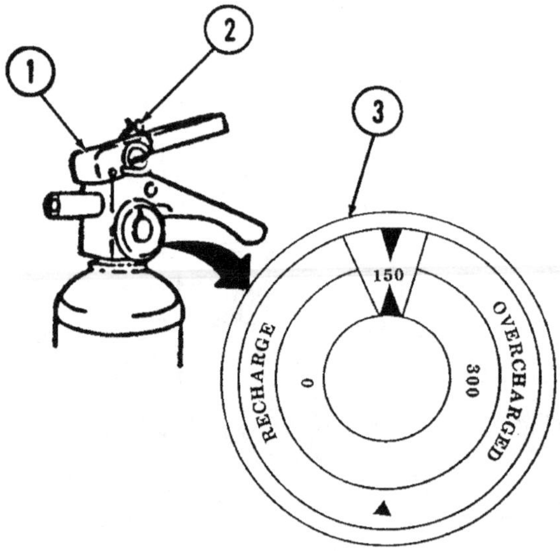
1
2
3
150
RECHARGE
0
300
OVERCHARGED

Table 2-3. Preventive Maintenance Checks and Services for Models M939/A1/A2 (Contd).

ITEM NO.	INTERVAL	LOCATION / ITEM TO CHECK/ SERVICE	CREWMEMBER PROCEDURE	NOT FULLY MISSION CAPABLE IF:
			DRIVER **NOTE** If leakage is detected, further investigation is needed to determine the location and cause of the leak (para. 2-9).	
11	Before	Left Front, Side Exterior	**a.** Check underneath vehicle for evidence of fluid leakage.	**a.** Class III leak of oil, fuel, or coolant.
			b. Visually check left side of vehicle for obvious damage that would impair operation.	**b.** Any damage will prevent operation.
			c. Check fuel tank (4) for leaks created by damage or expansion of fuel.	**c.** Class III leak of fuel.

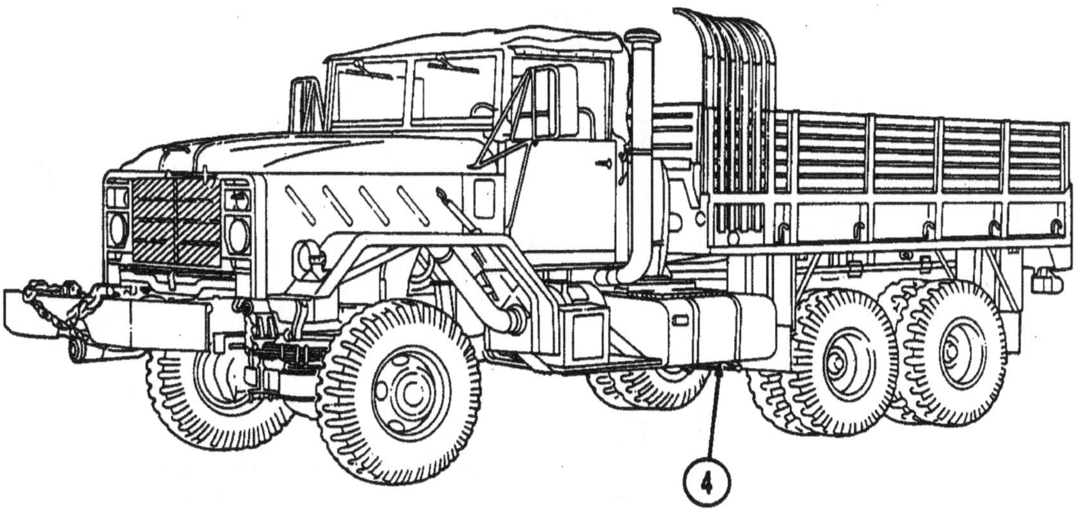
4

Table 2-3. Preventive Maintenance Checks and Services for Models M939/A1/A2 (Contd).

ITEM NO.	INTERVAL	LOCATION ITEM TO CHECK/ SERVICE	CREWMEMBER PROCEDURE	NOT FULLY MISSION CAPABLE IF:
			DRIVER **WARNING** If NBC exposure is suspected, all air filter media should be handled by personnel wearing protective equipment. Consult your unit NBC officer or NBC NCO for appropriate handling caution or disposal instructions.	
12	Before	Controls and Instru-ments	**a.** Check for missing or damaged seatbelts.	**a.** Seatbelt(s) are missing or damaged.
			b. Start engine (para. 2-12)..	**b.** Engine will not start.
			c. Check that ABS warning lamp (9) comes on for approximately three seconds and then goes out.	**c.** ABS warning lamp stays on.
			NOTE The engine must be running to perform the following checks.	
			d. Air cleaner indicator (1) should be in the green area.	**d.** Air cleaner is cracked, unser-viceable, or stays in the red.
			e. Tachometer (2) should read 600-650 rpm (M939/A1), 565-635 rpm (M939A2) at idle.	**e.** Tachometer (2) inoperable or reads less than 600 rpm or more than 650 rpm (M939/A1), 565-635 rpm (M939A2) at idle.
			f. Check engine water temperature gauge (3). (Normal range is 175° - 200°F (79 - 93°C) with engine warmed up.	**f.** Temperature gauge (3) reads less than 175° (79°C) or exceeds 220° F (104°C).
			g. Check primary air pressure gauge (4) and secondary air pressure gauge (6). Normal range is 90 to 130 psi (621 - 896 kPa). Make sure warning buzzer is operational.	**g.** Reads less than 60 psi (414 kPa). Warning buzzer stays on or does not operate.

Table 2-3. Preventive Maintenance Checks and Services for Models M939/A1/A2 (Contd).

ITEM NO.	INTERVAL	LOCATION / ITEM TO CHECK/ SERVICE	CREWMEMBER PROCEDURE	NOT FULLY MISSION CAPABLE IF:
12 (contd)	Before	Controls and Instru- ments (contd)	**h.** Check voltmeter (5). Needle should be in green area.	**h.** Readings above or below the green area or inoperative.
			i. Check transmission oil temperature gauge (7). Normal range is 120°F - 220°F (49°-104°C) (may not read at low temperature).	**i.** Oil tempera- ture exceeds 300°F (149°C).
			CAUTION If oil pressure gauge reads 0 psi, stop engine.	
			j. Engine oil pressure gauge (8) reads at least 15 psi (103 kPa) on M939/A1 or 10 psi (69 kPa) on M939A2 series vehicles at idle speed.	**j.** Engine oil pressure is less than 15 psi (103 kPa) on M939/A1 or 10 psi (69 kPa) on M939A2 series vehicles at idle speed.

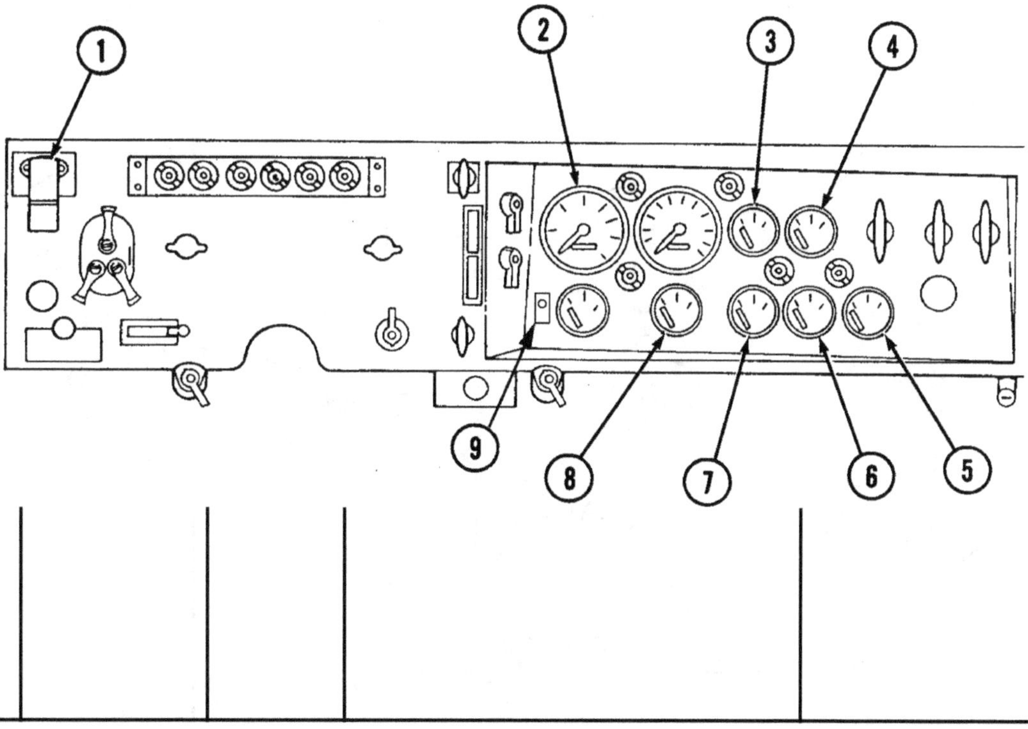
1
2
3
4
5
6
7
8
9

Table 2-3. Preventive Maintenance Checks and Services for Models M939/A1/A2 (Contd).

ITEM NO.	INTERVAL	LOCATION ITEM TO CHECK/ SERVICE	CREWMEMBER PROCEDURE	NOT FULLY MISSION CAPABLE IF:
12 (contd)	Before	Controls and Instruments (contd)	**j.** Check transmission selector (2) and transfer case shift lever (6). Shift in all ranges, observing unusual stiffness, abnormal operation, or binding.	**j.** Transmission selector (2) or transfer case shift lever (6) inoperative or binding.
			k. Put front-wheel drive lock-in switch (8) to IN position. Make sure axle lock-in light (1) is on (para. 2-3).	**k.** Front-wheel drive will not engage.
			l. Check steering response.	**l.** Steering binds or is unresponsive.
			m. Listen for leakage in exhaust system.	**m.** Any leak could cause injury to personnel.
			n. Determine parking brake (5) ability to hold vehicle. Depress override button (3) on dash, apply parking brake (5) and engage transmission in 1-5 drive. If vehicle moves at idle, adjust parking brake (para. 3-17).	**n.** Vehicle moves when parking brake is applied, and parking brake (5) cannot be adjusted.

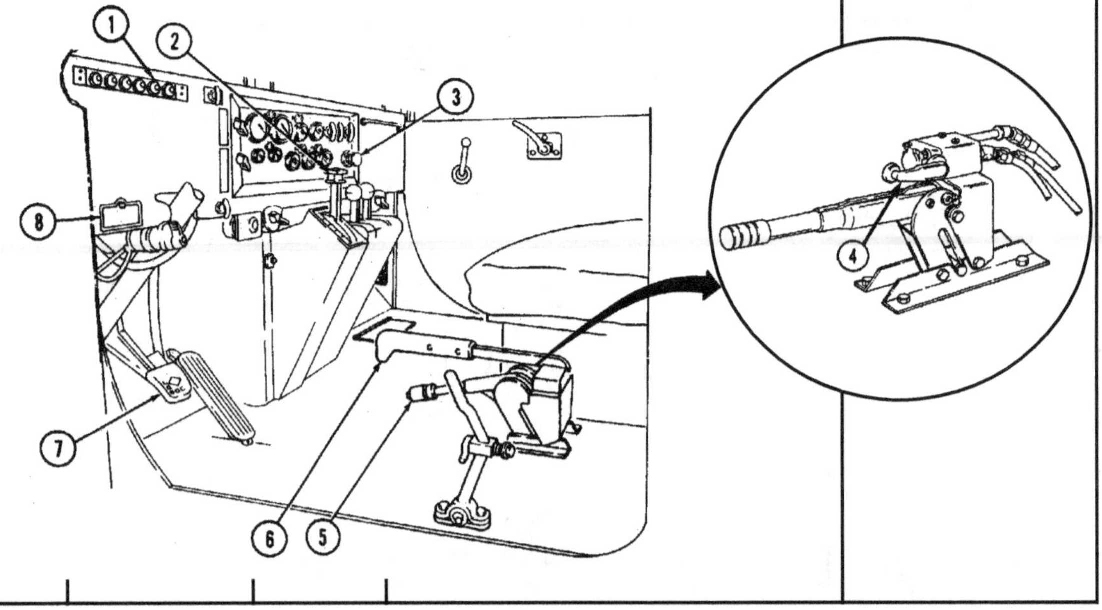
1
2
3
4
5
6
7
8

Table 2-3. Preventive Maintenance Checks and Services for Models M939/A1/A2 (Contd).

ITEM NO.	INTERVAL	LOCATION ITEM TO CHECK/ SERVICE	CREWMEMBER PROCEDURE	NOT FULLY MISSION CAPABLE IF:
12 (contd)	Before	Controls and Instruments (contd)	**o.** Determine spring brake ability to hold vehicle (para. 3-13). Apply service brake (7), place transmission in 1-5 drive, raise spring brake valve lever (4), release service brake (7), increase engine rpm until braking is felt — do not exceed 1000 rpm.	**o.** Vehicle moves with spring brake engaged.
13	Before	Central Tire Inflation System (CTIS)	DRIVER **NOTE** CTIS references apply only to M939A2 series vehicles. If CTIS is not operational, shut off power switch, disconnect electrical connector from ECU, and complete mission (para. 3-16). **a.** With engine running, depress RUN FLAT button (10). Check system for air leaks. Refer to table 3-2. **b.** With engine running, select one deflate and one inflate tire pressure mode on selector panel (9) and check that tires deflate or inflate (para. 2-14).	

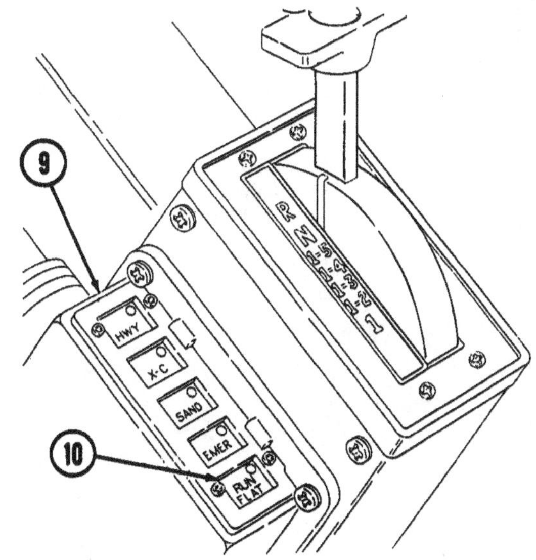
9
10
HWY
X-C
SAND
EMER
RUN FLAT

Table 2-3. Preventive Maintenance Checks and Services for Models M939/A1/A2 (Contd).

ITEM NO.	INTERVAL	LOCATION ITEM TO CHECK/ SERVICE	CREWMEMBER PROCEDURE	NOT FULLY MISSION CAPABLE IF:
			DRIVER **NOTE** • Air dryer references apply only to M939A2 series vehicles and those vehicles with air dryer kit or ABS kit. • Moisture ejector valve on M939A2 series vehicles without ABS kit must exceed 120 psi (827 kPa) for air to expel.	
14	Before	Air Dryer	**a.** Start engine and idle at 550-650 rpm (para. 2-12).	**a.** Engine is inoperable.
			b. Listen for unusual noises or vibrations.	**b.** Engine has unusual noises or vibrations.
			NOTE If unable to hear moisture ejector valve expulsion, set emergency brake, shift to park, and check right side of vehicle.	
			DRIVER **c.** Check for proper operation, listening for moisture ejector valve (1) to expel air.	**c.** Ejector valve is inoperable.
			WARNING Use care when checking heater; it may be hot to touch and cause injury to personnel.	
			d. Touch heater (2). It should be warm if operating properly and area temperature is below 40°F (2°C).	**d.** Heater is inoperable.

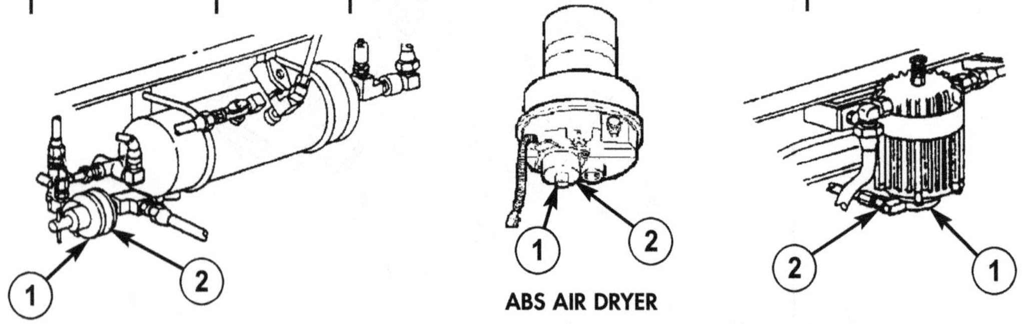
1
2
1
2
ABS AIR DRYER
2
1

Table 2-3. Preventive Maintenance Checks and Services for Models M939/A1/A2 (Contd).

ITEM NO.	INTERVAL	LOCATION ITEM TO CHECK/ SERVICE	CREWMEMBER PROCEDURE	NOT FULLY MISSION CAPABLE IF:
			NOTE Item 15 applies to M929/A1/A2 and M930/A1/A2 models only.	
			DRIVER	
15	Before	Hydraulic Hoist	**a.** Inspect all hydraulic lines and hoses (3) for signs of deterioration and leakage.	**a.** Class III leaks are evident.

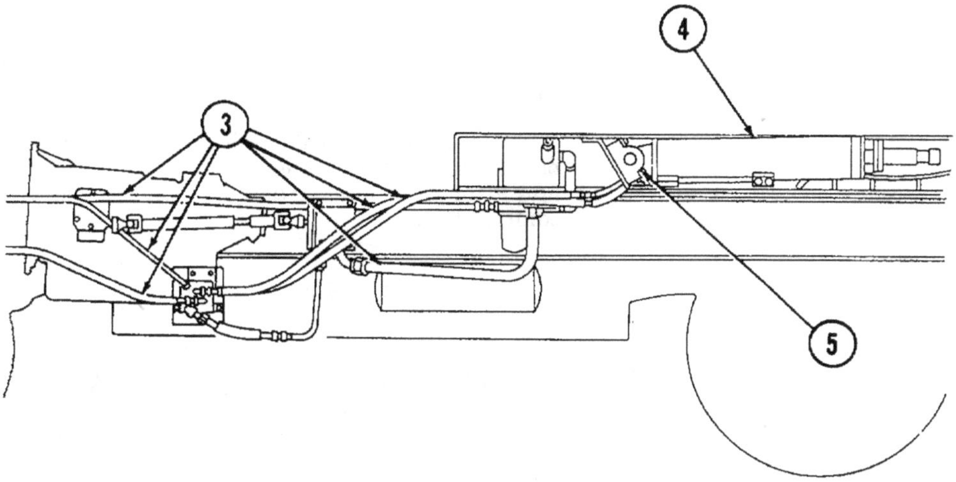
4
3
5

			b. Operate dump hoist (4) through raising, holding, and lowering position to check performance.	**b.** Class III leaks are evident.
			c. Inspect pivot points (5) for binding.	**c.** Dump hoist inoperative, or binding evident.
			d. Check safety catch operation.	**d.** Safety catch is inoperative.

Table 2-3. Preventive Maintenance Checks and Services for Models M939/A1/A2 (Contd).

ITEM NO.	INTERVAL	LOCATION ITEM TO CHECK/ SERVICE	CREWMEMBER PROCEDURE	NOT FULLY MISSION CAPABLE IF:
			DRIVER	
16	Before	Trailer Brakes	**a.** Couple and uncouple tractor (3) and trailer (1) to determine if fifth wheel (2) and locking mechanism (4) are working properly (para. 2-26).	**a.** Tractor (3) and trailer (1) will not couple properly or locking mechanism (4) fails to hold.
			NOTE Perform this check with the trailer empty and the trailer loaded after the tractor/trailer are coupled.	
			b. Check for air leaks at the intervehicular connecting hoses, relay valve, and air reservoirs (para. 3-12).	**b.** Any air leaks are present. Coupling or uncoupling action faulty.
			c. Using trailer brake/Johnnie bar (5) only, apply trailer brakes and attempt to move tractor/trailer combination.	**c.** Brakes fail to hold tractor/ trailer combination from moving.

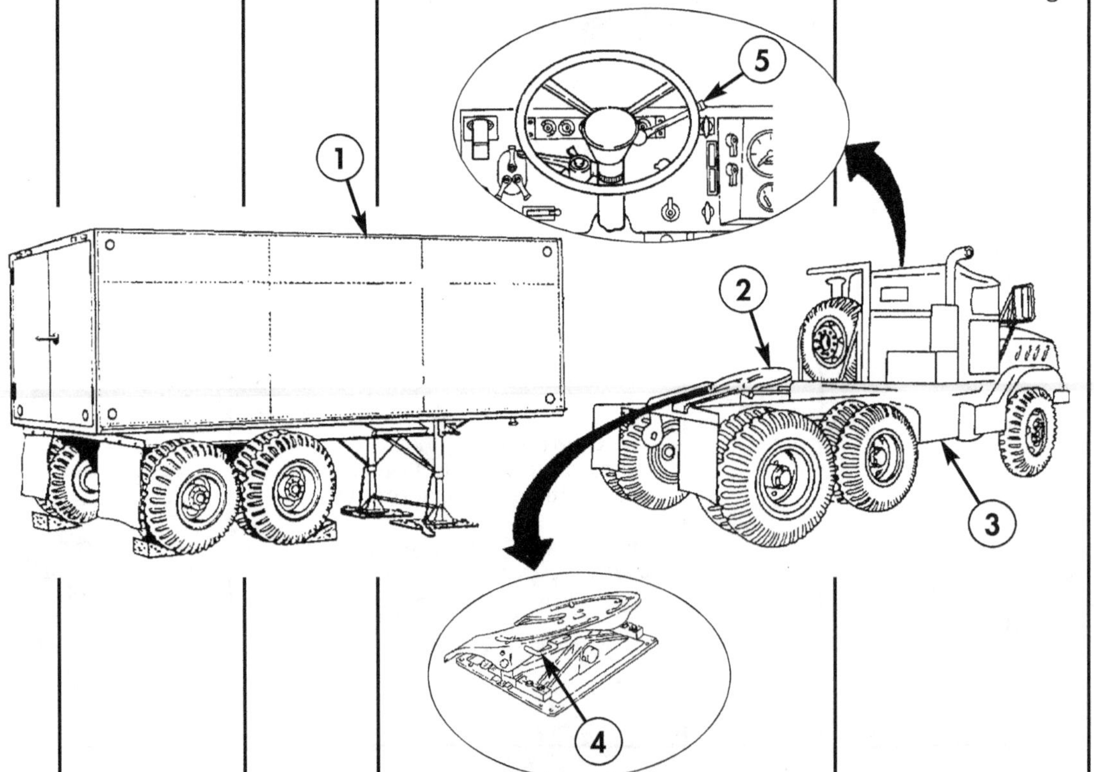
1
2
3
4
5

Table 2-3. Preventive Maintenance Checks and Services for Models M939/A1/A2 (Contd).

ITEM NO.	INTERVAL	LOCATION / ITEM TO CHECK/ SERVICE	CREWMEMBER PROCEDURE	NOT FULLY MISSION CAPABLE IF:
17	Before	Medium Wrecker	DRIVER **WARNING** • Vehicle will become charged with electricity if crane contacts or breaks high-voltage wire. Do not leave vehicle while high-voltage lines are in contact with crane or vehicle. Failure to do this will result in injury or death. Signal nearby personnel to have electrical power turned off. • Wear hand protection when handling cable. Do not handle cable with bare hands. Broken wires will cause injury. **a.** While operating crane, check filter (6) on front of crane oil reservoir (5). Filter indicator (7) should point to "clean".	

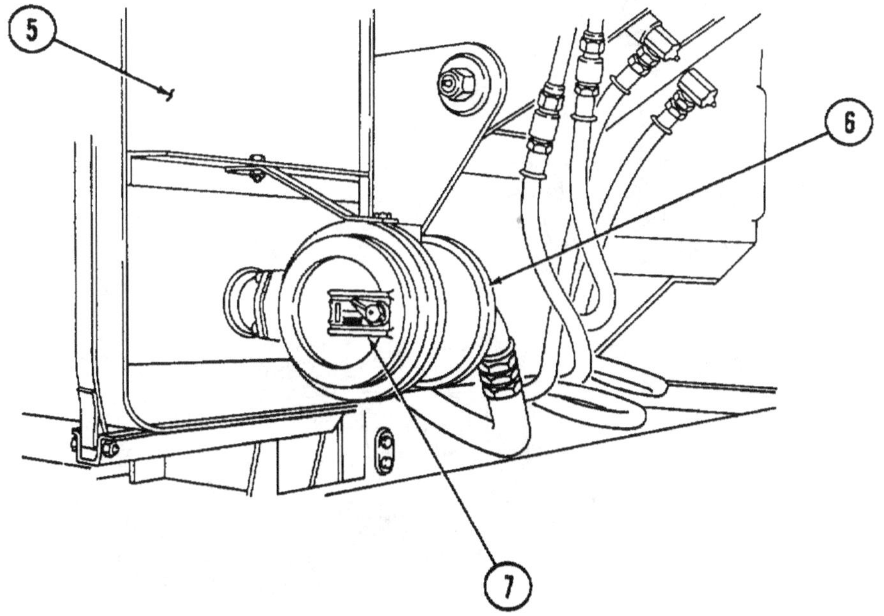
5
6
7

Table 2-3. Preventive Maintenance Checks and Services for Models M939/A1/A2 (Contd).

ITEM NO.	INTERVAL	LOCATION ITEM TO CHECK/ SERVICE	CREWMEMBER PROCEDURE	NOT FULLY MISSION CAPABLE IF:
17 (contd)	Before	Medium Wrecker (contd)	**b.** Operate crane through full range of elevation, rotation, and boom extension to determine performance of crane boom (2), hoist (1), and crane controls (5). Movement should be free and without hydraulic leaks.	**b.** Crane is inoperative or any class III leakage is evident.
			c. Inspect hoist cable (3) for breaks, kinks, and frays. Check sheaves (4) for damage.	**c.** Cable broken, kinked, frayed, or missing. Sheaves missing or damaged.
			DRIVER	
18	During	Steering/ Swaying	Check vehicle steering response for unusual free play, binding, wander, or shimmy.	Loose or binding steering action or steering wheel difficult to turn. Steering inoperative.

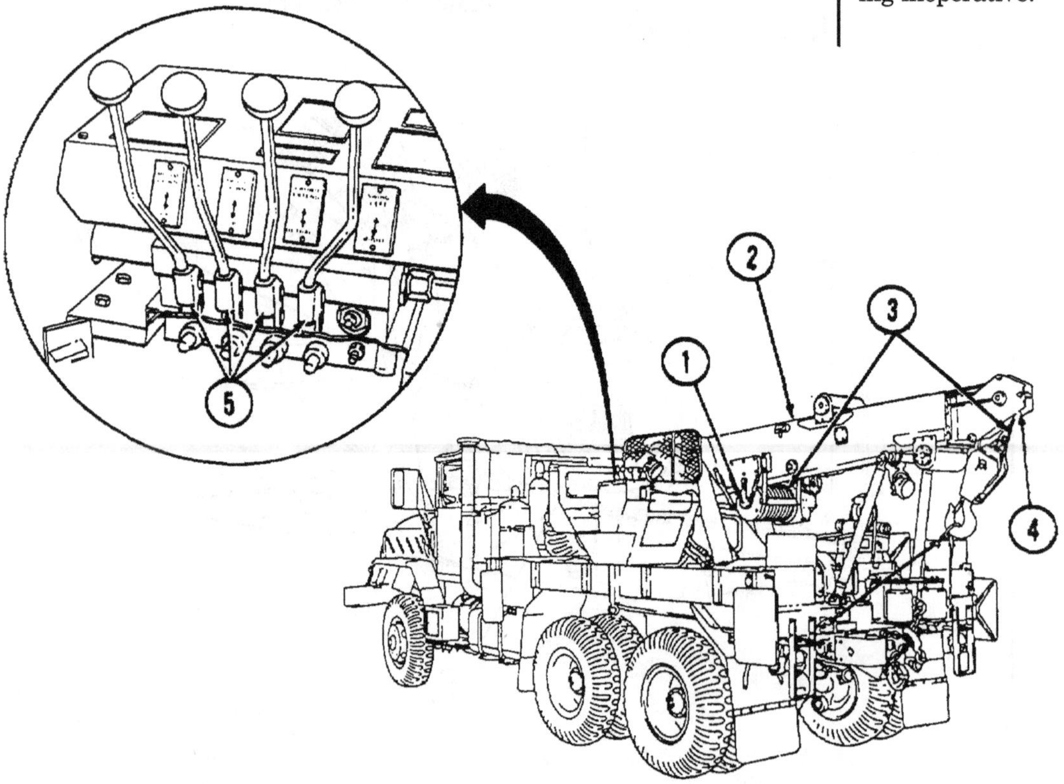

1
2
3
4
5

Table 2-3. Preventive Maintenance Checks and Services for Models M939/A1/A2 (Contd).

ITEM NO.	INTERVAL	LOCATION / ITEM TO CHECK/ SERVICE	CREWMEMBER PROCEDURE	NOT FULLY MISSION CAPABLE IF:
			DRIVER	
19	During	Gauges	**a.** Monitor all gauges during operation (para. 2-3). **CAUTION** If oil pressure gauge reads 0 psi, stop engine. Failure to do so may cause damage to internal engine components.	**a.** Speedometer is inoperative. Notify Unit Maintenance if speedometer needle does not move, jerks unevenly during sustained speeds, or appears stuck.
			b. Engine oil pressure gauge reads less than 15 psi (103 kPa) on M939/A1 series vehicles or 10 psi (68 kPa) on M939A2 series vehicles at idle; stop engine if pressure is zero.	**b.** Engine oil pressure is less than 15 psi (103 kPa) on M939/A1 or 10 psi (68 kPa) on M939A2 at idle.
			DRIVER	
20	During	CTIS	**CAUTION** Do not disconnect CTIS connector from ECU with power on. Damage to ECU will result. **NOTE** CTIS reference applies only to M939A2 series vehicles. If CTIS is not operational, shutoff power switch, disconnect electrical connector from ECU, and complete mission (para. 3-16). While driving, check that amber overspeed warning light (6) illuminates either in cross-country, sand, or emergency modes (para. 2-14). 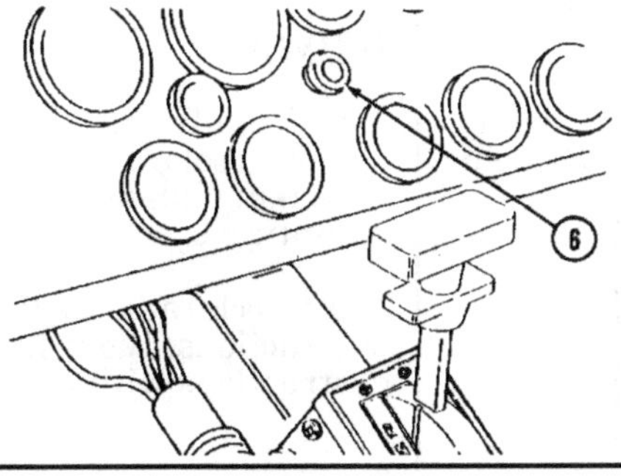6	

Table 2-3. Preventive Maintenance Checks and Services for Models M939/A1/A2 (Contd).

ITEM NO.	INTERVAL	LOCATION ITEM TO CHECK/ SERVICE	CREWMEMBER PROCEDURE	NOT FULLY MISSION CAPABLE IF:
			DRIVER	
21	During	Brake System	While driving, operate service brakes to determine ability to stop. Check for pulling, grabbing, or other abnormal operation (para. 3-13).	Service brakes do not operate properly.
			DRIVER **WARNING** Wear hand protection when handling winch cable. Do not handle with bare hands. Broken wires will cause injuries.	
22	After	Front Winch	**a.** Check winch hoses and lines for deterioration, leakage, and secure connections. **b.** Check all winch controls for proper operation (para. 2-22).	
			UNDER-HOOD CHECKS DRIVER **NOTE** Raise and secure hood at this time to complete the following checks (para. 2-19).	
23	After	Hood	Inspect tether cables, tether cable bolts, and washers.	
23.1	After	CV Boot	Check CV boot for leaks or tears. Failed boots may be repacked with grease until repair can be made by organization maintenance.	
			DRIVER	
24	After	Cooling System	**a.** Visually check radiator for obvious coolant leakage, damaged or leaking hoses, or damaged mounting brackets. **b.** Check radiator fins for obstructions. Blow out all such obstructions with compressed air.	**a.** Any class III leak or damaged mounting brackets are evident.

Table 2-3. Preventive Maintenance Checks and Services for Models M939/A1/A2 (Contd).

ITEM NO.	INTERVAL	LOCATION ITEM TO CHECK/ SERVICE	CREWMEMBER PROCEDURE	NOT FULLY MISSION CAPABLE IF:
24 (contd)	After	Cooling System (contd)	**c.** On M939A2 series vehicles check air aftercooler for coolant leakage.	**c.** Any class III coolant leak is evident.
			d. Check for cracking, fraying, obvious looseness, and breaks on alternator (3), fan (1), and power steering pump (2) drivebelts used on M939/A1 series vehicles, or serpentine drivebelt (4) used on M939A2 series vehicles.	**d.** Any drivebelt is missing, broken, cracked to the belt fiber, has more than one crack 1/8 in. (.32 cm) in depth or 50% of belt thickness, or has frays more than 2 in. (5.1 cm) long.

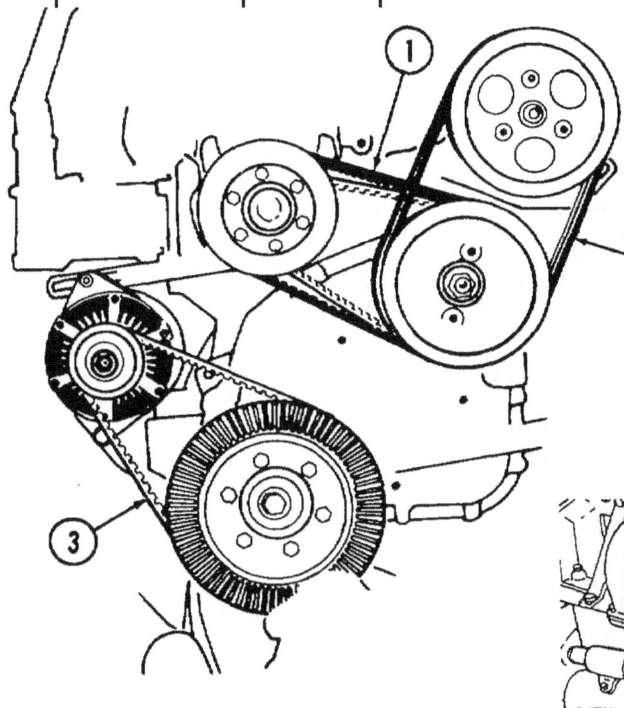
1
2
3

M939/A1 SERIES

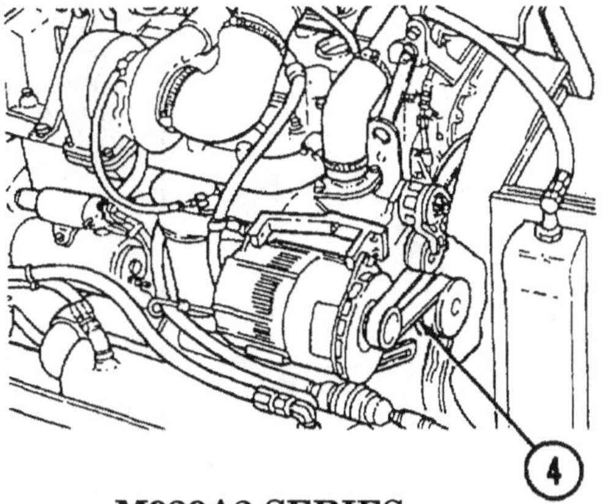
4

M939A2 SERIES

Table 2-3. Preventive Maintenance Checks and Services for Models M939/A1/A2 (Contd).

ITEM NO.	INTERVAL	LOCATION ITEM TO CHECK/ SERVICE	CREWMEMBER PROCEDURE	NOT FULLY MISSION CAPABLE IF:
24 (contd)	After	Cooling System (contd)	**e.** Check water pump for any obvious coolant leakage or damage.	**e.** Any class III coolant leak or damage is evident.
			f. Check pulleys for cracks and damage. DRIVER **WARNING** • Extreme care should be taken when removing surge tank filler cap if temperature gauge reads above 175° F (79° C). Steam or hot coolant under pressure will cause injury. • Accidental or intentional introduction of liquid contaminants into the environment is in violation of state, federal, and military regulations. Refer to Lubrication Order (para. 3-1) for information concerning storage, use, and disposal of these liquids. Failure to do so may result in injury or death. **NOTE** If surge tank on M939A2 series vehicles is found to be empty, open drainvalve on aftercooler and fill surge tank. Close drainvalve when coolant is observed flowing from drain, and continue to fill to approximately bottom end of fill tube.	**f.** Any pulley is cracked or damaged.

Table 2-3. Preventive Maintenance Checks and Services for Models M939/A1/A2 (Contd).

ITEM NO.	INTERVAL	LOCATION ITEM TO CHECK/ SERVICE	CREWMEMBER PROCEDURE	NOT FULLY MISSION CAPABLE IF:
25	After	Surge Tank	Check coolant level in surge tank (2). Tank should be filled to approximately bottom end of fill tube (1) before operation. Fill if necessary. 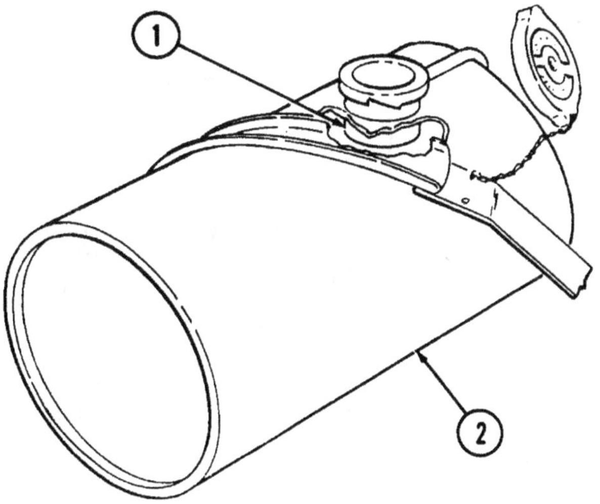	
26	After	Power-train	Check for oil leakage or damage (para. 2-9).	Any class III leak.

Table 2-3. Preventive Maintenance Checks and Services for Models M939/A1/A2 (Contd).

ITEM NO.	INTERVAL	LOCATION ITEM TO CHECK/ SERVICE	CREWMEMBER PROCEDURE	NOT FULLY MISSION CAPABLE IF:

LUBRICANTS	EXPECTED TEMPERATURES			
	+55°F TO -4°F (+13°C TO -40°C)	+110°F TO +10°F (+43°C TO -12°C)	+75°F TO -15°F (+24°C TO -26°C)	+40°F TO -65°F (+4°C TO -54°C)
OE/HDO 10, OE/HDO 15/40, DEXRON III	OE/HDO 10	OE/HDO 15/40	DEXRON III	OEA

DRIVER

WARNING

Accidental or intentional introduction of liquid contaminants into the environment is in violation of state, federal, and military regulations. Refer to Lubrication Order (para. 3-1) for information concerning storage, use, and disposal of these liquids. Failure to do so may result in injury or death.

CAUTION

- When checking transmission oil level, do not permit dirt, dust, or grit to enter transmission filler tube. Ensure dipstick handle and end of filler tube are clean. Serious internal transmission damage may result if transmission is contaminated.
- Do not overfill transmission. Internal transmission component damage will result.

NOTE

On M939A2 series vehicles, normal run level on transmission dipstick should show between ADD mark and FULL mark (para. 3-10).

ITEM NO.	INTERVAL	ITEM TO CHECK/ SERVICE	CREWMEMBER PROCEDURE	NOT FULLY MISSION CAPABLE IF:
27	After	Transmission Oil Level	**a.** Allow engine to idle. Shift transmission to neutral and apply parking brake (para. 2-16). **b.** Withdraw dipstick (2) slowly to prevent a false reading. If transmission oil temperature gauge reads 180°F (82°C) or below, level on dipstick (2) should show between marks designated for normal run (1). If transmission oil temperature is above 220°F (104°C), allow transmission oil to cool by turning engine off.	

Table 2-3. Preventive Maintenance Checks and Services for Models M939/A1/A2 (Contd).

ITEM NO.	INTERVAL	LOCATION ITEM TO CHECK/ SERVICE	CREWMEMBER PROCEDURE	NOT FULLY MISSION CAPABLE IF:
27 (contd)	After	Trans-mission Oil Level (contd)	**c.** If transmission oil level is low, add oil through filler tube (3). Return dipstick (2) to filler tube (3), tighten dipstick handle (4), and wipe away any oil spilled.	

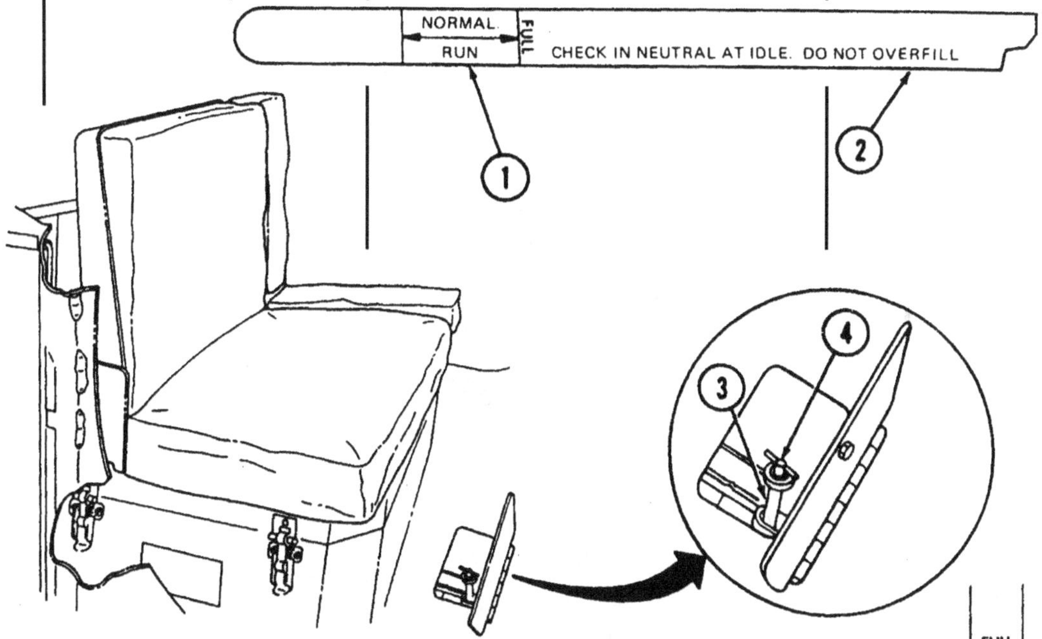
NORMAL
RUN
FULL
CHECK IN NEUTRAL AT IDLE. DO NOT OVERFILL
1
2
3
4

M939/A1 SERIES

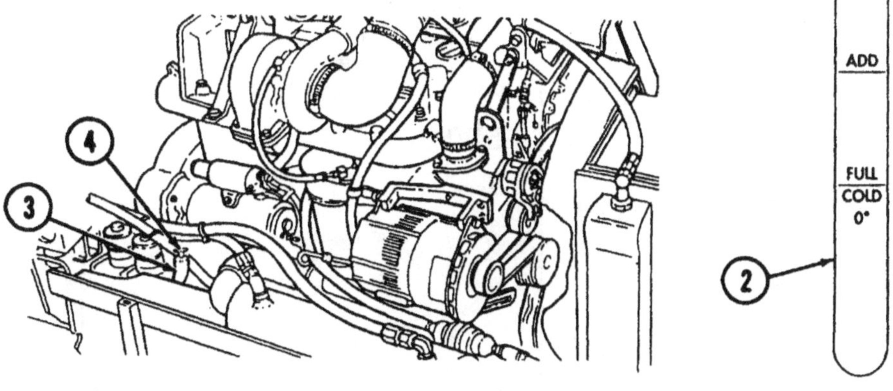
FULL
ADD
FULL COLD 0°
2
3
4

M939A2 SERIES

Table 2-3. Preventive Maintenance Checks and Services for Models M939/A1/A2 (Contd).

ITEM NO.	INTERVAL	LOCATION / ITEM TO CHECK/ SERVICE	CREWMEMBER PROCEDURE	NOT FULLY MISSION CAPABLE IF:

LUBRICANTS	EXPECTED TEMPERATURES		
	ABOVE 15°F (ABOVE -9°C)	+40°F TO -15°F (+4°C TO -26°C)	+40°F TO -65°F (+4°C TO -54°C)
OE/HDO 15/40	OE/HDO 15/40	OE/HDO 15/40	OEA

ITEM NO.	INTERVAL	LOCATION / ITEM TO CHECK/ SERVICE	CREWMEMBER PROCEDURE	NOT FULLY MISSION CAPABLE IF:
			DRIVER **WARNING** Accidental or intentional introduction of liquid contaminants into the environment is in violation of state, federal, and military regulations. Refer to Lubrication Order (para. 3-1) for information concerning storage, use, and disposal of these liquids. Failure to do so may result in injury or death. **CAUTION** Do not overfill. Damage to internal engine components will result. **NOTE** Engine oil level is checked after engine is stopped and dipstick is removed and wiped clean.	
28	After	Engine Oil Level	Withdraw dipstick (1) slowly to ensure an accurate reading. Check for proper oil level (para 3-8). Level should be between L (low) and H (high) marks. Add oil as necessary.	Engine is over filled.
			DRIVER	
29	After	Oil Filters	Check oil filter for obvious signs of leakage.	Any class III oil leak.

Table 2-3. Preventive Maintenance Checks and Services for Models M939/A1/A2 (Contd).

ITEM NO.	INTERVAL	LOCATION ITEM TO CHECK/ SERVICE	CREWMEMBER PROCEDURE	NOT FULLY MISSION CAPABLE IF:
			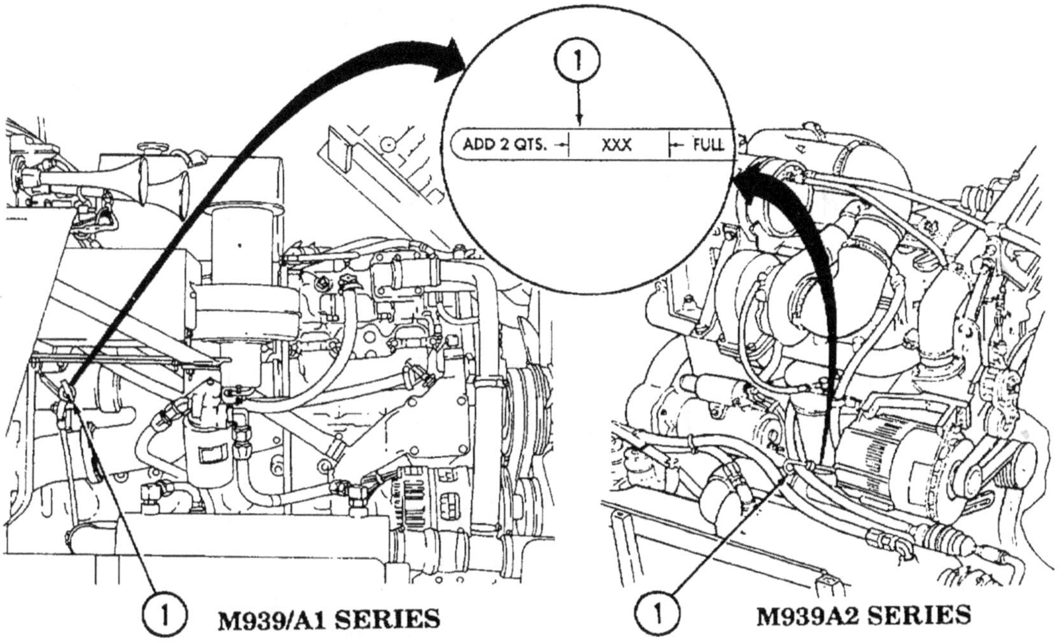1 ADD 2 QTS. XXX FULL 1 M939/A1 SERIES 1 M939A2 SERIES	
			DRIVER	
30	After	Power Steering Assist Cylinder	Check fluid lines of power steering assist cylinder (2) for damage, leaks, and looseness. 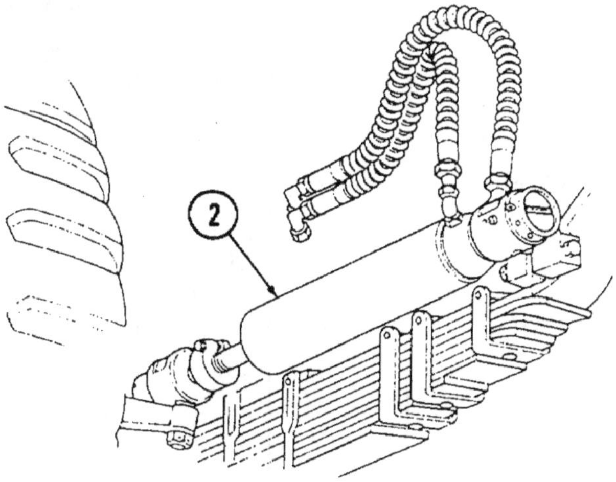2	Any cuts, breaks, or class III leakage.

Table 2-3. Preventive Maintenance Checks and Services for Models M939/A1/A2 (Contd).

ITEM NO.	INTERVAL	LOCATION ITEM TO CHECK/ SERVICE	CREWMEMBER PROCEDURE	NOT FULLY MISSION CAPABLE IF:

LUBRICANTS	EXPECTED TEMPERATURES		
	ABOVE 15°F (ABOVE -9°C)	+40°F TO -15°F (+4°C TO -26°C)	+40°F TO -65°F (+4°C TO -54°C)
OE/HDO 10	OE/HDO 10	OE/HDO 10	OEA

ITEM NO.	INTERVAL	ITEM TO CHECK/ SERVICE	CREWMEMBER PROCEDURE	NOT FULLY MISSION CAPABLE IF:
			DRIVER **WARNING** Accidental or intentional introduction of liquid contaminants into the environment is in violation of state, federal, and military regulations. Refer to Lubrication Order (para. 3-1) for information concerning storage, use, and disposal of these liquids. Failure to do so may result in injury or death. **CAUTION** • Before opening reservoir, make sure area around reservoir cap is clean. Do not allow dirt, dust, or water to enter reservoir. Failure to do this will cause damage to internal components. • Do not overfill power steering reservoir. Oil will overflow into vent system on the M939/A1 series or through the vent cap on the M939A2 series. **NOTE** Power steering reservoir oil level is checked with engine stopped (para. 3-8).	Oil in reservoir is contaminated.
31	After	Steering System	**a.** With engine cold, check power steering reservoir (1) with dipstick (2) on filler cap (3). If fluid is below COLD mark, add as necessary. **b.** If engine is at normal operating temperature, 175° F to 200° F (79° C to 93° C), use HOT FULL mark and add as necessary. **c.** Check power steering reservoir oil level when engine is stopped.	**c.** Class III leakage is evident.

Table 2-3. Preventive Maintenance Checks and Services for Models M939/A1/A2 (Contd).

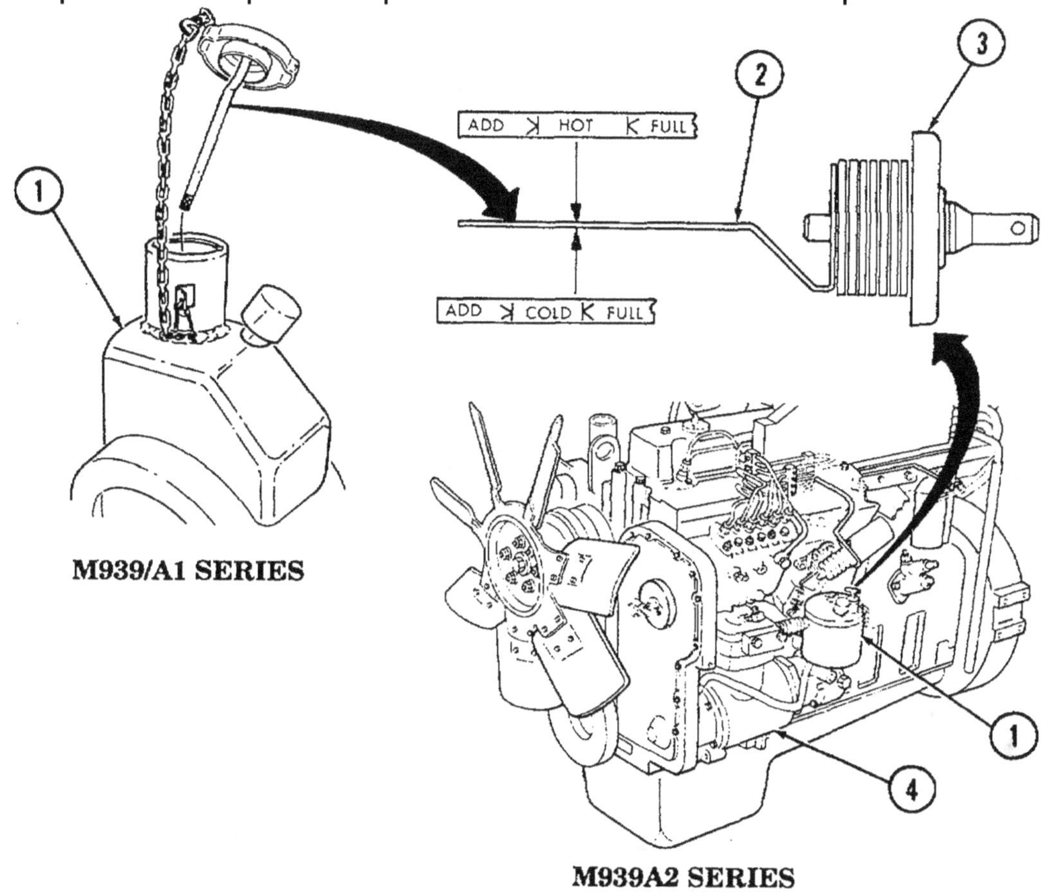
ADD HOT FULL
ADD COLD FULL
1
2
3
4
M939/A1 SERIES
M939A2 SERIES

ITEM NO.	INTERVAL	LOCATION ITEM TO CHECK/ SERVICE	CREWMEMBER PROCEDURE	NOT FULLY MISSION CAPABLE IF:
			DRIVER	
31 (contd)	After	Steering System (contd)	**d.** Visually check for oil leaks (para. 2-9).	**d.** Class III leakage is evident.
			e. Check steering arm for looseness and damage.	**e.** Loose or damaged arms.
			DRIVER	
32	After	Air Compressor	Check air compressor (4) for obvious signs of oil or air leakage or damage (oil in air tanks).	Any class III oil leak or damage is evident. Any air leakage is evident.

Table 2-3. Preventive Maintenance Checks and Services for Models M939/A1/A2 (Contd).

ITEM NO.	INTERVAL	LOCATION ITEM TO CHECK/ SERVICE	CREWMEMBER PROCEDURE	NOT FULLY MISSION CAPABLE IF:
			DRIVER **WARNING** • Do not perform fuel filter/water separator checks, inspections, or draining while smoking or near fire, flames, or sparks. Fuel could ignite, causing damage to vehicle, injury, or death. • Accidental or intentional introduction of liquid contaminants into the environment is in violation of state, federal, and military regulations. Refer to Lubrication Order (para. 3-1) for information concerning storage, use, and disposal of these liquids. Failure to do so may result in injury or death. **CAUTION** Do not overtighten plastic valve. Damaged valve will result in fuel leaks. **NOTE** If fuel is still unclear after draining one quart (0.946 l), notify unit maintenance.	
33	After	Fuel Filter/ Water Separator	**a.** Loosen valve (1) on bottom of fuel filter and allow the water to drain into a suitable container. Close the valve when clean fuel is visible. **b.** Upper valve (2) may have to be opened for fuel to drain. **c.** Prime fuel system (para. 3-8). **d.** Check for leaks.	**d.** Class III leakage is evident.

Table 2-3. Preventive Maintenance Checks and Services for Models M939/A1/A2 (Contd).

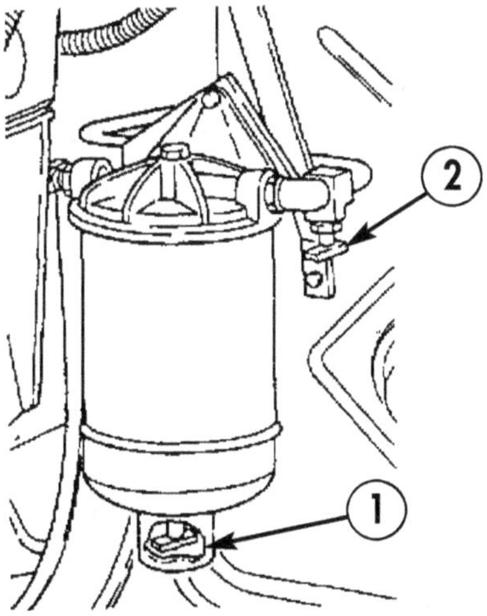
2
1

M939/A1 SERIES

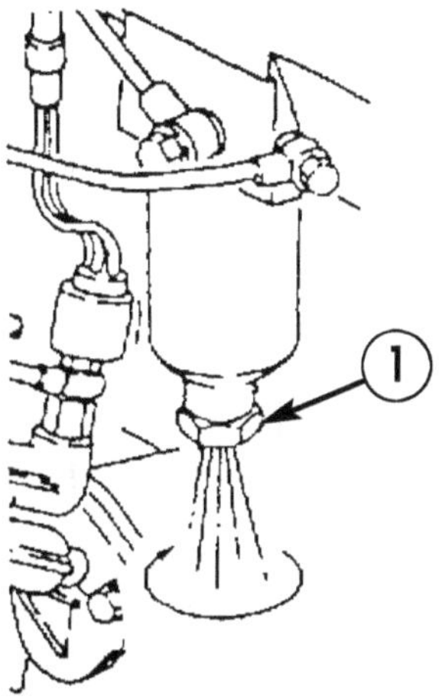
1

M939A2 SERIES

ITEM NO.	INTERVAL	LOCATION / ITEM TO CHECK/ SERVICE	CREWMEMBER PROCEDURE	NOT FULLY MISSION CAPABLE IF:
			EXTERIOR OF VEHICLE DRIVER **NOTE** Lower hood at this time to complete the following checks (para. 2-19).	
34	After	Tires (Right Side)	Visually check tires for under-inflation, cracks, gouges, or bulges. Remove all penetrating objects.	M939A1/A2 series has any tires missing or unserviceable. There is evidence of cuts, gouges, and bulges which would result in tire failure during operation (two or more tires for the M939 series).
			DRIVER	
35	After	Wheels, Studs and Nuts	Ensure all wheel stud nuts are tight using wheel stud nut wrench and handle (para. 3-11).	Any wheel stud or stud nuts are missing, loose, or damaged.

Table 2-3. Preventive Maintenance Checks and Services for Models M939/A1/A2 (Contd).

ITEM NO.	INTERVAL	LOCATION / ITEM TO CHECK/ SERVICE	CREWMEMBER PROCEDURE	NOT FULLY MISSION CAPABLE IF:
			DRIVER **CAUTION** Duel fuel tanks that remain unused may become contaminated with fungus.	Duel tank is contaminated.
36	After	Fuel Tank	Check fuel tank, lines, and fittings for leakage.	Class III leakage evident.
			DRIVER **WARNING** Do not touch hot exhaust pipes with bare hands. Severe burns will result.	
37	After	Exhaust System	Inspect exhaust stack (1) and muffler (2) for obvious damage and/or leaks and rusted-through conditions. Report all damage to unit maintenance.	Pipe, clamps, or hardware missing or damaged, and any leak which could cause injury to personnel.

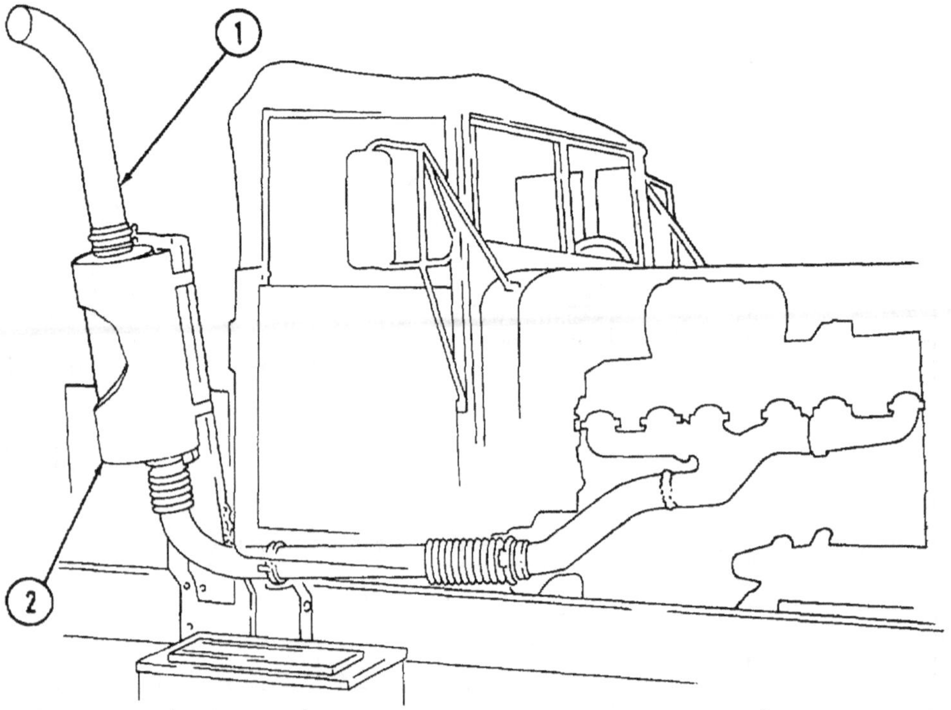
1
2

Table 2-3. Preventive Maintenance Checks and Services for Models M939/A1/A2 (Contd).

ITEM NO.	INTERVAL	LOCATION ITEM TO CHECK/ SERVICE	CREWMEMBER PROCEDURE	NOT FULLY MISSION CAPABLE IF:
			DRIVER	
38	After	Air Tank Drain	**NOTE** Drain moisture from tanks in the sequence listed below. After all moisture has been drained and only air is coming out, close drainvalves. **a.** Open drainvalve (5) and drain moisture from airbrake system wet tank reservoir. **b.** Open drainvalve (6) and drain moisture from spring brake air reservoir. **c.** Open drainvalve (3) and drain moisture from primary airbrake system air reservoir. **d.** Open drainvalve (4) and drain moisture from secondary airbrake system air reservoir.	

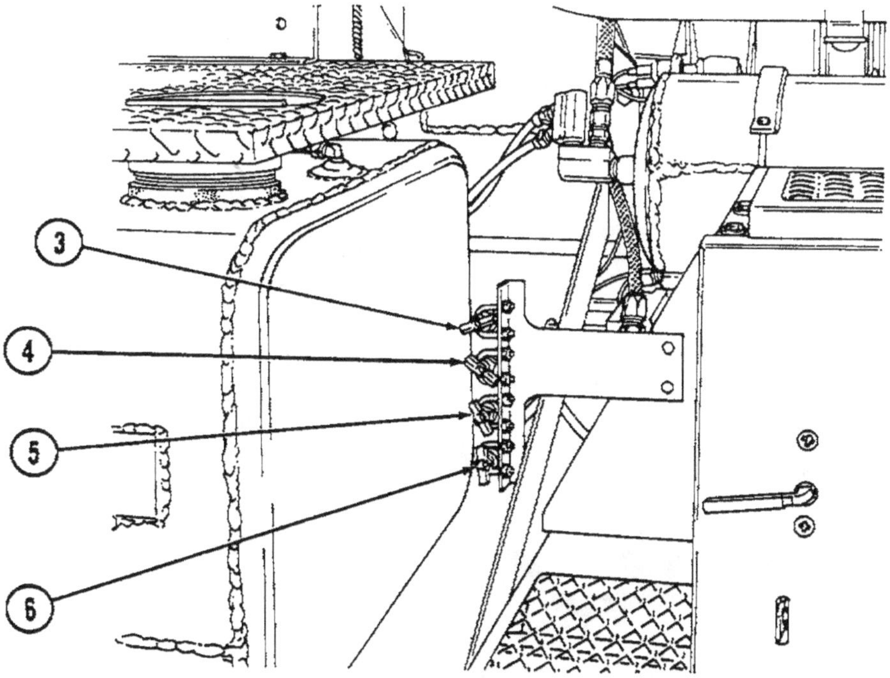
3
4
5
6

Table 2-3. Preventive Maintenance Checks and Services for Models M939/A1/A2 (Contd).

ITEM NO.	INTERVAL	LOCATION ITEM TO CHECK/ SERVICE	CREWMEMBER PROCEDURE	NOT FULLY MISSION CAPABLE IF:
			DRIVER	
39	After	Tires (Left Side)	Visually check tires (1) for under-inflation, cracks, gouges, or bulges. Remove all penetrating objects.	M939A1/A2 series have any tire missing or unserviceable. There is evidence of cuts, gouges, and bulges which would result in tire failure during operation (para. 3-11). (Two or more tires missing or unserviceable for the M939 series.)
			DRIVER	
40	After	Wheels, Studs, and Nuts	Ensure all wheel stud nuts (2) are tight using wheel stud nut wrench (3) and handle (4) (para. 3-11).	Any wheel stud or stud nut is missing, loose, or damaged.

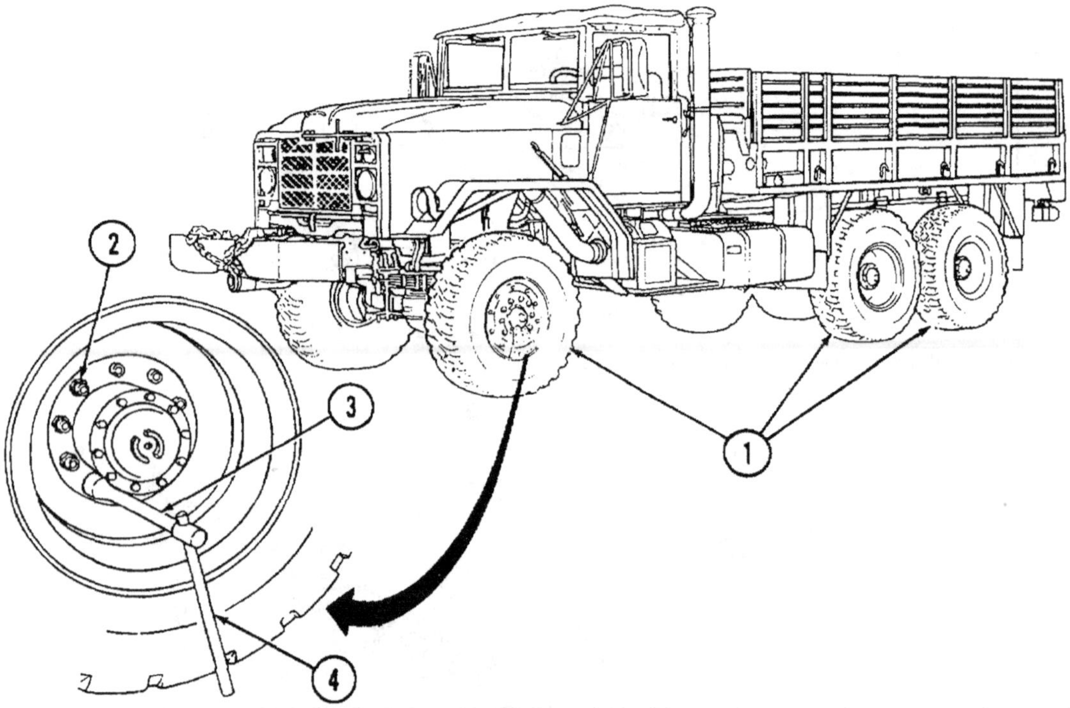
2
3
4
1

Table 2-3. Preventive Maintenance Checks and Services for Models M939/A1/A2 (Contd).

ITEM NO.	INTERVAL	LOCATION ITEM TO CHECK/ SERVICE	CREWMEMBER PROCEDURE	NOT FULLY MISSION CAPABLE IF:
			DRIVER **WARNING** If NBC exposure is suspected, all air filter media should be handled by personnel wearing protective equipment. Consult your unit NBC officer or NBC NCO for appropriate handling or disposal instructions.	
41	After	Air Intake System	**a.** Check clamps (5) and (9) for tightness and upper hump hose (6), tube (10), lower hump hose (7), and air cleaner assembly (8) for openings which would allow foreign material to enter engine.	**a.** Intake system has any obvious leaks.
			b. Check air cleaner (8) assembly for openings which would allow foreign material to enter engine.	**b.** Air cleaner missing or damaged that would allow dust or dirt into air intake.

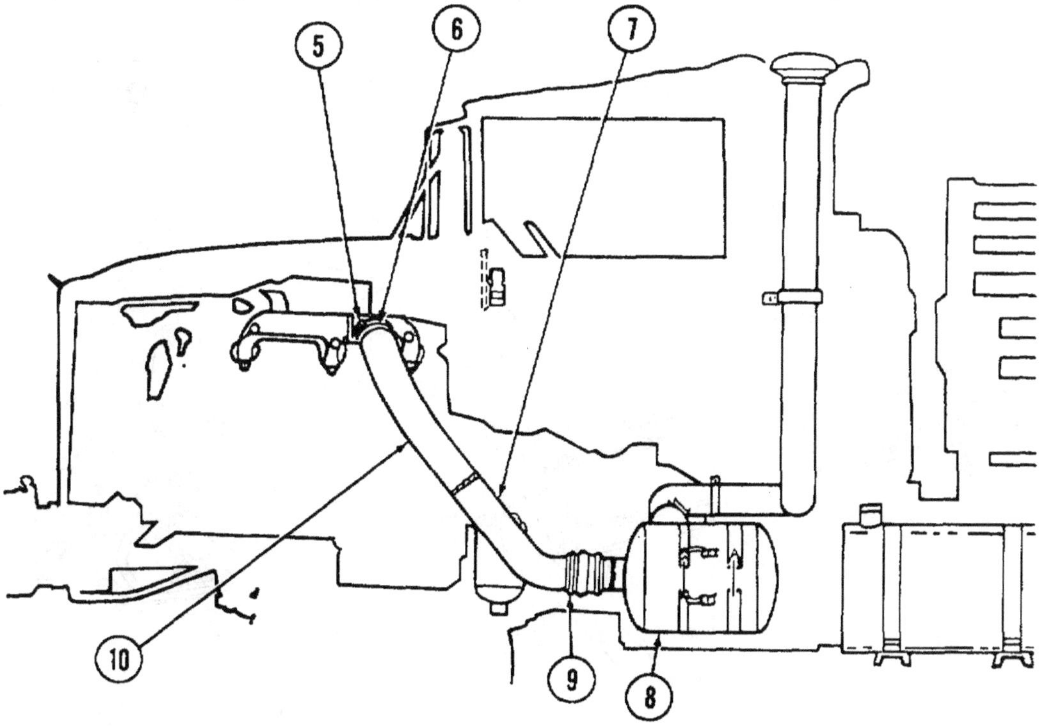
5
6
7
10
9
8

Table 2-3. Preventive Maintenance Checks and Services for Models M939/A1/A2 (Contd).

ITEM NO.	INTERVAL	LOCATION ITEM TO CHECK/ SERVICE	CREWMEMBER PROCEDURE	NOT FULLY MISSION CAPABLE IF:
			INTERIOR OF VEHICLE DRIVER **WARNING** Make sure companion seatbelt is not caught in battery box. This will cause belts to rot which may lead to injury of personnel. **NOTE** Missing, torn, or inoperative seat– belt may be in violation of AR 385-55.	
42	After	Seat and Seatbelts	**a.** Check driver's (2) and companion (1) seats for security of mounting. **b.** Check seatbelts for: 1. proper adjustment. 2. ability to lock. 3. security of mounting hardware. 4. belt material (3) for rips, tears, and exposure to electrolyte.	

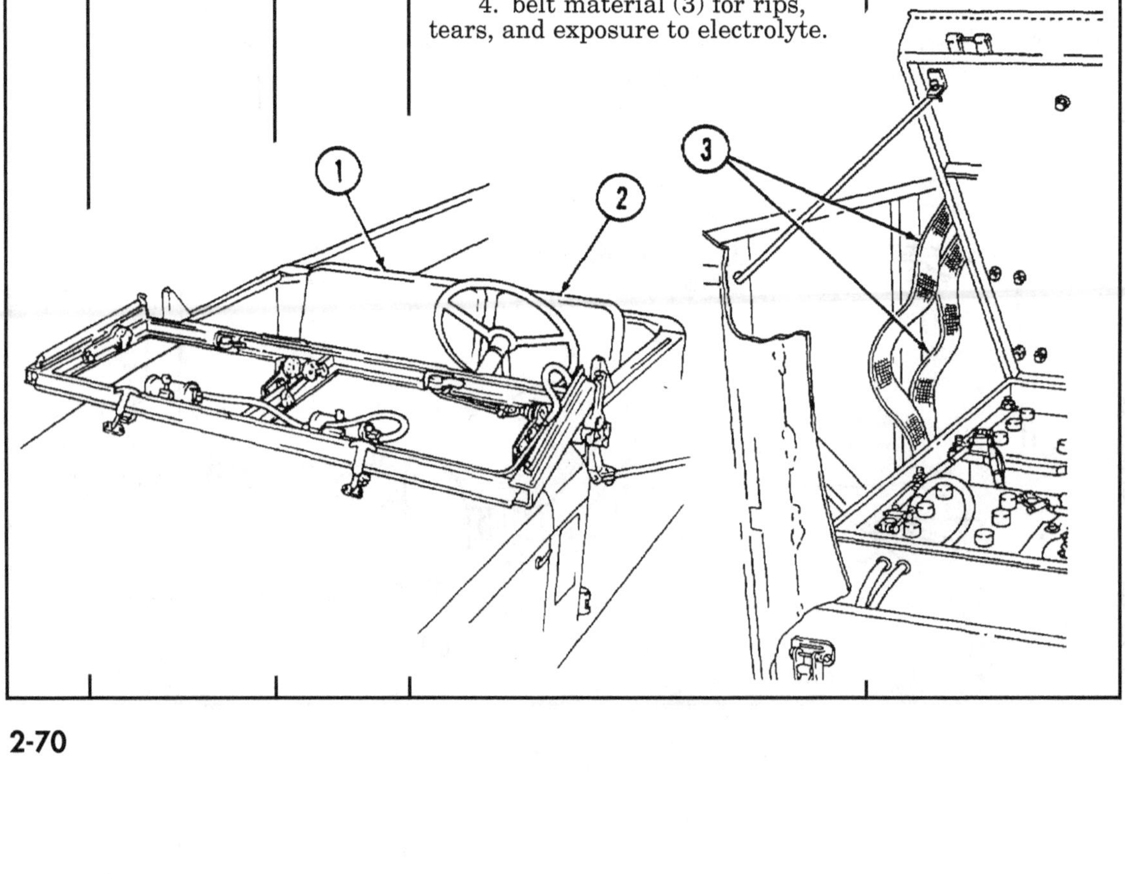
1
2
3

Table 2-3. Preventive Maintenance Checks and Services for Models M939/A1/A2 (Contd).

ITEM NO.	INTERVAL	LOCATION ITEM TO CHECK/ SERVICE	CREWMEMBER PROCEDURE	NOT FULLY MISSION CAPABLE IF:
			DRIVER **NOTE** CTIS reference applies only to M939A2 series vehicles. If CTIS is not operational, disable CTIS (para. 3-16).	
43	After	CTIS	**a.** With engine running, select RUN FLAT. Check system for air leaks (para. 2-14). **b.** With engine running, select one deflate and one inflate tire pressure mode on selector panel (4) and check that tires deflate or inflate (para. 2-14).	

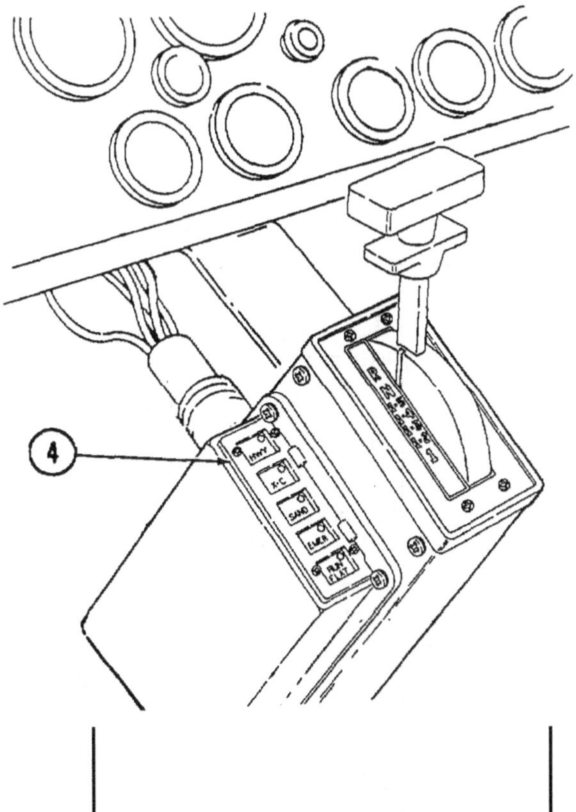
4
HWY
X-C
SAND
EMER
RUN FLAT

Table 2-3. Preventive Maintenance Checks and Services for Models M939/A1/A2 (Contd).

ITEM NO.	INTERVAL	LOCATION ITEM TO CHECK/ SERVICE	CREWMEMBER PROCEDURE	NOT FULLY MISSION CAPABLE IF:
			DRIVER **NOTE** Operation of vehicles with inoperative horn may violate AR 385-55.	
44	After	Horns	Check operation of horns if tactical situation permits.	
			DRIVER **NOTE** Operation of vehicle with malfunctioning lights may violate AR 385-55.	
45	After	Lights	Check operation of headlights, taillights, turn signals, brake, and blackout lights.	
			DRIVER	
46	After	Mirrors	Check for missing or cracked mirrors.	
			DRIVER	
47	After	Brake System	**a.** With the air system fully charged at 120 psi (827 kPa), engine off and parking brake applied, walk around vehicle and listen for leaks in the air system, air reservoirs, lines, and hoses. **b.** Visually check brake chamber and air reservoirs for obvious damage (para. 3-9 and 3-12). **c.** Visually check hoses and lines for cracks, breaks, etc. **d.** Check for presence of spring brake caging bolt and dust cover.	**a.** Any reservoir, line or hose is leaking (para. 3-9).

Table 2-3. Preventive Maintenance Checks and Services for Models M939/A1/A2 (Contd).

ITEM NO.	INTERVAL	LOCATION ITEM TO CHECK/ SERVICE	CREWMEMBER PROCEDURE	NOT FULLY MISSION CAPABLE IF:
			DRIVER	
48	After	Cargo Bed	**a.** Check condition of troop seat (4) and retainer pins (3). **b.** Check condition of troop seat latches (1). **c.** Check troop safety strap (2).	**a.** Troop seat retainer pins missing or damaged. **b.** Latches damaged or missing. **c.** Safety strap missing.

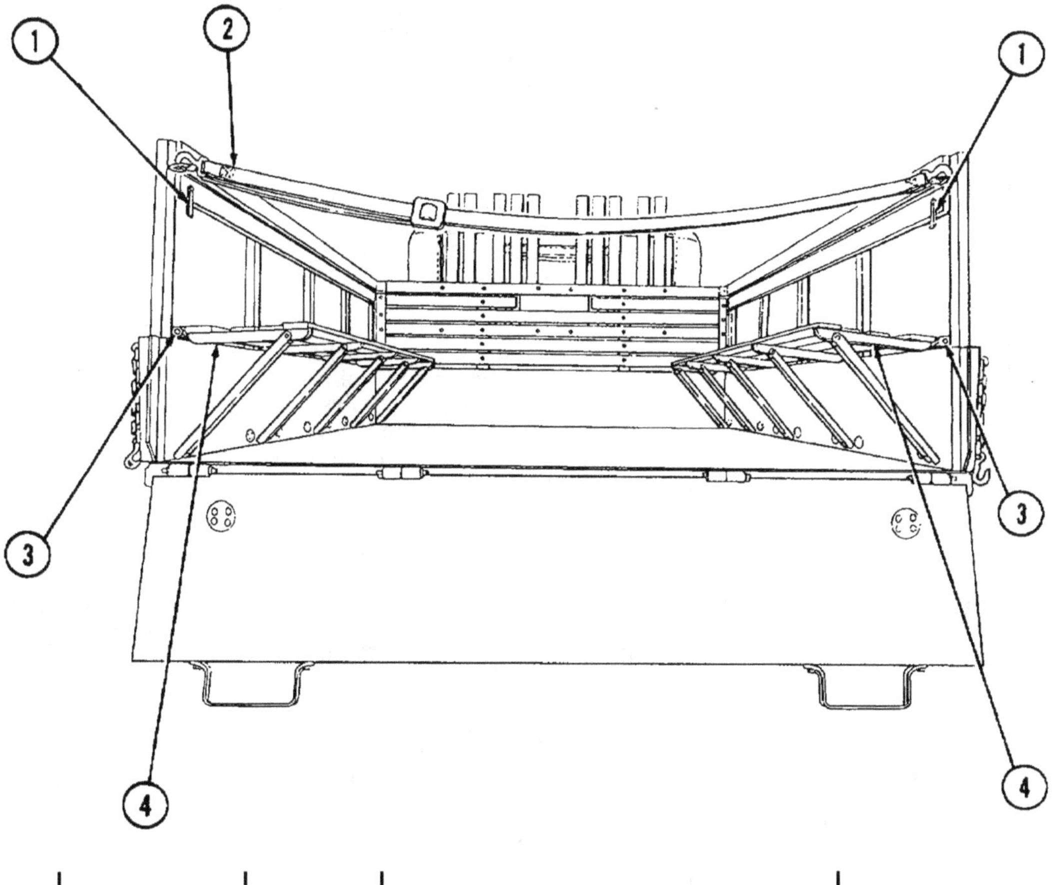
1
2
1
3
3
4
4

Table 2-3. Preventive Maintenance Checks and Services for Models M939/A1/A2 (Contd).

ITEM NO.	INTERVAL	LOCATION ITEM TO CHECK/ SERVICE	CREWMEMBER PROCEDURE	NOT FULLY MISSION CAPABLE IF:

LUBRICANTS	EXPECTED TEMPERATURES		
	ABOVE 15°F (ABOVE -9°C)	+40°F TO -15°F (+4°C TO -26°C)	+40°F TO -65°F (+4°C TO -54°C)
OE/HDO 10	OE/HDO 10	OE/HDO 10	OEA

ITEM NO.	INTERVAL	ITEM TO CHECK/ SERVICE	PROCEDURE	NOT FULLY MISSION CAPABLE IF:
			DRIVER **SPECIAL BODY EQUIPMENT** **CAUTION** Before opening reservoir, make sure area around reservoir filler cap is clean. Do not allow dirt, dust, or water to enter reservoir. Failure to do this will cause damage to internal components. **NOTE** • Hydraulic Tank Oil Level reference applies only to the M925/A1/A2, M928/A1/A2, M929/A1/A2, and M930/A1/A2 models. • If vehicle has positive locking device, inspect the attaching hardware for loose or missing bolts prior to stowing in up position. If loose or missing, do not stow bed in up position. • Make sure bed is in travel mode before checking oil level.	
49	After	Hydraulic Tank Oil Level	Check hydraulic reservoir level by removing filler cap (1) and pulling out dipstick (2). Oil level should be at the third mark from top of gauge with body down in traveling position. If oil level is low, fill to top line. After check, make sure filter cap is tight.	Any class III leak or cap missing.

Table 2-3. Preventive Maintenance Checks and Services for Models M939/A1/A2 (Contd).

ITEM NO.	INTERVAL	LOCATION ITEM TO CHECK/ SERVICE	CREWMEMBER PROCEDURE	NOT FULLY MISSION CAPABLE IF:
			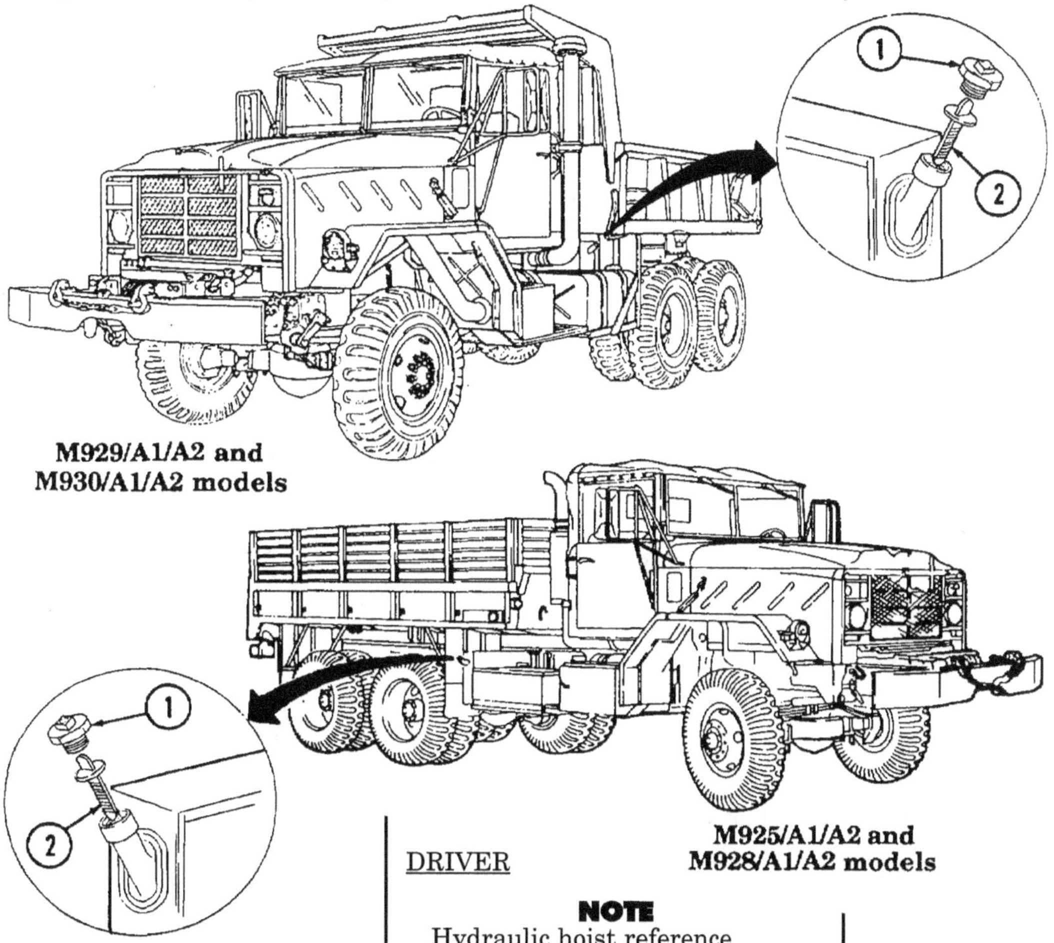1 2 1 2 **M929/A1/A2 and M930/A1/A2 models** **M925/A1/A2 and M928/A1/A2 models**	
			<u>DRIVER</u> **NOTE** Hydraulic hoist reference applies only to the M929/A1/A2 and M930/A1/A2 model vehicles.	
50	After	Hydraulic Hoist	**a.** Check PTO, drive shaft, hydraulic pump, and control valve for damage, leakage, and security of mounting.	**a.** Class III leakage is evident.
			b. Inspect cylinders for damage, leakage, and security mounting to sub-frame.	**b.** Class III leakage is evident.

Table 2-3. Preventive Maintenance Checks and Services for Models M939/A1/A2 (Contd).

ITEM NO.	INTERVAL	LOCATION ITEM TO CHECK/ SERVICE	CREWMEMBER PROCEDURE	NOT FULLY MISSION CAPABLE IF:
			DRIVER **NOTE** Van must be set up to perform the following check (para. 2-27).	
51	After	Expand-able Vans	Operate all switches in both AC and DC electrical systems to determine that all function properly and that power to ceiling lights (1) is on. Switch box (2) on right rear van wall.	Neither AC nor DC system will operate.

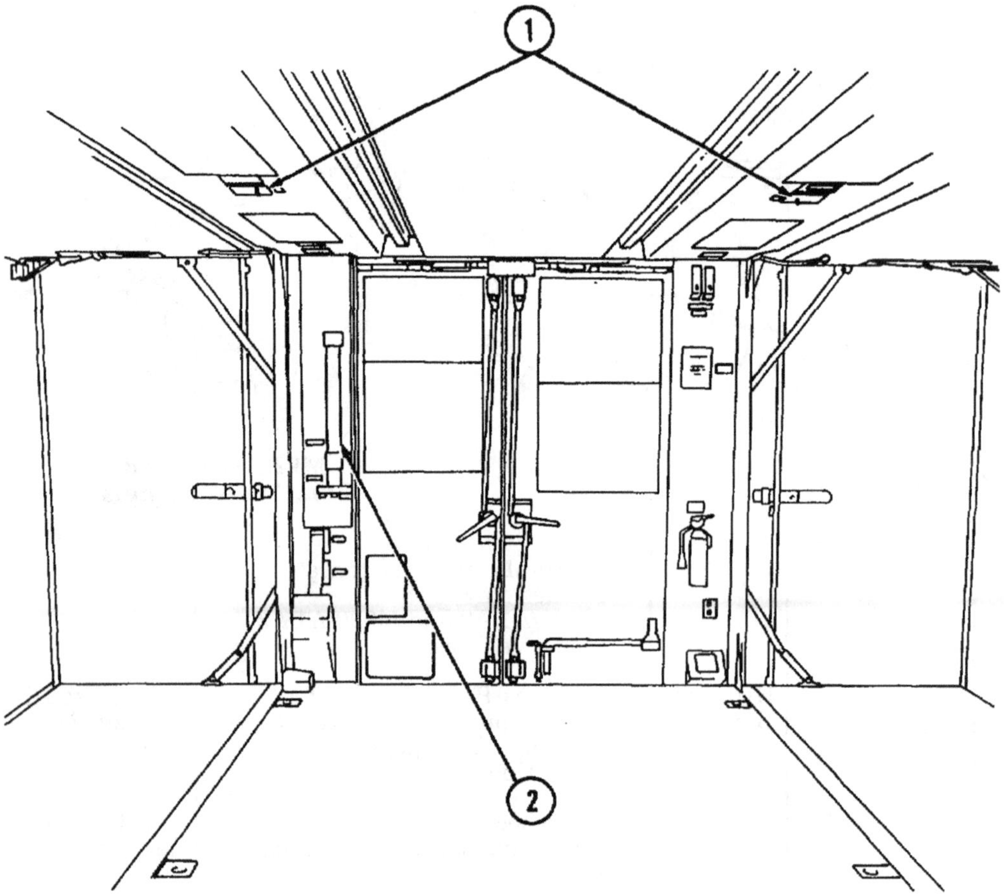
1
2

Table 2-3. Preventive Maintenance Checks and Services for Models M939/A1/A2 (Contd).

ITEM NO.	INTERVAL	LOCATION ITEM TO CHECK/ SERVICE	CREWMEMBER PROCEDURE	NOT FULLY MISSION CAPABLE IF:
			DRIVER **CAUTION** • Before opening reservoir, make sure area around the reservoir cap is clean. Do not allow dirt, dust, or water to enter reservoir. Failure to do so will cause damage to internal components. • Do not overfill hydraulic oil reservoir. Damage to internal components will result.	

LUBRICANTS	EXPECTED TEMPERATURES		
	ABOVE 15°F (ABOVE -9°C)	+40°F TO -15°F (+4°C TO -26°C)	+40°F TO -65°F (+4°C TO -54°C)
OE/HDO 10	OE/HDO 10	OE/HDO 10	OEA

ITEM NO.	INTERVAL	ITEM TO CHECK/ SERVICE	CREWMEMBER PROCEDURE	NOT FULLY MISSION CAPABLE IF:
52	After	Wrecker Crane	Check hydraulic oil level with dipstick in filler cap (3). Oil level should be at FULL mark. If low, add as necessary.	Any class III leak.

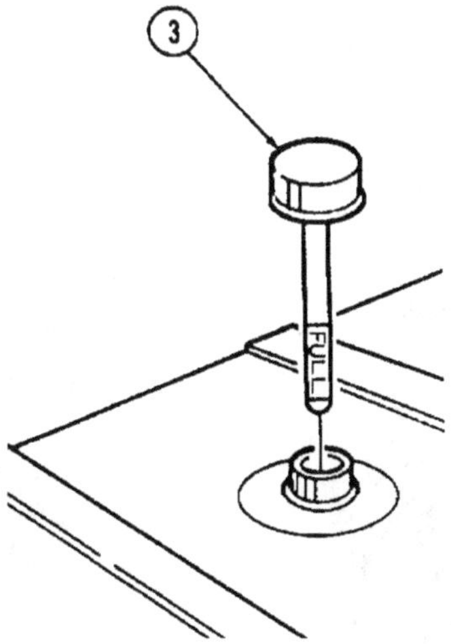
3
FULL

Table 2-3. Preventive Maintenance Checks and Services for Models M939/A1/A2 (Contd).

ITEM NO.	INTERVAL	LOCATION ITEM TO CHECK/ SERVICE	CREWMEMBER PROCEDURE	NOT FULLY MISSION CAPABLE IF:
52a	Weekly	Hood Support Rod, Locking Pins, Handle, Retaining Bracket, Stop Cables, Hinge and Mounting Hardware	**a.** Check hood holddown latches (1) for damage or loose or missing hardware. **b.** Check support rod (2) and locking pins (4) and (7) for damage or loose or missing hardware. **c.** Check handle (3) for damage or loose or missing hardware. **d.** Check retaining bracket (6) for damage or loose or missing hardware. **e.** Check stop cables (8) for damage and loose or missing hardware. **f.** Check hinge (5) for damage or loose or missing hardware. **g.** Inspect tether cables, tether cable bolts, and washers.	
52b	Weekly	Electrical Connectors, Receptacles, and Ground Strap	**a.** Check electrical connectors (10) and receptacles (11) on both right and left sides of hood for damage or loose or missing hardware. **b.** Check ground strap (9) for damage or loose or missing hardware.	

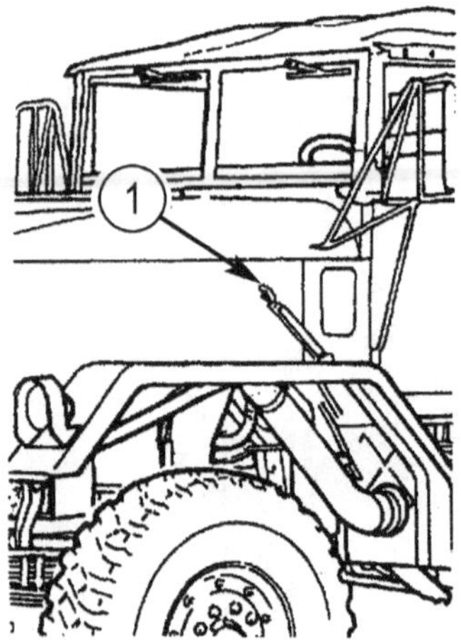
1

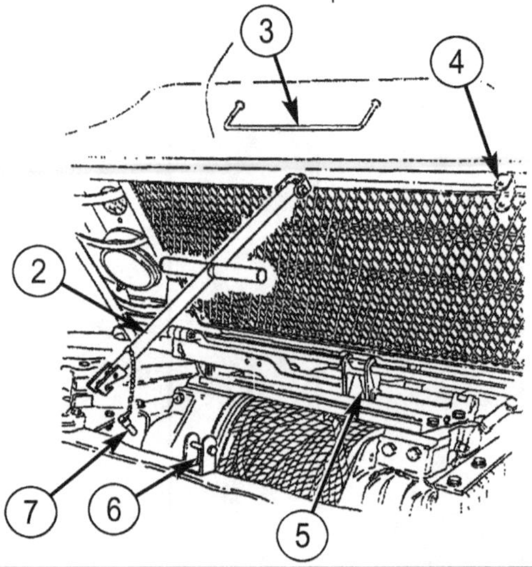
3
4
2
7
6
5

Table 2-3. Preventive Maintenance Checks and Services for Models M939/A1/A2 (Contd).

ITEM NO.	INTERVAL	LOCATION ITEM TO CHECK/ SERVICE	CREWMEMBER PROCEDURE	NOT FULLY MISSION CAPABLE IF:

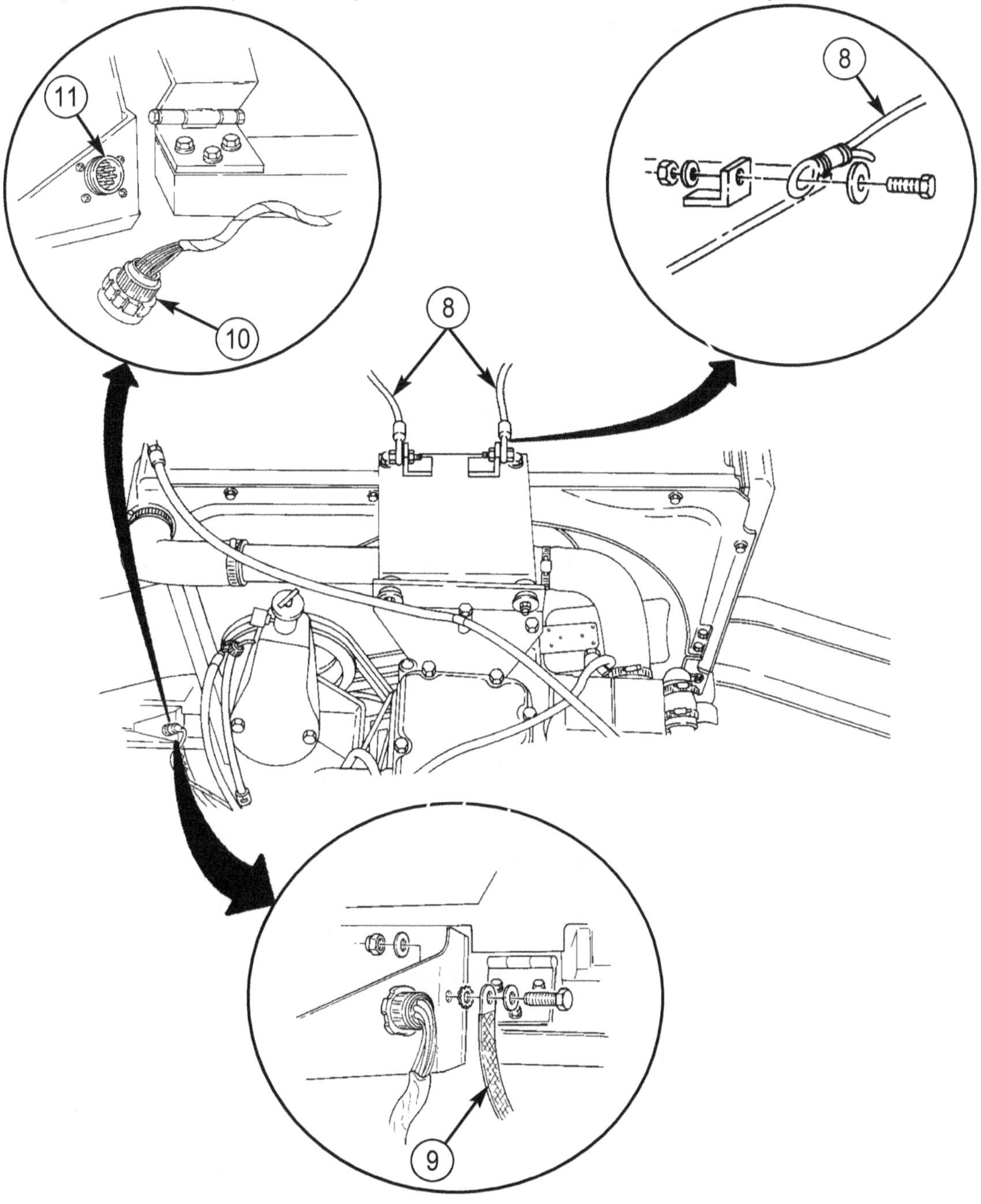
11
10
8
8
8
9

Table 2-3. Preventive Maintenance Checks and Services for Models M939/A1/A2 (Contd).

ITEM NO.	INTERVAL	LOCATION ITEM TO CHECK/ SERVICE	CREWMEMBER PROCEDURE	NOT FULLY MISSION CAPABLE IF:
53	Weekly	Batteries	DRIVER **WARNING** • Don't smoke, have open flames, or make sparks around the batteries, especially if the caps are off. Batteries can explode and cause injury or death. • Protective clothing, rubber gloves, and eye protection must be worn. • Remove all jewelry such as rings, dog tags, or bracelets. If jewelry or tools contact battery terminal, a direct short may occur resulting in instant heating, damage to equipment, and injury to personnel. • Ensure seatbelts and strapping do not come in contact with electrolyte. Damage to strapping material will result, leading to injury or death. • Ensure seatbelts are not caught inside battery box when closing cover. Failure to do so will result in injury or death. **a.** Check electrolyte level in battery (2). Electrolyte should be filled to the level/split ring (3) in the battery filler opening (vent). If fluid is low, fill with distilled water to the level ring. If fluid is gassing (boiling), notify unit maintenance. **b.** Inspect seatbelt straps (1) for damage from electrolyte.	Battery is cracked, unserviceable, missing, or leaking. Terminals or cables are loose or corroded, or hold downs are not secure.

Table 2-3. Preventive Maintenance Checks and Services for Models M939/A1/A2 (Contd).

ITEM NO.	INTERVAL	LOCATION ITEM TO CHECK/ SERVICE	CREWMEMBER PROCEDURE	NOT FULLY MISSION CAPABLE IF:
54	Weekly	Alcohol Evaporator	 1 2 3 DRIVER **WARNING** Alcohol used in alcohol evaporator is flammable, poisonous, and explosive. Do not smoke when adding fluid and do not drink fluid. Failure to do this will result in injury or death. **NOTE** • Use alcohol evaporator during cold weather operations only. • Alcohol evaporator is removed from vehicles equipped with ABS. **a.** Check fluid level. Fill with alcohol if bottle (4) is less than two-thirds full. **b.** Check bottle (4) for cracks or breaks. 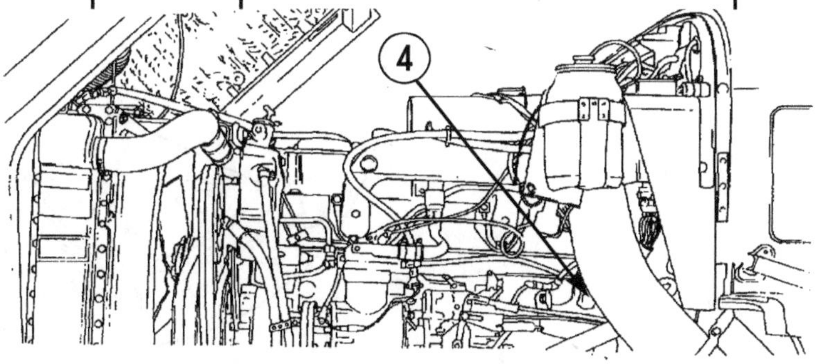 4	

Table 2-3. Preventive Maintenance Checks and Services for Models M939/A1/A2 (Contd).

ITEM NO.	INTERVAL	LOCATION ITEM TO CHECK/ SERVICE	CREWMEMBER PROCEDURE	NOT FULLY MISSION CAPABLE IF:
			DRIVER **WARNING** If NBC exposure is suspected, all air filter media should be handled by personnel wearing protective equipment. Consult your unit NBC officer or NBC NCO for appropriate handling or disposal instructions.	
55	Weekly	Air Cleaner	Empty automatic dust unloader (2) from air cleaner (1).	Dust unloader damaged.

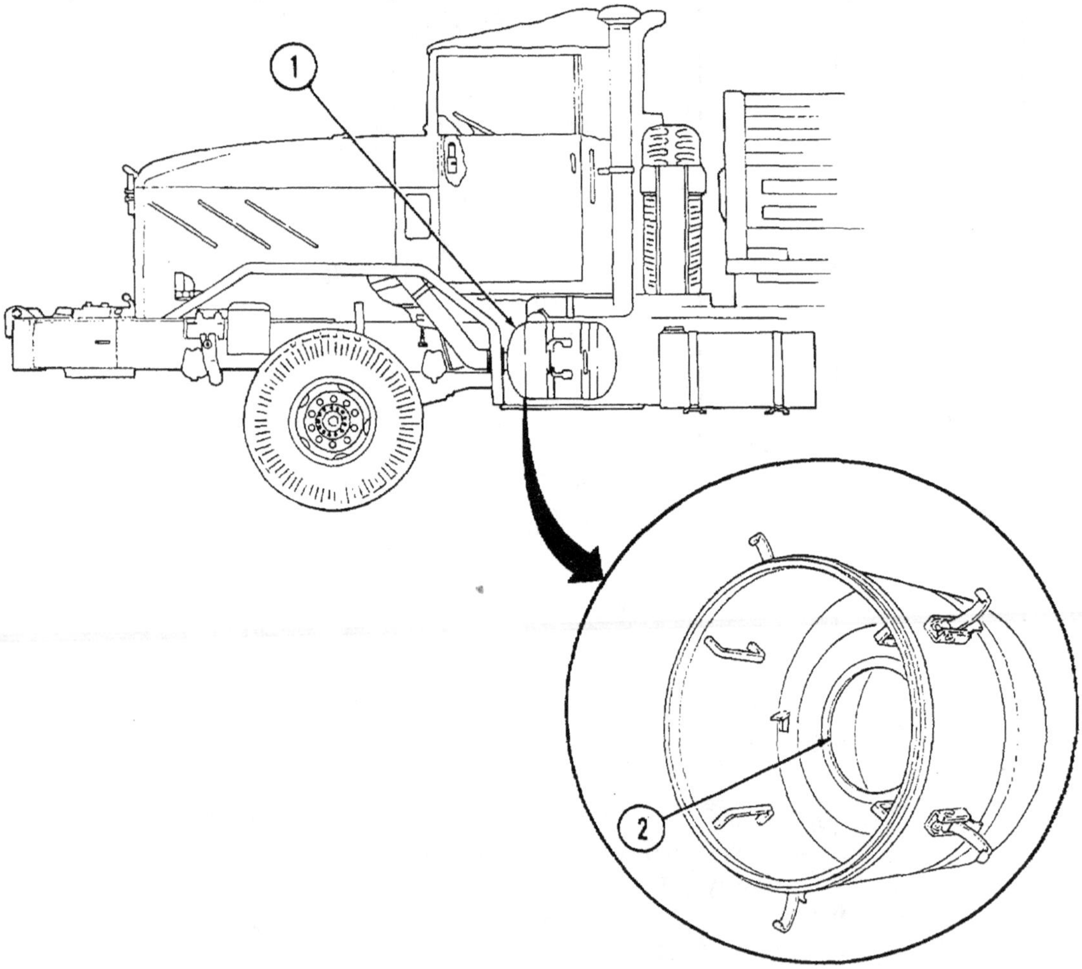
1
2

Table 2-3. Preventive Maintenance Checks and Services for Models M939/A1/A2 (Contd).

ITEM NO.	INTERVAL	LOCATION / ITEM TO CHECK/ SERVICE	CREWMEMBER PROCEDURE	NOT FULLY MISSION CAPABLE IF:
			DRIVER	
56	Weekly	Air Tank Check Valve	Build up pressure to normal range of between 90 to 130 psi (621-896 kPa). Open wet tank drain (5) and observe primary (3) and secondary (4) air gauges.	Primary/secondary gauge pressure falls.

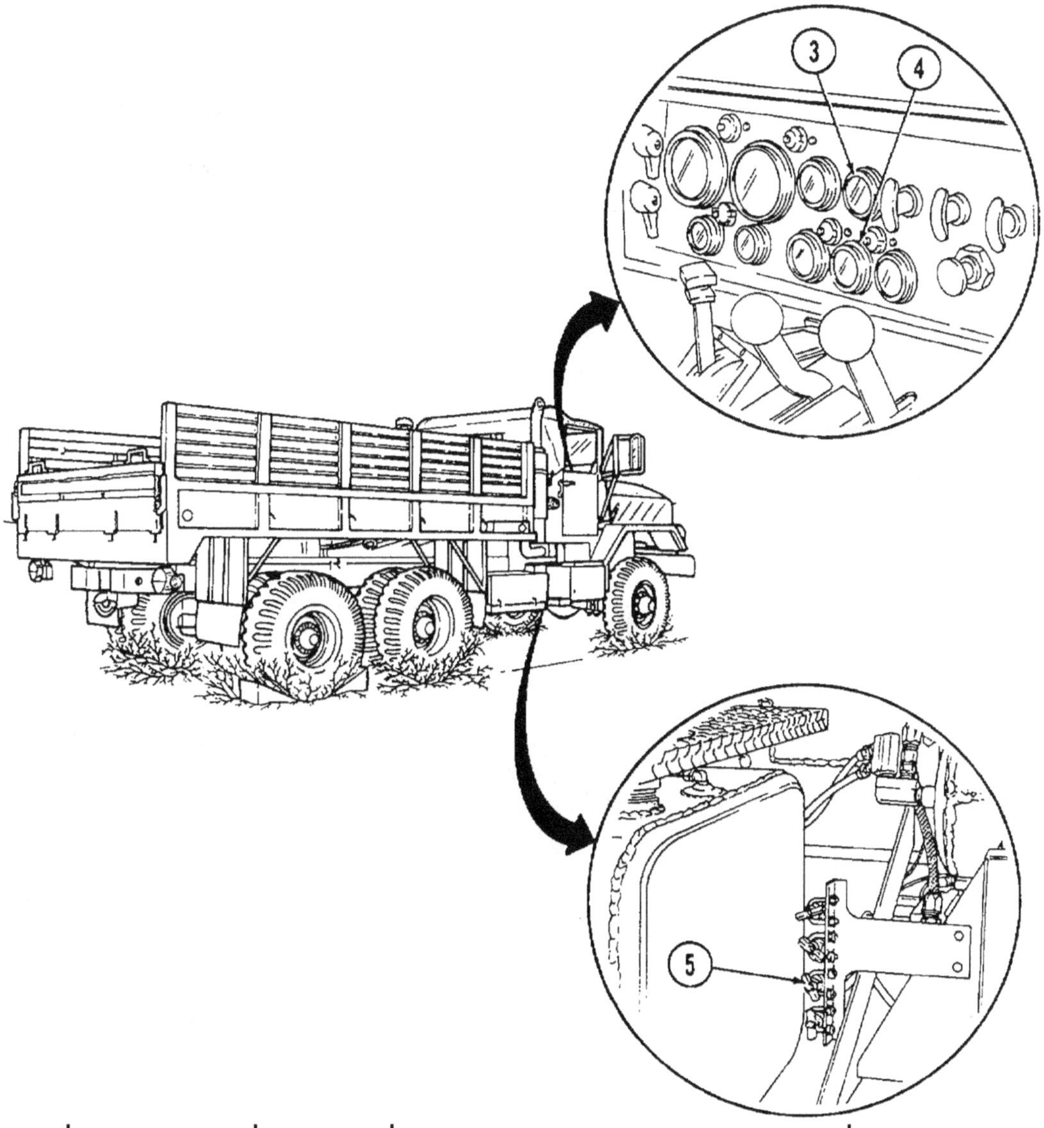
3
4
5

Table 2-3. Preventive Maintenance Checks and Services for Models M939/A1/A2 (Contd).

ITEM NO.	INTERVAL	LOCATION ITEM TO CHECK/ SERVICE	CREWMEMBER PROCEDURE	NOT FULLY MISSION CAPABLE IF:
			DRIVER	
57	Weekly	Cab	Visually inspect the cab mounts (1) for cracks that penetrate the mounting brackets, breaks that loosen the cab mounts, and damage or missing cab mounts.	Cab mounts missing or welds cracked through brackets or broken loose cab mounts.

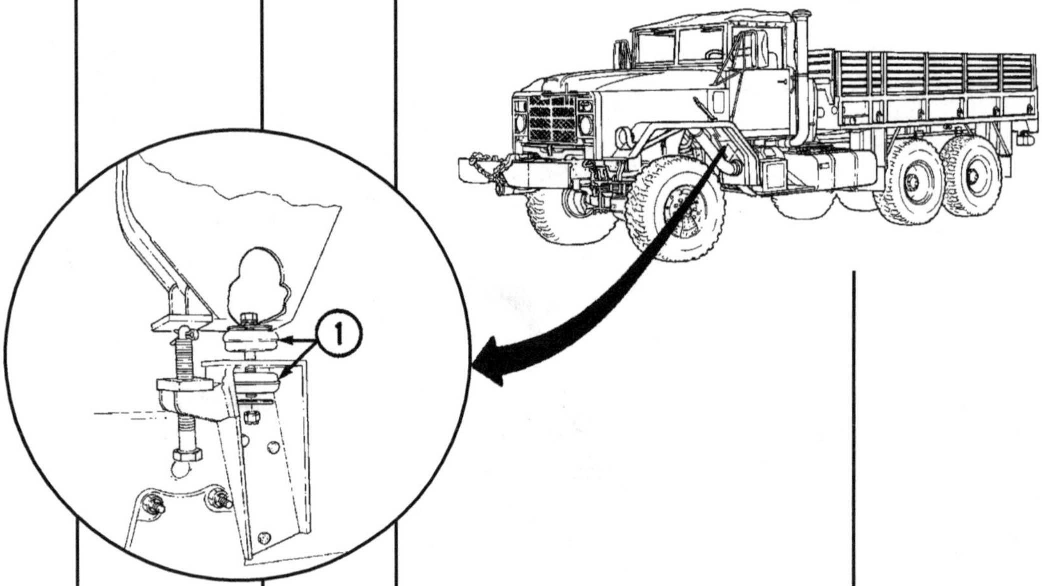
1

ITEM NO.	INTERVAL	LOCATION ITEM TO CHECK/ SERVICE	CREWMEMBER PROCEDURE	NOT FULLY MISSION CAPABLE IF:
58	Weekly	Van Electrical System	**a.** Operate all switches to determine that all function properly and that ceiling lights illuminate. **b.** Open doors to determine reliability of blackout switches. **c.** Visually check cables and harnesses for breaks or loose connections.	**a.** Electrical system will not operate. **b.** Blackout switch is inoperative.

Table 2-3. Preventive Maintenance Checks and Services for Models M939/A1/A2 (Contd).

ITEM NO.	INTERVAL	LOCATION ITEM TO CHECK/ SERVICE	CREWMEMBER PROCEDURE	NOT FULLY MISSION CAPABLE IF:
			DRIVER **CAUTION** Do not disconnect CTIS connector from ECU with power on. Damage to ECU will result. **NOTE** CTIS reference applies only to M939A2 series vehicles.	
59	Weekly	Tires	**a.** Check tire tread depth. When worn to 4/32 in. (3 mm), change tire (para. 3-11). **b.** Check right, left, and spare for correct tire pressure per Tire Inflation Data (table 1-10).	**a.** Tread worn beyond 4/32 in. (3 mm).

Table 2-3. Preventive Maintenance Checks and Services for Models M939/A1/A2 (Contd).

ITEM NO.	INTERVAL	LOCATION ITEM TO CHECK/ SERVICE	CREWMEMBER PROCEDURE	NOT FULLY MISSION CAPABLE IF:
59 (contd)	Weekly	Tires (contd)	**NOTE** On M931A2 and M932A2 models equipped with CTIS, the CTIS must be neutralized at unit maintenance if towing a 5,000-gallon semitrailer (M131 series, M967/A1, M969/A1/A2, and M970/A1) (table 1-10) **c.** Check all tires with inflation gauge with CTIS set at each setting (para. 2-14). Start with HWY through EMG. **d.** Check that tire pressure is inflated back to HWY.	

	CTIS TIRE PRESSURES			
Vehicle	**Highway** Standard (psi) Metric (kPa)	**Cross Country** Standard (psi) Metric (kPa)	**Sand/Snow** Standard (psi) Metric (kPa)	**Emergency** Standard (psi) Metric (kPa)
M923A2, M925A2, M927A2, M928A2, M929A2, M930A2, M931A2, M932A2, M934A2	60/414	35/241	25/172	12/83
All Models: Spare	Maximum Highway Pressure			
M936A2	80/551	35/241	25/172	12/83

Table 2-3. Preventive Maintenance Checks and Services for Models M939/A1/A2 (Contd).

ITEM NO.	INTERVAL	LOCATION ITEM TO CHECK/ SERVICE	CREWMEMBER PROCEDURE	NOT FULLY MISSION CAPABLE IF:
			NOTE Body sides reference applies only to M923/A1/A2, M925/A1/A2, M927/A1/A2, and M928/A1/A2 model vehicles.	
60	Weekly	Body Sides	**a.** Check cargo body sides for damage, broken welds, and rusted-through conditions.	
			b. Check cargo body side racks for cracks and breaks.	
			c. Check dropside T-bolts for presence and security (M923/A1/A2 and M925/A1/A2 model vehicles only).	**c.** Any T-bolts missing.
			d. Check condition of safety strap eyelets.	**d.** Safety strap eyelets are missing or damaged.
			e. Check cargo tie-down brackets for presence and damage.	**e.** Cargo tiedown brackets missing or damaged.
			f. Check dropside hinges and pins for presence and damage (M923/A1/A2 and M925/A1/A2 model vehicles only).	**f.** Dropside hinges and pins are missing or damaged.

Table 2-3. Preventive Maintenance Checks and Services for Models M939/A1/A2 (Contd).

ITEM NO.	INTERVAL	LOCATION ITEM TO CHECK/ SERVICE	CREWMEMBER PROCEDURE	NOT FULLY MISSION CAPABLE IF:
			DRIVER	
61	Weekly	Dump Trucks	**a.** Inspect dump body and cab protector for cracks, broken welds, loose or broken bolts, and rusted-through conditions. Ensure all bolts securing cab protector to dump body are secure.	**a.** Cab protector missing.
			b. Check dump body support braces (1) for presence and damage.	**b.** Support braces (1) are bent, broken, or damaged.

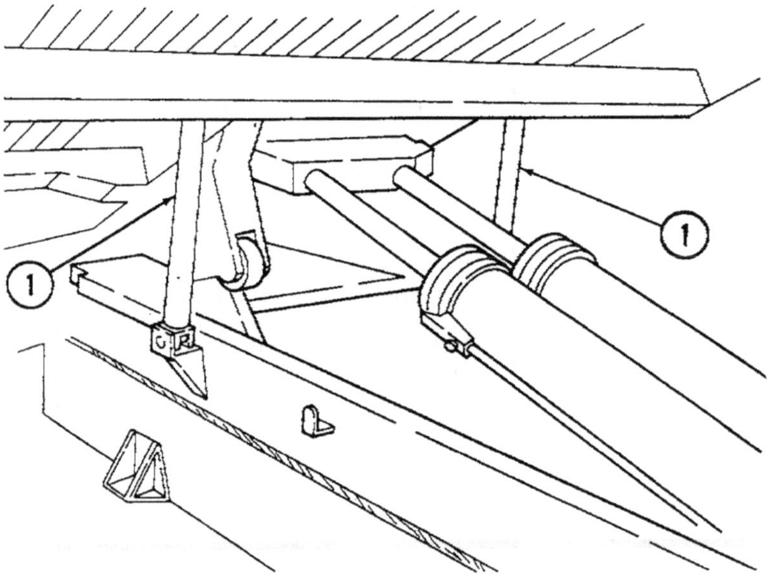
1
1

Table 2-3. Preventive Maintenance Checks and Services for Models M939/A1/A2 (Contd).

ITEM NO.	INTERVAL	LOCATION ITEM TO CHECK/ SERVICE	CREWMEMBER PROCEDURE	NOT FULLY MISSION CAPABLE IF:
			DRIVER	
62	Weekly	Tailgate	**a.** Inspect tailgate for damage, security, and ease of operation. **b.** Check tailgate chains for security, presence, and damage. **c.** Check security of latches, brackets, and retaining pins for presence and damage. **d.** Check security of chains, tailgate wings, and harness hooks for presence and damage.	**a.** Tailgate is inoperative.
			DRIVER	
63	Weekly	Transfer Case and Transmission Bolts	Check all transfer case and transmission bolts for looseness.	Any loose or missing bolts.
			DRIVER	
64	Weekly	Spare Tire Davit	**a.** Check spare tire davit boom assembly for proper operation (para. 3-11). **b.** Secure spare tire to davit/tire carrier by using BII chain and 3/8 in. bolt, NSN 5305-00-725-2317 and nut, NSN 5310-00-732-0558.	

Table 2-3. Preventive Maintenance Checks and Services for Models M939/A1/A2 (Contd).

ITEM NO.	INTERVAL	LOCATION ITEM TO CHECK/ SERVICE	CREWMEMBER PROCEDURE	NOT FULLY MISSION CAPABLE IF:
			DRIVER	
65	Weekly	Underbody Frame	Visually inspect frame side rails, crossmembers, and underbody supports for broken bolts, cracks, breaks, broken welds, rivets, and rusted-through conditions.	Any side rail or crossmember is obviously broken; any weld, bolt, or rivet broken or rusted through.
			DRIVER	
66	Weekly	Differentials	Visually inspect rear differentials for oil leaks (para. 2-9).	Class III leakage is evident.
			DRIVER	
67	Weekly	Tow Pintle	Check for presence and condition. Ensure safety pin and chain are present.	Safety pin is missing.
			DRIVER	
68	Weekly	Van Body Exterior	**a.** Check for condition and proper function of panels and doors.	**a.** Panels or doors do not function properly.
			b. Check for presence and condition of ladders, stabilizers, receptacles, and power cables.	**b.** Ladders, stabilizers, receptacles, or power cables are missing.
			DRIVER	
69	Weekly	Wrecker Crane	**a.** Check PTOs, drive shafts, hydraulic pumps, and control valves for damage, leakage, and security mounting.	Class III leakage is evident.
			b. Check hydraulic tank oil level.	

Table 2-3. Preventive Maintenance Checks and Services for Models M939/A1/A2 (Contd).

LUBRICANTS	EXPECTED TEMPERATURES		
	ABOVE 15°F (ABOVE -9°C)	+40°F TO -15°F (+4°C TO -26°C)	+40°F TO -65°F (+4°C TO -54°C)
Oil, lubricating, multipurpose (MIL-L-2105) GO-80/90	GO-80/90	GO-80/90	
Oil, lubricating, multipurpose (MIL-L-2105) GO-75			GO-75

ITEM NO.	INTERVAL	LOCATION ITEM TO CHECK/ SERVICE	CREWMEMBER PROCEDURE	NOT FULLY MISSION CAPABLE IF:
70	Weekly	Rear Winch and Controls (M936/ A1/A2)	DRIVER **a.** Remove oil level plug (1) from winch gearcase. If level is below level plug hole, fill to bottom of hole.	
			b. Check for secure connections.	**b.** Winch inoperable; mount loose or damaged.

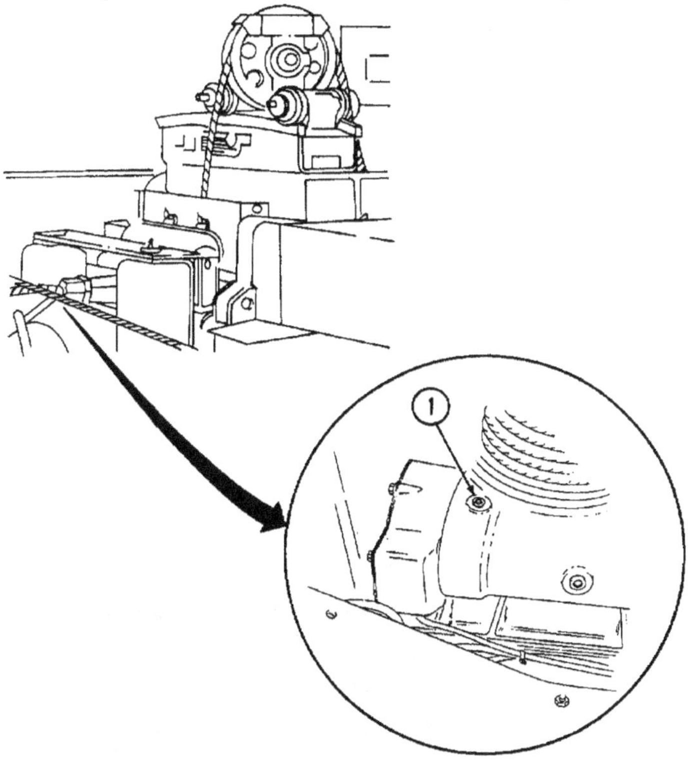
1

Table 2-3. Preventive Maintenance Checks and Services for Models M939/A1/A2 (Contd).

ITEM NO.	INTERVAL	LOCATION ITEM TO CHECK/ SERVICE	CREWMEMBER PROCEDURE	NOT FULLY MISSION CAPABLE IF:
			DRIVER **SPECIAL PURPOSE KITS**	
71	Weekly	Rifle Mount Kit	Check stock brace (2) for looseness or damage. Check catch (1) assembly for excessive looseness, binding, or damage. 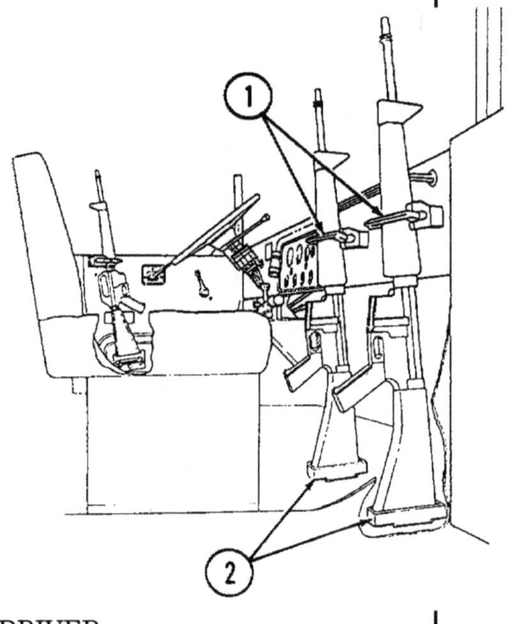1 2	
			DRIVER	
72	Weekly	Machine-gun Mount	Check for damage to cab and security of mount and ring (TM 9-1005-245-14).	
			DRIVER	
73	Weekly	Deep-water Fording Kit	**a.** Tighten fuel tank filler cap(s) (para. 2-40). **b.** Make sure all battery filler caps are present and secure (para. 2-40). **c.** Make sure transmission dipstick is secured in filler tube (para. 2-40). **d.** Check operation of control handle. Make sure fording valves open and close (para. 2-40).	**a.** Requires deepwater fording kit operation and kit is inoperative. **d.** Control handle on fording valves are inoperative.

Table 2-3. Preventive Maintenance Checks and Services for Models M939/A1/A2 (Contd).

ITEM NO.	INTERVAL	LOCATION ITEM TO CHECK/ SERVICE	CREWMEMBER PROCEDURE	NOT FULLY MISSION CAPABLE IF:
			DRIVER	
74	Weekly	Arctic Winterization Kits	**a.** Check fuel burning personnel and engine heater air intake and exhaust tubes for damage, obstructions, and leakage (para. 2-44). **b.** Ensure both engine coolant heater shutoff valves (3) are open. **c.** Check fuel burning and engine coolant heater controls by depressing indicator lamps to make sure they illuminate (para. 2-44).	**a.** Any exhaust leakage or class III fuel leak is evident.

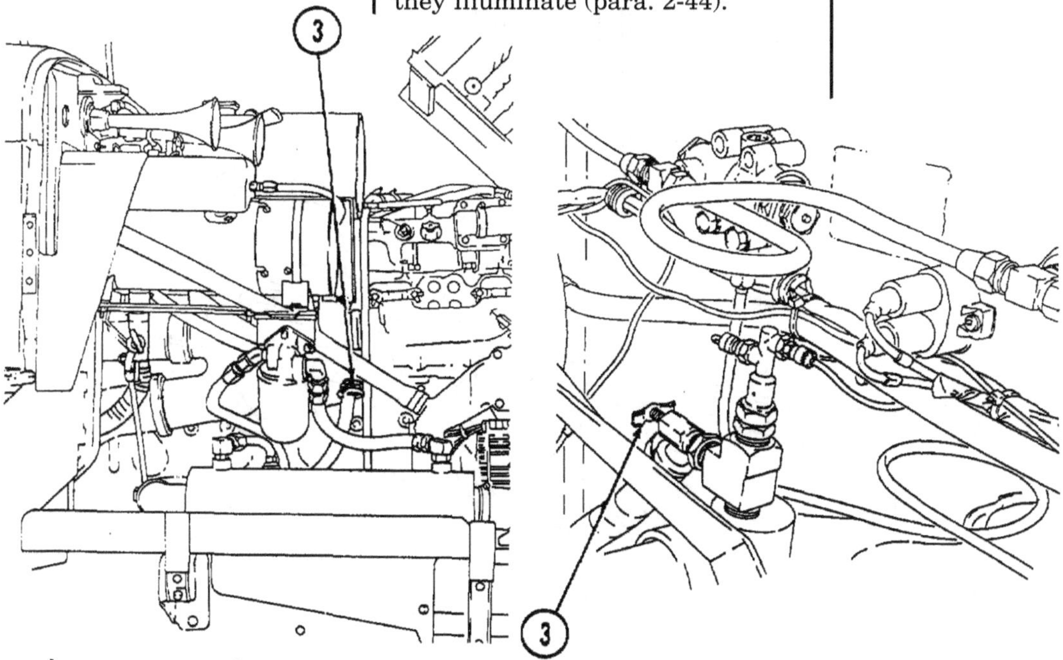
3
3

ITEM NO.	INTERVAL	LOCATION ITEM TO CHECK/ SERVICE	CREWMEMBER PROCEDURE	NOT FULLY MISSION CAPABLE IF:
			DRIVER	
75	Weekly	M-8 Chemical Alarm	Refer to TM 3-6665-225-12 for Preventive Maintenance Checks and Services.	
			DRIVER	
76	Weekly	M-11 Decontamination Unit	Refer to TM 3-4230-204-12&P for Preventive Maintenance Checks and Services.	

Table 2-3. Preventive Maintenance Checks and Services for Models M939/A1/A2 (Contd).

ITEM NO.	INTERVAL	LOCATION ITEM TO CHECK/ SERVICE	CREWMEMBER PROCEDURE	NOT FULLY MISSION CAPABLE IF:
			DRIVER	
77	Weekly	Troop Seat Kit	Check for broken or splintered side racks (1) and troop seats (2) (para. 2-41).	

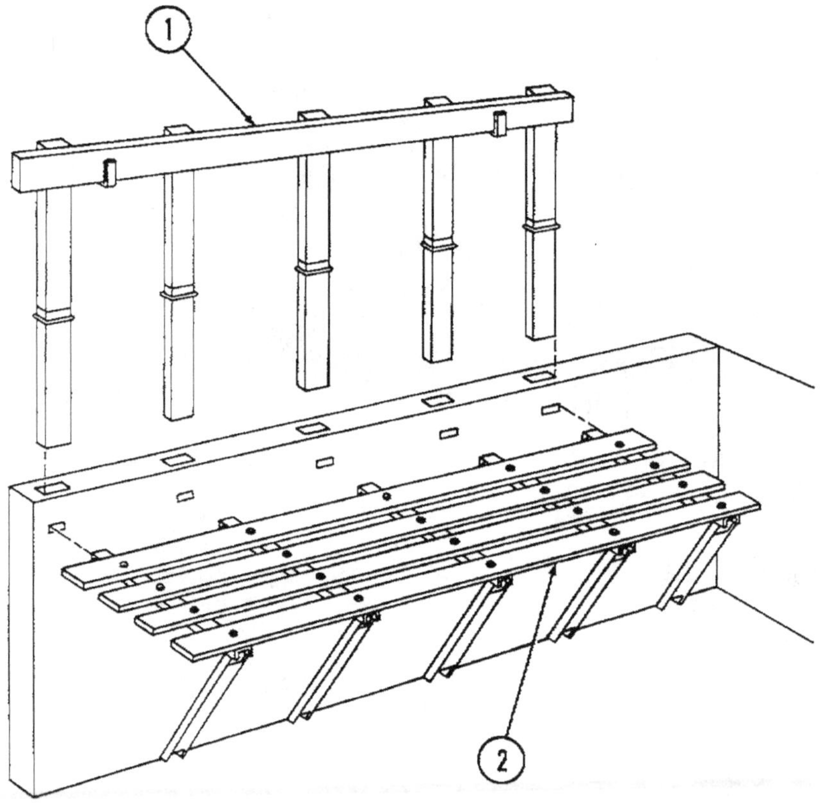
1
2

Table 2-3. Preventive Maintenance Checks and Services for Models M939/A1/A2 (Contd).

ITEM NO.	INTERVAL	LOCATION ITEM TO CHECK/ SERVICE	CREWMEMBER PROCEDURE	NOT FULLY MISSION CAPABLE IF:
			DRIVER	
78	Weekly	Bow and Tarp Kit	Check staves (3), crossbows (4), ropes (5), rear end curtain (6), and tarpaulin (7) for damage (para. 2-42).	

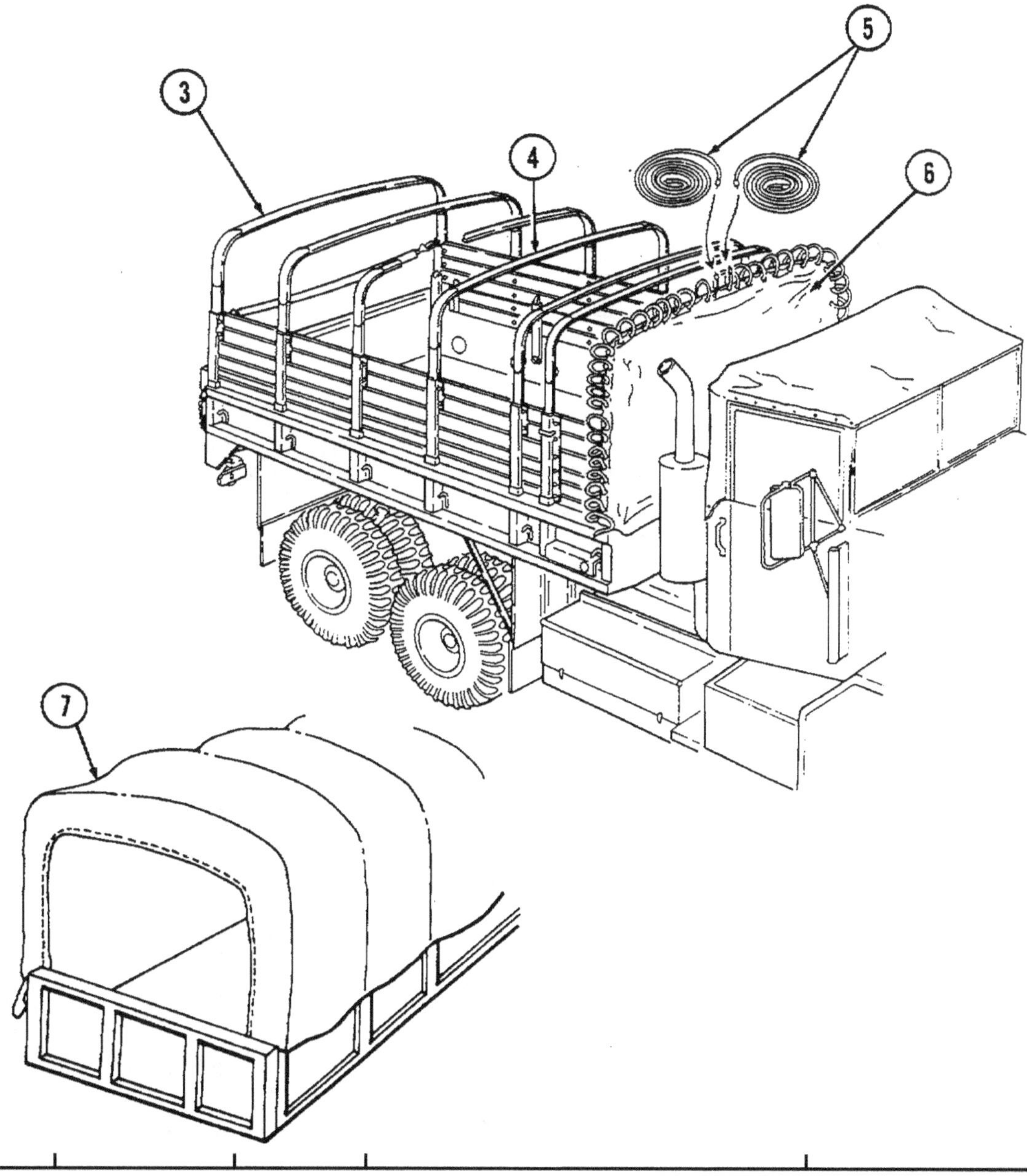

3
4
5
6
7

Table 2-3. Preventive Maintenance Checks and Services for Models M939/A1/A2 (Contd).

ITEM NO.	INTERVAL	LOCATION ITEM TO CHECK/ SERVICE	CREWMEMBER PROCEDURE	NOT FULLY MISSION CAPABLE IF:
			DRIVER	
79	Weekly	Radiator and Hood Cover Kit	Clean and inspect radiator cover flap (1) and tie rope (2) for damage (para. 2-43).	

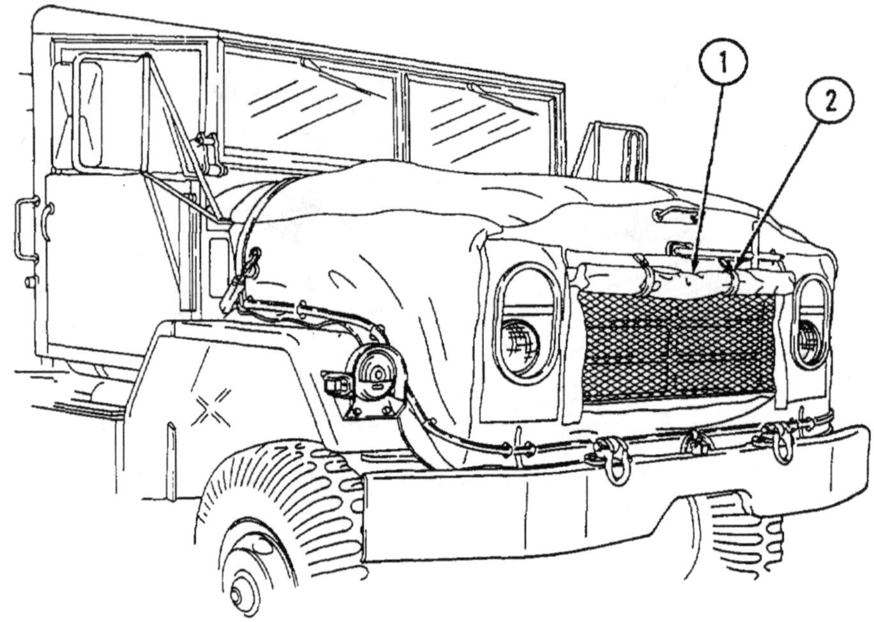
1
2

Table 2-3. Preventive Maintenance Checks and Services for Models M939/A1/A2 (Contd).

ITEM NO.	INTERVAL	LOCATION / ITEM TO CHECK/ SERVICE	CREWMEMBER PROCEDURE	NOT FULLY MISSION CAPABLE IF:
			DRIVER **NOTE** Make sure primary air pressure gauge on instrument panel reads 120 psi (827 kPa).	
80	Monthly	Air Compressor	**a.** Check air compressor (3) for air leakage.	**a.** Any air leakage is evident.
			b. Check air lines and fittings for air leakage.	**b.** Any air leakage is evident.

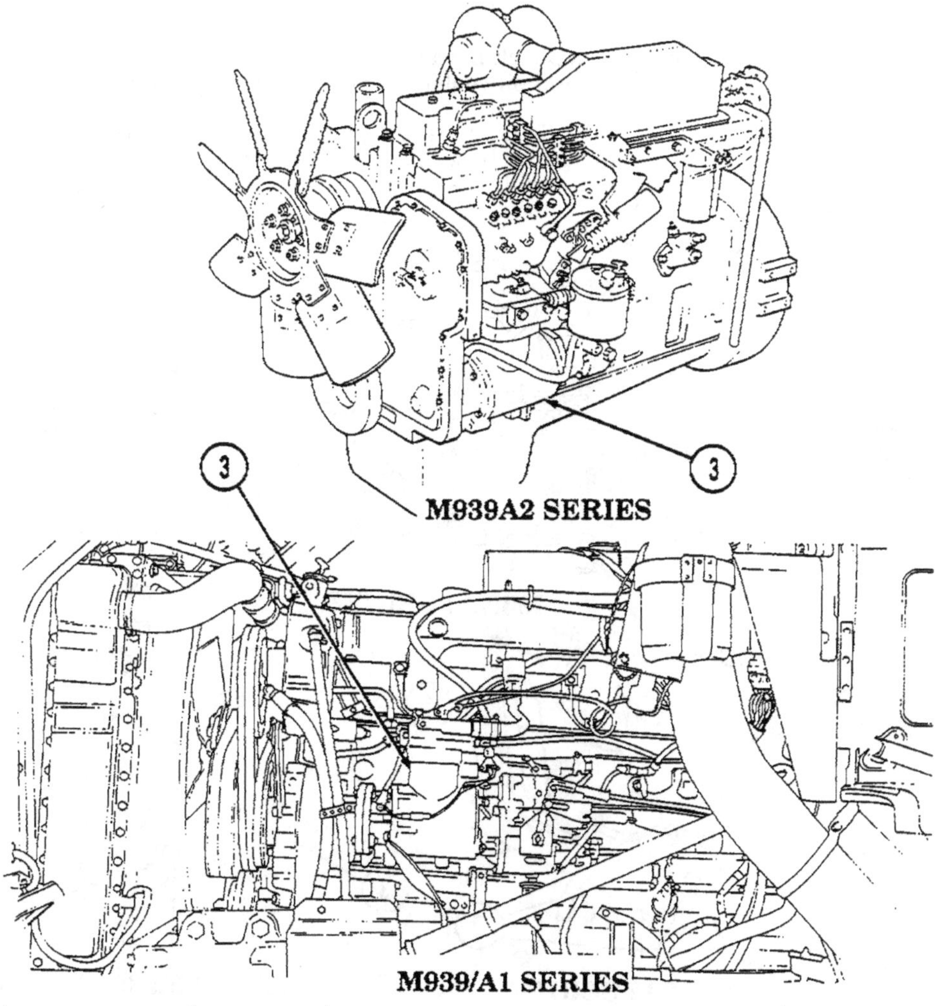
3
3
M939A2 SERIES
M939/A1 SERIES

Table 2-3. Preventive Maintenance Checks and Services for Models M939/A1/A2 (Contd).

ITEM NO.	INTERVAL	LOCATION ITEM TO CHECK/ SERVICE	CREWMEMBER PROCEDURE	NOT FULLY MISSION CAPABLE IF:
			DRIVER **CAUTION** Before opening reservoir, make sure area around reservoir cap is clean. Do not allow dirt, dust, or water to enter reservoir to prevent damage to steering system internal components. **NOTE** Power steering reservoir oil level is checked with engine stopped.	
81	Monthly	Steering System	**a.** Check power steering pump (1) and oil cooler (3) for leakage. **b.** Visually inspect power steering pump hoses (2) for deterioration and leaks.	**a.** Class III leakage is evident. **b.** Class III leakage is evident or hoses are cut or broken.

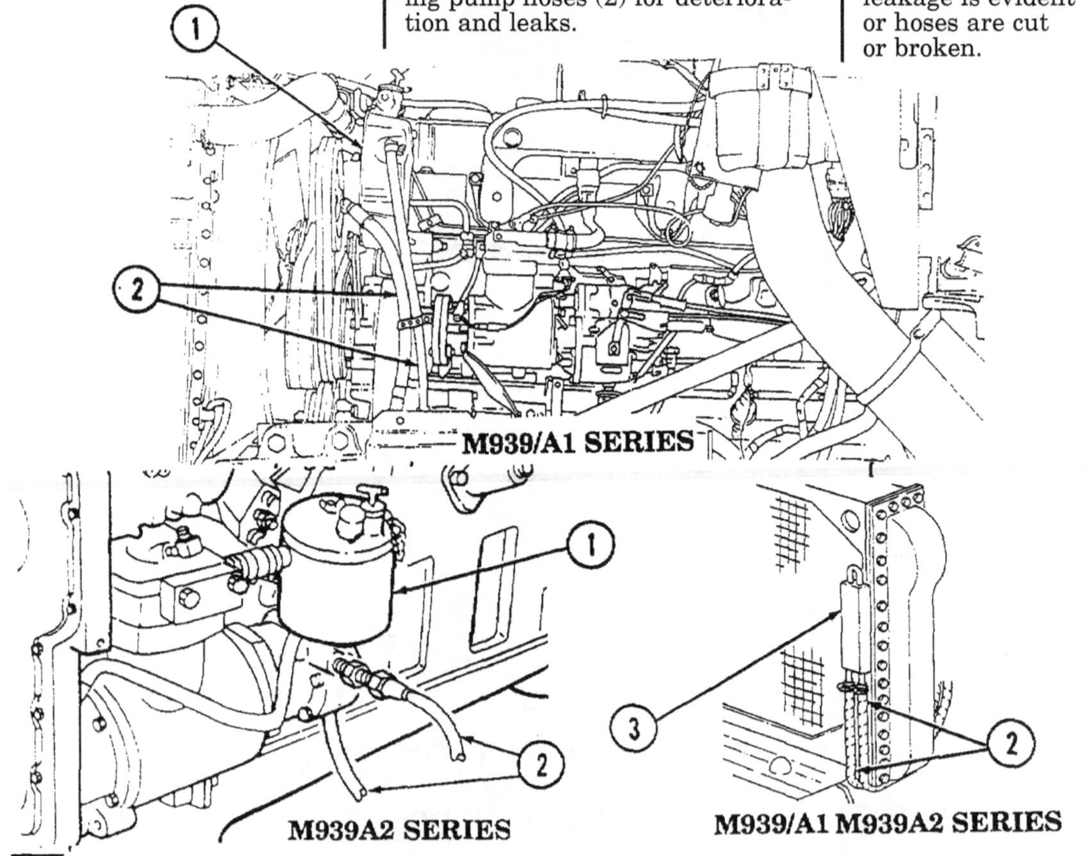
1
2
M939/A1 SERIES
1
2
M939A2 SERIES
3
2
M939/A1 M939A2 SERIES

Table 2-3. Preventive Maintenance Checks and Services for Models M939/A1/A2 (Contd).

ITEM NO.	INTERVAL	LOCATION ITEM TO CHECK/ SERVICE	CREWMEMBER PROCEDURE	NOT FULLY MISSION CAPABLE IF:
82	Monthly	Surge Tank	DRIVER **WARNING** If temperature gauge reads above 175° F (79° C), use care when removing surge tank filler cap. Pressurized steam or hot coolant will cause injury to personnel. **a.** Check all hoses (4) for deterioration and/or leakage. Tighten loose or leaking hose connections.	Class III leakage is evident, or hoses (4) are cut or broken.

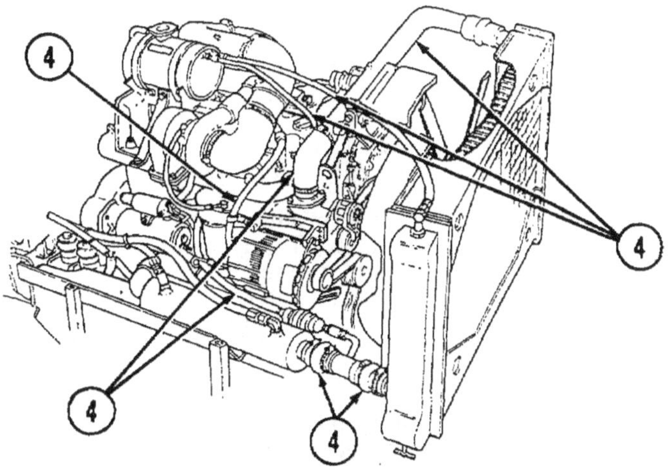
4
4
4
4
4

ITEM NO.	INTERVAL	LOCATION ITEM TO CHECK/ SERVICE	CREWMEMBER PROCEDURE	NOT FULLY MISSION CAPABLE IF:
83	Monthly	Wrecker Boom	DRIVER Check the date of the last boom load test. If more than one-year-old, notify unit maintenance (refer to TB 9-352).	Load test is more than one year old.

Table 2-3. Preventive Maintenance Checks and Services for Models M939/A1/A2 (Contd).

ITEM NO.	INTERVAL	LOCATION ITEM TO CHECK/ SERVICE	CREWMEMBER PROCEDURE	NOT FULLY MISSION CAPABLE IF:
			DRIVER	
84	Monthly	Frame Inspection	**a.** Check the chassis for loose or missing screws (3) and rivets (4) securing fifth wheel (2) to side rails (5), and side rails (5) to vehicle (TB 9-2300-247-40).	
			b. Using a .001–inch–thick feeler gauge, check for space between rivet head and the riveted frame members. Penetration of the feeler gauge between the rivet head and the riveted member is reason to suspect that the riveted connection and/or rivet should be replaced.	**b.** Loose or missing rivets.
			c. Thoroughly clean rivet and riveted connection of all dirt, grease, and oil. Using an oil can, apply lubricating oil around the suspect rivet and riveted connection. Allow approximately 10 to 20 seconds for the oil to penetrate. Wipe rivet and riveted connection free of oil. Tap rivet with an eight-pound hammer. Any indication of oil around the rivet indicates a loose rivet. Notify unit maintenance to replace all loose rivets. Check all riveted connections for signs of movement, such as bare or shiny spots, or other indications of movement between rivet and framing member. If movement is indicated, rivet and connection are loose.	**c.** Loose or missing rivets.
			CAUTION Axle breathers must be cleaned before servicing to prevent damage to axle from contamination. Remove, clean, and lubricate axle breathers every 1000 miles (1600 km) or monthly, whichever occurs first.	
84a	Monthly	Axle Breather	Check all axle breathers (6) for damage or dirt.	

Table 2-3. Preventive Maintenance Checks and Services for Models M939/A1/A2 (Contd).

ITEM NO.	INTERVAL	LOCATION ITEM TO CHECK/ SERVICE	CREWMEMBER PROCEDURE	NOT FULLY MISSION CAPABLE IF:

LUBRICANTS	EXPECTED TEMPERATURES		
	ABOVE 15°F (ABOVE -9°C)	+40°F TO -15°F (+4°C TO -26°C)	+40°F TO -65°F (+4°C TO -54°C)
GAA-GREASE, AUTOMOTIVE AND ARTILLERY (MIL-G-10924)	ALL TEMPERATURES		

ITEM NO.	INTERVAL	LOCATION ITEM TO CHECK/ SERVICE	CREWMEMBER PROCEDURE	NOT FULLY MISSION CAPABLE IF:
			DRIVER	
85	Monthly	Fifth Wheel	**a.** Thoroughly clean base plate of fifth wheel (2) and approach plate (1) of all dirt, grease, and oil. Coat approach plate (1) and base plate of fifth wheel (2) with grease.	

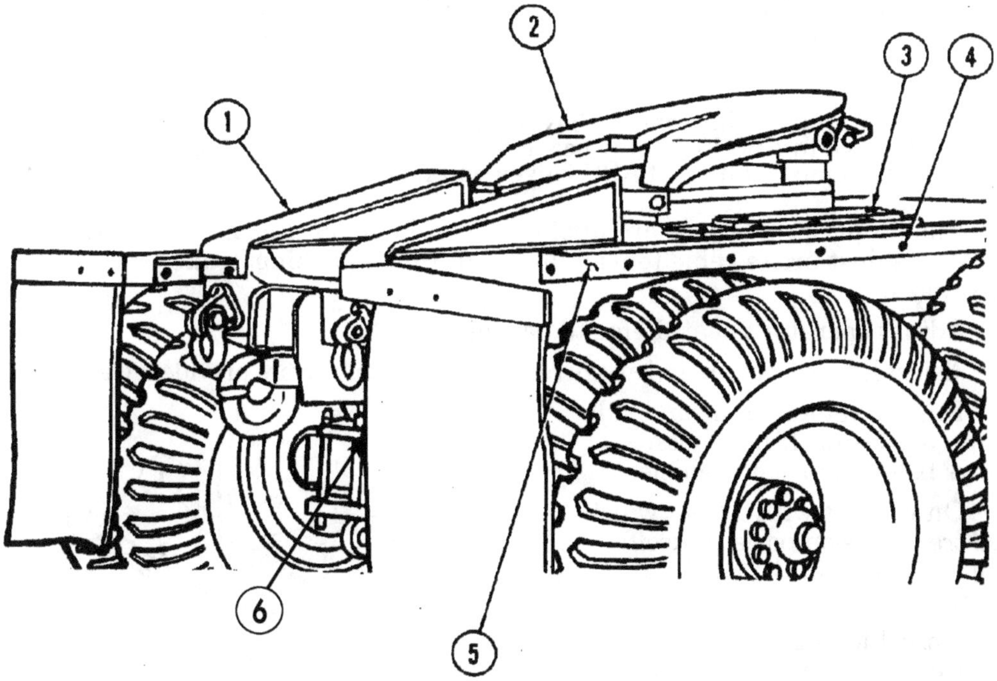
1
2
3
4
5
6

Section III. OPERATION UNDER USUAL CONDITIONS

2-11. GENERAL

This section provides instructions for vehicle operations under moderate temperature, humidity, and terrain conditions. For vehicle operations under unusual conditions, refer to Section IV of this chapter.

WARNING

This vehicle has been designed to operate safely and efficiently within the limits specified in this TM. Operation beyond these limits is prohibited IAW AR 70-1 without written approval from the Commander, U.S. Army Tank-automotive and Armaments Command, ATTN: AMCPEO-CM-S, Warren, MI 48397-5000.

NOTE

Before you attempt to operate your vehicle, be sure to perform the preventive maintenance checks and services shown in table 2-3.

2-12. STARTING THE ENGINE (ABOVE +32°F) (0°C)

CAUTION

Start-up procedure should be strictly adhered to, otherwise damage to ABS ECU may occur if vehicle is so equipped and may also induce faults and ABS valves will not function.

a. Ensure parking brakes are applied. Turn knob on the end of parking brake lever (4) to adjust brake cable tension and pull up on parking brake lever (4) to apply brakes.

b. Adjust operator's seat. Refer to paragraph 2-3.

c. Adjust left and right rearview mirrors. Ensure both mirrors provide a clear rearview.

d. Ensure vehicle front and side windows are clean. If not, clean windows before starting vehicle.

e. On vehicles with a front winch, ensure transmission power takeoff control lever (2) is in DISENGAGE position.

f. On vehicles equipped with transfer case power takeoff control lever (3), ensure lever (3) is locked in neutral (full forward) position.

g. Place automatic transmission selector lever (1) in N (neutral).

h. Ensure EMERGENCY ENGINE STOP control (8) is pushed in all the way.

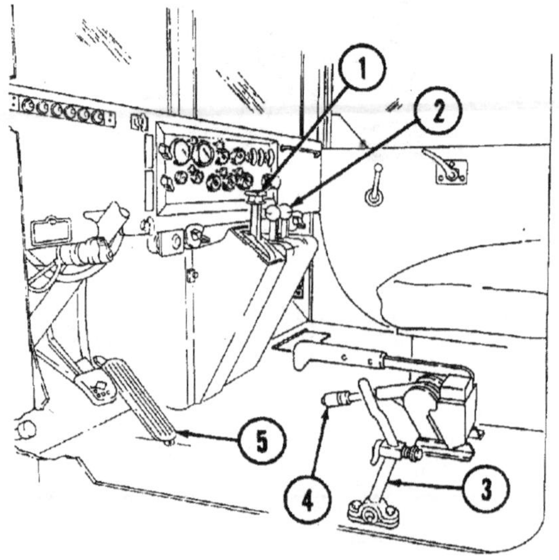
1
2
5
4
3

2-12. STARTING THE ENGINE (ABOVE +32°F) (0°C) (Contd)

i. Check air cleaner indicator (6). If red appears, indicating air restriction, notify your supervisor.

j. Place battery switch (10) in ON position.

NOTE

- Perform steps k. and l. for M939A1 vehicles.
- For M939A2 vehicle, crank engine first, then depress accelerator pedal all the way down. When engine fires, release pedal to partial travel.

k. Depress accelerator pedal (5).

CAUTION

Do not operate starter continuously for more than 10 seconds at a time, or with headlights on. Wait 10-15 seconds between periods of starter operation.

l. Place ignition switch (11) in START position. Release switch (11) after engine starts.

m. Check your instruments:

(1) Air pressure gauges (14) and (15) must read 50-60 psi (345-414 kPa) before warning light (7) goes out and warning buzzer stops. Normal operating pressures for both gauges (14) and (15) is 90-130 psi (621-896 kPa).

(2) Oil pressure gauge (13) should read 15 psi (103 kPa) on M939/A1 series vehicles, 10 psi (69 kPa) on M939A2 series vehicles, or higher.

(3) Voltmeter (16) should read in green area.

(4) Fuel gauge (17) indicates fuel level in fuel tank(s).

(5) ABS warning lamp (18) must not be illuminated for more than three secounds.

n. If necessary, pull out hand throttle control (9) until tachometer (12) indicates an operating range of 700-800 rpm (800-1000 rpm on M939A2 vehicles).

o. Allow engine to warm up approximately five minutes. If engaged, disengage hand throttle control (9) by rotating handle and pushing in to allow engine speed to drop to idle after warmup period.

p. Pull out EMERGENCY ENGINE STOP control (8) if any of the following conditions occur:

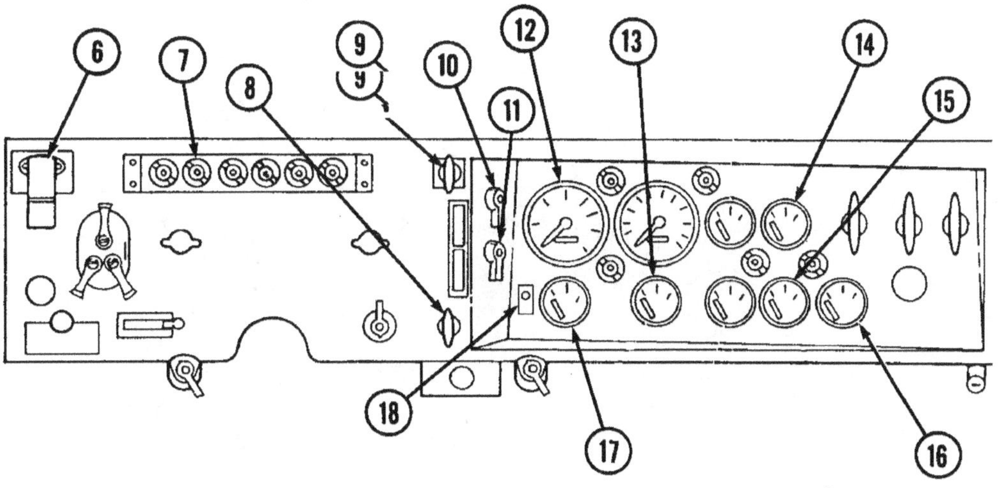
6
7
8
9
10
11
12
13
14
15
16
17
18

2-12. STARTING THE ENGINE (ABOVE +32°F) (0°C) (Contd)

(1) Noisy engine and/or excessive engine vibration.

(2) Oil pressure does not register, or suddenly drops to less than 15 psi (103 kPa) on M939A1 series; 10 psi (69 kPa) on M939A2 series, as indicated by engine oil pressure gauge (6).

(3) Sudden increase in coolant temperature beyond normal operating temperature, 175°-200°F (79°-93°C), as indicated by engine coolant temperature gauge (5).

(4) Engine continues to run after ignition switch (1) and battery switch (3) are turned to OFF positions.

q. After an emergency shutdown, the engine will not restart until unit maintenance resets the fuel shutoff valve. On M939A2 series vehicles, position EMERGENCY ENGINE STOP control (7) to reset.

r. Turn vehicle light switch (9) to desired position (para. 2-18).

2-13. COLD WEATHER STARTING (BELOW +32°F) (0°C)

a. Perform steps a. through k. in paragraph 2-12.

b. Press ether start switch (8) during cranking. Allow three seconds for ether to discharge into system after releasing switch (8).

CAUTION

Do not operate starter continuously for more than 10 seconds at a time or with headlights on. Wait 10-15 seconds between periods of starter operation.

NOTE

If engine cranks but will not start, turn battery switch to OFF position. See troubleshooting, malfunction 3.

c. Release engine ignition switch (1) after engine starts.

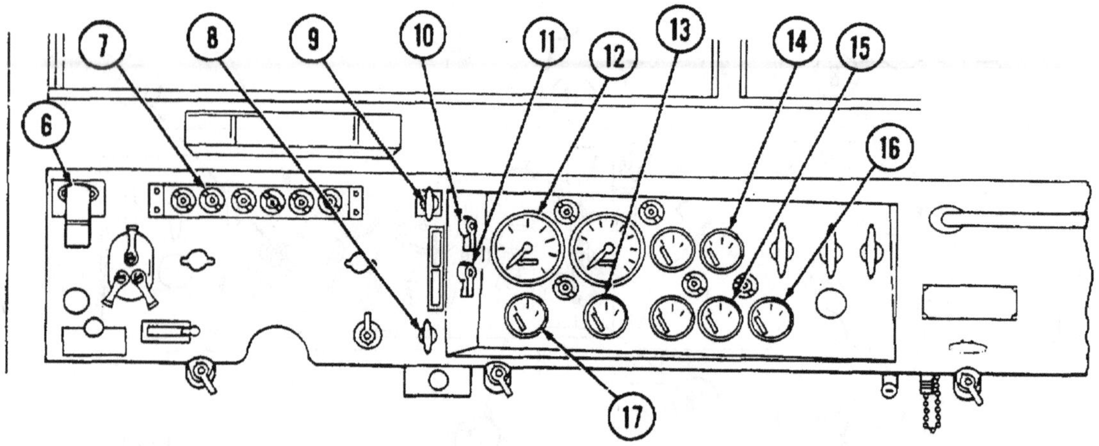
7
8
9
10
11
12
13
14
15
6
16
17

2-13. COLD WEATHER STARTING (BELOW +32°F) (0°C) (Contd)

NOTE

- If engine needs additional ether to prevent stalling, perform step b. again.
- At temperatures below 0°F (-18°C), M939A2 series vehicles will need to be repeated up to six times before engine will start. The engine start switch on the M939A2 series vehicles must be held in the START position to inject ether.

d. Check your instruments again (para. 2-12).

e. If necessary, pull out hand throttle control (2) until tachometer (4) indicates an operating range of 700-800 rpm (800-1000 rpm on M939A2 vehicles).

f. Allow engine to warm up approximately ten minutes. If engaged, disengage hand throttle control (2) by rotating handle and pushing in to allow engine speed to drop to idle after warmup period.

g. Stop engine immediately by pulling out EMERGENCY ENGINE STOP control (7) if at any time one or more of the following conditions arise:

(1) Noisy engine and/or excessive engine vibration.

(2) Engine oil pressure does not register, or suddenly drops to less than 15 psi (103 kPa) on M939/A1 series vehicles or 10 psi (69 kPa) on M939A2 series vehicles, as indicated by engine oil pressure gauge (6).

(3) Sudden increase in engine coolant temperature as indicated by engine coolant temperature gauge (5). Normal operating temperature is 175°-200°F (79°-93°C).

(4) Engine continues to run after ignition switch (1) and battery switch (3) are turned to OFF positions.

h. After an emergency shutdown, the engine will not restart until unit maintenance resets the fuel shutoff valve. On M939A2 series vehicles, position EMERGENCY ENGINE STOP control (7) to reset.

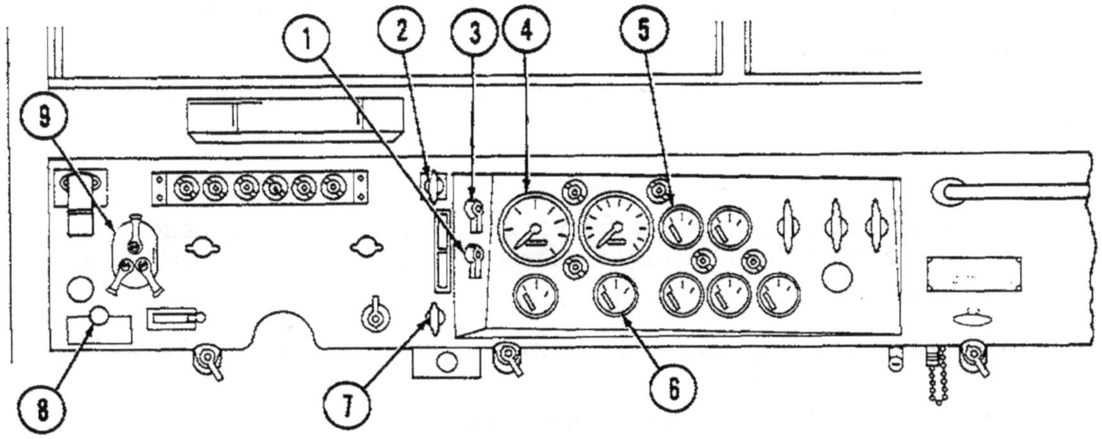
1
2
3
4
5
6
7
8
9

2-14. CTIS OPERATION

NOTE

If CTIS is not operational, refer to paragraph. 3-16.

a. Hwy Mode. CTIS highway selection is automatically programmed Hwy (4) when you start the engine.

b. X-C Mode. When the mission requires off-road driving, select X-C (cross-country) by depressing X-C (3) on the selector panel (1).

c. Sand Mode. When the mission requires driving in sand, snow, or mud, select SAND by depressing SAND (2) on the selector panel (1).

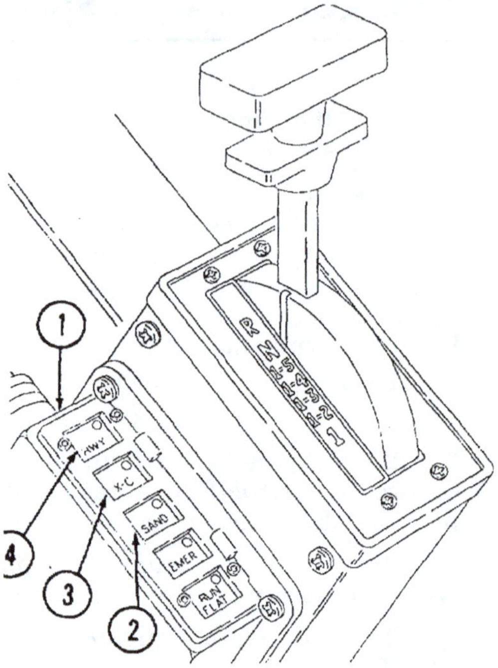
1
4
3
2
HWY
X-C
SAND
EMER
RUN FLAT

2-15. PLACING VEHICLE IN MOTION

WARNING

Do not put vehicle in motion until warning light goes out and alarm (buzzer) stops sounding. Air pressure gauge should indicate at least 90 psi (621 kPa). If warnings continue beyond three minutes, and/or pressure gauge does not reach 90 psi (621 kPa), turn ignition switch and battery switch to OFF positions and notify your supervisor. Failure to do this may cause injury or death.

a. Be sure all auxiliary equipment and tools are stored and locked.

b. Start engine. Refer to paragraph 2-12 for instructions.

c. Set vehicle lights for operating conditions. Refer to paragraph 2-18 for light switch operating instructions.

CAUTION

Do not shift transfer case shift lever from high range to low range, or low range to high range, unless transmission selector lever is in N (neutral).

d. With transmission selector lever (2) in N (neutral), select transfer case driving range:

(1) Depress lockout switch (1) and place transfer case shift lever (3) down to HIGH range for normal driving conditions.

2-15. PLACING VEHICLE IN MOTION (Contd)

(2) Depress lockout switch (1) and place transfer case shift lever (3) up to LOW range if vehicle is heavily loaded, facing a steep grade, and/or operating off-road.

e. Apply service brake pedal (6).

f. Release parking brake lever (4) by pushing forward to the floor.

g. Engage transmission with transmission shift lever (2).

(1) Select **1-5** (drive) if operation is on good roads and/or is on minimal grades.

(2) Select **1-4** (fourth) setting if operation is over moderately hilly road grades and/or is on restricted road speeds.

(3) Select **1-3** (third) setting if speed limits are low.

(4) Select **1-2** (second) setting if operation is over steep grades and/or is on rough terrain.

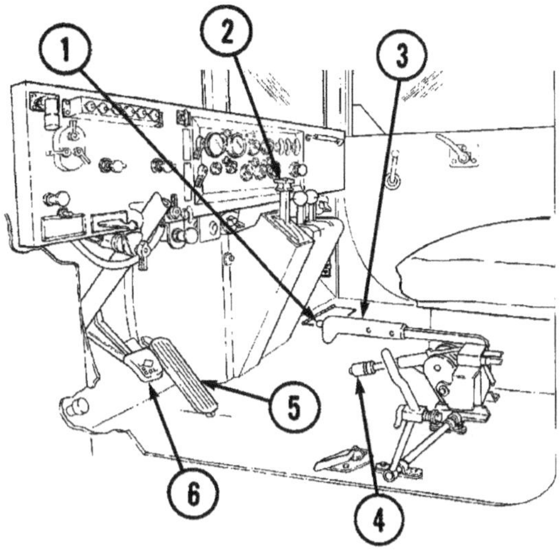
1
2
3
4
5
6

(5) Select **1** (first) setting if operation is under heavy loads, on extreme grades, and/or is on rough terrain.

h. Release brake pedal (6) and depress accelerator pedal (5). Accelerate at a safe, steady speed.

WARNING

- Do not use hand throttle while driving. The hand throttle will not disengage when brakes are applied. Failure to do this will result in injury or death.
- Do not drive too fast for road or weather conditions. The maximum safe speed limit for highway is 55 mph (88 km/h) for vehicles equipped with Antilock Brake System (ABS) and 40 mph (64 km/h) for vehicles not equipped with ABS.

CAUTION

- Do not allow engine speed to exceed 2100 rpm in any transmission gear ratio.
- Do not accelerate at full power when downshifting or upshifting to and from forward driving range **1** (first).

i. Upshift or downshift transmission selector lever (2), as necessary, whenever driving conditions change.

2-15. PLACING VEHICLE IN MOTION (Contd)

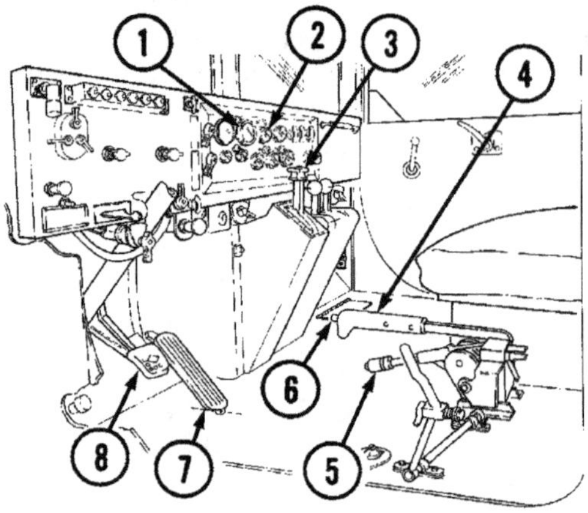
1
2
3
4
5
6
7
8

CAUTION

- Do not shift transfer case shift lever when transmission is in gear. Transmission selector lever must be in N (neutral) before shifting transfer case shift lever from high range to low range, or low range to high range.
- Never shift transfer case shift lever from high range to low range until vehicle is slowed down to 22 mph (35 km/h) or less.

j. Shift transfer case shift lever (4) as required by vehicle speed and changes in driving conditions.

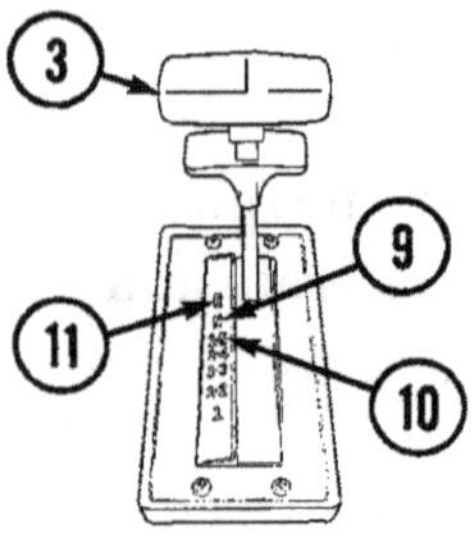
3
9
11
10

(1) To shift transfer case shift lever (4) from high range to low range when vehicle is in motion:

(a) Slow vehicle to 22 mph (35 km/h) or less as indicated by speedometer (1).

(b) Shift transmission selector lever (3) to N (neutral) (9).

NOTE

If gears do not mesh smoothly, return transfer case shift lever to neutral and attempt to re-engage low range.

(c) Press lockout switch (6) with thumb and shift transfer case shift lever (4) from high range to low range.

(d) Release transfer case shift lever (4) and shift transmission selector lever (3) into 1-5 (drive) (10) position.

(2) To shift transfer case shift lever (4) from low range to high range when vehicle is in motion:

(a) Shift transmission selector lever (3) to N (neutral) (9).

(b) Press lockout switch (6) and shift transfer case shift lever (4) to high.

(c) Release transfer case shift lever (4) and shift transmission selector lever (3) into 1-5 (drive) (10).

k. Apply brake pedal (8) as needed when going down a grade.

2-15. PLACING VEHICLE IN MOTION (Contd)

l. To drive in reverse:

WARNING

Do not back up without a ground guide. Failure to do this may result in damage to vehicle, injury, or death.

(1) Stop vehicle (para. 2-16).

CAUTION

Do not back up with transfer case shift lever in low range.

(2) Place transmission selector lever (3) in R (reverse) (11).

(3) Have ground guide direct backup operation.

2-16. STOPPING THE VEHICLE AND ENGINE

a. Release accelerator pedal (7).

NOTE

This warning applies to vehicles not equipped with ABS. To stop a vehicle equipped with ABS, perform step C.

WARNING

Pump brakes gradually when slowing or stopping vehicle on ice, snow, or wet pavement. Sudden stop will cause vehicle wheels to lock, engine to stall, and loss of power steering. Failure to pump brakes may result in injury or death.

b. Apply brake pedal (8) to bring vehicle to a gradual stop.

WARNING

Do not pump brakes that are locking on a vehicle equipped with ABS when stopping. ABS will automatically release wheels that are locking and apply pressure to the other wheels. Failure to do so may result in damage to vehicle or injury or death to personnel.

c. Apply firm steady pressure to brake pedal(s) to bring vehicle to a gradual stop.

d. Move transmission selector lever (3) to N (neutral) (9).

NOTE

Park on hard surface if possible.

e. Apply parking brake by pulling up on parking brake lever (5).

2-16. STOPPING THE VEHICLE AND ENGINE (Contd)

CAUTION

Do not shut down engine if engine coolant temperature gauge reads above 200°F (93°C).

f. Let engine idle for five minutes if engine coolant temperature gauge (4) reads above 195°F (91°C).

CAUTION

Shut down procedure should be strictly adhered to, otherwise damage to ABS ECU may occur if vehicle is so equipped and may also induce faults and ABS valves will not function.

g. Turn vehicle light switch (1) and ignition switch (3) to OFF position. Wait for engine to completely stop before turning battery switch (2) to OFF position.

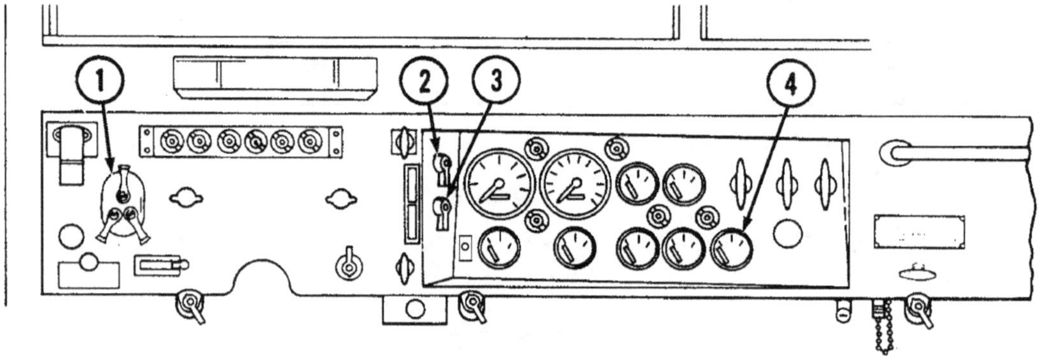
1
2
3
4

CAUTION

Pull out EMERGENCY ENGINE STOP control if engine continues to run after ignition and battery switches are in OFF position. Do not attempt to restart M939/A1 series vehicle engine until unit maintenance has reset fuel cutoff valve.

h. Perform AFTER operation checks and services (table 2-3).

2-17. USING SLAVE RECEPTACLE TO START ENGINE

a. Position right sides of both vehicles together.

b. Stop slaving vehicle engine.

c. Pull covers (6) from slave receptacles (7) of disabled vehicle and slaving vehicle. Receptacle (7) is located below grab handle (5).

CAUTION

Always connect slave cable to disabled vehicle first. Damage to batteries or cable may result from improperly connecting cables.

d. Connect slaving cable (8) between disabled vehicle and slaving vehicle.

2-17. USING SLAVE RECEPTACLE TO START ENGINE (Contd)

NOTE

Ensure all unused electrical switches in both vehicles are off.

e. Start slaving vehicle engine (para. 2-12).

f. Start disabled vehicle engine.

g. After disabled vehicle engine starts, disconnect slaving cable (8) from both vehicles.

h. Put covers (6) back over receptacles (7).

i. Clean and stow slaving cable (8).

j. If voltmeter (4) is not in green area, notify unit maintenance.

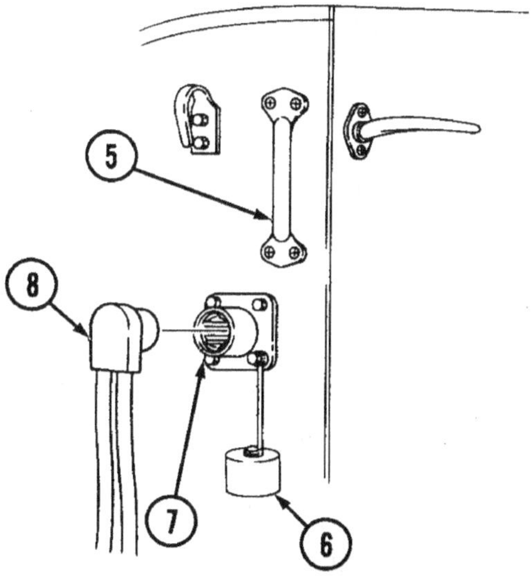
5
8
7
6

2-18. OPERATION OF VEHICLE SERVICE LIGHTS

Table 2-4. Main Light Switch Logic Table.

MAIN LIGHT SWITCH OPERATION						
LEVER POSITION						
SER DRV	PARK	STOP TURN	OFF	BO MKR	BO DRV	SYSTEMS OPERATED
X				X	X	PANEL LIGHT
X		X				SERVICE STOP LIGHTS
					X	BLACKOUT DRIVING LIGHTS
				X	X	BLACKOUT MARKERS AND TAILLIGHTS
X		X				SERVICE TURN INDICATOR (LEFT AND RIGHT)
X		X				STOP LIGHTS
X	X					PARKING LIGHTS
X						SERVICE HEADLIGHTS
				X	X	BLACKOUT STOP LIGHTS

2-18. OPERATION OF VEHICLE SERVICE LIGHTS (Contd)

a. Service Lights.

NOTE

Unlock lever must be in UNLOCK position to move main switch to any position other than BO MARKER.

(1) To illuminate instrument panel, turn main switch (4) to either STOP LIGHT, SERVICE DRIVE, BO MARKER, or BO DRIVE position.

(2) To brighten or dim instrument panel illumination, move auxiliary switch (6) to either PANEL BRT or DIM position.

(3) For normal daylight driving, turn main switch (4) to STOP LIGHT position.

(4) For night driving, turn main switch (4) to SERVICE DRIVE position.

(5) In blackout operation:

(a) Turn main switch (4) to BO DRIVE position before driving vehicle.

(b) Turn main switch (4) to BO MARKER position after stopping vehicle.

(6) To illuminate parked vehicle at night, turn main switch (4) to SERVICE DRIVE position and auxiliary switch (6) to PARK position.

b. Turn Signal and Hazard Warning Lights Tab Button.

NOTE

- Turn signal control lever must be moved to center position after completing turn.
- When the hazard warning light/emergency flashers are in use, they override the brake light/stop light operation. When driving, exercise caution and be prepared to use hand signals to indicate a stop.

(1) For right turns, move turn signal control lever (1) upward towards windshield (3). For left turns, move lever (1) downward away from windshield (3).

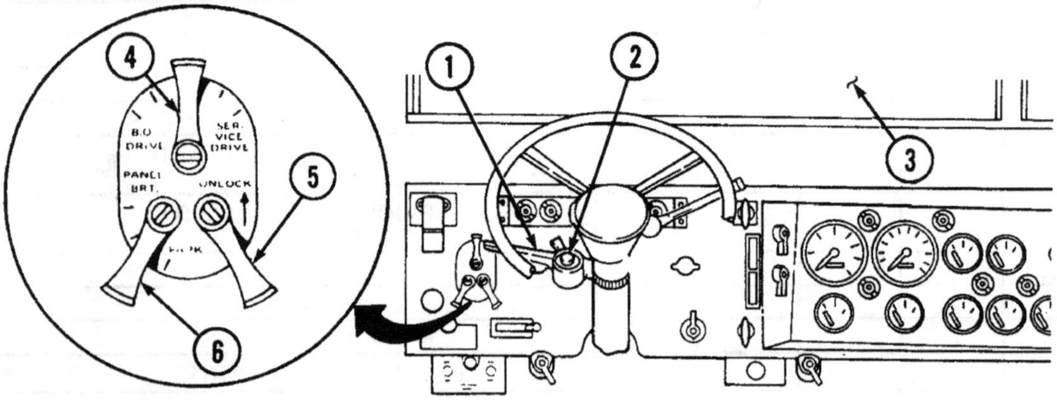
4
B.O DRIVE
SER-VICE DRIVE
PANEL BRT.
UNLOCK
5
6
1
2
3

2-18. OPERATION OF VEHICLE SERVICE LIGHTS (Contd)

(2) For hazard warning lights (blinking lights):

(a) Turn main switch lever (4) to STOP LIGHT position.

(b) Depress hazard tab button (2) and move turn signal control lever (1) up to lock tab button (2) in position.

(c) To deactivate, move turn signal control lever (1) back to neutral. Hazard tab button (2) will automatically disengage.

2-19. RAISING AND SECURING CAB HOOD

a. General. All M939/A1/A2 series vehicles are equipped with a tilt-forward hood which provides easy access to the engine compartment.

b. Raising and Securing Hood.

(1) Release left and right hood latches (7).

(2) Remove pin (11) from hood bracket (9) and swing bar (12) out.

(3) Pull hood forward by grasping hood handle (8).

WARNING

Ensure pin is placed in end of retaining bar. Failure to do so may damage vehicle, or cause injury or death.

(4) Once hood is raised, secure bar (12) to bumper bracket (10) with pin (11).

c. To lower and secure hood, reverse steps of task b, steps 1 through 4.

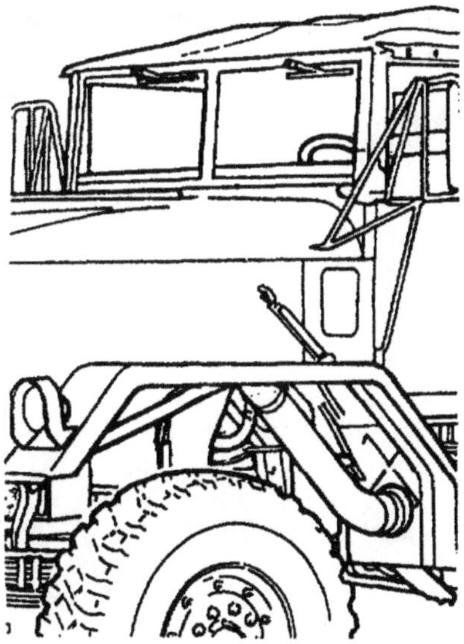

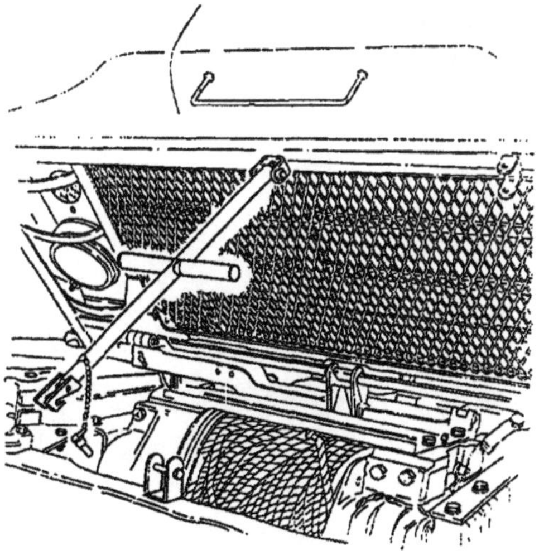

2-20. TOWING WITH TOWBAR

WARNING

Personnel must not occupy vehicle in tow. Injury or death may result.

CAUTION

- Do not use towing as a means to start engine of vehicle with automatic transmission.
- Damage to automatic transmission of disabled M939/A1/A2 series vehicles will result from towing unless transmission, transfer case, and PTO are all in neutral. Refer to disabled vehicle operator's manual for towing instructions.
- When towing M939/A1/A2 series vehicles with inoperative compressed air system, spring brakes must be caged prior to towing. Refer to paragraph 3-13.
- Maximum towing speed shall not exceed 35 mph (56 km/h) on paved highway or 15 mph (24 km/h) on off-highway surfaces.

NOTE

When towing M939/A1/A2 series vehicles, normal towing procedures require removal of propeller shaft.

(a) Remove two lifting shackles (3) from front bumper (5) of vehicle to be towed and store in safe place.

(b) Install clevis (2) end of towbar (7) on front bumper shackle brackets (4) and secure in place with clevis bolt (6) and safety pin (1).

2-20. TOWING WITH TOWBAR (Contd)

(c) Install yoke (12) end of towbar (7) to pintle hook (13) of towing vehicle.

(d) Loosely install a utility chain (9) through front springs (8) of disabled vehicle and around frame (11) of towing vehicle. Make sure utility chain (9) is clear of any light brackets or wiring.

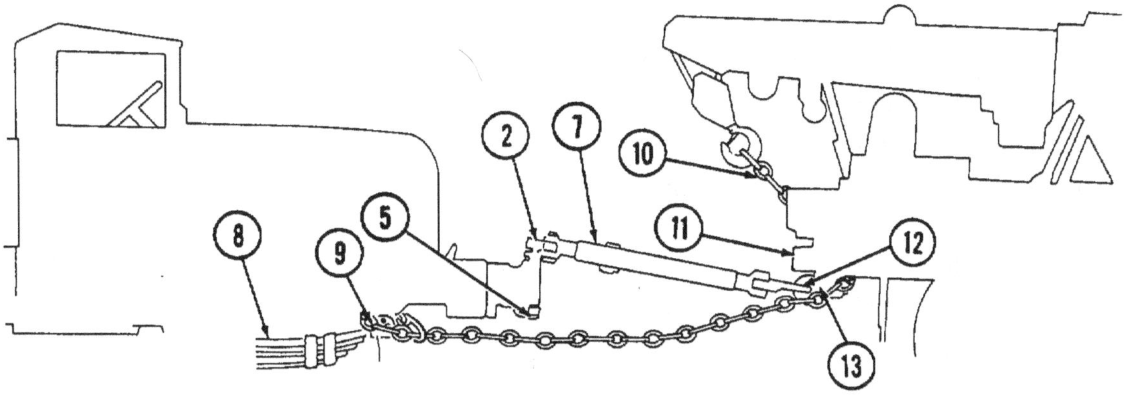
2
7
10
5
11
8
9
12
13

WARNING

If vehicle being towed has inoperative compressed air system, emergency air and service air lines must not be connected between vehicles. Failure to do this may result in damage to equipment, or cause injury or death.

(e) Connect emergency air line (15) and service air line (17) to half couplings (14) and (16) on each vehicle.

(f) Release parking brake (20) and place transmission selector lever (18), transfer case shift lever (19), and power takeoff lever (21) in neutral on disabled vehicle.

(g) Turn on hazard warning lights on both towing and disabled vehicles (para. 2-18).

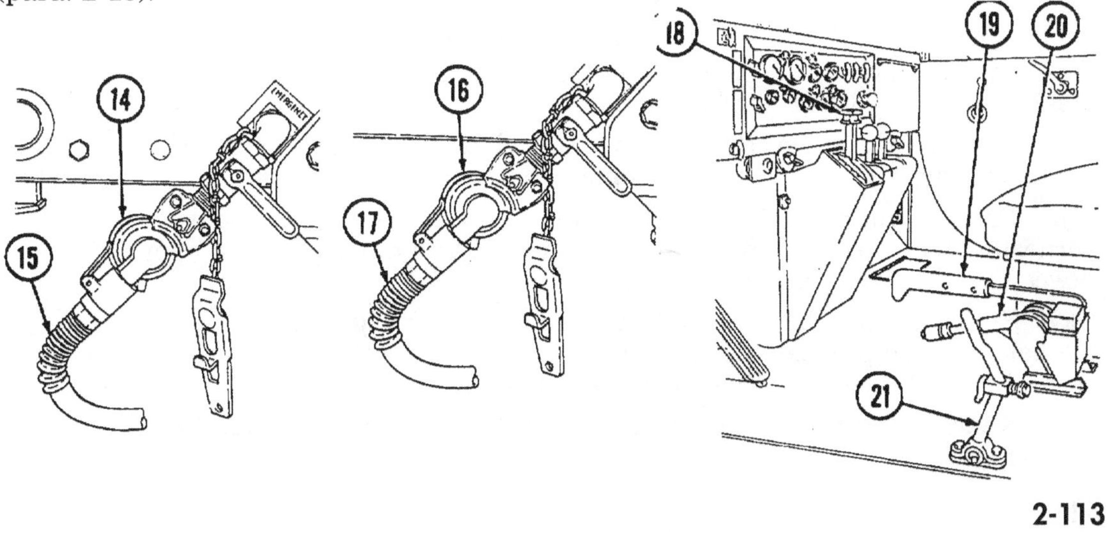
14
15
16
17
18
19
20
21

2-21. RAISING WINDSHIELD AND INSTALLING CAB TOP

NOTE

This operation is best accomplished by the operator and one crewmember.

a. Release windshield catches (3) and raise windshield frame (1) to vertical position.

b. Tighten knobs (4) on left and right sides of windshield frame (1).

c. Secure windshield catches (3) to hood (2) mounts.

d. Lower cab windows.

e. Install two pillar posts (5) in rear corners of cab (10).

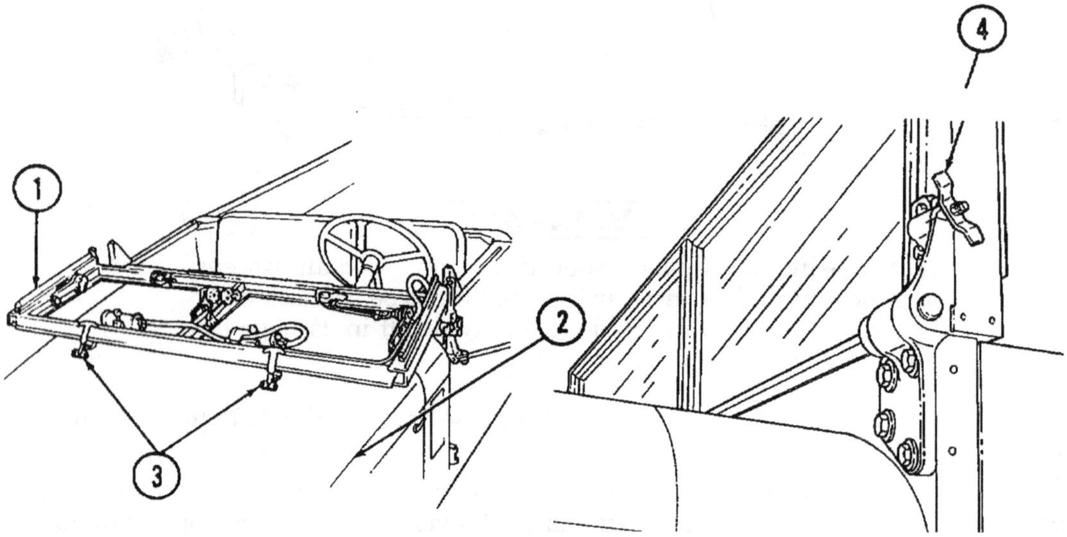
4
1
2
3

f. Insert crossbow (8) in roof rail bows (6) and crossbow (7) with stave holes in curved portion of pillar posts (5).

g. Insert roof rails (12) on pillar posts (5) and windshield frame (1) and push roof rail catch (11) into windshield frame (1) to lock catch (11).

h. Install overhead staves (14) by placing washer end of staves (14) in windshield frame (1) and other end in top crossbow (7).

i. Place tarpaulin top (13) on hood (2) of vehicle, and slide front edge of cab top (13) sideways into windshield channel (15) from either side of windshield frame (1).

j. Pull tarpaulin top (13) towards rear of cab (10) over windshield frame (1), overhead staves (14), and crossbows (7) and (8). Place inner flap of cab top (13) between seats (9) and inside of cab (10).

k. Slide right and left edge of tarpaulin top (13) in right and left pillar post channels (20) and pull cab top (13) down to back of cab (10). Make sure inner flap of tarpaulin top (13) slides behind seats (9) evenly.

l. Secure top edge of tarpaulin top (13) to roof rail (12) with turnbutton fasteners (16).

2-21. RAISING WINDSHIELD AND INSTALLING CAB TOP (Contd)

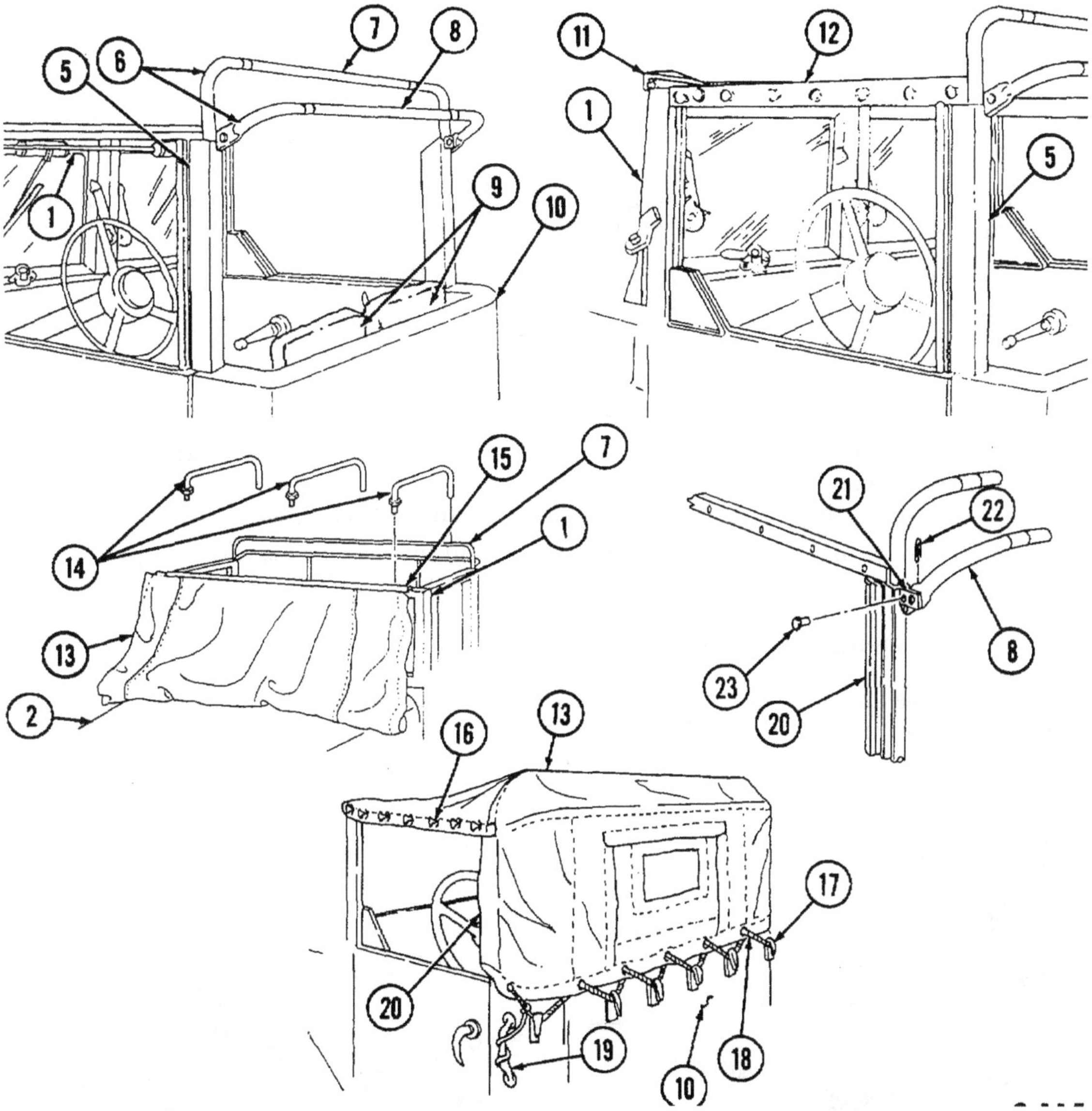
5
6
7
8
1
9
10
11
12
1
5
15
7
14
1
13
2
21
22
8
23
20
16
13
17
20
19
10
18

m. Remove retaining clip (22) and pin (23) from crossbow (8) and push movable crossbow (8) outward to take up slack of cab top (13). Push pin (23) through crossbow (8) and bracket (21) holes and push retaining clip (22) through hole in pin (23).

n. Thread rope (18) through tarpaulin top (13) holes and around lashing hooks (17). Tie ends of rope (18) to grab handles (19) on each side of cab (10).

o. To remove cab top, reverse steps e. through n.

p. Clean and fold tarpaulin top (13). Do not fold or stow when wet. Refer to paragraph 2-8 for tarpaulin cleaning procedures.

q. Store tarpaulin top (13), overhead staves (14), crossbows (7) and (8), and pillar posts (5).

2-22. OPERATION OF FRONT WINCH

NOTE

All winching and recovery operations will be performed IAW FM 20-22.

a. Preparation for Use.

(1) Park vehicle directly facing object to be winched, if possible. If vehicle cannot be parked in a direct line with object to be winched, refer to task e.

(2) Place transmission selector lever (3) in N (neutral).

(3) Apply parking brake (4).

(4) Turn ignition switch (1) and battery switch (2) to OFF position.

CAUTION

- Before opening reservoir, make sure area around reservoir filler cap is clean. Do not allow dirt, dust, or water to enter reservoir. Failure to do this may cause damage to internal components.
- Do not proceed with winch operation if oil level is less than halfway from end of dipstick to FULL mark on wrecker models or in red area of dipstick on all other models. Damage to internal components may result.

(5) Check oil level in hydraulic oil reservoir (5). Refer to paragraph 2-4 for locations of hydraulic oil reservoir.

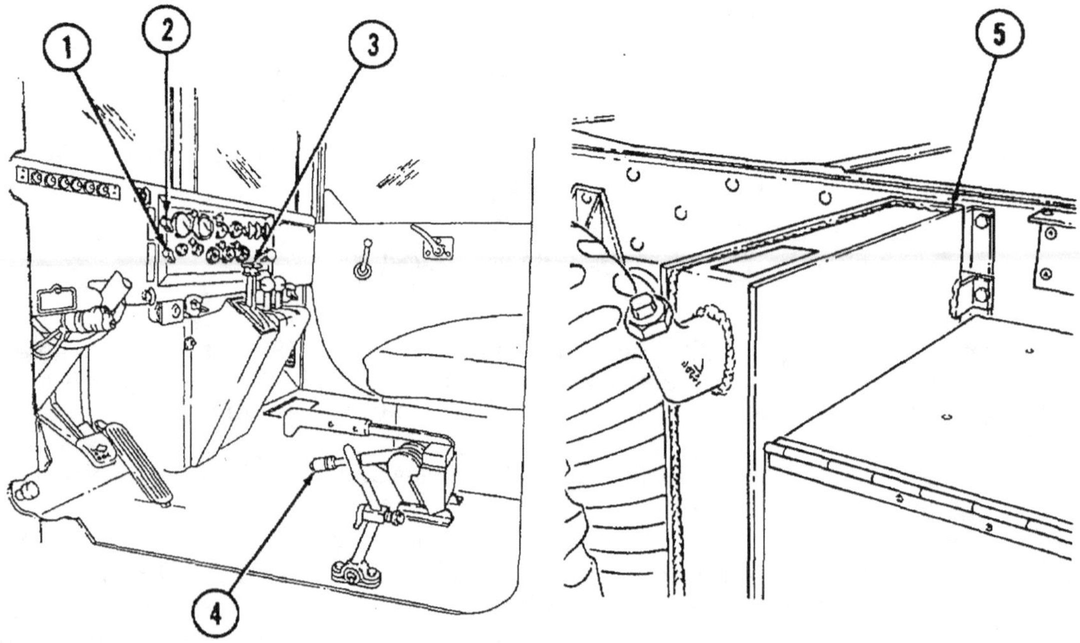
1
2
3
4
5

2-22. OPERATION OF FRONT WINCH (Contd)

b. Unwinding Winch Cable.

(1) Free winch cable chain (6) and hook from vehicle.

(2) Pull out drum lock knob (8), rotate 90 degrees, and release.

WARNING

Wear hand protection when handling winch cable. Do not handle cable with bare hands. Broken wires may cause injury.

CAUTION

- Do not wind out winch cable when attached to load. Load must be wound in only, except when using A-frame kit. Failure to do this may cause damage to winch brakedrum.
- Leave at least four turns of cable on winch drum. Refer to table 1-2 for winch load capacities. Failure to do this may cause damage to winch.

NOTE

M936A2 model vehicles do not have a level wind.

(3) On M936/A1 model vehicles:

(a) Pull out level wind lock knob (7), rotate 90 degrees, and release.

(b) Pull out cable tensioner lock knob (10) with left hand and push tensioner lever (9) toward the left side of the vehicle with right hand. Release lock knob (10).

(4) Pull out required length of cable. Do not allow cable to knot or kink.

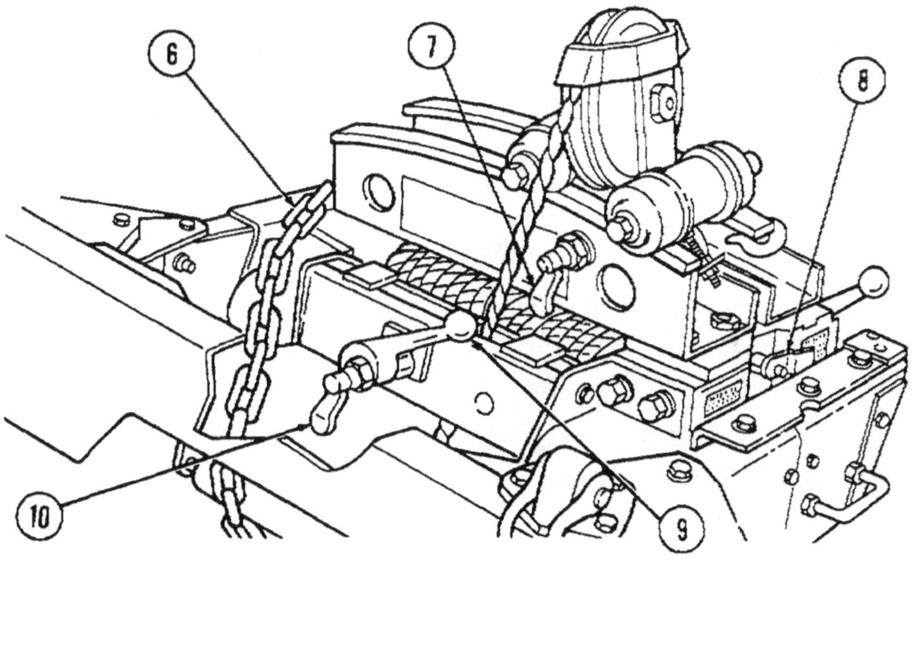
6
7
8
10
9

2-22. OPERATION OF FRONT WINCH (Contd)

c. Rigging the Load.

(1) Attach utility chain (4) to lifting shackles (5) or pintle hook of load.

(2) If load is very heavy or deeply mired, install a snatch block (3) to increase winch pulling power.

(3) To rig a snatch block (3):

(a) Unwind enough cable (7) to reach the load and back to the front winch. Attach cable chain hook (2) to lifting shackle (1).

(b) Turn snatch block hook (8) to the right. Lift up rear of snatch block (3) and open support link (6). Insert cable (7). Lift up rear of snatch block (3) to lower and lock support link (6) to snatch block hook (8). Return hook (8) to original position.

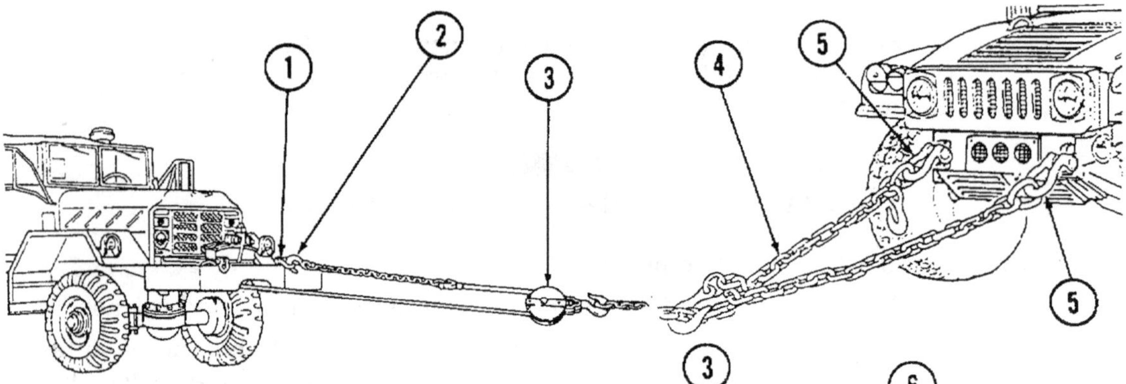
1
2
3
4
5
5

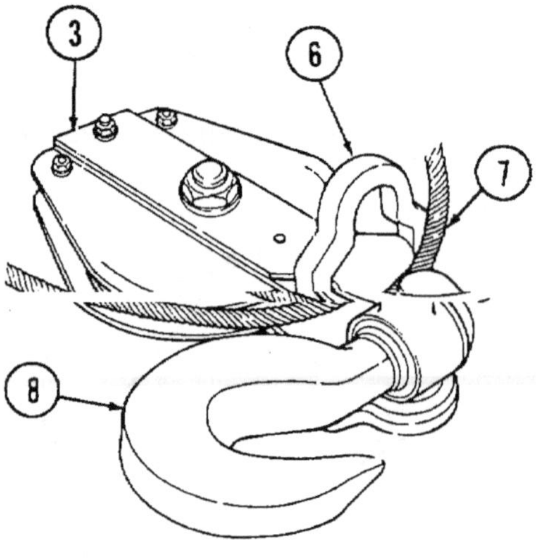
3
6
7
8

(c) Attach utility chain (4) to lifting shackles (5) or pintle hook of load. Attach snatch block (3) to utility chain (4).

NOTE

- M936A2 wreckers do not have front anchors for field chocks.
- M936 wreckers are equipped with field chocks for heavy recovery operations. Refer to paragraph 2-24 for field chock installation.

(4) Disengage brakes, transmission, and transfer case of vehicle being retrieved.

d. Pulling Load.

NOTE

This operation requires two crewmembers.

2-22. OPERATION OF FRONT WINCH (Contd)

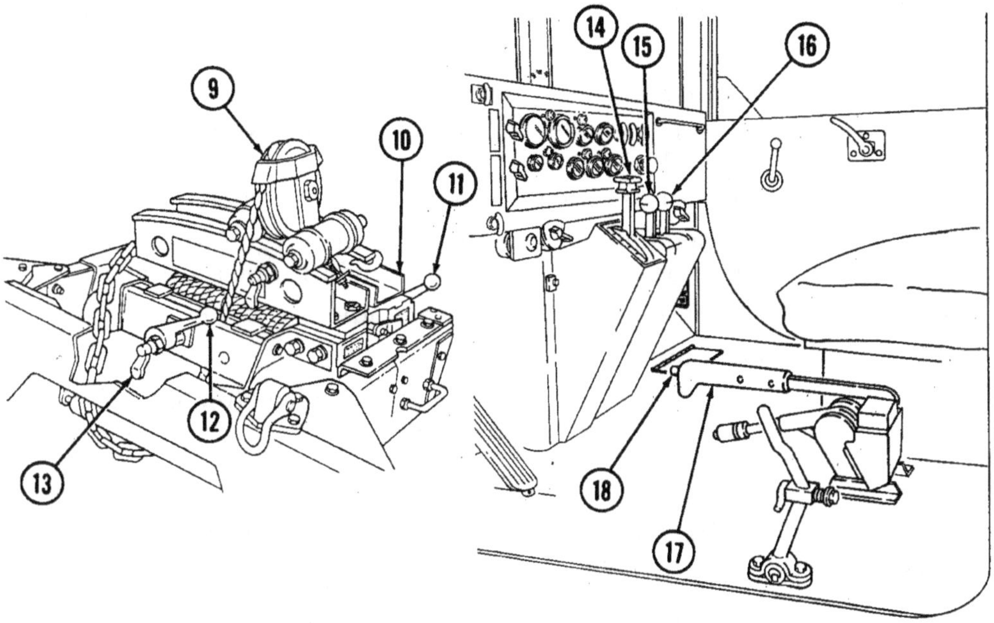
9
10
11
12
13
14
15
16
17
18

(1) Start engine (para. 2-12).

(2) Release hinged latch (10) and pull clutch lever (11) as far back as it will go.

(3) On vehicles with level wind device (9), pull out lock knob (13) and tensioner lever (12) back. Align lock knob (13) with hole in housing and release.

(4) Press lockout switch (18) and shift transfer case shift lever (17) into high range.

WARNING

- Direct all personnel to stand clear of winch cable and vehicle when engaging transmission or transfer case. Failure to do so may result in injury or death.
- Do not operate winch erratically. Erratic winding will result in a snapped cable, causing injury or death.

CAUTION

- If temperature is above 70°F (21°C), stop winding operation for six minutes every 100 ft (30.5 m) of cable winched in, and leave engine and power takeoff engaged. Failure to do so may cause damage to winch.
- Do not operate winch when engine is running over 1800 rpm. Damage to equipment will occur.

NOTE

2-22. OPERATION OF FRONT WINCH (Contd)

Ensure each layer of cable winds evenly, if not equipped with level wind.

(5) With parking brake applied, place transmission selector lever (2) in 1-5 (drive) and pull transmission power takeoff control lever (3) back to ENGAGE. Return transmission selector lever (2) to N (neutral).

(6) Pull front winch control lever (4) back to WIND and hold.

(7) Winding speed and pulling capacity of winch is regulated by engine rpm. To increase, depress accelerator pedal (5) or adjust hand throttle control (1).

(8) Release winch control lever (4) to stop winding.

e. Pulling Indirect Loads.

(1) If vehicle (11) cannot be lined up straight with load (6), line vehicle (11) up with a reliable go-between such as a large tree (10).

CAUTION

Do not perform front winch operation if direct pull or use of a go-between object is unavailable.

(2) Unwind enough cable (7) to reach tree (10) and load (6). Refer to task b. for instructions on unwinding cable (7).

(3) Attach snatch block (8) to cable (7) (task c).

(4) Rig chain (9) from tree (10) to the snatch block (8). Attach cable chain to pintle hook or lifting shackles of load (6) (task c).

(5) Wind cable (7) until cable chain reaches snatch block (8) (task d).

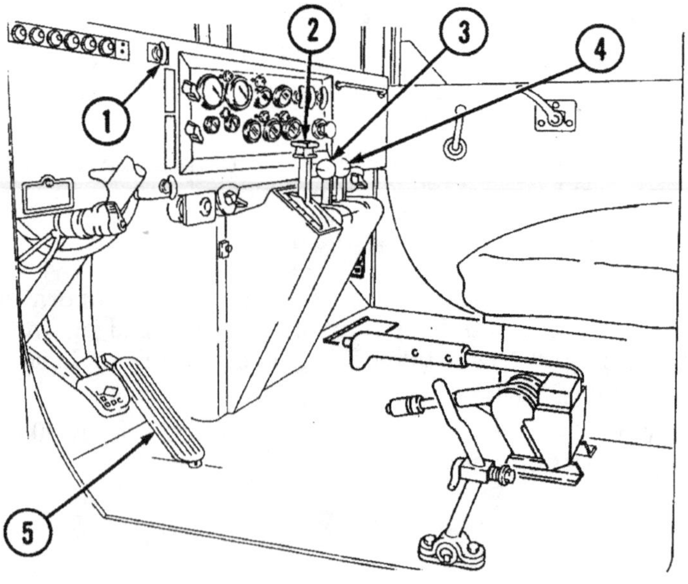
1
2
3
4
5

2-22. OPERATION OF FRONT WINCH (Contd)

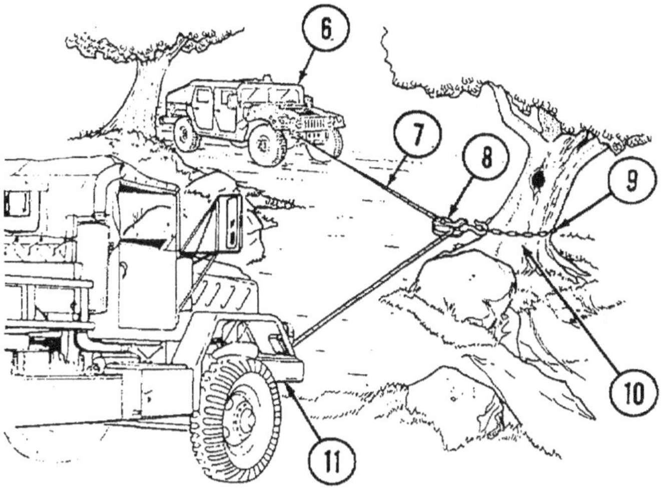
6
7
8
9
10
11

NOTE

If load is on a slope, block wheels of lead before loosening cable.

(6) Briefly push front winch control lever (4) forward to WINCH. Cable (7) will unwind (loosen) to permit removal of snatch block (8).

(7) Remove snatch block (8) and utility chain (9).

(8) Continue winding operation.

f. Lifting and Lowering Loads.

(1) A-frame kit is installed and rigged by unit maintenance.

WARNING

- Vehicle will become charged with electricity if A-frame contacts or breaks high-voltage wire. Do not leave vehicle while charged with high-voltage. Notify nearby personnel to have electrical power turned off. Failure to do this may result in injury or death.
- Do not lower load without a ground guide. Direct all personnel to stand clear of lifting operation. Swinging loads may cause injury or death.

CAUTION

- Do not winch out line loads for distances greater than 10 ft (3 m), as this may result in damage to winch brakedrum.
- Do not attempt to lift loads heavier than 3,000 lb (1,362 kg), as this may result in damage to the A-frame kit.

(2) Rig winch to load (task c).

(3) To lift load, follow same winding instructions as in task d.

2-22. OPERATION OF FRONT WINCH (Contd)

(4) To lower load:

(a) Push front winch control lever (2) forward to WINCH.

(b) Observe directions of ground guide.

(c) After load has been lowered, release front winch control lever (2) to NEUTRAL.

(d) Direct ground guide to maintain tension on cable while unrigging load.

g. After Winch Operation.

(1) Direct ground guide to maintain tension on cable.

(2) Pull front winch control lever (2) back to WIND.

(3) Watch ground guide for signal indicating cable chain coupling is approaching drum.

(4) Release winch control lever (2) when signalled by ground guide.

(5) Direct crewmember to:

(a) Disengage drum clutch by pushing clutch control lever (4) toward the winch.

CAUTION

Do not force clutch control lever. If lever does not easily disengage, slightly engage winch control lever in WIND until clutch control lever returns without force.

(b) Swing hinged latch (3) down to lock clutch control lever (4) in disengaged position.

(c) Pull out drum lock knob (5), rotate 90 degrees, and release. If necessary, rotate drum by hand to allow drum lock plunger to engage.

2-22. OPERATION OF FRONT WINCH (Contd)

(d) On winches with level wind device, manually push level wind (12) completely to the operator's side of level wind (12) frame. Pull out level wind lock knob (13), rotate 90 degrees, and release. If necessary, adjust level wind (12) to assure lock plunger engages.

(6) Push transmission power takeoff control lever (1) forward to DISENGAGE.

h. Preparing Winch for Travel.

(1) On winches without level wind:

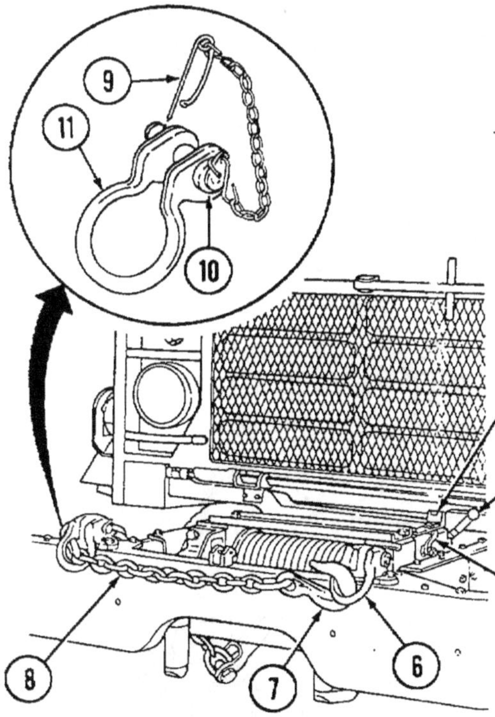
9
11
10
3
4
5
8
7
6

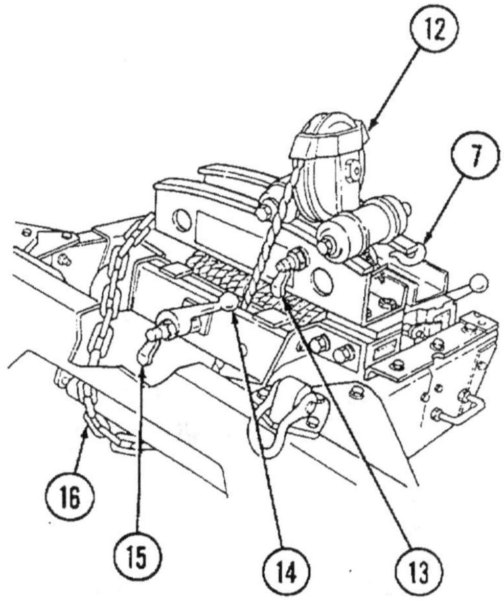
12
7
16
15
14
13

(a) Put cable chain (8) under and over right frame extension, then across top of bumper. Attach cable hook (7) to left lifting shackle (6).

(b) Remove right lifting shackle (11) by unsnapping pin lock (9) and removing shackle bolt (10). Place chain (8) through lifting shackle (6) and reinstall shackle (11).

(2) On winches with level wind:

(a) Pull cable chain (16) up through the space between bumper and winch.

(b) Wind cable chain (16) around level wind (12) frame and attach hook (7) to frame.

(c) Pull out cable tensioner lock knob (15) with left hand and push tensioner lever (14) toward left of vehicle with right hand. Release lock knob (15).

2-23. OPERATION OF CARGO TRUCKS

a. General. M923/A1/A2 and M925/A1/A2 cargo trucks have 7 x 14 ft (2.1 x 4.3 m) cargo beds. M927/A1/A2 and M928/A1/A2 cargo trucks have 7 x 20 ft (2.1 x 6.1 m) beds. All can be equipped with bow and tarp kit.

b. Lowering and Raising Tailgate.

WARNING

On dropside trucks, make sure forward end of dropsides are secured before lowering tailgate. Failure to do this may result in injury or death.

(1) On fixed-side vehicles, remove hooks (2) from retainer slots on both sides of tailgate (1). On dropside trucks, turn locking handles (3) on both sides of tailgate (1) counterclockwise to loosen. Grasp ring (4) and turn T-bolt (7) 90°. Remove locking handles (3).

(2) Lower top of tailgate (1). Do not drop tailgate (1).

(3) Reverse steps 1 and 2 to raise tailgate (1).

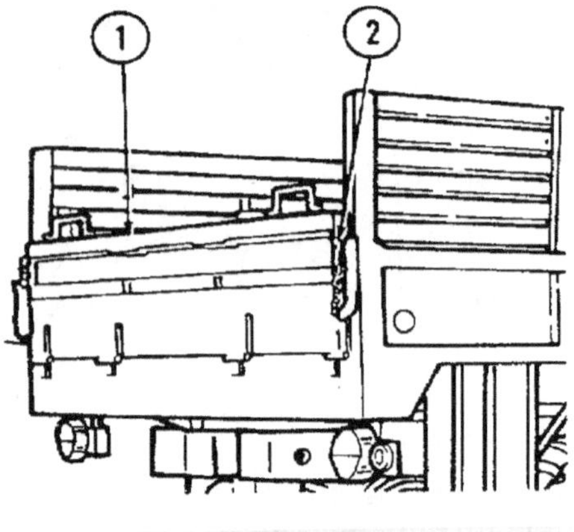
1
2

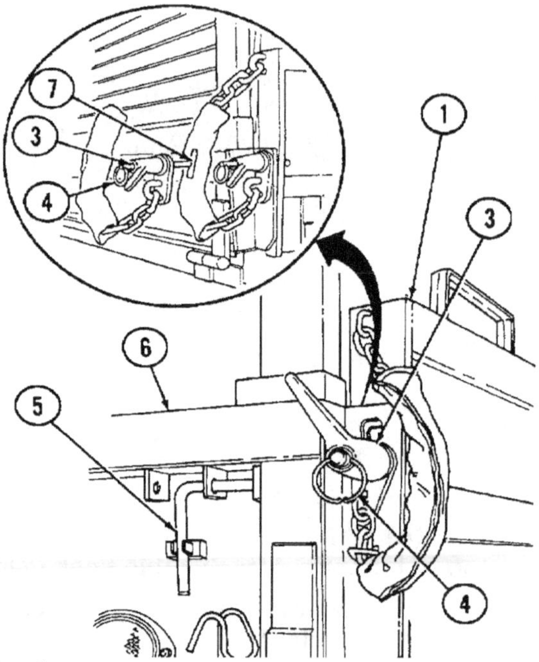
7
3
4
1
3
6
5
4

c. Lowering and Raising Troop Seats.

(1) To lower troop seats (15), pull troop seat supports (10) forward 45 degrees, release latches (9), and lower seats (15).

(2) Adjust troop seat supports (10) to contact both side and floor of vehicle.

(3) On dropside trucks, install troopseat locking rods (11) in hole (14) near tailgate (1). Locking rod (11) can be shortened or lengthened. To adjust locking rod (11):

(a) Loosen locknut (12).

(b) Turn end (13) clockwise to shorten; counterclockwise to lengthen.

(c) Tighten locknut (12).

2-23. OPERATION OF CARGO TRUCKS (Contd)

(4) To raise troop seat (15), reverse steps 1 through 3.

d. Removing Front and Side Racks.

(1) Lower tailgate (1). Refer to task b.

(2) If troop seats are lowered, raise troop seats (15) and secure in place with latches (9). Refer to task c.

NOTE

This operation requires two crewmembers.

(3) On dropside trucks:

(a) Remove troop seat locking rods (10) from holes (14) near tailgate (1) and secure to side rack clip (8).

(b) Raise tailgate (1) (task b).

(c) Pull back four troop seat securing pins (5) from corners of dropsides (6).

(d) Remove retaining clip (16) from anchor pin (17).

(e) Lift and remove side rack (19).

(f) Repeat steps c and d for opposite side rack (19).

(4) Lift and remove front rack (18).

(5) Lift and remove side racks (19).

e. Installing Front and Side Racks.

NOTE

When installing front rack, be sure front rack retainer clip is inserted in side rack rings.

(1) Reverse steps 1 through 5 of task d.

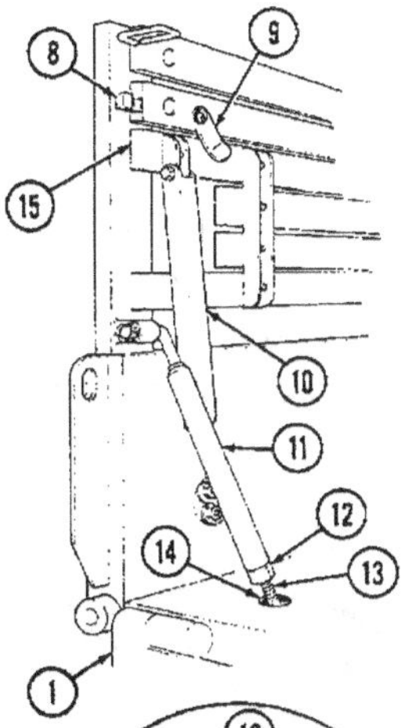
8
9
15
10
11
12
14
13
1

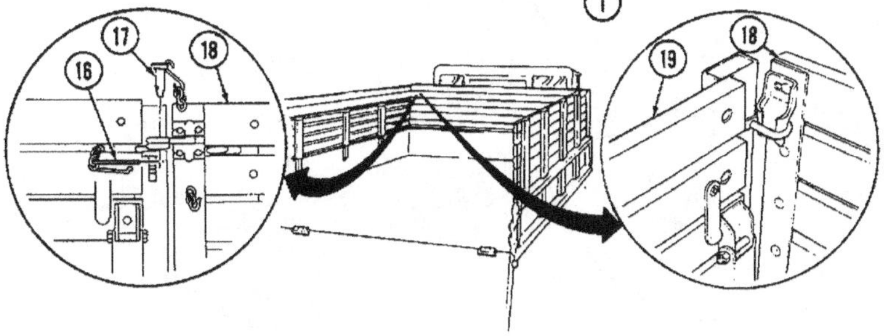
17
18
16
19
18

2-23. OPERATION OF CARGO TRUCKS (Contd)

f. Lowering and Raising Dropsides.

(1) Park vehicle where it can best be loaded or unloaded. Turn ignition switch and battery switch to OFF and apply parking brake.

(2) Turn locking handle (4) counterclockwise to loosen grasp ring (5) and turn T-bolt (7) 90°. Remove locking handle (4) and repeat operation on opposite end of dropside (2).

WARNING

- Troop seats, side rack braces, bows, side racks, and troop seat securing pins must be secured in stowed position before lowering dropside. Failure to do so may result in injury or death.
- Make sure side panel front locks are secured before lowering tailgate or dropsides will fall. Failure to do so may result in injury or death.

NOTE

This operation requires two crewmembers.

(3) Grasp side rack (3) and carefully lower dropside (2).

(4) To load from both sides:

(a) Lower tailgate (1) (task b).

(b) Remove locking handles (4) securing dropside (2) and lower dropside (2). Repeat operation for other side.

(5) To raise dropside (2):

(a) Raise dropside (2) and install T-bolt (7) at forward end. Place T-bolt (7) in slot, turn 90 degrees and hold in position with ring (5). Turn locking handle (4) clockwise to secure.

(b) Raise opposite dropside (2) and secure it in position.

(6) Raise tailgate (1) and install left rear and right rear locking handles (6).

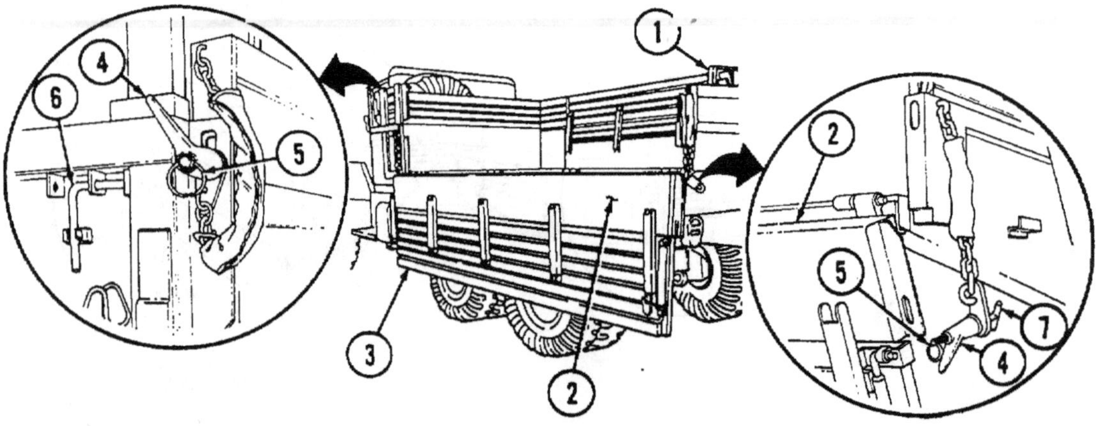
4
6
5
1
3
2
2
5
7
4

2-24. OPERATION OF MEDIUM WRECKER

a. General. The medium wrecker (M936/A1/A2) has a hydraulic crane and front and rear winches. The vehicle's winch and towing capacities are adequate for recovering all wheeled vehicles. The medium wrecker can also remove and replace engines, power packs, and gun tubes.

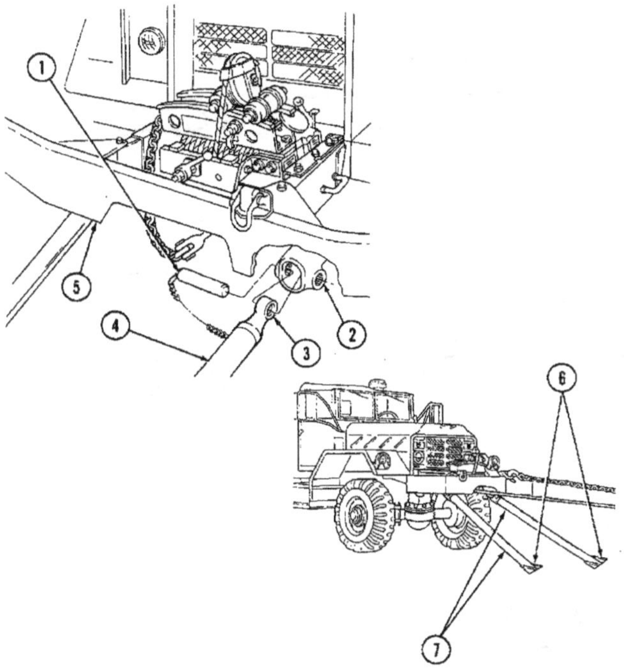
1
2
3
4
5
6
7

NOTE

- M936A2 vehicles do not have a level wind or front anchors for field chocks.
- All winching and recovery operations will be performed IAW FM 20-22.

b. Front Winch Operation.

(1) Refer to paragraph 2-22 for operating instructions.

(2) Install field chocks (4) for heavy recovery operations or operations on slippery terrain. To install chocks (4):

(a) Remove chocks (4) from storage area at left rear of wrecker body.

(b) Insert chocks (4) in left and right brackets (2) below the front bumper (5). Insert pin (1) through bracket (2) and yoke (3) to secure chock (4) in place.

(c) Dig two 12 in. (30 cm) holes at spade end (6) of chocks (7). Insert spade ends (6) in holes.

c. Rear Winch Operation.

WARNING

Do not wind out winch cable when attached to load. Load must be wound in only. Failure to do this may result in injury or death.

(1) Position rear of wrecker in direct line with load to be winched if possible.

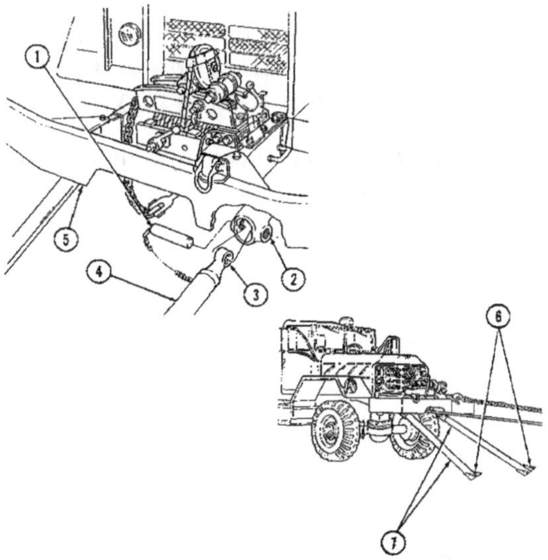
1
2
3
4
5
6
7

2-24. OPERATION OF MEDIUM WRECKER (Contd)

(2) Engage parking brake lever (9) and turn ignition switch (3) and battery switch (2) to OFF.

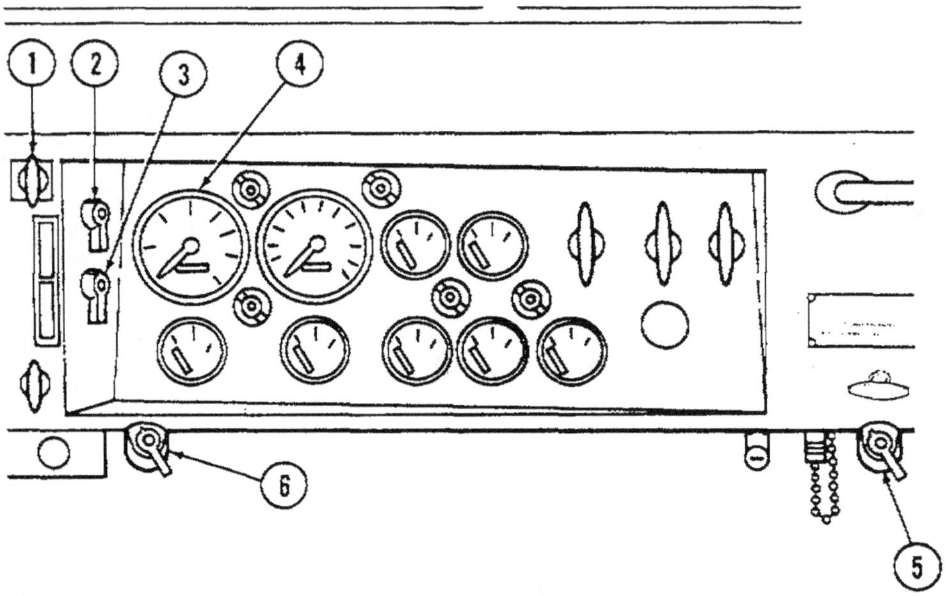
1
2
3
4
5
6

CAUTION

- Before opening hydraulic oil reservoir, make sure area around reservoir filler cap is clean. Do not allow dirt, dust, or water to enter reservoir. Failure to do so may cause damage to internal components.
- Do not proceed with winch operation if oil level is less than halfway from end of dipstick to full mark. Fill as needed. Failure to do so may cause damage to internal components.

(3) Check oil level in hydraulic oil reservoir (12). Refer to LO 9-2320-272-12.

2-24. OPERATION OF MEDIUM WRECKER (Contd)

(4) For heavy pulls, install field chocks (15). Perform the following:

(a) Remove field chocks (15) from storage area at left rear of wrecker body.

(b) For direct pulls, install chocks (15) facing the load in left and right rear chock brackets (13) below bumperettes. Insert pin (16) through bracket (13) and yoke (17) to secure each field chock (15) in place.

(c) For indirect pulls, install field chocks (15) in left side and rear brackets (13) for left side pulls or in right side and rear brackets (13) for right side pulls.

(d) Dig two 12 in. (30 cm) holes at spade end (14) of chocks (15). Insert chock spade ends (14) in holes.

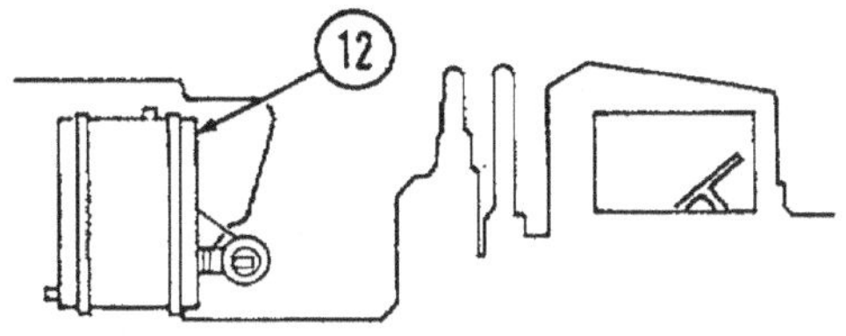
12

(5) Start engine (para. 2-12).

(6) Place transfer case shift lever (8) in NEUTRAL.

(7) Unlock safety latch (11) and push transfer case power take-off lever (10) back to engage.

(8) Place transmission selector lever (7) in 1-5 (drive).

NOTE

Vehicles equipped with automatic throttle kit/MWO will automatically increase engine rpms to the proper range when the PTO is engaged.

(9) Pull out hand throttle control (1) and set engine speed between 1250 and 1300 rpm as indicated by tachometer (4).

(10) Turn on floodlight control switch (5) if operation is at night and tactical situation permits.

(11) Turn on amber warning light switch (6), if required, and if tactical situation permits.

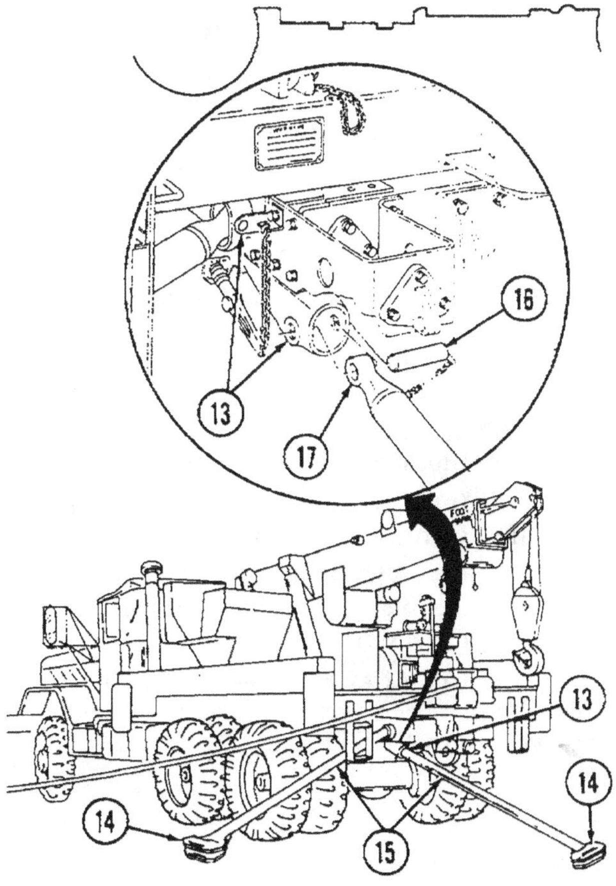
16
13
17
13
14
14
15

2-24. OPERATION OF MEDIUM WRECKER (Contd)

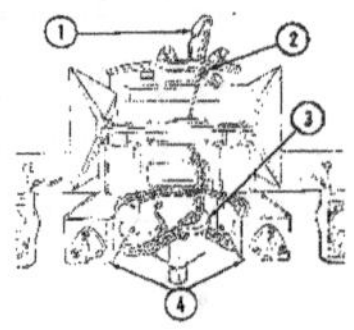
1
2
3
4

(12) Release level wind (1) by pulling out lock knob (2), rotating it 90 degrees, and releasing knob (2).

(13) Turn on floodlights (5) for night operation if tactical situation permits.

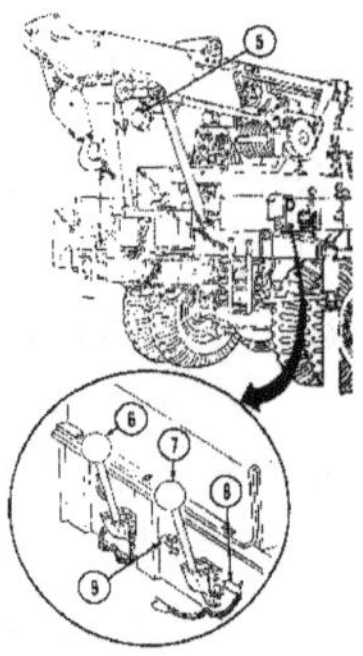
5
6
7
8
5

NOTE

Torque control lever must be in HIGH or LOW before directional control lever can be operated.

(14) Remove travel pin (8) and pull torque control lever (7) outward to HIGH.

WARNING

Wear hand protection when handling winch cable. Do not handle cable with bare hands. Broken wires may result in injury.

NOTE

Cable and snatch block ratings on level surface are 14,500 lb (6,583 kg) for 3/4 in. (19 mm) cable; 22,500 lb (10,215 kg) for single sheave snatch block; 27,500 lb (12,485 kg) for double-sheave snatch block.

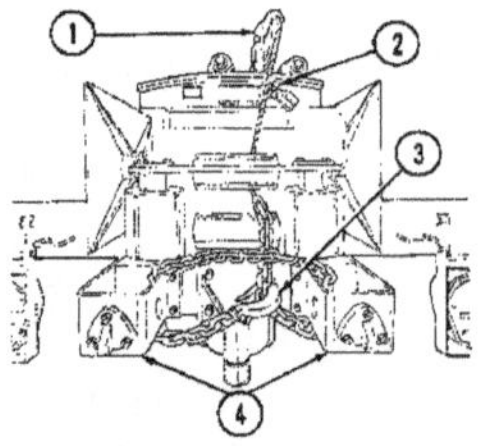
1
2
3
4

(15) To unwind winch cable:

(a) Release cable tensioner switch (9) if engaged.

(b) Remove travel pin (8) and pull directional control lever (6) outward to UNWIND until winch cable hook and chain (3) are loosened from bumperettes (4).

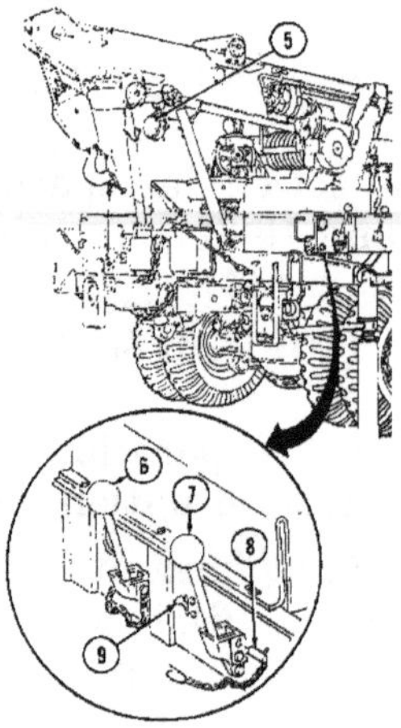
5
6
7
8
9

(c) Direct crewmember to free cable hook and chain (3) from rear bumperettes (4).

(d) With crewmember maintaining manual tension on cable, pull directional control lever (6) outward to UNWIND.

(e) After required length of cable (14) has been unwound, return directional control lever (6) to NEUTRAL.

2-24. OPERATION OF MEDIUM WRECKER (Contd)

(16) To rig the load:

(a) Attach utility chain (10) to lifting shackles (11) or pintle hook of load (12).

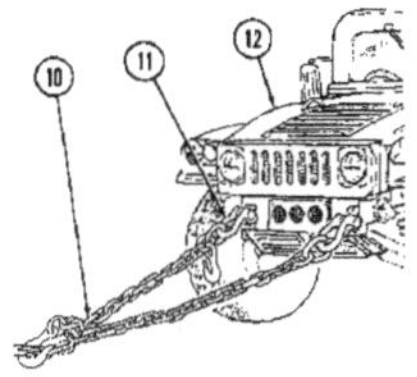
10
11
12

NOTE

M939A2 vehicles do not have front anchors.

(b) If load (12) is very heavy or deeply mired, install snatch block (16) or combination of snatch blocks (16) to increase winch pulling power. To rig a snatch block (16):

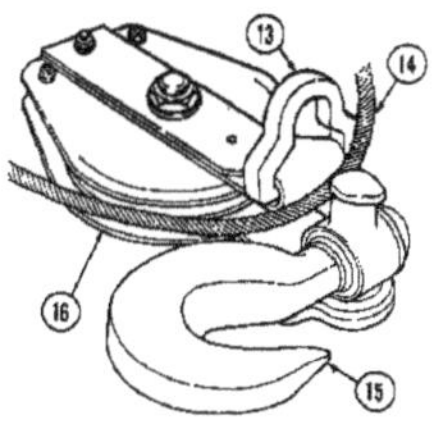
13
14
16
15

1. Unwind enough cable (14) to reach the load (12) and back to rear winch. Rig cable hook and chain (3) to rear bumperettes (4).

2. Turn hook (15) to right. Lift up rear of snatch block (16) and open support link (13). Insert cable (14). Lift up rear of snatch block (16) and lower and lock support link (13) to hook (15). Return hook (15) to original position.

3. Rig hook (15) to load (12).

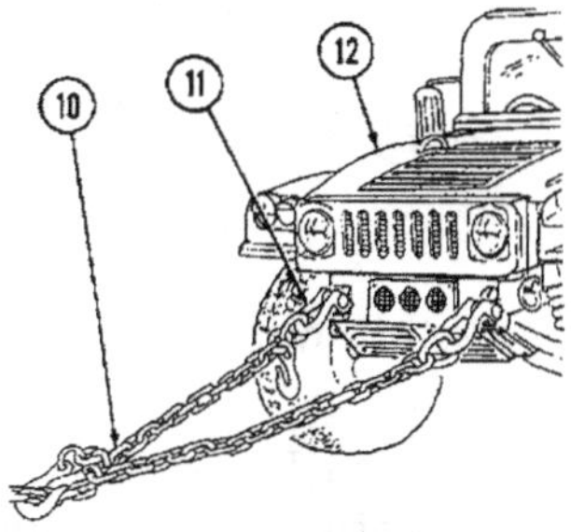
10
11
12

(17) Release brakes, transmission, and transfer case of vehicle being retrieved.

WARNING

Direct all personnel to stand clear of winch cable during winch operation. A snapped winch cable may result in injury or death.

NOTE

If using the wrecker's rear winch for self-recovery, release parking brake and set spring brake override.

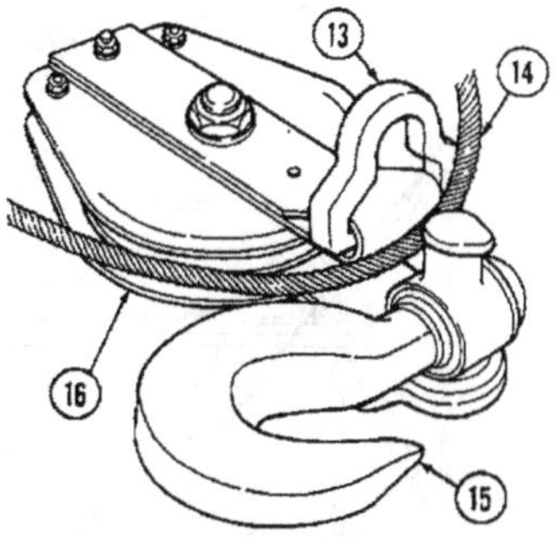
13
14
16
15

(18) To pull load:

(a) Position lever (7) to LOW for heavy loads or HIGH for light loads.

(b) Push tensioner switch (9) down to engage tensioner.

2-24. OPERATION OF MEDIUM WRECKER (Contd)

WARNING

Do not operate winch erratically. Erratic winding may result in a snapped cable, causing injury or death.

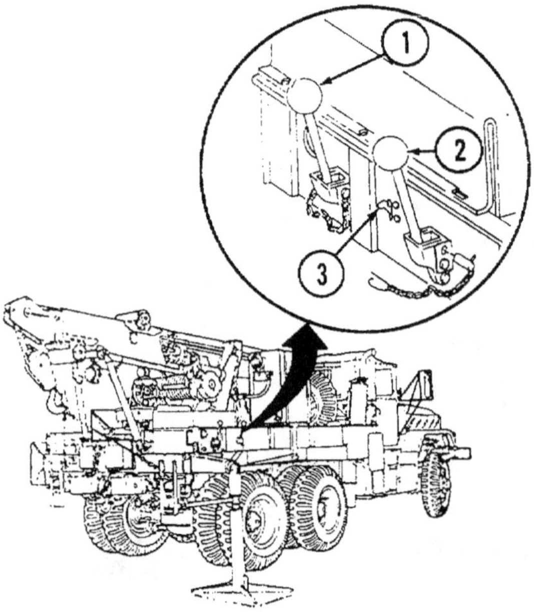
1
2
3

(c) Push directional control lever (1) into WIND.

(19) To shift torque control lever (2) from LOW to HIGH or HIGH to LOW:

(a) Pull directional control lever (1) out to NEUTRAL.

(b) Shift torque control lever (2).

(c) Push directional control lever (1) in to resume winding.

(20) Place directional control lever (1) in NEUTRAL to stop winding.

(21) Briefly pull directional control lever (1) outward to UNWIND to loosen cable for unrigging.

(22) Remove winch cable chain (7) and hook (9) from load. Remove snatch blocks if used.

(23) Push directional control lever (1) in to WIND. Release directional control lever (1) to stop winding operation when cable chain (7) approaches guide rollers (4).

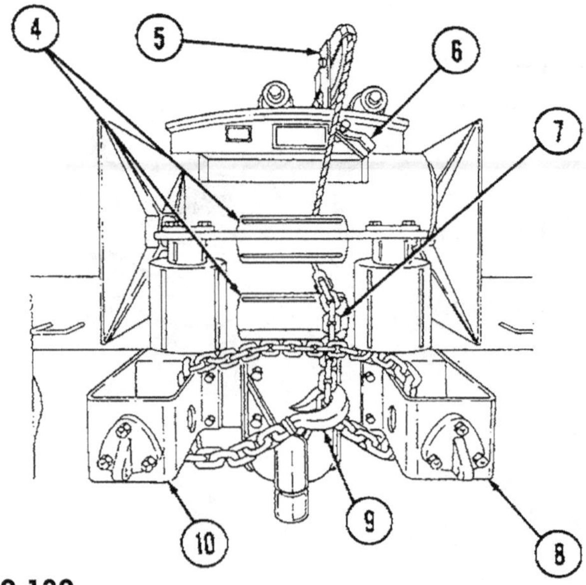
4
5
6
7
10
9
8

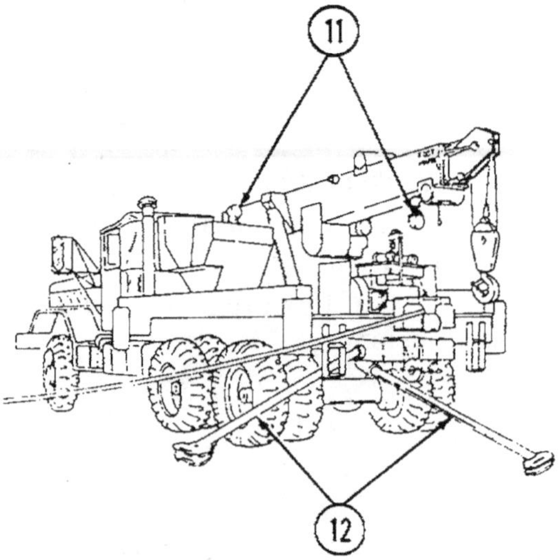
11
12

2-24. OPERATION OF MEDIUM WRECKER (Contd)

(24) To prepare rear winch for travel:

(a) Manually push level wind (5) completely to the right. Pull out drum lock knob (6), rotate 90 degrees, and release. If necessary, adjust level wind (5) to ensure lock plunger engages.

(b) Place cable chain (7) and hook (9) up through right bumperette (8) and down through left bumperette (10).

(c) Place cable chain (7) and hook (9) around chain (7) between bumperettes (8) and (10).

(d) Push directional control lever (1) into WIND. Stop when cable is snug and replace travel pin.

(25) Pull cable tensioner switch (3) up to disengage tensioner.

(26) Place torque control lever (2) in NEUTRAL and replace travel pin.

(27) Turn off floodlights (11) if used.

(28) Disengage hand throttle control (13) by rotating handle and pushing in to allow engine speed to drop to idle.

(29) Pull transfer case power takeoff control lever (18) forward to disengage, and lock in position.

(30) Place transmission selector lever (15) in N (neutral).

(31) Turn off main floodlight switch (16) and amber warning light switch (14), if used.

(32) If field chocks (12) were used:

(a) Make sure transfer case shift lever (17) is in high range.

(b) Release parking brake lever (19).

(c) Place transmission selector lever (15) in 1-5 (drive) to move vehicle ahead far enough to free chocks (12).

(d) Stop vehicle, shut down engine, and engage parking brake lever (19) (para. 2-16).

(e) Remove field chocks (12).

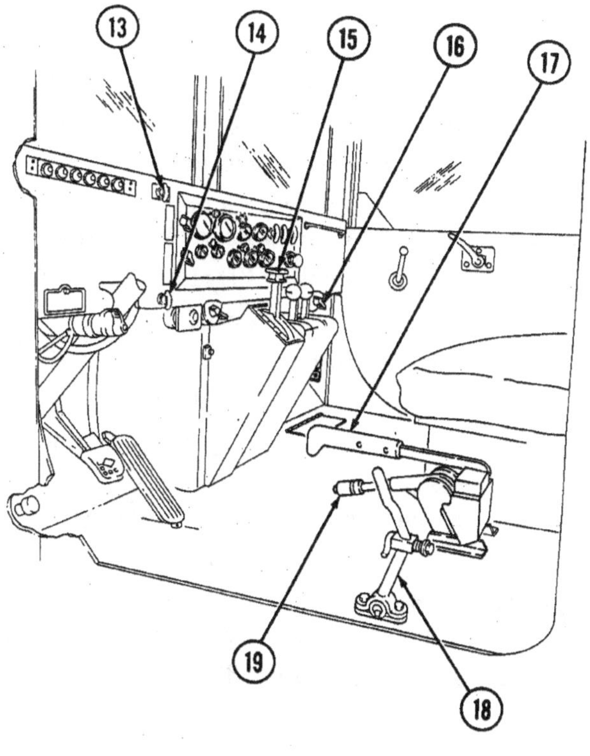
13
14
15
16
17
19
18

2-24. OPERATION OF MEDIUM WRECKER (Contd)

d. Crane Operation. The medium wrecker crane is capable of rotating 360°, extending its boom 18 feet (5.5 meters) and elevating the boom to 45°. A data plate above crane controls lists variations in safe load crane extension and how it is rigged. Maximum capacity is 20,000 pounds (9,080 kilograms) with three-part line.

WARNING

Gondola safety guard must be in place prior to crane operation.

NOTE

- Whenever possible, position wrecker for a direct rear lift.
- This operation requires two crewmembers.
- Vehicle equipped with auto throttle will automatically idle between 1350 and 1400 rpms.
- Set CTIS to SAND mode after outrigging (M936A2 model vehicles).

(1) Park wrecker on a level, hard surface if possible. Wrecker position depends upon type of lifting operation such as rear lift, side lift, or lift and swing.

(2) Position outriggers (10) as follows:

(a) Remove retaining clip (15) holding L-shaped retaining pin (14) at corner of outrigger frame tube (16) and remove pin (14).

(b) Pull outrigger (10) out until it stops and lower to a vertical position.

(c) Insert outrigger handle (12) into hole in collar (9).

(d) With crewmember holding collar (9), grasp outrigger base (11) and turn counterclockwise until base (11) makes contact with ground.

(e) Turn collar (9) clockwise until base (6) seats against ground.

(f) Repeat procedure to lower remaining outriggers (10).

(3) Start engine (para. 2-12).

(4) Place transfer case shift lever (4) in neutral.

(5) Unlock safety latch (5) and push transfer case power takeoff lever (6) back to engage.

(6) Place transmission selector lever (2) in 1-5 (drive).

(7) Pull out hand throttle control (1) on M936/A1 to last stop.

(8) If tactical situation permits, turn on amber warning light switch (7), if required, and floodlight control switch (3) if operation is at night.

(9) To obtain required lift, adjust shipper braces (17):

(a) Remove retaining clips (19) securing T-shaped retaining pins (18) to shipper braces (17).

(b) Raise boom (8) to required height by pulling boom control lever (13) back. Adjust height until holes in male and female sections of shipper braces (17) are aligned.

2-24. OPERATION OF MEDIUM WRECKER (Contd)

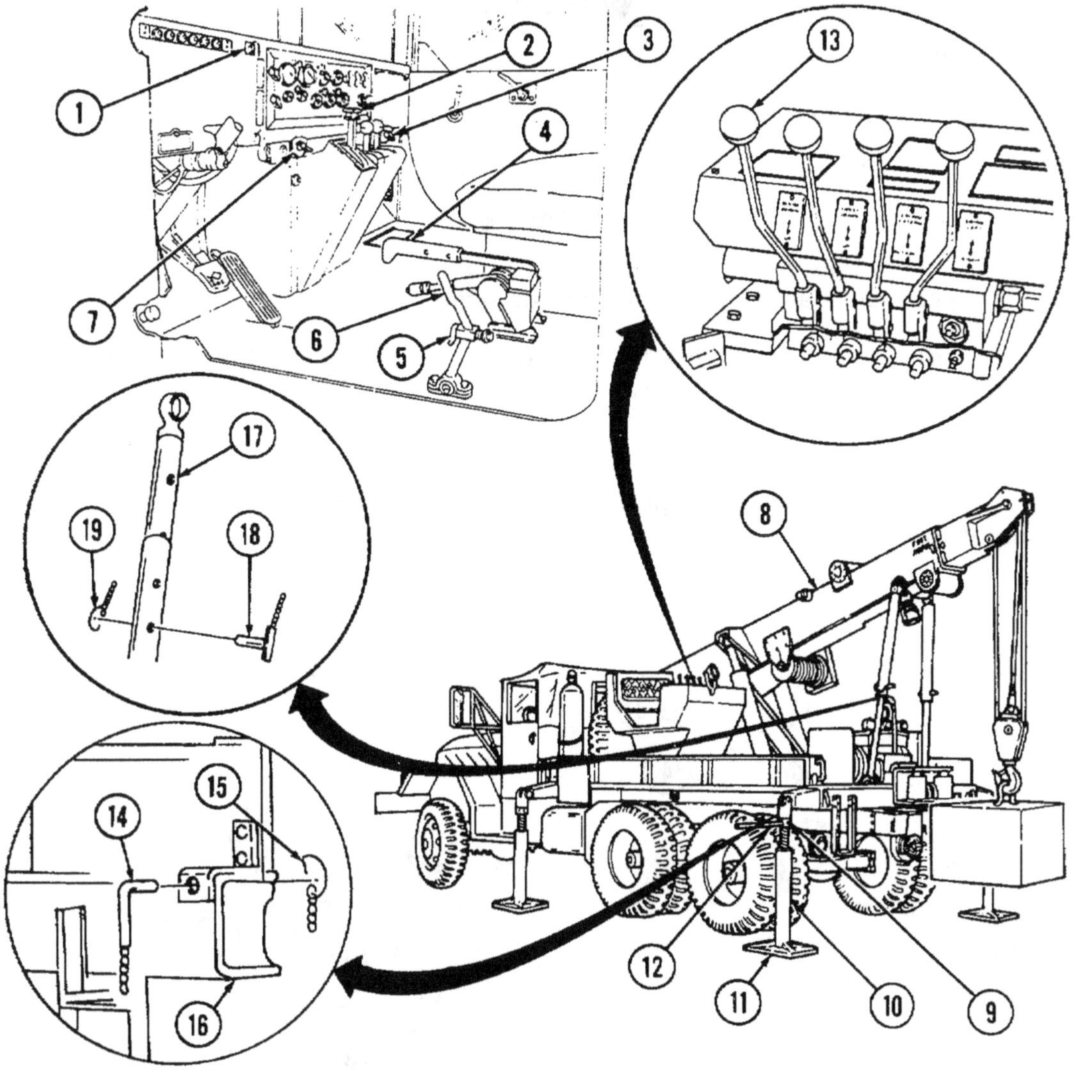
1
2
3
4
5
6
7
8
9
10
11
12
13
14
15
16
17
18
19

WARNING

Do not get underneath the wrecker boom when raised unless properly secured. Failure to do this may result in injury or death to personnel.

NOTE

If more height is needed than full length of shipper braces allow, use boom jacks. Refer to paragraph d, step 13 for installation of boom jacks.

(c) Reinsert T-shaped retaining pins (18) in lined-up holes and secure with retaining clips (19).

(d) Lower boom (8) to support boom (8) weight on shipper braces (17).

2-24. OPERATION OF MEDIUM WRECKER (Contd)

WARNING

Direct all personnel to stand clear of crane or load during crane operation. A snapped cable, shifting, or swinging load may cause injury or death.

(10) To lift load:

CAUTION

Do not allow crane block to jam boom sheaves in raising operation. Failure to do this will damage boom sheaves.

(a) Lower crane block (3) to load (5) by pushing hoist control lever (7) to DOWN position, and releasing it to stop.

(b) Rig load (5) with utility chains, cable slings, or attach hook (4) directly to lifting devices on the load (5).

(c) Raise load (5) by pulling hoist control lever (7) to UP position, and releasing it to stop.

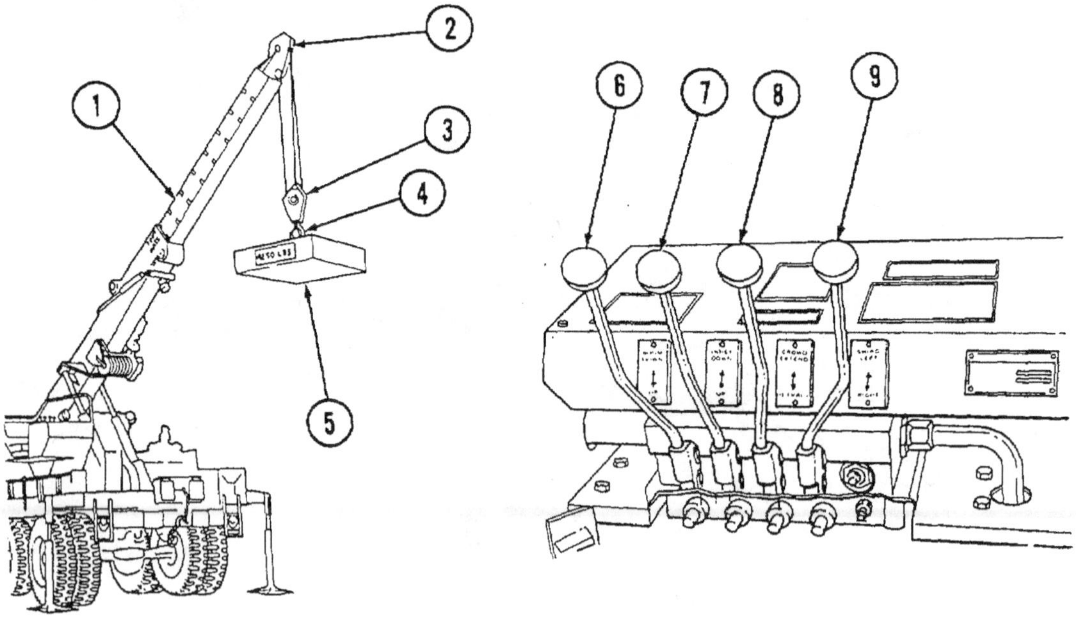
1
2
3
4
5
6
7
8
9

(11) To lower load:

(a) Lower load (5) by pushing hoist control lever (7) to DOWN position, and releasing it to stop.

(b) Block load (5) to prevent tipping or shifting.

2-24. OPERATION OF MEDIUM WRECKER (Contd)

(12) To lift and swing load:

NOTE

This operation requires use of all outriggers.

(a) Perform steps 1 through 9.

(b) Slightly pull boom control lever (6) back to UP to take boom (1) weight off shipper braces (12).

(c) Remove retaining clips (16) from L-shaped shipper brace retaining pins (14) and remove pins (14) from shipper brace brackets (13).

(d) Swing brace (12) upward to brace retaining brackets (10) on each side of shipper (11). Secure braces (12) on brackets (10) with retaining clips (15).

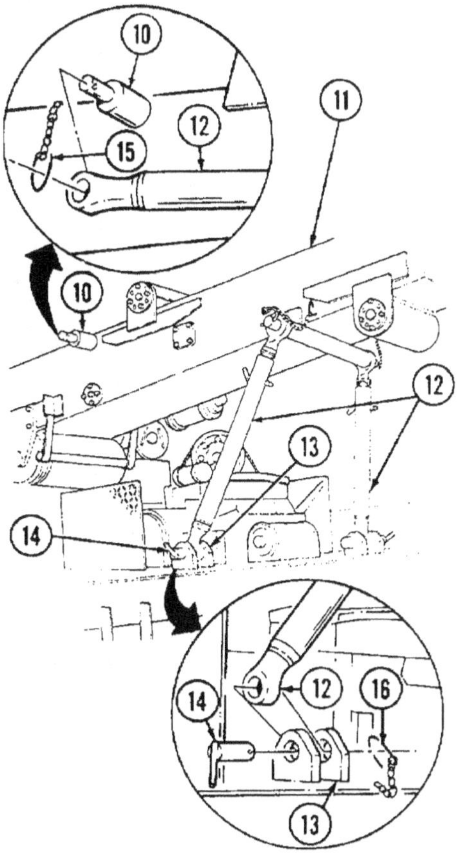
10
11
12
15
10
12
13
14
14
12
16
13

CAUTION

When extending boom, move HOIST and CROWD levers at the same time. Failure to do so may result in boom damage.

(e) Extend boom (1) as necessary by pushing crowd control lever (8) forward to DOWN. Maintain an even distance between crane block (3) and boom sheaves (2).

(f) Elevate boom (1) to desired height by pulling boom control lever (6) back to UP.

(g) Lift load (5). Refer to step 10.

CAUTION

Boom rotates 360°. Make sure area is clear of obstacles, and caution is used when operating boom over cab area. Damage to windshield, exhaust stack, air intake, and cab may result.

(h) Push swing control lever (9) forward to LEFT to swing load (5) left. Pull swing control lever (9) back to RIGHT to swing load (5) right.

(i) Lower load. Refer to step 11.

2-24. OPERATION OF MEDIUM WRECKER (Contd)

(13) For heavy rear lift:

NOTE

This operation requires use of front outriggers.

(a) Perform steps 1 through 9.

(b) Slightly pull boom control lever (2) back to UP to take boom (1) weight off shipper braces (8).

(c) Remove retaining clips (11) form L-shaped shipper brace retaining pins (9) and remove retaining pins (9) from shipper brackets (8).

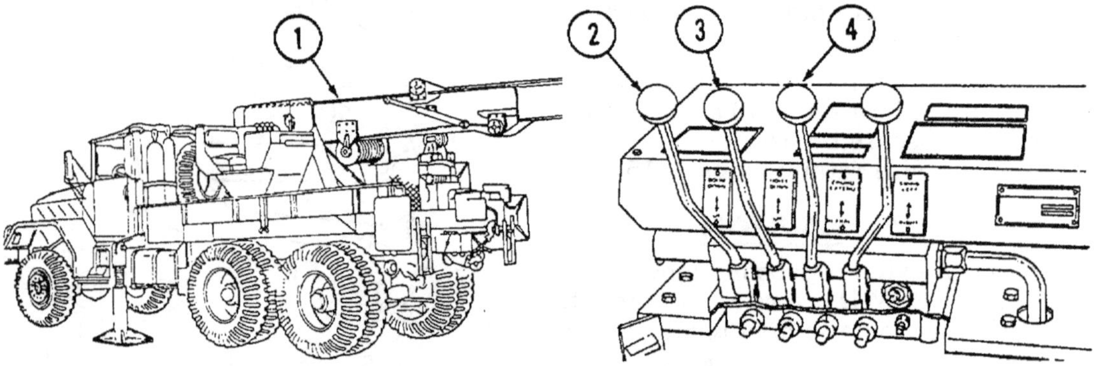
1
2
3
4

(d) Swing shipper braces (7) upward to brace retaining brackets (5) on each side of shipper (6). Secure braces (7) on brackets (5) with retaining clips (10).

(e) Push boom control lever (2) forward to DOWN to lower boom (1) to horizontal position.

CAUTION

When extending boom, move HOIST and CROWD levers at the same time. Failure to do this will result in boom damage.

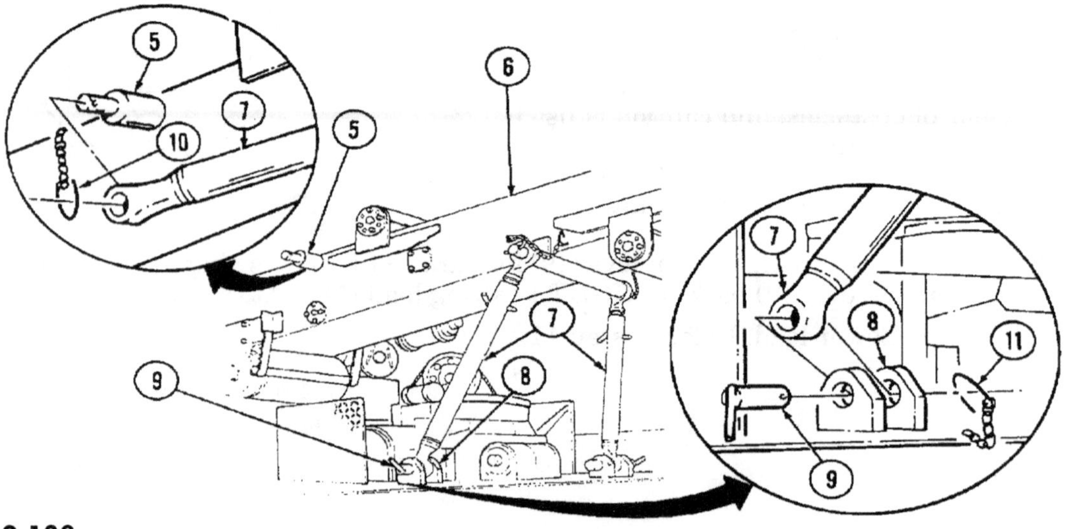
5
6
7
10
5
7
8
9
7
8
11
9

2-24. OPERATION OF MEDIUM WRECKER (Contd)

(f) Extend boom (1) to desired operational length by pushing crowd control lever (4) forward to EXTEND and hoist control lever (3) forward to DOWN. Maintain even distance between crane block (14) and boom sheaves (13).

(g) Remove boom jacks (15) from wrecker bed.

(h) Remove both retaining pins (16) and jack pin (20). Insert jack pin (20) in hole of 12-ft mark (21) on boom (1).

(i) Insert boom jack yoke end (17) on jack pin (20) and secure with retaining pin (16).

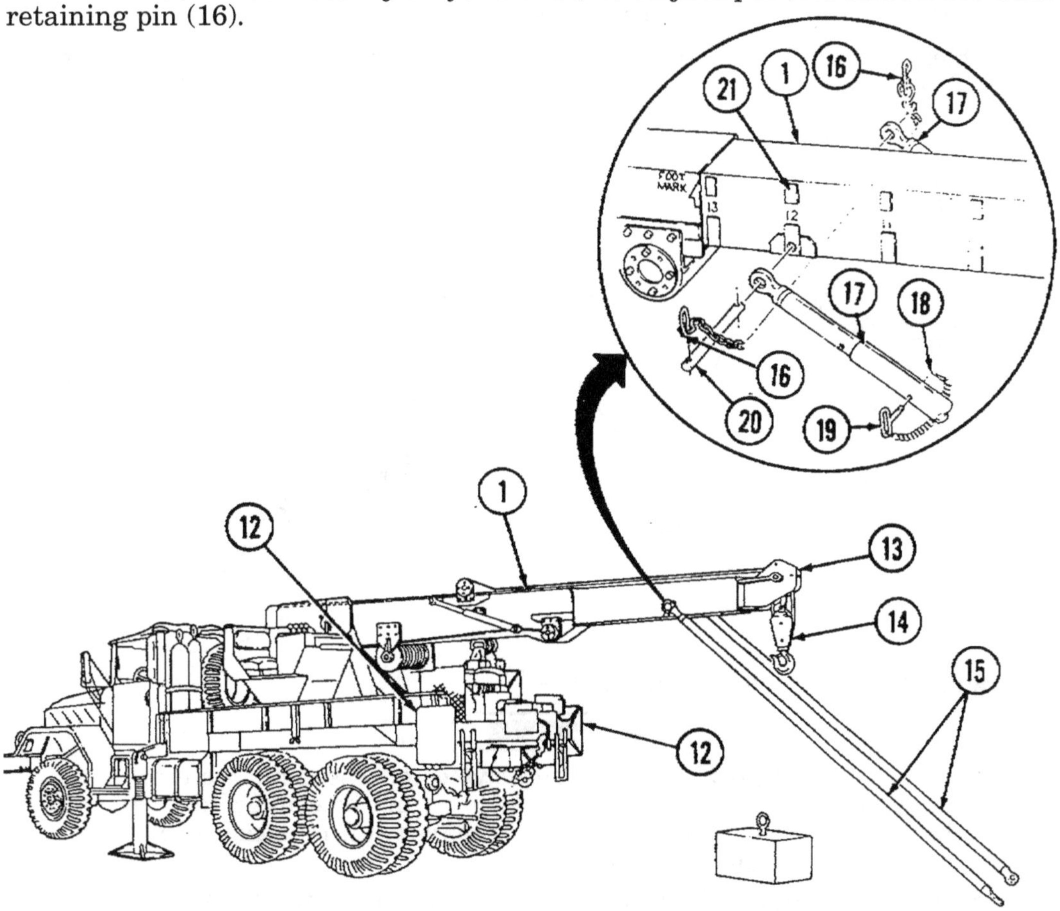
21
1
16
17
FOOT MARK
13
12
17
18
16
20
19
1
12
13
14
15
12

(j) Obtain ring-handled pin (19) and extend boom jacks (15) to required length. Insert ring-handled pin (19) when required length is obtained and secure with retaining clip (18).

(k) Repeat steps i and j on opposite side of boom (1) with second boom jack (15).

WARNING

Direct all personnel to stand clear of crane or load during crane operation. A snapped cable, shifting, or swinging load may result in injury or death.

(l) Pull boom control lever (2) to UP position until boom jacks (15) are off the ground.

(m) Remove boom jack base plates (12) from wrecker bed.

2-24. OPERATION OF MEDIUM WRECKER (Contd)

(n) Install boom jack (1) in base plate (3) and secure in place with retaining pin (7) and retaining clip (8).

(o) Obtain tie bar (2) from wrecker bed and install between boom jacks (1). Secure in place with retaining pin (5) and retaining clip (6).

(p) Push boom control lever (9) to DOWN position, and release it when boom jacks (1) make firm contact with ground.

(q) Perform lifting and lowering operation. Refer to steps 10 and 11.

(14) Heavy side lifts are similar to heavy rear lifts except that front and rear outriggers (4) from the lifting side of the vehicle will be positioned.

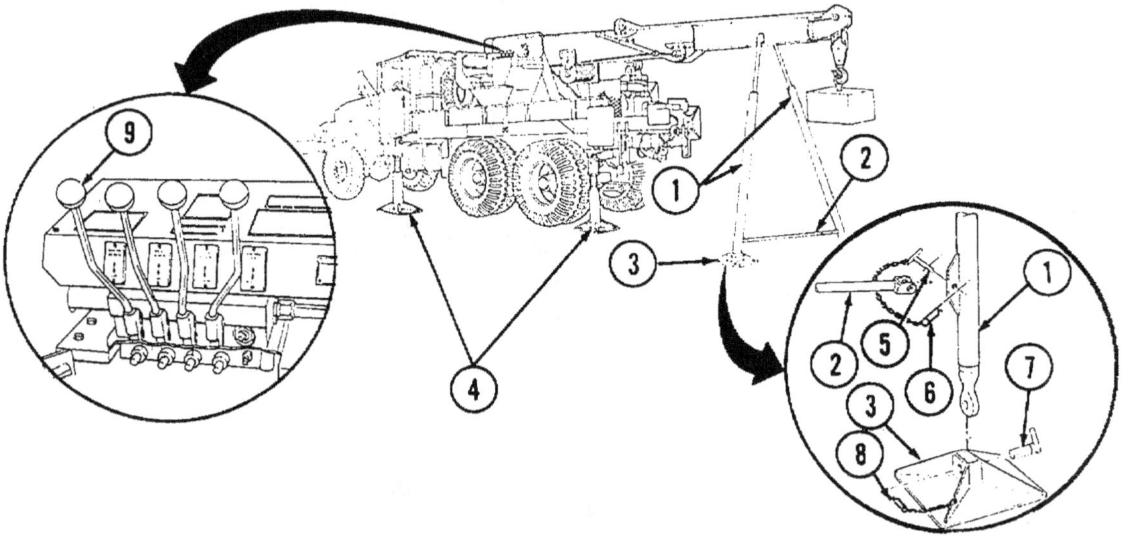
9
1
2
3
4
1
5
2
6
7
3
8

e. Towing With Wrecker Crane.

WARNING

- If vehicle being towed has inoperative compressed air system, emergency air and service air lines must not be connected between vehicles. Failure to do this may result in damage to equipment, or cause injury or death.
- If the compressed air system of the vehicle being towed has emergency air and service air lines, they must not be connected.

CAUTION

- Do not use towing as a means to start engine of vehicle with automatic transmission. Refer to disabled vehicle operator's manual for towing instruction. Failure to so may will result in damage to vehicle.
- Maximum towing speed shall not exceed 35 mph (56 km/h) on paved highway or 15 mph (24 km/h) on off-highway surfaces.

NOTE

When towing M939/A1/A2 series vehicles with inoperative compressed air system, the spring brakes must be caged.

2-24. OPERATION OF MEDIUM WRECKER (Contd)

(1) Position rear of wrecker directly in front of disabled vehicle.

(2) If tires, transmission, and steering of disabled vehicles are serviceable, proceed to step 4 and install towbar (10) to front bumper shackle brackets (14).

(3) If disabled vehicle has no shackle brackets (14), or if tires, transmission, and/or steering of disabled vehicle are unserviceable, proceed to step 5d and use lift-tow procedure.

(4) To tow a disabled vehicle using front bumper shackle brackets:

(a) Remove lifting shackles (13) from bumper (15) of disabled vehicle.

(b) Install clevis (12) end of towbar (10) on front bumper shackle brackets (14) and secure in place with clevis bolt (16) and safety pin (11).

(c) Install yoke (18) end of towbar (10) to wrecker pintle hook (19).

WARNING

Crisscross and connect utility chains between vehicles in the event towbar breaks or becomes disconnected. Failure to do this will result in injury or death.

(d) Crisscross and connect utility chains (17) to spring hangers (20) on towed vehicle and secure to towing vehicle.

(e) Connect emergency and service air line (22) to respective half coupling (21) on each vehicle.

(f) Release parking brake and place transmission, transfer case, and PTO of disabled M939/A1/A2 series vehicles in neutral.

(g) Reverse steps a through f after completion of towing.

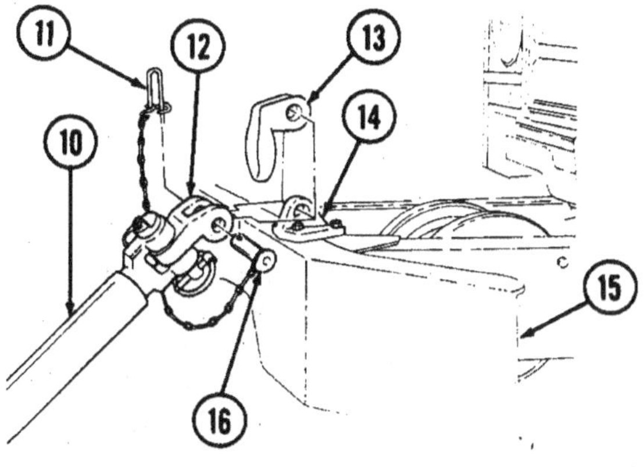
11
12
13
14
10
15
16

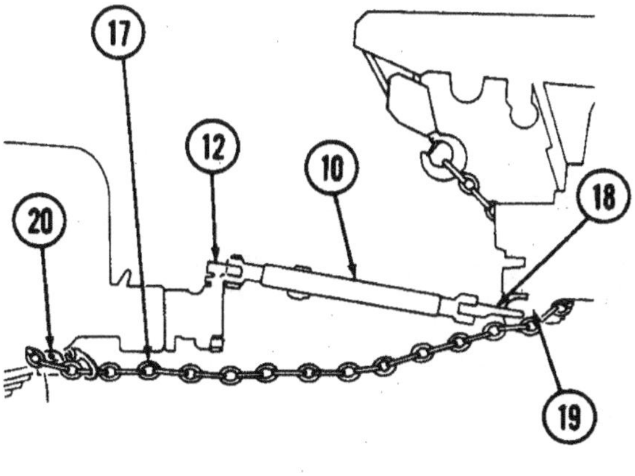
17
12
10
18
20
19

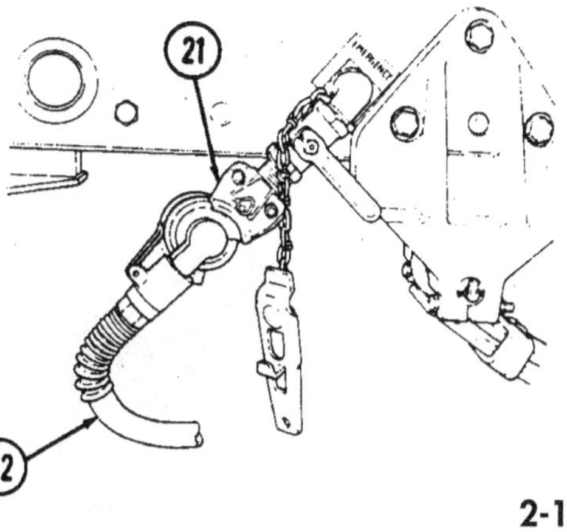
21
22

2-24. OPERATION OF MEDIUM WRECKER (Contd)

(5) For lift-tow operations:

(a) Attach whiffletree (3) to lifting shackles (4) of disabled vehicle. Whiffletree (3) is attached the same way as a towbar (12). Refer to step 4c.

(b) Lower crane block (2) and insert hook (5) through center hole of whiffletree (3). Refer to crane operation (para. 2-24d).

(c) Remove two retaining pins (11) and towbar adjustment pins (10). Slide out leg extensions (9).

(d) Attach two clamps (8) to clevis (13) of leg extensions (9) and secure in place with clevis bolts (14) and attached safety pins (15).

(e) Loosen two tension adjusting nuts (7) on clamps (8).

(f) Position clamps (8) against front axle (16) of disabled vehicle and wrap clamp chain (6) around axle and back up through clamp (8). Lock in place by turning tension adjusting nut (7) until secure.

(g) Repeat step f with second clamp (8) in position against axle (16) on opposite side of forward axle.

NOTE

If necessary, adjust length of towbar while performing step h. Yoke end of towbar should extend one foot or more beyond front bumper of disabled vehicle.

(h) After both clamps (8) are secured to disabled vehicle, slide leg extensions (9) back into towbar (12) and secure in place with adjustment pins (10) and retaining pins (11).

(i) Install yoke end of towbar (12) on wrecker pintle hook (16).

(j) Disengage brakes and remove propeller shafts between transmission and transfer and place transfer in high range of disabled M939/A1/A2.

(k) Raise crane block (2) and lift front wheels of disabled vehicle off ground until towbar is level (parallel to ground).

(l) To secure from lift-tow operation, reverse steps a through k.

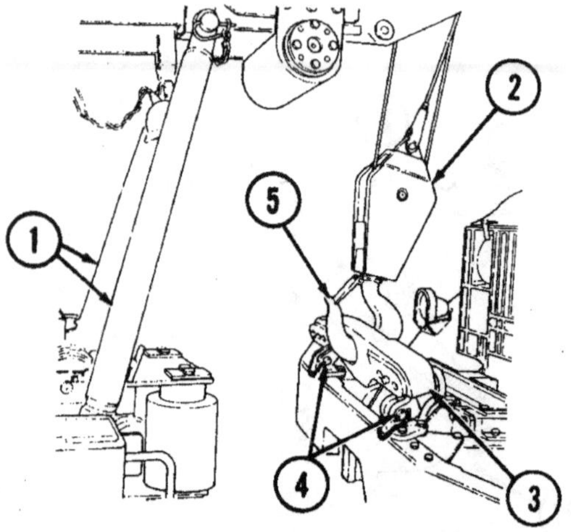
1
2
3
4
5

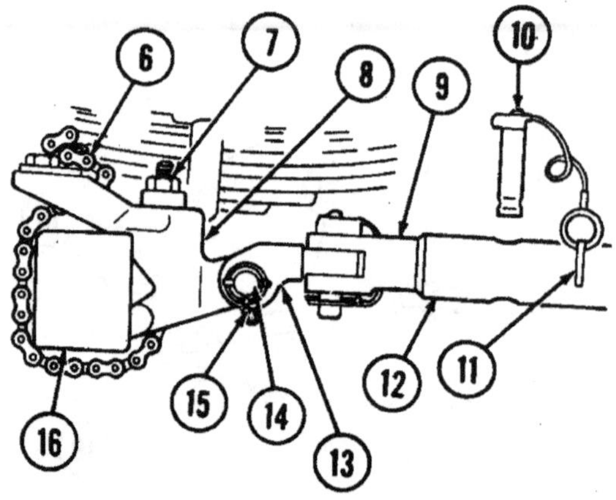
6
7
8
9
10
11
12
13
14
15
16

2-24. OPERATION OF MEDIUM WRECKER (Contd)

f. Securing Crane After Operation.

(1) Secure outriggers (4), two boom jacks (3), tie bar (6), and boom jack base plates (5). Refer to step d.

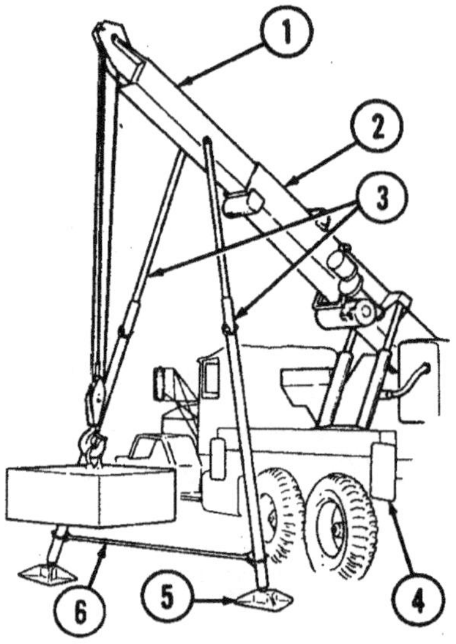
1
2
3
4
5
6

NOTE

When retracting boom, pull crowd control lever back to RETRACT and hoist control lever back to UP to prevent block from becoming tangled.

(2) Fully retract boom (1) into shipper (2) and center boom (1) to rear of wrecker.

(3) Remove retaining clip (11) from upper retaining brackets (7) and swing shipper braces (8) to shipper brace brackets (9) on wrecker body. Secure shipper braces (8) in place with L-shaped retaining pins (10) and retaining clip (12).

(4) Push boom control lever (17) forward to DOWN to allow shipper braces (8) to support boom (1) and weight of shipper (2).

(5) Install lifting sling (15) onto crane block hook (14). Attach hook ends of sling (15) to brackets (16) on outrigger (4).

(6) Pull hoist control lever (18) back to UP and remove all slack from lifting sling (15) and boom cable (13).

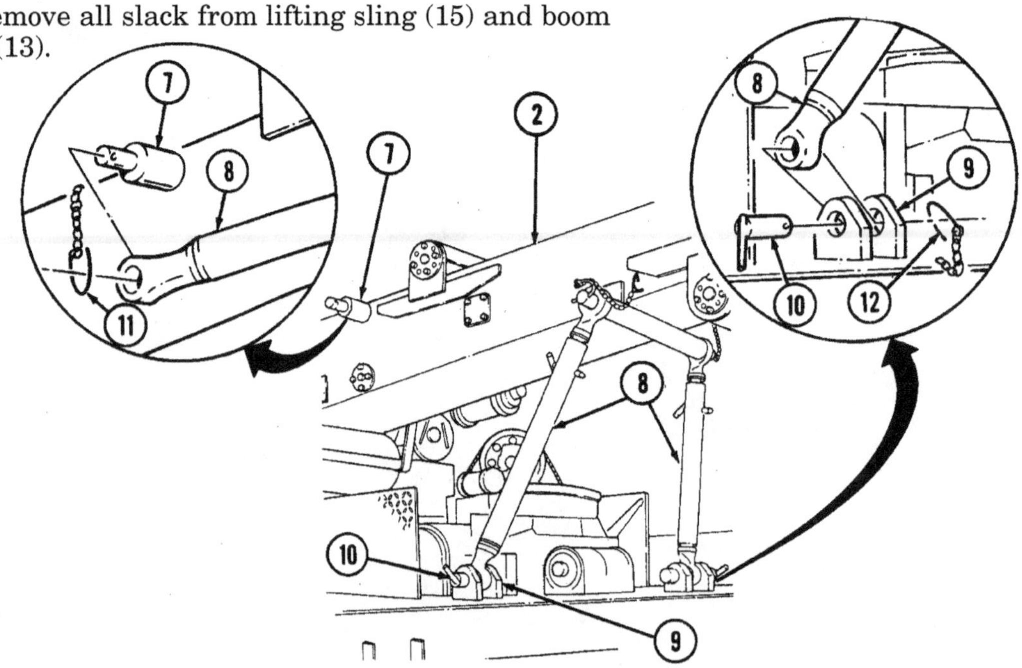
7
8
11
7
2
8
9
10
12
8
10
9

2-24. OPERATION OF MEDIUM WRECKER (Contd)

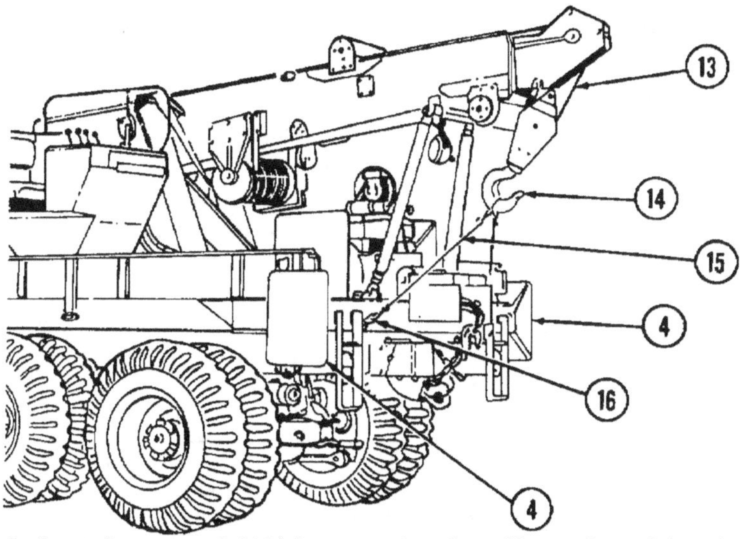
13
14
15
4
16
4

(7) Release hand throttle control (19) by rotating handle and pushing in to allow engine speed to drop to idle.

(8) Push transfer case power takeoff lever (23) forward to disengage.

(9) Place transmission selector lever (20) in N (neutral).

(10) Apply parking brake lever (24).

(11) Turn off amber warning light (25) and floodlight control switch (21) if used during crane operation.

(12) Place transfer case shift lever (22) in desired operating range.

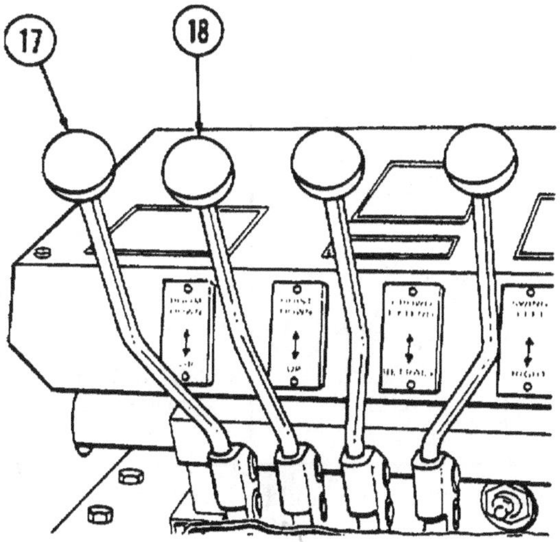
17
18

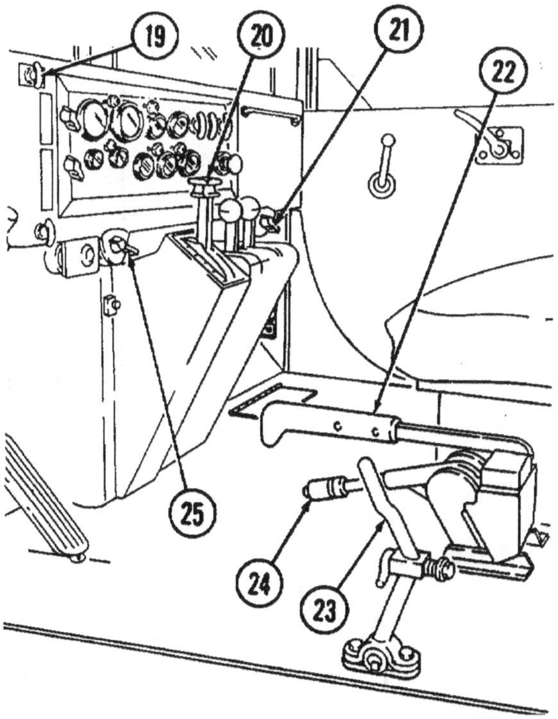
19
20
21
22
25
24
23

2-25. OPERATION OF DUMP TRUCKS

a. Payload Capacities. M929/A1/A2 and M930/A1/A2 dump trucks can carry 10,000 lb (4,540 kg) of material cross-country. Table 2-5 lists typical material weights.

WARNING

Stay clear of dump body and cab protector at all times during loading and unloading operations. Dump body can unexpectedly raise when a heavy load is dropped into dump body and will cause injury or death.

Table 2-5. Typical Material Weights.

	MATERIAL WEIGHT		Capacity level full 5.0 cu-yd (3.8 cu-M)	Capacity Heaping full 7.5 cu-yd (5.7 cu-M)
	lb per cu-ft	lb per cu-yd (kg per cuM)	lb (approx) (kg)	lb (approx) (kg)
Ashes	43	1,161 (1,518.5)	5,805 (2,635.5)	8,708 (3,953.4)
Cinders	46	1,242 (1,624.4)	6,210 (2,819.3)	9,315 (4,229.0)
Clay (dry and loose)	77	2,079 (2,719.1)	*10,395 (4,719.3)	*15,593 (7,079.2)
Clay (wet)	110	2,970 (3,884.5)	*14,850 (6,741.9)	†22,275 (10,113.0)
Clay and gravel (dry)	100	2,700 (3,513.3)	*13,500 (6,129.0)	†20,250 (9,193.5)
Clay and gravel (wet)	65	1,755 (2,295.4)	8,775 (3,983.4)	*13,163 (5,976.0)
Coal, anthracite (hard)	54	1,458 (1,906.9)	7,290 (3,309.7)	*10,935 (4,964.5)
Coal, bituminous (soft)	81	2,187 (2,860.4)	*10,935 (4,964.5)	*16,403 (7,447.0)
Coke	28	756 (988.8)	3,780 (1,716.1)	5,670 (2,587.2)
Concrete	138	3,726 (4,873.2)	*18,630 (8,58.0)	†27,945 (12,678.2)
Concrete mix (wet)	124	3,348 (4,379.1)	*16,740 (7,600.0)	†25,110 (11,400.1)
Earth (dry and loose)	75	2,025 (2,648.8)	*10,125 (4,596.8)	*15,188 (6,895.4)
Earth (moist and packed)	95	2,565 (3,345.8)	*12,825 (5,822.6)	*19,238 (8,734.1)

* Over rated cross-country payload

† Over rated cross-country and highway payload

2-25. OPERATION OF DUMP TRUCKS (Contd)

Table 2-5. Typical Material Weights (Contd).

	MATERIAL WEIGHT		Capacity level full 5.0 cu-yd or (3.8 cu-M)	Capacity Heaping full 7.5 cu-yd or (5.7 cu-M)
	lb per cu-ft	lb per cu-yd (kg per cuM)	lb (approx) (kg)	lb (approx) (kg)
Earth and gravel (dry and loose)	100	2,700 (3,531.3)	*13,500 (6,129.0)	†20,250 (9,193.5)
Garbage (dry)	37	999 (1,306.6)	4,995 (2,267.7)	7,493 (3,401.8)
Garbage (wet)	47	1,269 (1,659.7)	6,345 (2,880.6)	9,518 (4,321.2)
Gravel	110	2,970 (3,884.5)	14,850 (6,741.9)	22,275 (10,112.9)
Gravel and sand (dry and loose)	95	2,565 (3,354.8)	12,825 (5,822.6)	19,238 (8,734.1)
Gravel and sand (wet)	120	3,240 (4,237.6)	16,200 (7,354.8)	24,300 (11,032.2)
Limestone (crushed)	100	2,700 (3,531.3)	13,500 (6,129.0)	20,250 (9,193.5)
Mud (wet)	120	3,240 (4,237.6)	16,200 (7,354.8)	24,300 (11,032.2)
Rock and stone	95	2,565 (3,354.8)	12,825 (5,822.6)	19,238 (8,734.1)
Salt (fine)	50	1,350 (1,765.7)	6,750 (3,064.5)	10,125 (4,596.8)
Sand (dry and loose)	98	2,646 (3,460.7)	13,230 (6,006.4)	19,845 (9,009.6)
Sand (dry and packed)	110	2,970 (3,884.5)	14,850 (6,741.9)	22,275 (10,112.9)
Sand (moist and loose)	120	3,240 (4,237.6)	16,200 (7,354.8)	24,300 (11,032.2)
Slag (crushed)	75	2,025 (2,648.5)	10,125 (4,596.8)	15,188 (6,895.4)
Snow (moist and packed)	50	1,350 (1,765.7)	6,750 (3,064.5)	10,125 (4,596.8)
Stone (crushed)	100	2,700 (3,531.3)	13,500 (6,129.0)	20,250 (9,193.5)
Stone (loose)	95	2,565 (3,354.8)	12,825 (5,822.6)	19,238 (8,734.1)

* Over rated cross-country payload

† Over rated cross-country and highway payload

2-25. OPERATION OF DUMP TRUCKS (Contd)

b. Regular Dump Operation.

NOTE

Transfer case shift lever should be in H (high) range.

(1) Start engine (para. 2-12) and position vehicle for dumping. Apply parking brake lever (7), place transmission selector lever (1) in N (neutral), and transfer case shift lever (3) to HIGH.

(2) Check chains (16) to ensure they will not restrict tailgate (14) opening.

(3) Unhook safety chain and unlock tailgate (14) by pulling control lever (8) forward and down.

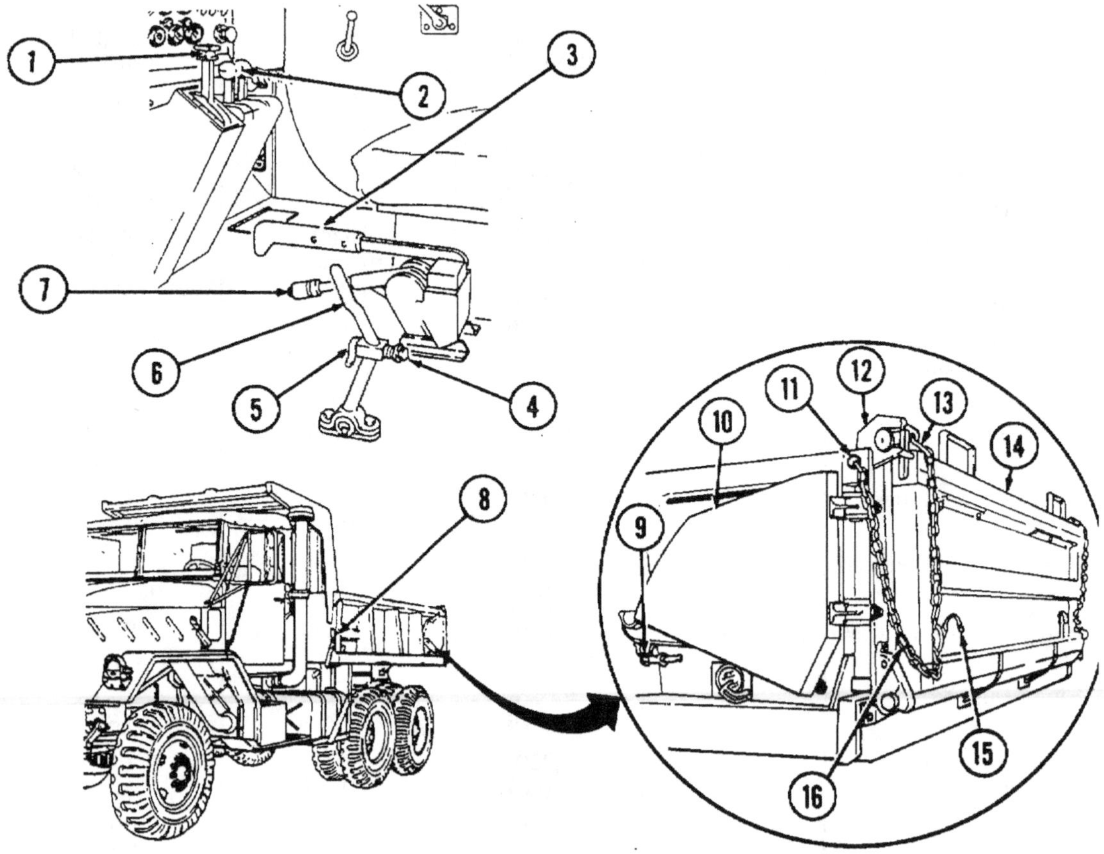
1
2
3
4
5
6
7
8
9
10
11
12
13
14
15
16

WARNING

Direct all personnel to stand clear of vehicle when engaging transmission or transfer case. Failure to do this will cause injury or death.

(4) Apply parking brake (7) and shift transmission selector lever (1) in 1-5 (drive).

2-25. OPERATION OF DUMP TRUCKS (Contd)

(5) Pull transmission power takeoff control lever (2) back to ENGAGE.

(6) Return transmission selector lever (1) to N (neutral).

(7) Depress spring (4) to release safety latch (5) and push dump body control lever (6) back to raise dump body.

NOTE

- Dump body will stop automatically when fully raised.
- Engine rpm should not exceed 1000 rpm during dumping operation.

(8) To lower dump body, pull dump body control lever (6) full forward to lower dump body.

CAUTION

To prevent dump body from raising during vehicle operation, dump body control lever must remain locked in N (neutral) position.

(9) Return dump body control lever (6) to N (neutral) when dump body is completely lowered. Secure lever (6) with control lever safety latch (5).

(10) Push power takeoff control lever (2) forward to DISENGAGE.

(11) Push tailgate control lever (8) up and back as far as it will go to lock tailgate (14) in closed position.

c. Rocker-Type Dump Operation.

NOTE

This operation requires two crewmembers. Perform steps 2 through 6 on left side of vehicle first.

(1) Position vehicle for dumping and apply parking brake lever (7). Place transmission selector lever (1) in N (neutral).

(2) Remove chain (16) from upper chain slot (11).

(3) Thread chain (16) through chain bracket (15) at corner of tailgate (14).

(4) Remove bracket pin (13) in upper hinged bracket (12).

(5) Unfasten retaining hook (9) and swing tailgate wing (10) fully to the rear of vehicle.

(6) Insert bracket pin (13) in upper hinged bracket (12).

(7) Repeat steps 2 through 5 for right side of tailgate (14).

(8) Remove bracket pin (13) from upper hinged bracket (12).

(9) Lower tailgate (14) and insert bracket pins (13) from upper hinged brackets (12) on both sides of the vehicle.

2-25. OPERATION OF DUMP TRUCKS (Contd)

(10) Install chains (6) in lower chain slots (7).

(11) Raise and lower dump body as required. Refer to task 6, steps 4 through 10.

(12) After dump operation, remove two bracket pins (2) and raise tailgate (4). Swing two tailgate wings (3) to sides of dump body and reinsert bracket pins (2) into upper hinged brackets (1).

(13) Remove chains (6) from lower chain slots (7).

(14) Secure tailgate wings (3) to retaining hooks (8).

(15) Insert chains (6) in upper chain slots (9).

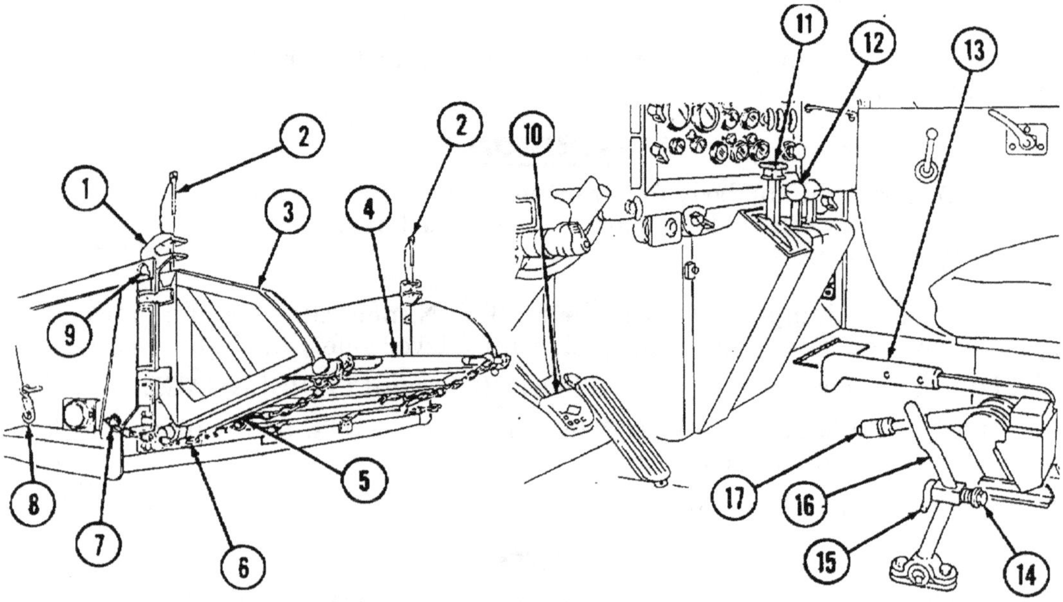
1
2
2
3
4
5
6
7
8
9
10
11
12
13
14
15
16
17

d. Spreader-Type Dump Operation.

NOTE

This operation requires two crewmembers. Perform steps 2 through 6 on left side first.

(1) Position vehicle for dumping and apply parking brake lever (17). Place transmission selector lever (11) in N (neutral).

(2) Remove chain (6) from upper chain slot (9).

(3) Thread chain (6) through chain brace (5) at corner of tailgate (4).

(4) Unfasten retaining hook (8) from tailgate wing (3) and swing wing (3) fully to the rear of vehicle.

(5) Loop chain (6) under tailgate (4), take up slack, and insert link of chain (6) into lower chain slot (7).

(6) Return tailgate wing (3) to side of dump body and secure with retaining hook (8).

(7) Repeat steps 2 through 6 for other side of tailgate (4).

2-25. OPERATION OF DUMP TRUCKS (Contd)

(8) Pull tailgate control lever (18) forward and down to unlock tailgate.

(9) Raise and lower dump body as required. Refer to task b, steps 4 through 6.

(10) Depress spring (14) to release safety latch (15), and pull dump body control lever (16) back to raise dump body. When dump body lifts 2 or 3 ft (.6 or .9 m), move dump control lever (16) back to neutral position to lock dump body.

(11) Shift transfer case shift lever (13) up to low range.

(12) Push brake pedal (10) and release parking brake lever (17).

(13) Place transmission selector lever (11) in 1 (first).

(14) Release brake pedal (10) and accelerate.

(15) Raise dump body at intervals by pulling dump body control lever (16) back to raise and then forward to neutral position as required.

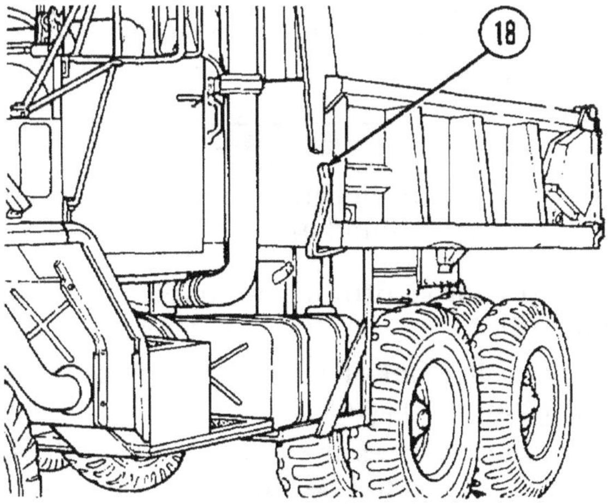
18

CAUTION

Do not exceed 5 mph (8 km/h) in 1 (first). If more speed is required, refer to caution data plate for correct transmission gear range.

(16) After vehicle has been unloaded:

(a) Stop vehicle and apply parking brake lever (17).

(b) Place transmission selector lever (11) in N (neutral).

(c) Shift transfer case shift lever (13) into desired position.

(d) Pull dump body control lever (16) forward to lower position.

(e) Return dump body control lever (16) to neutral position and secure lever with safety latch (15).

(f) Push power takeoff control lever (12) forward to DISENGAGE.

(g) Unhook safety chain and push tailgate control lever (18) up and back as far as it will go to lock tailgate (4) in closed position.

2-26. OPERATION OF TRACTOR AND FIFTH WHEEL

a. General. A fifth wheel, or semitrailer coupler, is mounted on the rear of M931/A1/A2 and M932/A1/A2 tractor trucks. When connected to a semitrailer, the fifth wheel pivots up, down, and sideways to allow for changes in road conditions. The fifth wheel is rated at 37,500 lb (17,025 kg) cross-country.

b. Wedge Adjustment.

(1) Position fifth wheel wedges (13) fully below walking beam (10) for highway operations.

(2) Position wedges (13) back and away from walking beam (10) for cross-country operations.

(3) To position wedges:

(a) Remove screws (14) from center wedges (13).

(b) Remove wedge (13) and reverse position.

(c) Install screws (14) on center wedges (13) and tighten.

c. Coupling Semitrailer.

WARNING

When backing up, maintain centerline of tractor with centerline of semitrailer and use ground guide.

(1) Back up tractor so fifth wheel coupler jaws (11) are directly in line with semitrailer kingpin (3).

(2) Stop tractor in front of semitrailer, place transmission shift lever (8) in N (neutral), and apply parking brake (9).

(3) Turn landing gear crank (4) to adjust semitrailer height to tractor. Semitrailer approach plates (2) should be slightly lower than tractor fifth wheel (1).

(4) Block semitrailer wheels with chocks (5).

(5) Pull plunger handle (12) forward, then out, to open fifth wheel coupling jaws (11).

(6) Release parking brake (9) and slowly back tractor under semitrailer. Place transmission selector (8) in N (neutral) and apply parking brake (9).

WARNING

Make sure to connect service hose to service coupling and emergency hose to emergency coupling. Hoses not properly connected will cause brake failure.

(7) Connect tractor air coupling (15) to semitrailer air couplings (16).

2-26. OPERATION OF TRACTOR AND FIFTH WHEEL (Contd)

WARNING

Airbrake hose shutoff valves must be open at all times during normal operation of tractor truck and trailer, and brakes should be functional. Failure to follow these precautions may cause injury or death to personnel.

(8) Open shutoff valves (7) by placing handles (6) in alignment with valves (7).

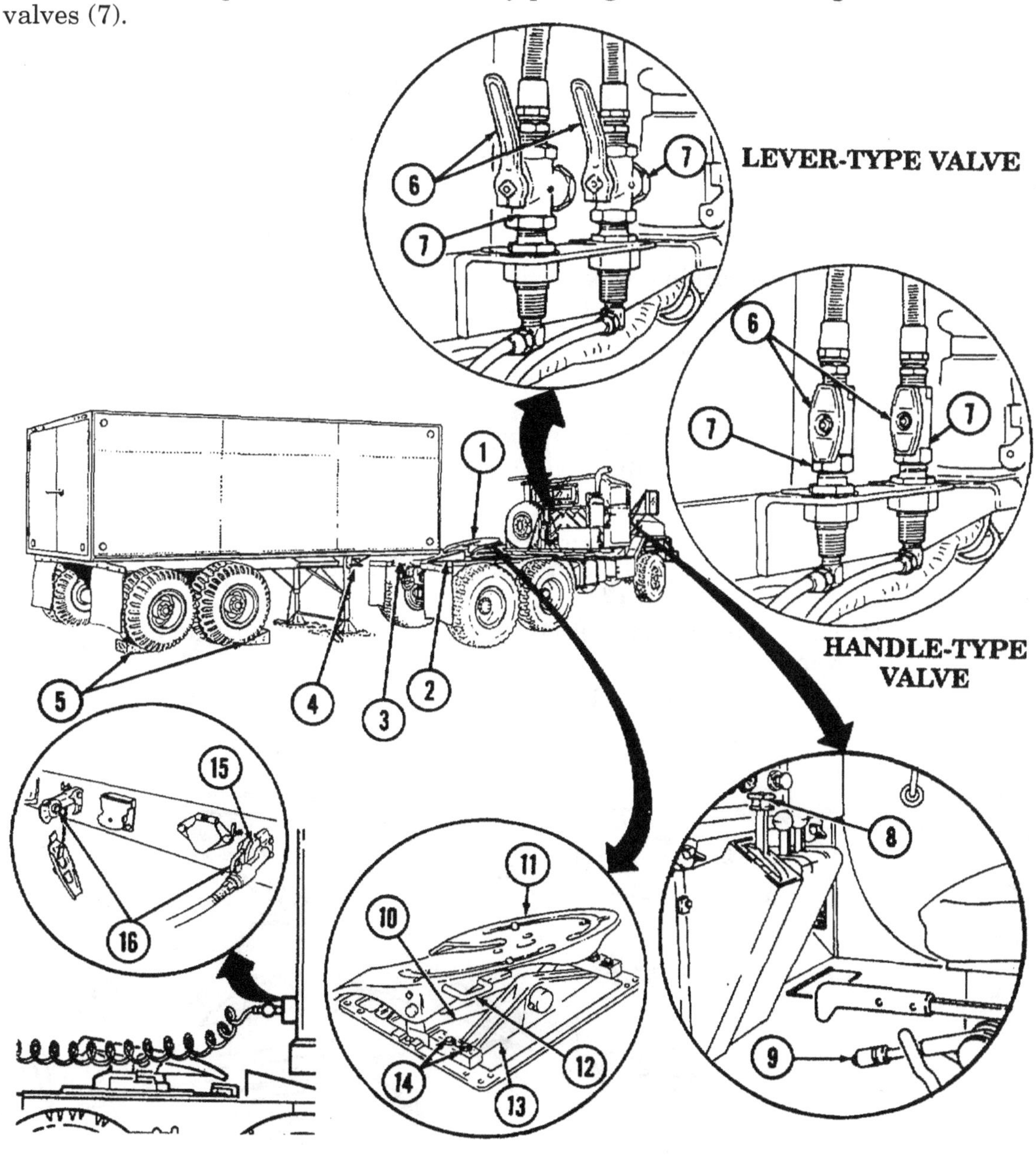
LEVER-TYPE VALVE
HANDLE-TYPE VALVE
1
2
3
4
5
6
7
8
9
10
11
12
13
14
15
16

2-26. OPERATION OF TRACTOR AND FIFTH WHEEL (Contd)

(9) Press in trailer air supply valve (3) and hold in place for 15 seconds. Release valve (3). Valve (3) should remain in engaged position indicating semitrailer airbrake system has proper air pressure. If valve (3) does not remain in engaged position, disconnect couplings (9) and notify your supervisor.

(10) Pull down trailer air brake hand control lever (2) to engage semitrailer brakes.

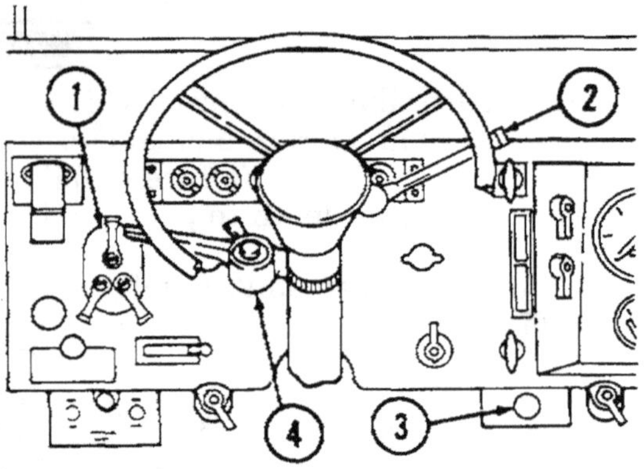
1
2
4
3

WARNING

Do not back up without a ground guide. Doing so may result in damage to vehicle, injury, or death.

(11) Release parking brake lever (6), place transmission selector lever (5) in R (reverse), and resume backing up.

(12) Stop vehicle when coupling jaws (7) close around semitrailer kingpin (8). Visually check to make sure jaws (7) have completely closed.

(13) With trailer air brake control handle (2) engaged, place transmission selector lever (5) in 1 (first) and slightly depress accelerator pedal. Tractor will not move forward if fifth wheel is properly connected to semitrailer.

CAUTION

Stop vehicle immediately if tractor moves forward and repeat task c, steps 9 through 12.

(14) Place transmission selector lever (5) in N (neutral) and apply parking brake lever (6).

(15) Connect electrical cable (11) to electric receptacle (10) on semitrailer.

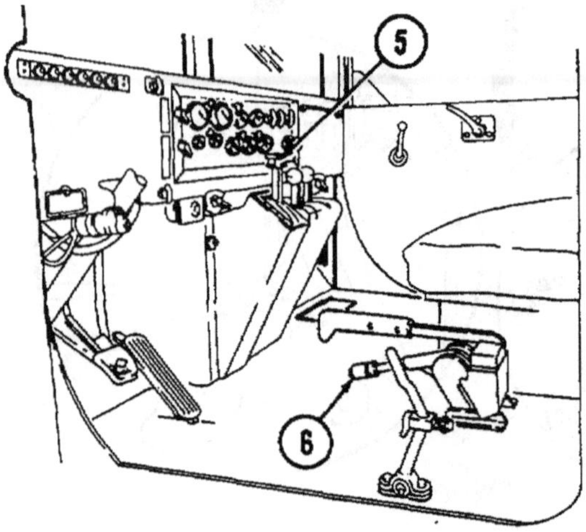
5
6

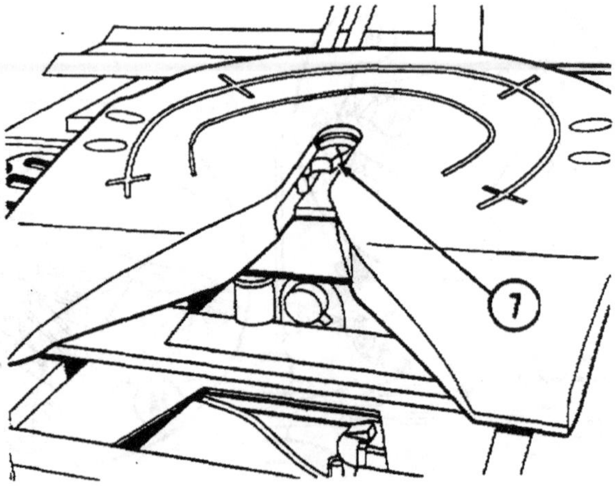
7

2-26. OPERATION OF TRACTOR AND FIFTH WHEEL (Contd)

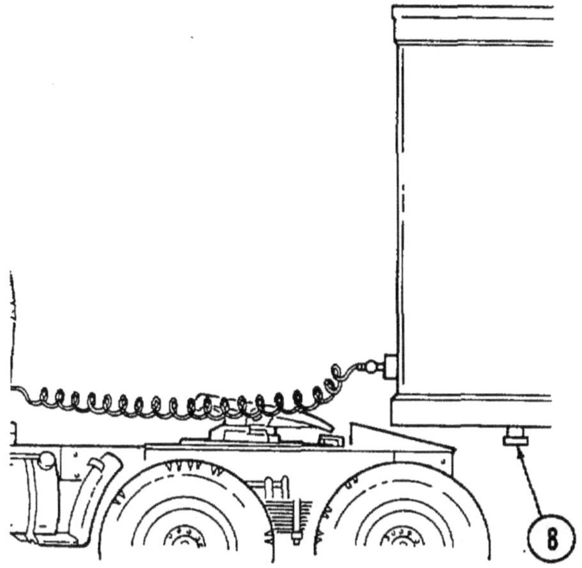
8

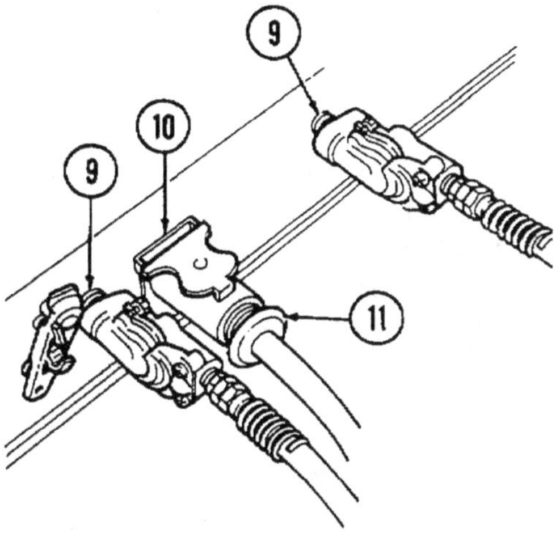
9
10
9
11

(16) Check semitrailer lights:

(a) Turn light switch (1) to STOP TURN position or SER DRIVE (service drive) position.

(b) Operate turn signal switch (4) and direct ground guide to check for proper operation of semitrailer signal lights.

(c) Depress brake pedal and direct ground guide to check for proper operation of semitrailer stoplights.

(17) Turn crank (12) to raise landing gear (14) on semitrailer.

(18) Stow landing gear float pads (15) in racks (13), and remove wheel chocks (16).

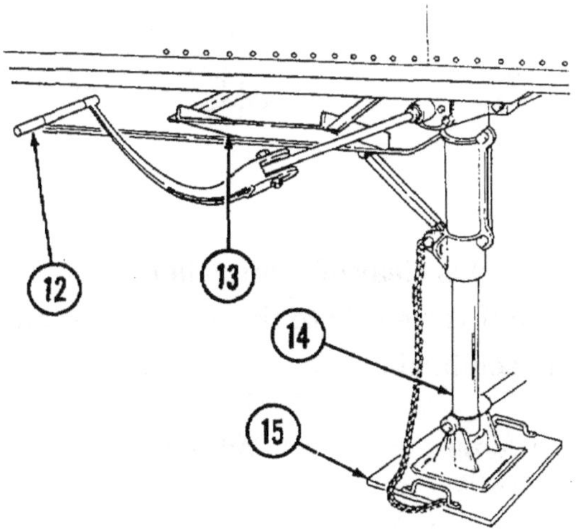
12
13
14
15

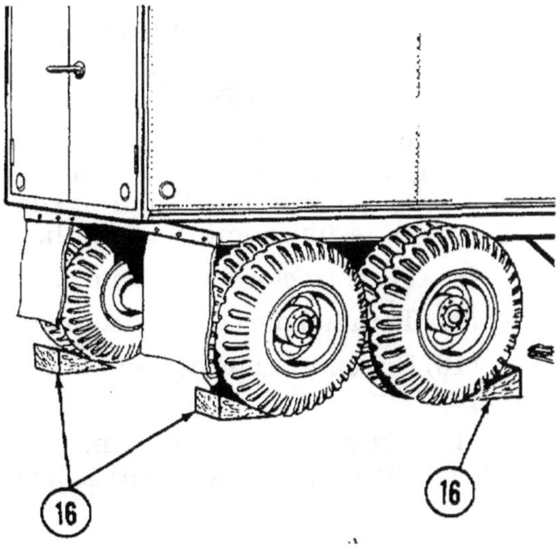
16
16

2-26. OPERATION OF TRACTOR AND FIFTH WHEEL (Contd)

d. Uncoupling Semitrailer.

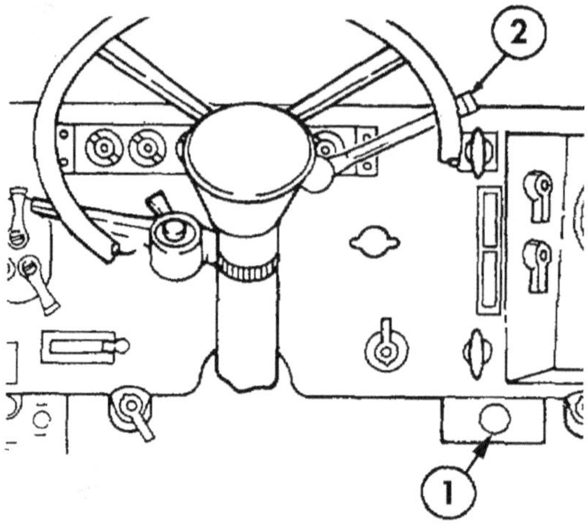
2
1

WARNING

Use ground guide when backing up to park semitrailer. Failure to do this will result in damage to vehicle, injury, or death.

(1) Push trailer air supply valve (M931/A1/A2 and M932/A1/A2 (1) in.

(2) Place semitrailer in proper location and engage airbrake hand control lever/Johnnie bar (2) and parking brake lever (3).

(3) Place wheel chocks (12) in front and behind semitrailer wheels.

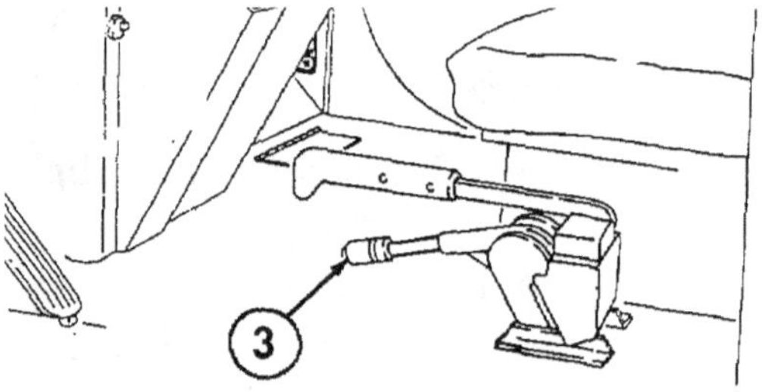
3

(4) Place landing gear float pads (6) on ground under semitrailer landing gear (5).

(5) Turn crank (4) until landing gear (5) makes firm contact with float pads (6).

CAUTION

Ensure all lights are turned to OFF position. Failure to do this will result in damage to vehicles electrical system.

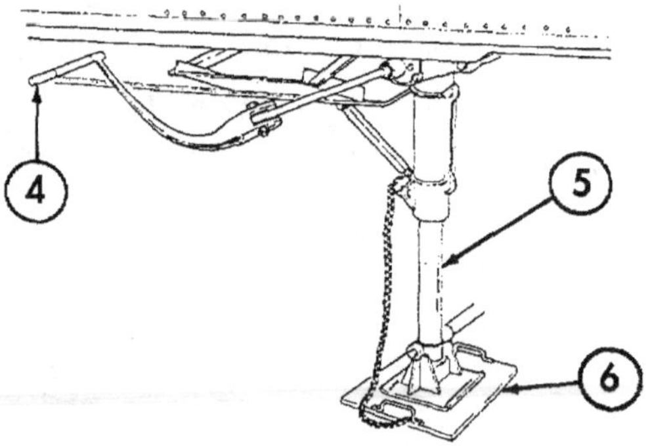
4
5
6

(6) Disconnect and remove electrical cable (8) from semitrailer and secure cable (8) on tractor.

(7) To release semitrailer kingpin (11), pull plunger handle (14) forward, then out, to open fifth wheel coupling jaws (13).

WARNING

Do not pull tractor forward beyond approach ramps until all air lines are disconnected. Failure to do this will result in injury or death.

(8) Move tractor forward until fifth wheel (9) is clear of semitrailer.

(9) Turn handle (16) of airbrake hose shutoff valve (15) to the closed position.

(10) Disconnect airhose couplings (7) from semitrailer and secure airhose couplings (7) on tractor.

(11) Disengage airbrake hand control lever/Johnnie bar (2) and parking brake lever (3), and continue operations without trailer.

2-26. OPERATION OF TRACTOR AND FIFTH WHEEL (Contd)

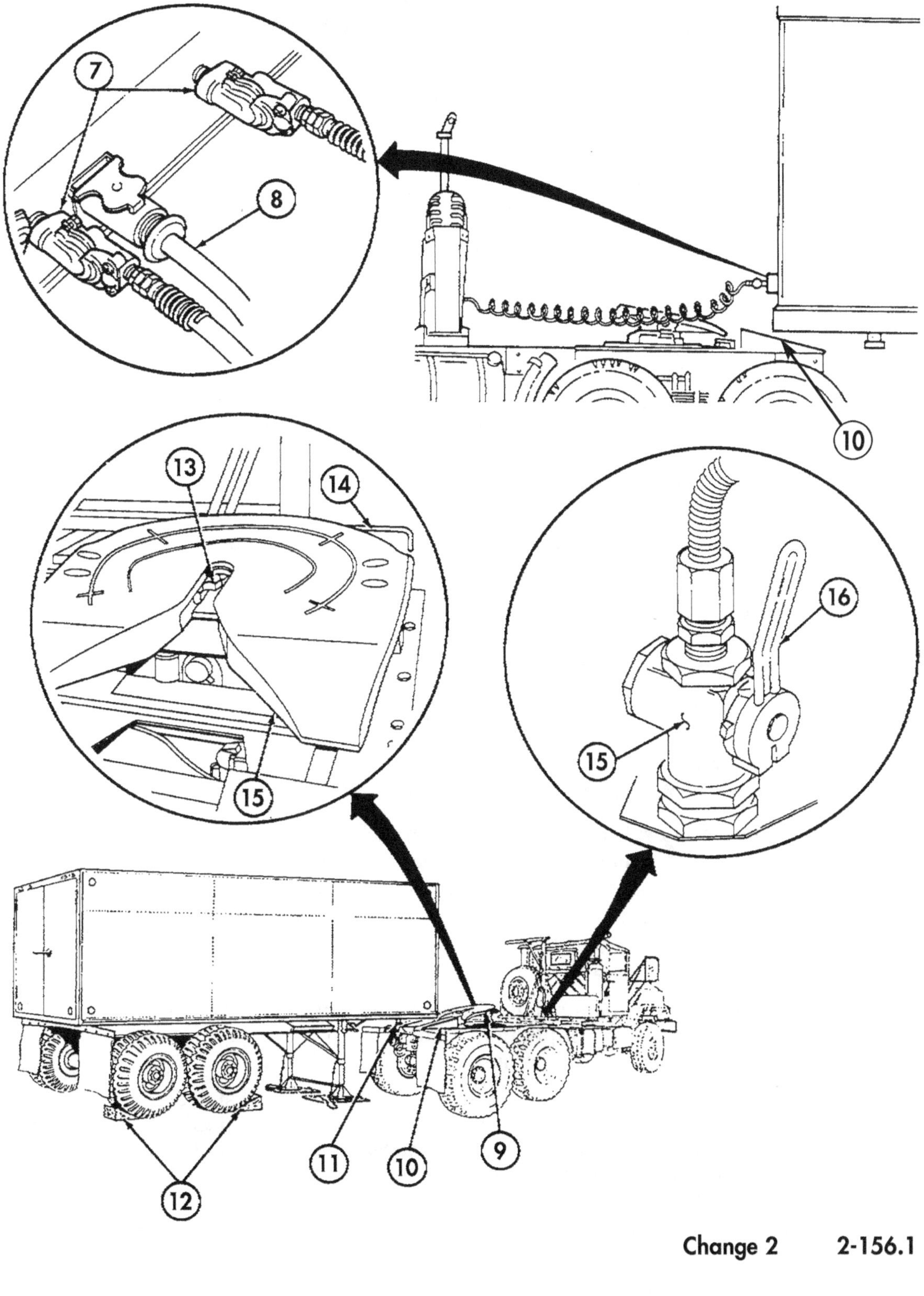
7
8
10
13
14
15
16
15
11
10
9
12

2-26.1. OPERATION OF M931/A1/A2 W/O ABS AND M932/A1/A2 W/O ABS TRACTORS WHEN TOWING M131 SERIES, M967/A1, M969/A1/A2, AND M970/A1 5,000-GALLON FUEL TANKERS

NOTE

This paragraph is for vehicles without ABS.

a. Braking.

WARNING

- M939 series vehicles have a conventional air brake system, which is very sensitive. Drivers of these vehicles must be well-trained in operating tactical vehicles with air brakes. Air brakes are unique because braking force is proportional to pedal travel, but the driver does not experience resistance from the brake pedal. An inexperienced driver may respond to lack of resistance by applying too much force to brake pedal. This may cause brakes to lock up and vehicle to become uncontrollable, resulting in damage to vehicle or injury or death to personnel.
- Apply brakes gradually. Panic stops may cause wheel lockup, stalled engine, and loss of power steering. Failure to comply may cause damage to vehicle or injury or death to personnel.
- Hard braking while turning may cause wheels to lock up and vehicle to skid out of control, resulting in damage to vehicle or injury or death to personnel.
- Do not drive too fast for total weight of vehicle, amount of fuel in tanker, length and angle of grade, road conditions, and weather. Failure to do so may result in damage to vehicle or injury or death to personnel.
- The use of brakes on a long or steep downgrade supplements braking effect of engine. Ensure vehicle is in correct low gear before starting down a grade, and follow correct braking techniques described in steps 1 through 5. Failure to comply may result in damage to equipment or injury or death to personnel.
- Comply with warning signs indicating length and angle of grade. Failure to comply may result in damage to equipment or injury or death to personnel.
- Ensure vehicle is moving 10–15 mph (16–24 km/h) slower than posted ramp speed for entrance or exit ramps. Failure to comply may result in vehicle rollover, causing damage to vehicle or injury or death to personnel.

2-26.1. OPERATION OF M931/A1/A2 W/O ABS AND M932/A1/A2 W/O ABS TRACTORS WHEN TOWING M131 SERIES, M967/A1, M969/A1/A2, AND M970/A1 5,000-GALLON FUEL TANKERS (Contd)

WARNING

- Operation of 5,000-gallon semitrailers (M131 series, M967/A1/A2, and M970/A1) can carry 5,000 gallons of fuel (but not water) when towed with M931/A1/A2 or M932/A1/A2 tractors. This applies to operating on prepared surfaces, such as paved, gravel, or dirt roads. Failure to comply may result in damage to vehicle or injury or death to personnel.
- On cross-country terrain, payload is limited to 3,000 gallons of fuel if the prime mover is an M931/A1/A2 or M932/A1/A2 series 5-ton tractor. Failure to comply may result in damage to vehicle or injury or death to personnel.

1. When making a normal stop, push brake pedal down, controlling pressure so vehicle comes to a smooth and safe stop. For an emergency stop, brake so vehicle can be steered and controlled to maintain a straight line. Use one of the following methods.

WARNING

If wheels lock up, remain off of brakes until wheels begin turning again. It may take approximately one second for wheels to begin turning. Failure to comply may cause vehicle to stray off course of travel, resulting in damage to vehicle or injury or death to personnel.

(a) Controlled Braking. Without turning steering wheel, apply brakes as hard as possible without locking wheels. If large steering adjustments are necessary, or wheels are sliding, release brakes. Brake again when tires gain traction.

(b) Stab Braking. Apply brakes are hard as possible. Release brakes when wheels lock up. Once wheels begin turning, apply brakes fully again.

2-26.1. OPERATION OF M931/A1/A2 W/O ABS AND M932/A1/A2 W/O ABS TRACTORS WHEN TOWING M131 SERIES, M967/A1, M969/A1/A2, AND M970/A1 5,000-GALLON FUEL TANKERS (Contd)

WARNING

Liquid surge results from liquid's movement in partially-filled tanks, which may result in loosing control of 5,000 gallon fuel tankers causing vehicle damage, injury or death to personnel.

2. Because of vehicle's high center of gravity and liquid surge, start, slow down, and stop smoothly. If a quick stop is necessary to avoid a crash, use controlled braking.

3. Know how much space is required to stop vehicle. More space may be necessary to stop an empty tanker vehicle than a full tanker vehicle.

4. Do not tailgate.Always maintain a safe distance from the vehicle ahead of you. Drive far enough behind other vehicles to allow at least three vehicle lengths.

5. Look far enough down the road to avoid unexpected obstacles and perform necessary lan changes. At night, drive slowly enough to observe obstacles with headlights, avoiding the need for sudden lan changes or abrupt stops.

b. Steering.

WARNING

Avoid steering more than necessary to clear an obstacle. Oversteering may cause a skid, jackknife, or rollover. Failure to comply may cause damage to vehicle or injury or death to personnel.

1. After clearing an obstacle, prepare to countersteer (turn steering wheel in opposite direction) to recenter vehicle.

2. Slow down before curves, then accelerate slightly through the curve. Posted speeds for a curve may be too fast for tanker vehicle.

2-26.1. OPERATION OF M931/A1/A2 W/O ABS AND M932/A1/A2 W/O ABS TRACTORS WHEN TOWING M131 SERIES, M967/A1, M969/A1/A2, AND M970/A1 5,000-GALLON FUEL TANKERS (Contd)

c. Speed Limits.

WARNING

- Do not drive faster than road or weather conditions permit. Maximum safe speed limit for normal highway driving is 55 mph (88 km/h).
- Stopping can be adversely affected by poor road/weather conditions. Drive at a slow, safe speed in poor conditions to avoid excessive braking. Failure to comply may result in damage to equipment or serious injury or death to personnel.

1. At unit-level maintenance, all operators must adhere to the following maximum driving speeds:

Prepared surfaces (paved, gravel, or dirt roads) . . . .40 mph (64 km/h)

Cross-country . . . . .35 mph (56 km/h)

Sand, snow, mud . . . . .25 mph (40 km/h)

Icy conditions . . . . .12 mph (19 km/h)

2. When necessary to slow down vehicle to maintain posted safe speed limit, perform the following steps:

(a) Apply brakes just hard enought to feel a definite slowdown of vehicle. This brake application should last for roughly three seconds.

(b) When speed has been reduced to approximately 5 pmh (8 km/h) below posted speed limit, release brakes.

2-26.2. OPERATION OF M931/A1/A2 W/ABS AND M932/A1/A2 W/ABS TRACTORS WHEN TOWING M131 SERIES, M967/A1, M969/A1/A2, AND M970/A1 5,000-GALLON FUEL TANKERS

WARNING

Death or serious injury to soldiers, or damage to army equipment will occur if the instructions in this procedure are not followed.

a. Payload.

WARNING

- Operation of 5,000-gallon semitrailers (M131 series, M967/A1/A2, and M970/A1) can carry 5,000 gallons of fuel (but not water) when towed with M931/A1/A2 or M932/A1/A2 tractors. This applies to operating on prepared surfaces, such as paved, gravel, or dirt roads. Failure to comply may result in damage to vehicle or injury or death to personnel.
- On cross-country terrain, payload is limited to 3,000 gallons of fuel if the prime mover is an M931/A1/A2 or M932/A1/A2 series 5-ton tractor. Failure to comply may result in damage to vehicle or injury or death to personnel.
- Liquid surge results from liquid's movement in partially-filled tanks, which may result in loosing control of 5,000 gallon fuel tankers causing vehicle damage, injury or death to personnel.

b. Mountain Operations.

The general rule-of-thumb for determining proper gear range and speed is to use the same speed descending a grade as the speed achieved ascending the grade. A typical descent speed for this combination on a 9 percent grade should be limited to 20 mph (32 km/h). The brake system provides more than adequate brake capacity for safe mountain terrain operations, if properly operated and maintained.

c. Braking.

WARNING

Stopping distance is generally reduced by ABS technology. ABS technology is designed to perform a conventional braking technique called "stab" braking automatically using wheel speed sensors. Drivers must understand they should not pump the brakes on an ABS-equipped vehicle, as this will deactivate the ABS. Drivers must also understand that by removing pressure from the brake pedal, drivers can also deactivate the ABS. Failure to comply may result in damage to vehicle or injury or death to personnel.

2-26.2. OPERATION OF M931/A1/A2 W/ABS AND M932/A1/A2 W/ABS TRACTORS WHEN TOWING M131 SERIES, M967/A1, M969/A1/A2, AND M970/A1 5,000-GALLON FUEL TANKERS

WARNING

- When the ABS senses impending wheel lockup, the ECU will modulate the relays which will repeat a “release and recharge” cycle of air in the brake chambers. Unlike a car’s ABS, where you can feel this modulation on the brake pedal, you will not feel any modulation of the brake pedal on an air brake system. When the ABS does modulate, you will feel a jerking sensation of the vehicle as the brakes rapidly release and lock. Failure to comply may result in damage to vehicle or injury or death to personnel.
- M939 series vehicles have a conventional air brake system, which is very sensitive. Drivers of these vehicles must be well-trained in operating tactical vehicles with air brakes. Air brakes are unique because braking force is proportional to pedal travel, but the driver does not experience resistance from the brake pedal. An inexperienced driver may respond to lack of resistance by applying too much force to brake pedal. Operators can be confident that M939 series trucks equipped w/ABS brakes have more than adequate brake capacity for safe mountain terrain operations.
- Do not drive too fast for total weight of vehicle, amount of fuel in tanker, length and angle of grade, road conditions, and weather. Failure to do so may result in damage to vehicle or injury or death to personnel.
- Comply with warning signs indicating length and angle of grade. Failure to comply may result in damage to equipment or injury or death to personnel.
- Ensure vehicle is moving 10–15 mph (16–24 km/h) slower than posted ramp speed for entrance or exit ramps. Failure to comply may result in vehicle rollover, causing damage to vehicle or injury or death to personnel.

d. Convoy Integrity. For convoys that include medium tractors transporting petroleum tankers, convoy speeds will be established in advance of movement so as not to exceed the maximum safe speed of the slowest transport equipment operating within each convoy. Convoy briefings will set forth the speed restrictions with a clear statement to all convoy participants as to the purpose and reasons for such speed restrictions.

2-26.2. OPERATION OF M931/A1/A2 W/ABS AND M932/A1/A2 W/ABS TRACTORS WHEN TOWING M131 SERIES, M967/A1, M969/A1/A2, AND M970/A1 5,000-GALLON FUEL TANKERS

e. Safe Operation.

1. ABS allows the wheels to roll while the driver maintains full brake pressure on the brake pedal. The rolling action helps to regain traction control (stability) on the rear wheels.

2. Because of vehicle's high center of gravity and liquid surge, start, slow down, and stop smoothly. If a quick stop is necessary to avoid a crash, maintain full brake pressure.

3. Know how much space is required to stop vehicle. More space may be necessary to stop an empty tanker vehicle than a full tanker vehicle.

4. Do not tailgate. Always maintain a safe distance from the vehicle ahead of you. Drive far enough behind other vehicles to allow at least three vehicle lengths.

5. Look far enough down the road to avoid unexpected obstacles and perform necessary lane changes. At night, drive slowly enough to observe obstacles with headlights, avoiding the need for sudden lane changes or abrupt stops.

f. Steering.

WARNING

- ABS technology is designed to maintain rolling traction and steering. The rolling action may produce longer stopping distances on some surfaces, such as freshly fallen snow or loose gravel. The ABS steering advantage outweighs any braking disadvantage on these surfaces. Evasive steering techniques are designed to allow the driver to steer the vehicle clear of damage. By maintaining a speed reduction without wheel lockup, ABS increases steerability of the vehicle. The driver should use just enough steering movement to adjust the vehicle to a clear space on the roadway.
- Avoid steering more than necessary to clear an obstacle. Oversteering may cause a skid, jackknife, or rollover. Failure to comply may cause damage to vehicle or injury or death to personnel.

Slow down before curves, then accelerate slightly through the curve. Posted speeds for a curve may be too fast for tanker vehicle.

2-26.2. OPERATION OF M931/A1/A2 W/ABS AND M932/A1/A2 W/ABS TRACTORS WHEN TOWING M131 SERIES, M967/A1, M969/A1/A2, AND M970/A1 5,000-GALLON FUEL TANKERS

g. Speed Limits.

WARNING

- Do not drive faster than road or weather conditions permit. Maximum safe speed limit for normal highway driving is 55 mph (88 km/h).
- Stopping can be adversely affected by poor road/weather conditions. Drive at a slow, safe speed in poor conditions to avoid excessive braking. Failure to comply may result in damage to equipment or serious injury or death to personnel.

All operators must adhere to the following maximum driving speeds:

Prepared surfaces (paved, gravel, or dirt roads) . . . .55 mph (88 km/h)

Cross-country . . . . .40 mph (64 km/h)

Sand, snow, mud . . . . .25 mph (40 km/h)

Icy conditions . . . . .5-12 mph (8-19 km/h)

2-27. OPERATION OF EXPANSIBLE VAN TRUCK

a. General. Expansible van trucks transport communication equipment into the field. Van bodies are 8 ft (2.4 m) wide in travel position. In the field, van sides expand to nearly 14 ft (4.3 m).

b. Selecting Operating Site. Whenever possible, position van on level, firm ground.

WARNING

Block vehicle wheels if operating site is on a grade, no matter how slight. Failure to do this will result in injury or death.

c. Leveling Van Body.

(1) Remove four adjustable leveling jacks (4) and foot pads (5) from stowage compartment in rear of van body.

(2) Attach foot pad (5) to bottom of each jack (4). Assemble and install inner and outer tubes of jack (4) and adjust length to approximate height of brackets (1) (marked H) at each corner of van. Secure jack (4) with chained pin (7).

(3) Insert upper foot (2) of jack (4) into bracket (1). Install jack handle (3) and unscrew jack (4) until foot pads (5) are in firm contact with ground. Anchor each jack foot pad (5) with two jack spikes (6). Do not attempt to raise entire van off of ground with leveling jacks.

(4) Repeat procedure at each corner of van until body is level.

d. Expanding Van Body.

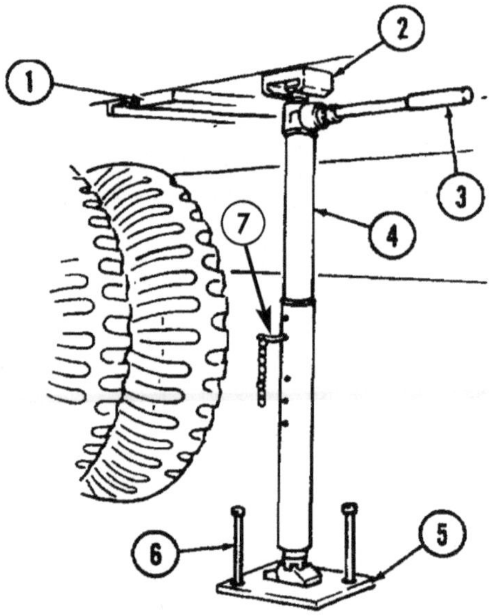
1
2
3
4
5
6
7

WARNING

Open van door slowly. Personnel may be on ladder. Use caution when using ladder.

CAUTION

Vehicle must be approximately level for expansion or retraction of van body.

NOTE

This operation requires two crewmembers.

(1) Ladder setup:

(a) Release toggle clamp (11) securing right ladder (10).

(b) Lift ladder (10) up to remove.

(c) Install ladder (10) in brackets below rear doors.

(2) Remove chained pin (16) from lock handle (15). Pull handle (15) out and disengage handle (15) end from retaining bracket (17). Repeat procedure at all four van corners.

2-27. OPERATION OF EXPANSIBLE VAN TRUCK (Contd)

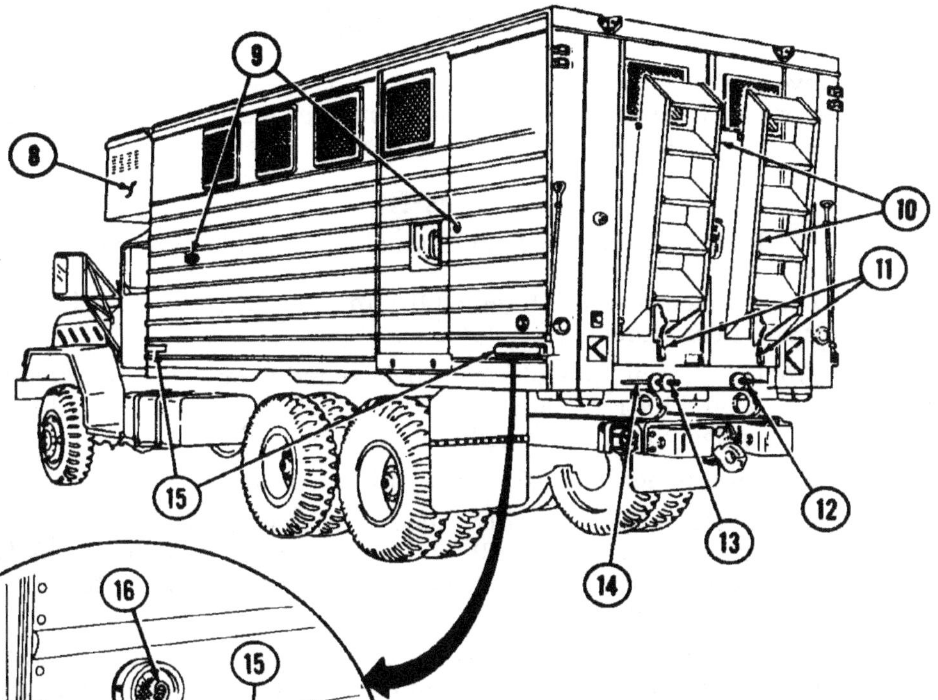
9
8
10
11
15
12
13
14

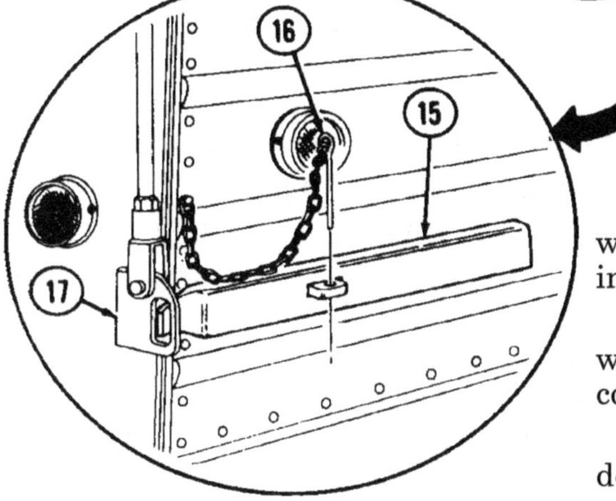
16
15
17

(3) Remove side panel lock wrench and ratchet wrench from holders inside left rear door.

(4) Using side panel lock wrench, turn four side locks (9) (marked A) counterclockwise.

(5) Push locking plunger (14) downward to release left ratchet (13).

(6) Turn left ratchet (13) (marked B) counterclockwise with ratchet wrench to expand left side panel (8) on left side of vehicle. Turn right ratchet (12) (marked B) to expand right side panel (8) on right side of vehicle. Crank both sides fully out.

(7) Unfold two end panels (18) on each side panel (8). Unclip and use side lock rod (19) (marked C) to keep end panel door open and out of the way while roof (21) and floor panels (20) are being raised and lowered.

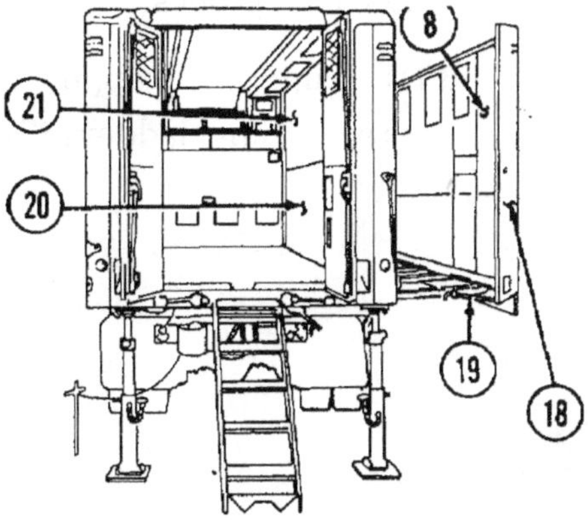
8
21
20
19
18

2-27. OPERATION OF EXPANSIBLE VAN TRUCK (Contd)

WARNING

Have crewmember support raised floor panel when operator turns hinged roof lock handle. Floor panel may fall, resulting in injury or death.

NOTE

Push open roof and floor panel from inside van only.

(8) Turn hinged roof lock handle (10) (marked D) counterclockwise to unlock roof panel (3) and floor panel (4).

(9) From inside van, push hinged roof (3) and floor panel (4) outward, step out onto floor panel (4) and lift roof panel (3) until swivel hooks (8) (marked E) can be turned at right angles. Support hinged roof (3) on three swivel hooks (8).

(10) Slide end panel bolt (15) (marked F) into corner post guide (14).

(11) Crank both sides (1) in with ratchet wrench (17) until swivel eyes (7) on toggle clamps (9) (marked G) on van roof (3) can be attached to swivel hooks (8). Left ratchet (16) is turned clockwise to retract left side of van. Right ratchet (18) is turned counterclockwise to retract right side of van.

(12) Pull side panel (2) straight by partially closing toggle clamps (9). While doing this, push up on hinged roof (3) and out on end panel (1) to ensure seal alignment.

(13) Stand on hinged floor panel (4) to relieve any binding.

(14) Adjust left ratchet (16) and right ratchet (18) to ensure a tight seal.

(15) Close three toggle clamps (9) on each side (2), closing the center clamp first.

(16) Remove side swing rod (6) from retaining clip (11). Swing rod (6) down and engage end of swing rod (6) with lock handle assembly (12). Push assembly closed and secure with chained pin (13). Repeat procedure at all four van corners.

NOTE

Make sure sliding end panel bolts are fully extended into corner post guides.

(17) Set up van as applicable:

(a) Mount ladders (5) to rear and/or side doors.

2-27. OPERATION OF EXPANSIBLE VAN TRUCK (Contd)

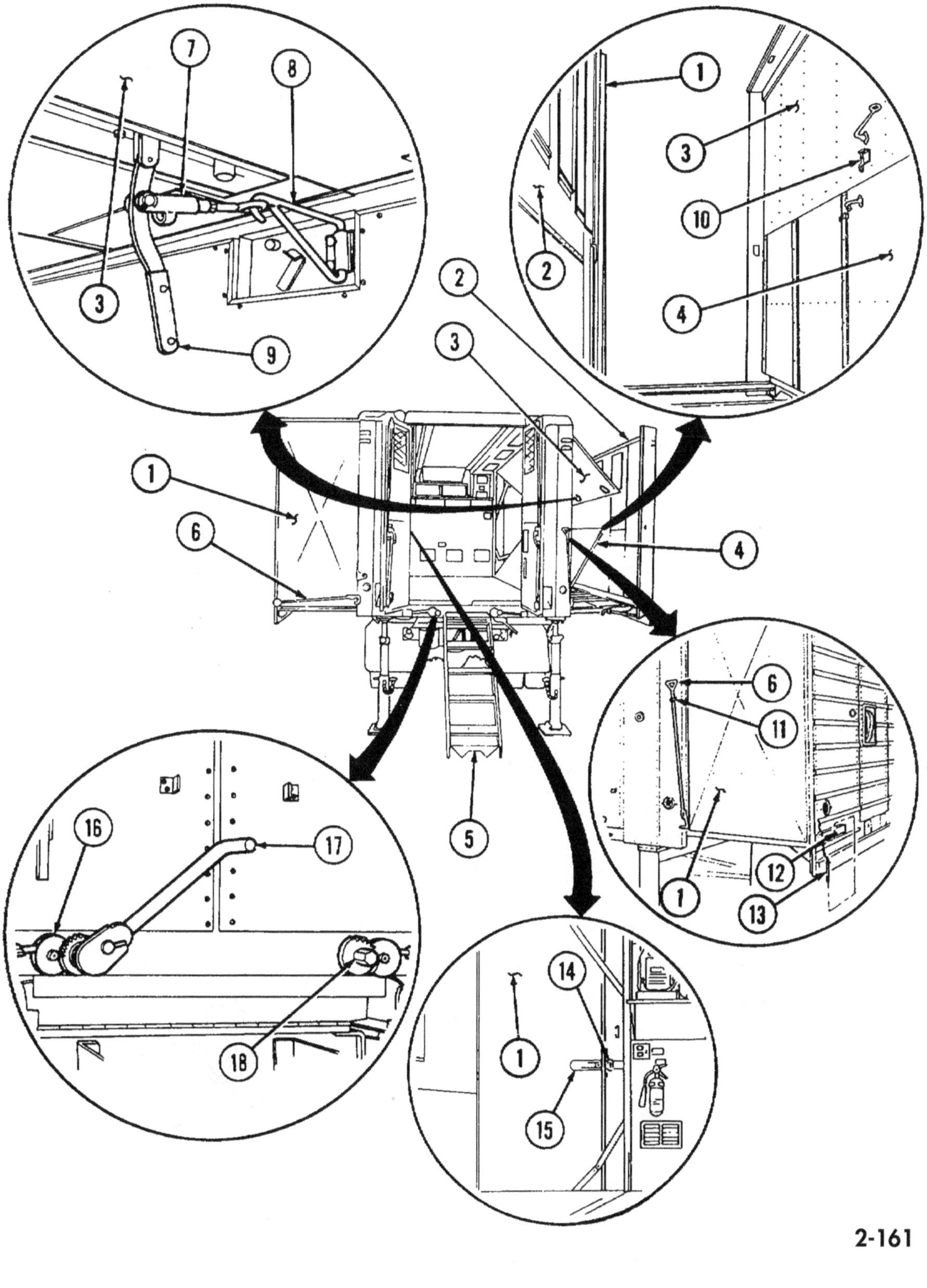

7
8
3
9
1
3
10
2
4
2
3
1
6
4
6
11
12
13
1
5
16
17
18
14
1
15

2-27. OPERATION OF EXPANSIBLE VAN TRUCK (Contd)

(b) Remove ground spike (6) and spike cable (4) from storage location.

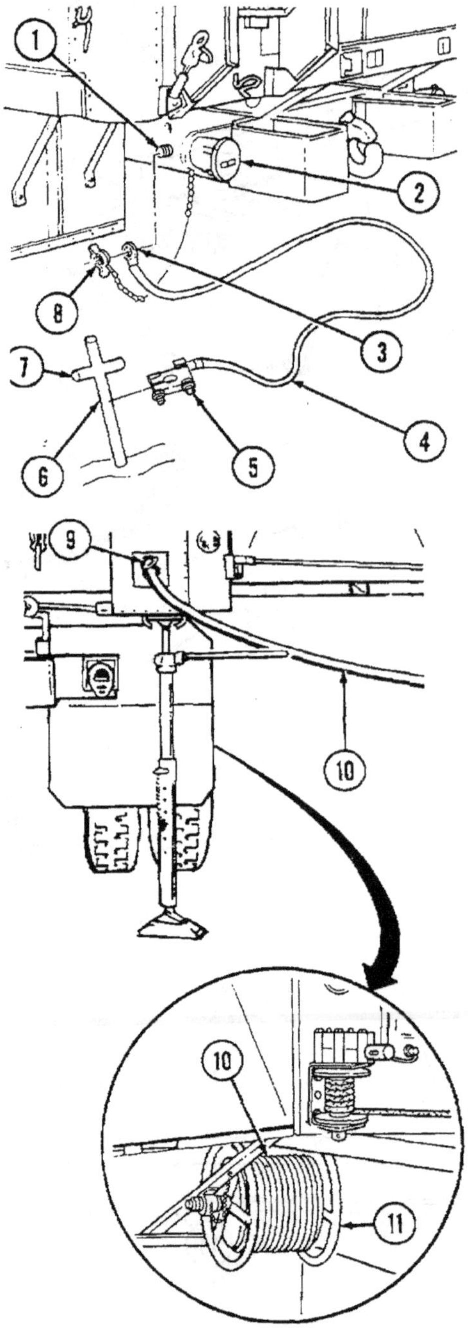
1
2
3
4
5
6
7
8
9
10
10
11

WARNING

- Ground spike must be driven into ground 18-24 in. (46-61 cm) and spike cable connected to the chassis, ensuring that all terminals make good contact with bare metal before connecting power from outside source. If necessary, scrape dirt, paint, or rust from contact area. Failure to do this will result in electrical damage, injury, or death.
- Open van door slowly. Personnel may be on ladder. Use caution when using ladder.

(c) Remove wingnut (8) and connect ring terminal (3) on spike cable (4) to chassis stud (1) behind left-rear stoplight (2). Connect spike cable terminal clamp (5) to ground spike (6) and slide up to T-handle (7).

(d) Drive ground spike (6) 18-24 in. (46-61 cm) into ground.

(e) Remove canvas cover on cable reel (11).

(f) Remove power cable (10) on cable reel (11) using ratchet wrench.

NOTE

It may be necessary to use electric auxiliary cable ring to connect to auxiliary power source.

(g) Connect power cable (10) to appropriate auxiliary power source.

(h) Connect other end of power cable (10) to power receptacle (9).

2-27. OPERATION OF EXPANSIBLE VAN TRUCK (Contd)

e. Operating Van Electrical System.

(1) Inspect ground spike (6), spike cable (4), and chassis stud (1) for loose or damaged components.

(2) Turn to connect switch box (12) to outside power source.

(3) Turn on ceiling light switch (switch 14).

(4) Turn on receptacle switches (switches 1, 3, 4, 6, 7, 9, 10, 12, 13).

(5) Turn on switch 5 if left heater (13) is to be used. Turn on switch 8 if right heater (15) is to be used. Refer to step f. for left and right heater operating instructions.

(6) Turn air conditioner switch if air conditioning unit (14) is to be used. Refer to step h. for air conditioner operating instructions.

(7) Turn on switch 2 if blackout switch (16) is to be used. Refer to step i. for blackout operating instructions.

(8) Turn on emergency light switch (17) if outside power source fails.

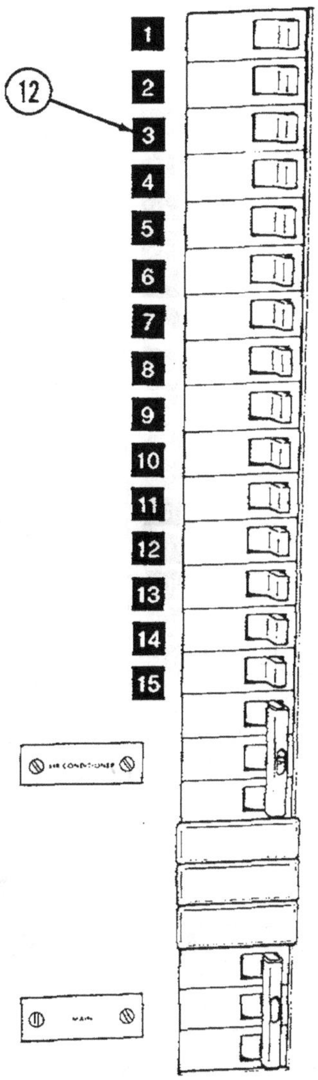
12
1
2
3
4
5
6
7
8
9
10
11
12
13
14
15

f. Operating Van Heaters.

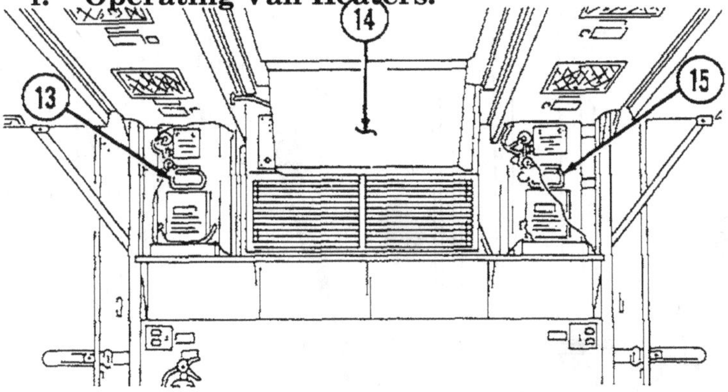
14
13
15

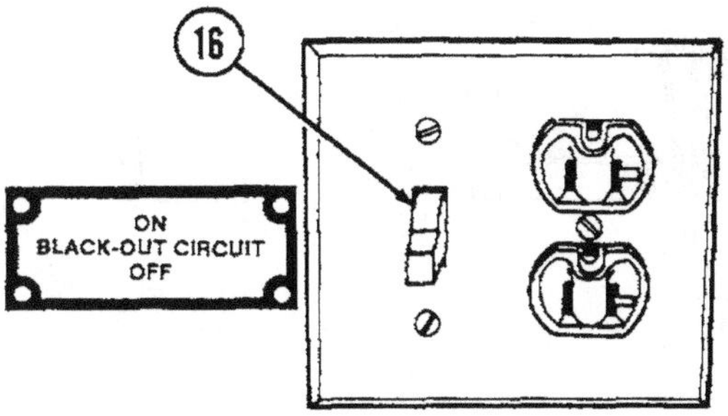
16
ON
BLACK-OUT CIRCUIT
OFF

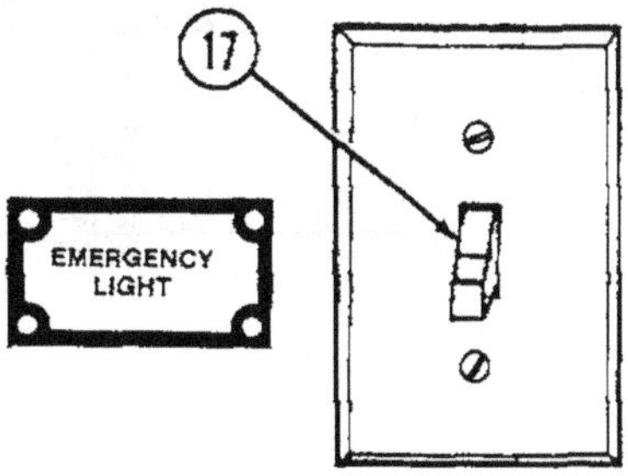
17
EMERGENCY
LIGHT

2-27. OPERATION OF EXPANSIBLE VAN TRUCK (Contd)

WARNING

Ground spike must be driven into ground 18-24 in. (46-61 cm) and spike cable connected to the chassis before power can be taken from outside source. Failure to do this will result in electrical damage, injury, or death.

(1) Turn on main switch in circuit breaker switch box (1).

(2) Turn on left heater switch (switch 5) and/or right heater switch (switch 8) in circuit breaker switch box (1).

(3) Set heater thermostat (2) to desired temperature.

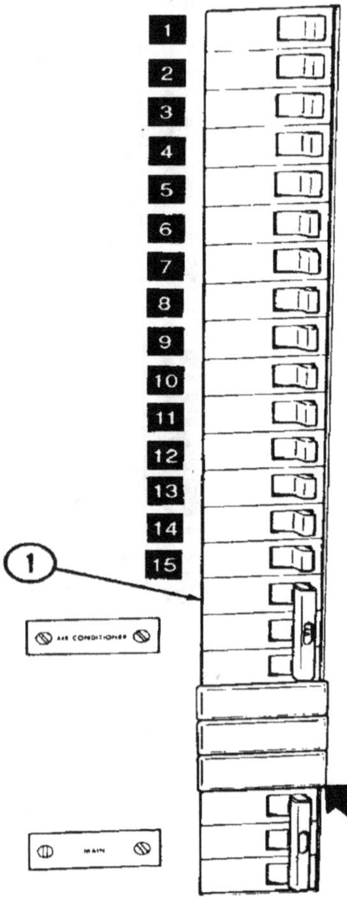
1
2
3
4
5
6
7
8
9
10
11
12
13
14
15
1

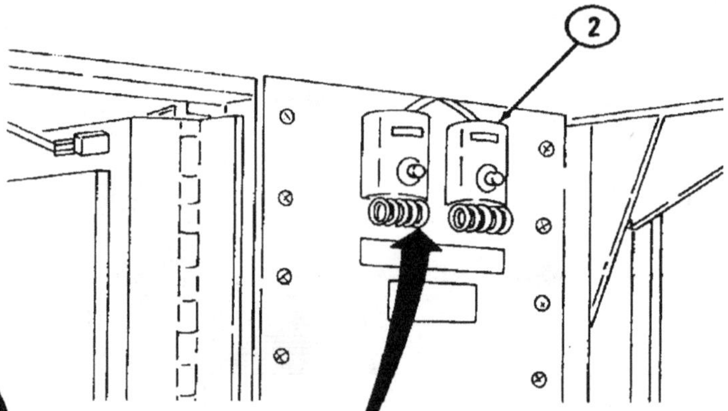
2

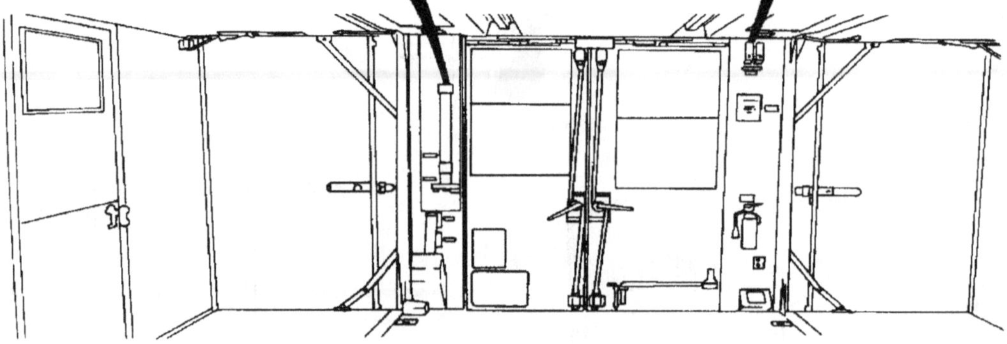

2-27. OPERATION OF EXPANSIBLE VAN TRUCK (Contd)

(4) Open heater fuel shutoff valve (5).

(5) Set heater switch (3) to HEATER (for heated air) or FAN (for unheated air) as desired. White indicator light (4) should come on when heater is working properly.

(6) Set louver operating handles (4) on each side to control mix of outside air with recirculated air.

(7) Open heat registers (7) below heaters.

(8) To stop heaters, turn off heater control switch (3).

CAUTION

Do not turn off heater circuit breaker switches (switches 5 and 8) until white indicator light goes off. Damage to heater will result.

(9) To secure heater for transit, turn heater fuel shutoff (5) and turn off left heater switch (switch 5) and right heater switch (switch 8), as required.

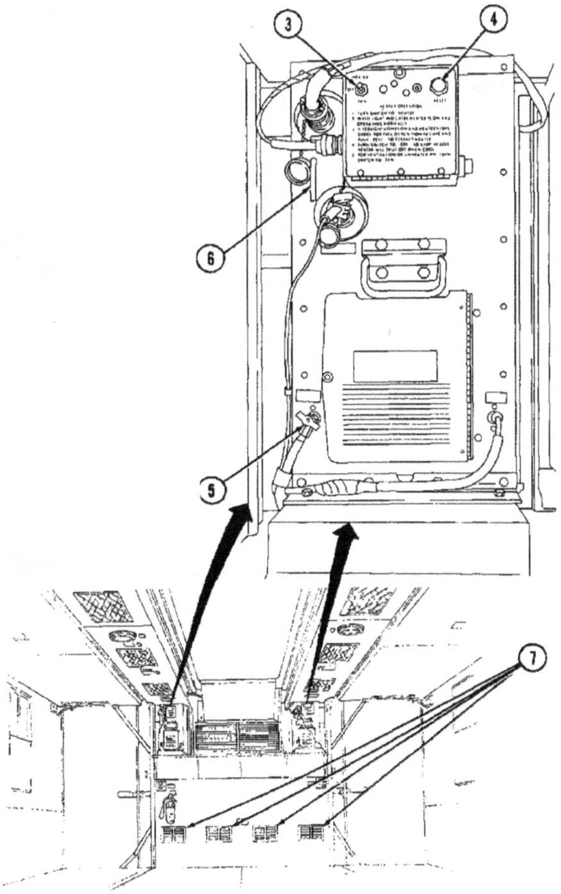
3
4
6
5
7

2-27. OPERATION OF EXPANSIBLE VAN TRUCK (Contd)

g. Operating Van Air Conditioner.

(1) Push bonnet door control rod (9) forward to open bonnet door (11).

(2) Turn on main circuit breaker switch in circuit breaker switch box (1).

(3) Turn on air conditioner switch in circuit breaker switch box (1).

(4) Turn on power input switch (3) and compressor circuit breaker (4).

(5) Set air conditioner control (5) to COOL for cold air or VENT for ventilation of outside air into van.

(6) Turn compressor switch (6) to HIGH when starting air conditioning unit. Turn switch (6) to LOW after desired temperature is obtained.

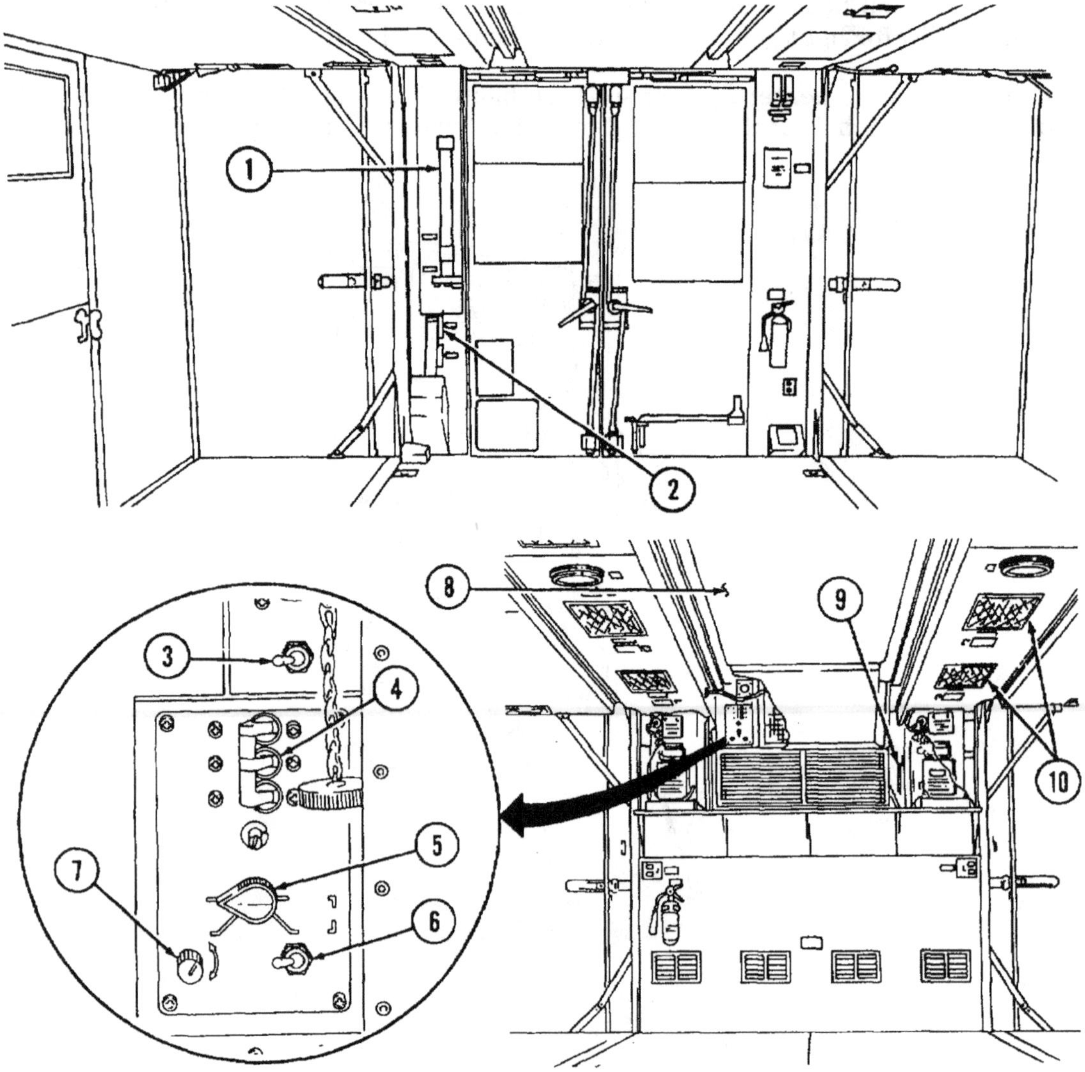
1
2
3
4
5
6
7
8
9
10

2-27. OPERATION OF EXPANSIBLE VAN TRUCK (Contd)

(7) Adjust temperature selector (7). Cooler temperatures are obtained when temperature selector (7) is turned counterclockwise.

(8) Open air conditioner vents (8).

(9) To shut off air conditioner:

(a) Turn air conditioner control (5) to VENT.

(b) Turn compressor circuit breaker (4) to OFF.

(c) Turn power input switch (3) to OFF.

(d) Turn off air conditioner switch in circuit breaker switch box (1).

(e) Pull bonnet door control rod (9) back to close bonnet door (11).

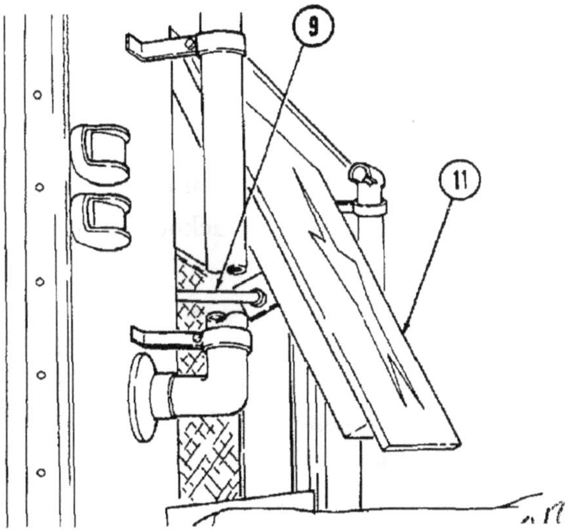
9
11

h. Blackout Operations.

(1) Push up blackout panels (12) on van sides and rear doors to block in all interior light.

(2) Turn on blackout circuit switch (switch 2) in circuit breaker box (1).

(3) Turn on main circuit breaker switch in circuit breaker box (1).

(4) Turn on blackout switch (2). Ceiling lights (10) will cut off automatically when van door is opened.

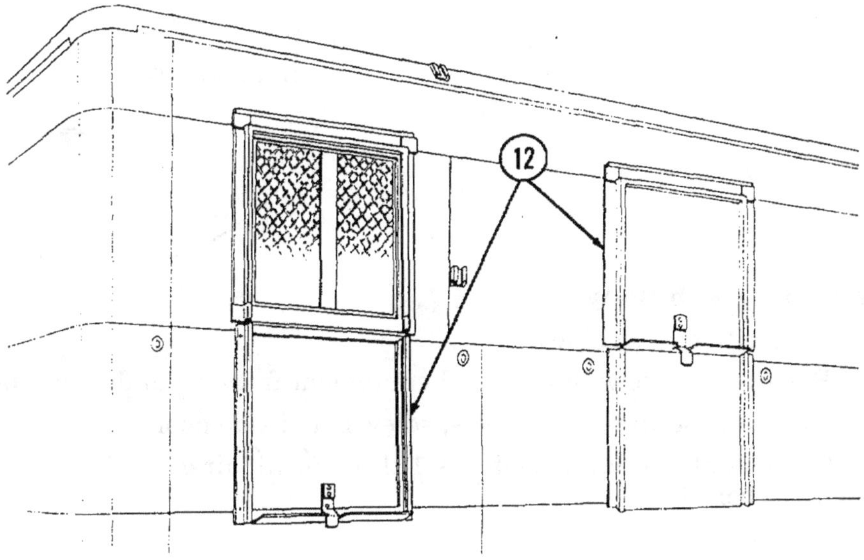
12

2-27. OPERATION OF EXPANSIBLE VAN TRUCK (Contd)

NOTE

Leave switches servicing machines that must operate without interruption during blackout conditions in OFF positions. Lights to operate these machines should be plugged into a separate overhead receptacle with blackout switch in ON position.

(5) After blackout operation:

(a) Turn all overhead receptacle blackout switches (6) to OFF.

(b) Turn off main blackout switch (2).

(c) Turn off blackout circuit breaker (switch 2) in circuit breaker box (1).

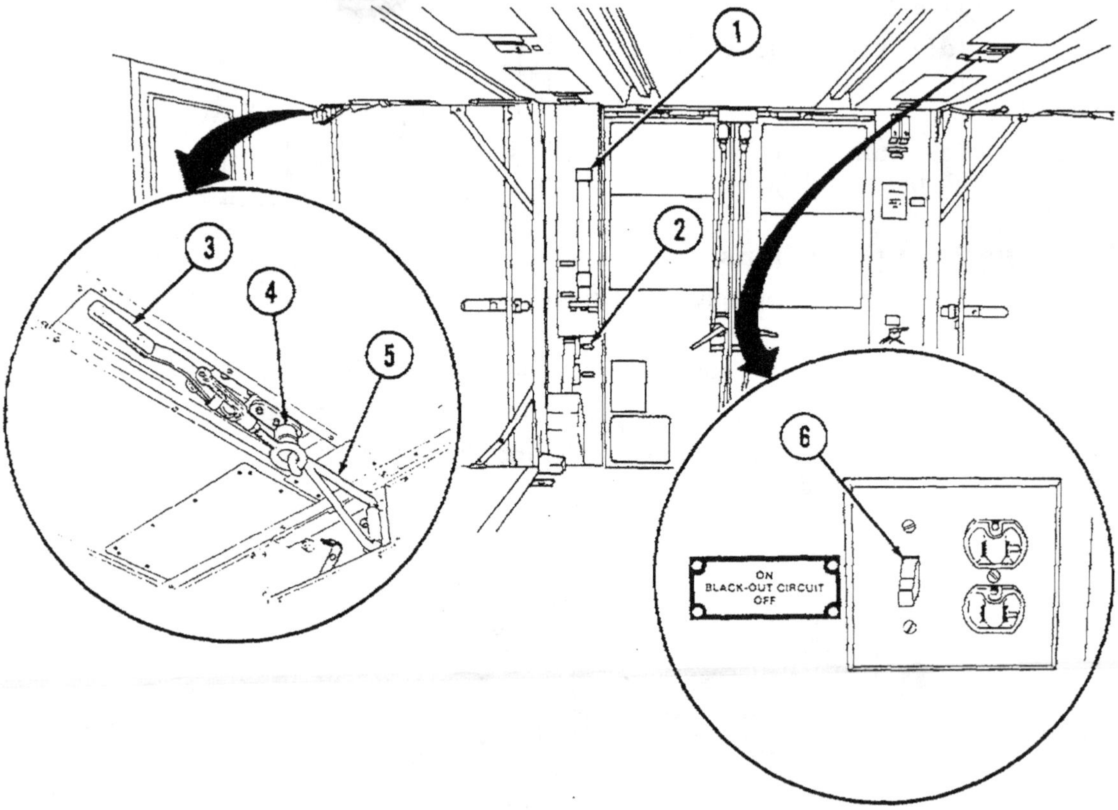
1
2
3
4
5
6
ON
BLACK-OUT CIRCUIT
OFF

i. Retracting Van Body.

(1) Turn off van machines.

(2) Remove and stow all gear and equipment from expanded van floor.

(3) Close and secure all windows, screws, and side doors.

(4) Turn off all switches (switches 1 through 15, air conditioner, and main switch) in circuit box (1).

2-27. OPERATION OF EXPANSIBLE VAN TRUCK (Contd)

(5) Release and unhook six toggle clamps (3). Do not place swivel hooks (5) in stowed position.

(6) Disconnect field telephone lead-in (7), if used.

(7) Disconnect power cable (8) from van auxiliary power entrance receptacle and auxiliary power source.

(8) Disengage side lockrods (9) at all four corners from lock handle assemblies (11) and place rods (9) in retaining clips (10).

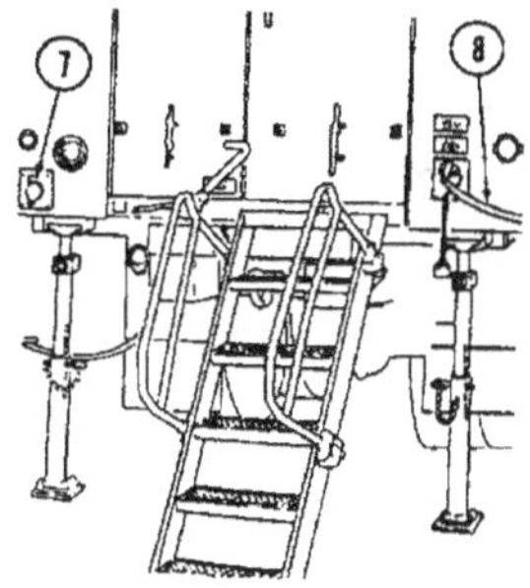
7
8

(9) Make sure locking plunger is unlocked to release right ratchet (13).

(10) Turn right ratchet (13) clockwise with ratchet wrench (14) to expand right side panel (12). Crank side panel (12) until fully expanded.

(11) Retract two end sliding bolts (15).

(12) Push up on hinged roof (16) to free end panels (18). Push out on end panels (17) and hold panels open with holding rod (18).

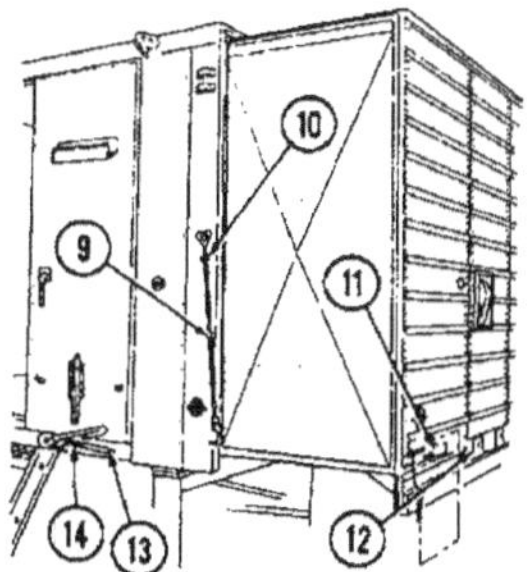
10
9
11
14
13
12

(13) Push up on hinged roof (16) and swing swivel hooks (5) into stored position. Lower hinged roof (16).

(14) Engage eye of each toggle clamp (3) with anchor post (4) in stored position and close toggle clamp (3).

(15) From outside van, push floor panel (19) upward and inward with roof panel (16) downward until fully closed.

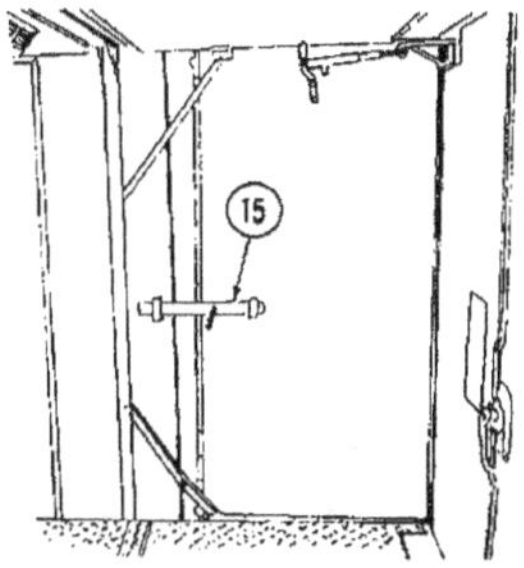
15

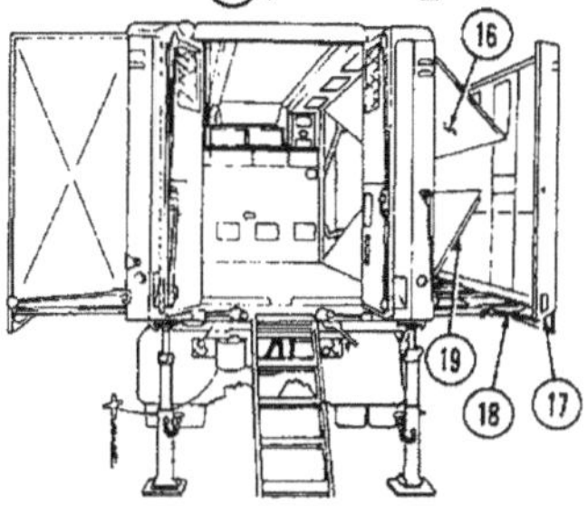
16
19
18
17

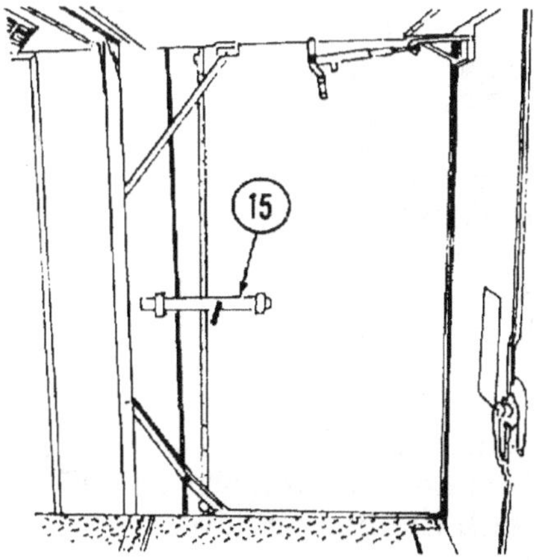
15

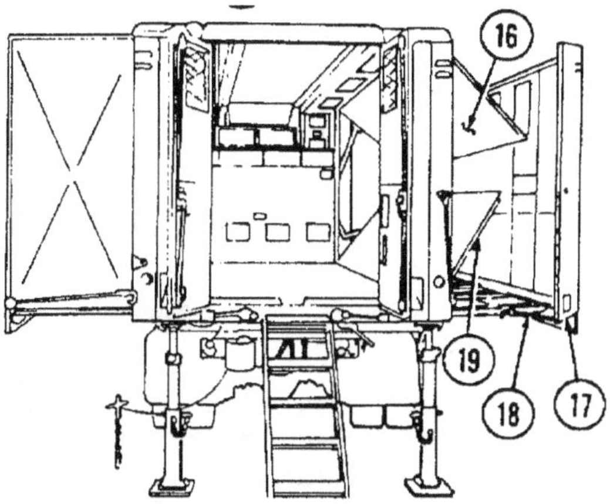
16
19
18
17

2-27. OPERATION OF EXPANSIBLE VAN TRUCK (Contd)

(16) Turn ratchet wrench (8) clockwise to lock floor (5) and roof (1) panels in position.

(17) Remove holding rods (3) from end panels (2) and insert each rod (3) into retaining clips on beam (4).

(18) Close end panels (2) at all corners.

(19) Turn right ratchet (7) counterclockwise with ratchet wrench (8) to retract right side panel (11). Crank side (11) until fully retracted. Push locking plunger (6) upward to lock ratchet (7).

(20) Repeat steps 9 through 19 for left side.

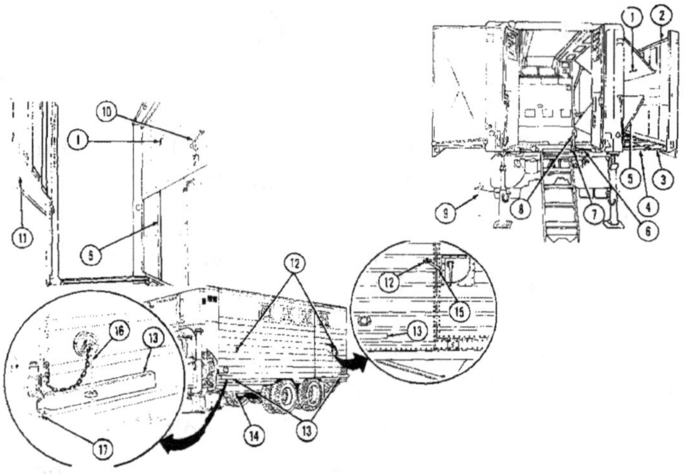

NOTE

Left ratchet operates opposite that of the right ratchet. Reverse direction of rotation for operation in steps 10 and 19.

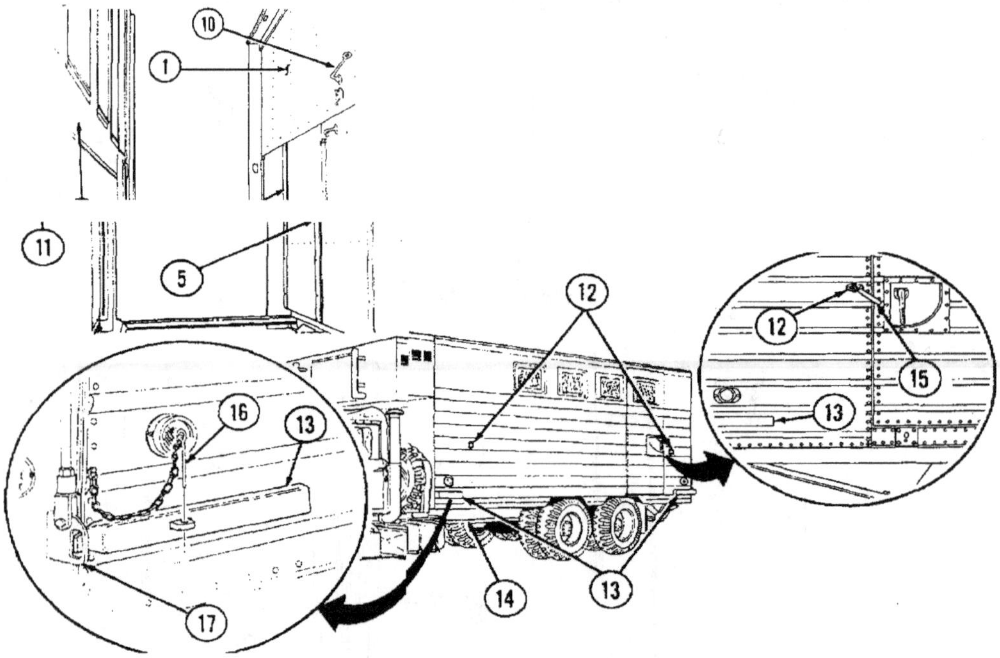
10
1
11
5
12
12
15
16
13
13
17
14
13

(21) Turn four side locks (12) clockwise with side panel lockwrench (15) and insert all four corner ends of lock handle assembly (13) into retaining bracket (17). Close lock handle assembly (13) and secure in place with chained pin (16).

2-27. OPERATION OF EXPANSIBLE VAN TRUCK (Contd)

(22) Remove ground spike (9) and store in storage location (14).

(23) Remove two jack spikes (22) from footpad (21) and install jack handle (19) in leveling jack (20). Rotate jack handle (19) clockwise until footpad (21) clears ground.

(24) Remove leveling jack (20) from van body bracket (18) and detach footpad (21) from jack (20).

(25) Remove leveling jack eyebolt (23) and telescope leveling jack (20). Reinsert eyebolt (23).

(26) Repeat steps 23 through 25 with three remaining van body leveling jacks (20).

(27) Stow two leveling jacks (20), footpads (21), and jack spikes (22) in compartment at rear of van body.

(28) Stow all tools and equipment used in van operation.

(29) Close and secure rear doors.

(30) Install ladders (24) to rear doors (26) and secure for travel with toggle clamps (25).

(31) Start with van end of auxiliary power cable (28) and, using handle (29), wind up power cable (28) on reel (27).

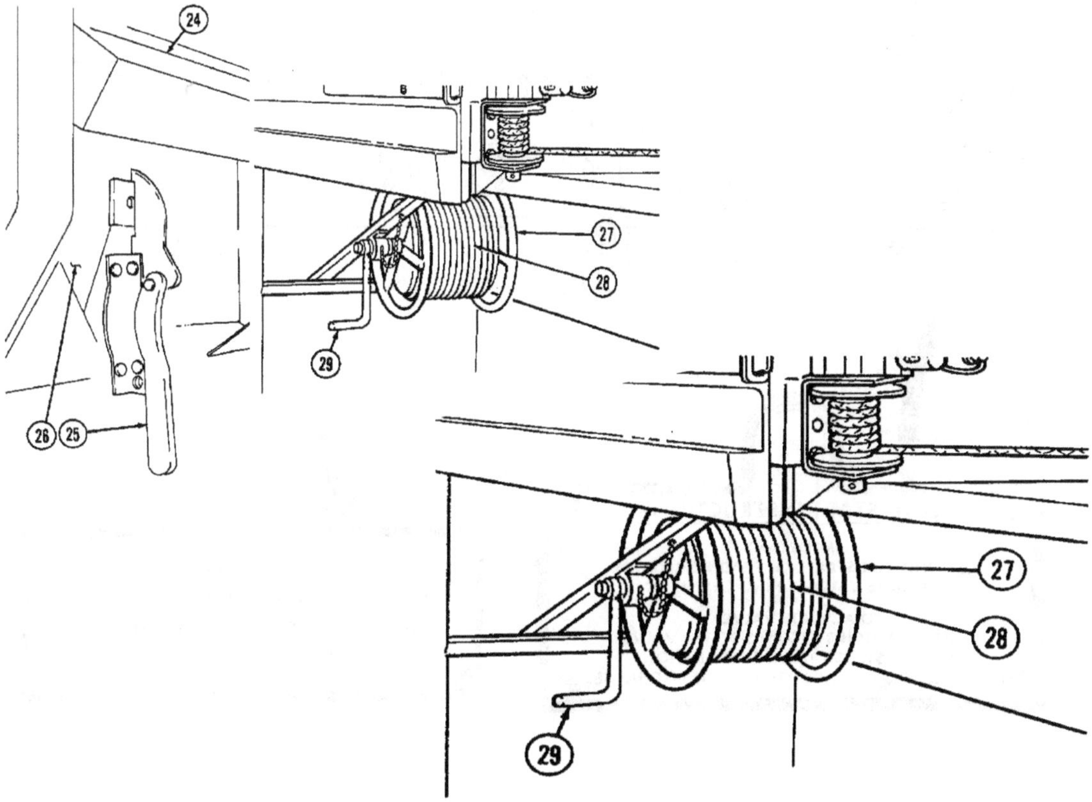
24
27
28
29
26
25
27
28
29

2-28. DECALS AND INSTRUCTION PLATES

a. The location and contents of decals and instruction plates are provided in this paragraph. A complete list and location of all decals and instruction plates is in TM 9-2320-272-20P. If any of these plates are worn, broken, painted over, missing, or unreadable, they must be replaced.

b. Below are those decals and plates that are located inside the cab. These decals and plates are common to one or more models covered in this manual.

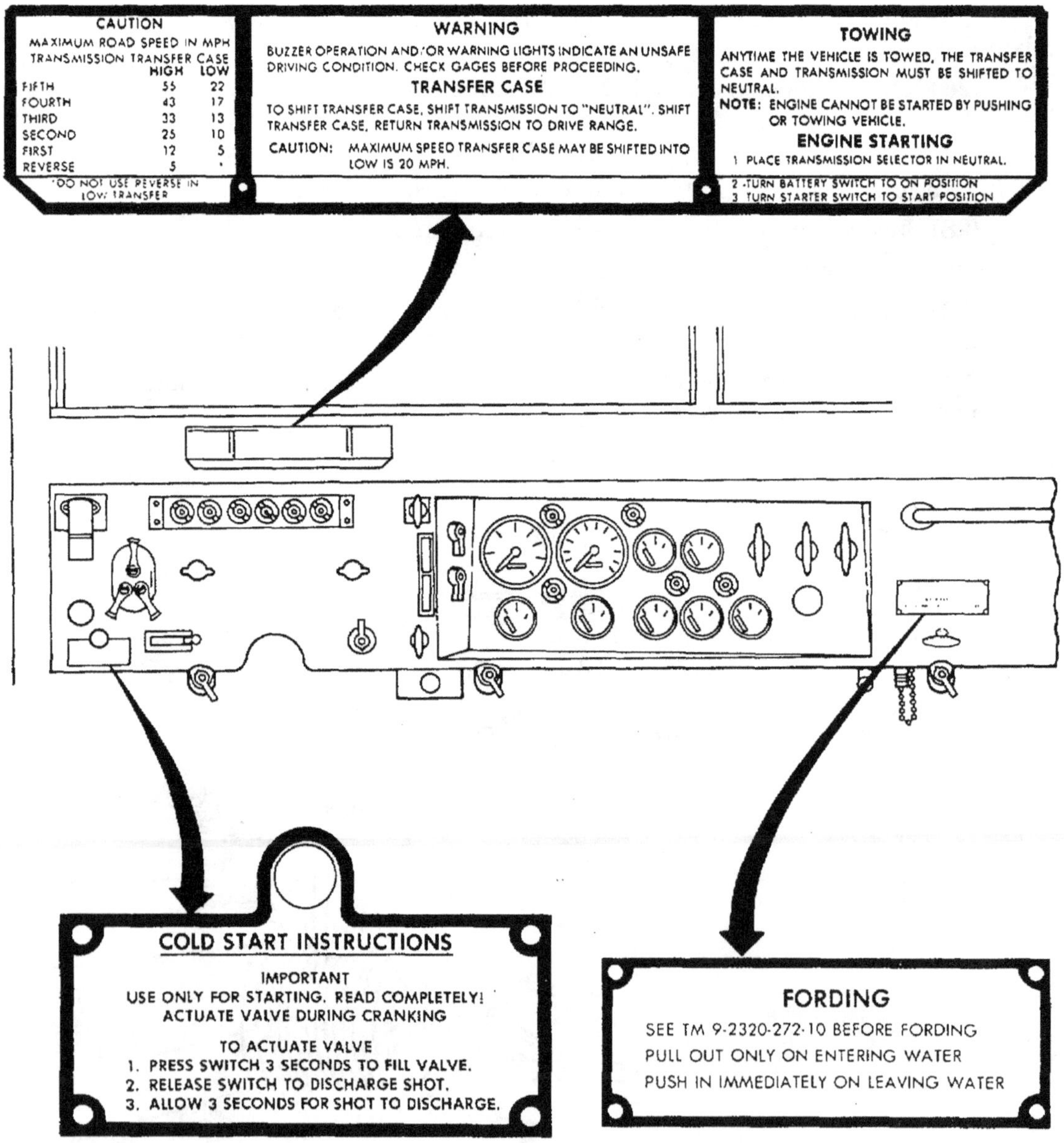
CAUTION
MAXIMUM ROAD SPEED IN MPH
TRANSMISSION TRANSFER CASE
HIGH LOW
FIFTH 55 22
FOURTH 43 17
THIRD 33 13
SECOND 25 10
FIRST 12 5
REVERSE 5 *
*DO NOT USE REVERSE IN LOW TRANSFER
WARNING
BUZZER OPERATION AND/OR WARNING LIGHTS INDICATE AN UNSAFE DRIVING CONDITION. CHECK GAGES BEFORE PROCEEDING.
TRANSFER CASE
TO SHIFT TRANSFER CASE, SHIFT TRANSMISSION TO "NEUTRAL". SHIFT TRANSFER CASE. RETURN TRANSMISSION TO DRIVE RANGE.
CAUTION: MAXIMUM SPEED TRANSFER CASE MAY BE SHIFTED INTO LOW IS 20 MPH.
TOWING
ANYTIME THE VEHICLE IS TOWED, THE TRANSFER CASE AND TRANSMISSION MUST BE SHIFTED TO NEUTRAL.
NOTE: ENGINE CANNOT BE STARTED BY PUSHING OR TOWING VEHICLE.
ENGINE STARTING
1 PLACE TRANSMISSION SELECTOR IN NEUTRAL.
2 TURN BATTERY SWITCH TO ON POSITION
3 TURN STARTER SWITCH TO START POSITION
COLD START INSTRUCTIONS
IMPORTANT
USE ONLY FOR STARTING. READ COMPLETELY!
ACTUATE VALVE DURING CRANKING
TO ACTUATE VALVE
1. PRESS SWITCH 3 SECONDS TO FILL VALVE.
2. RELEASE SWITCH TO DISCHARGE SHOT.
3. ALLOW 3 SECONDS FOR SHOT TO DISCHARGE.
FORDING
SEE TM 9-2320-272-10 BEFORE FORDING
PULL OUT ONLY ON ENTERING WATER
PUSH IN IMMEDIATELY ON LEAVING WATER

2-28. DECALS AND INSTRUCTION PLATES (Contd)

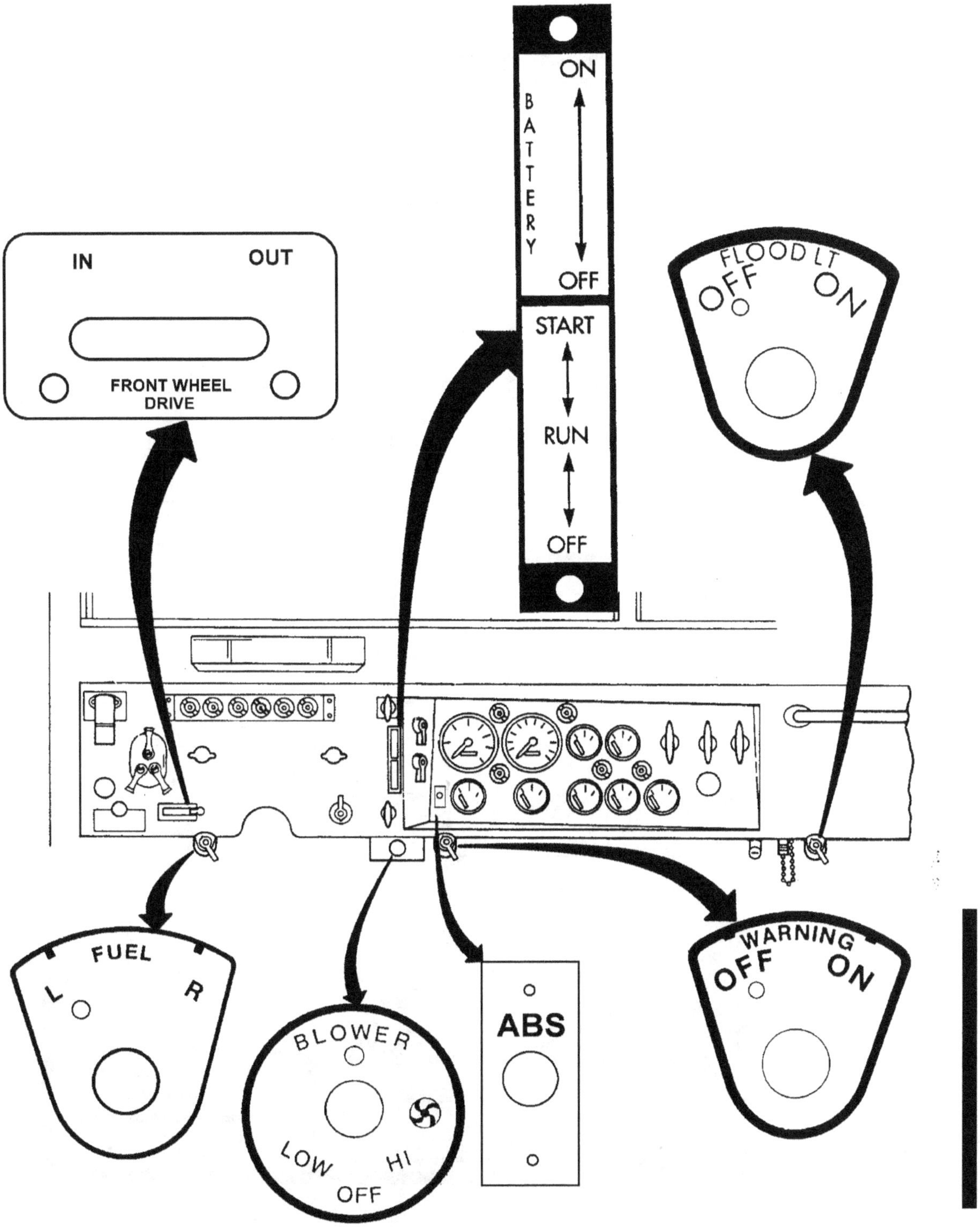
IN
OUT
FRONT WHEEL DRIVE
BATTERY
ON
OFF
START
RUN
OFF
FLOOD LT
OFF
ON
FUEL
L
R
BLOWER
LOW
OFF
HI
ABS
WARNING
OFF
ON

2-28. DECALS AND INSTRUCTION PLATES (Contd)

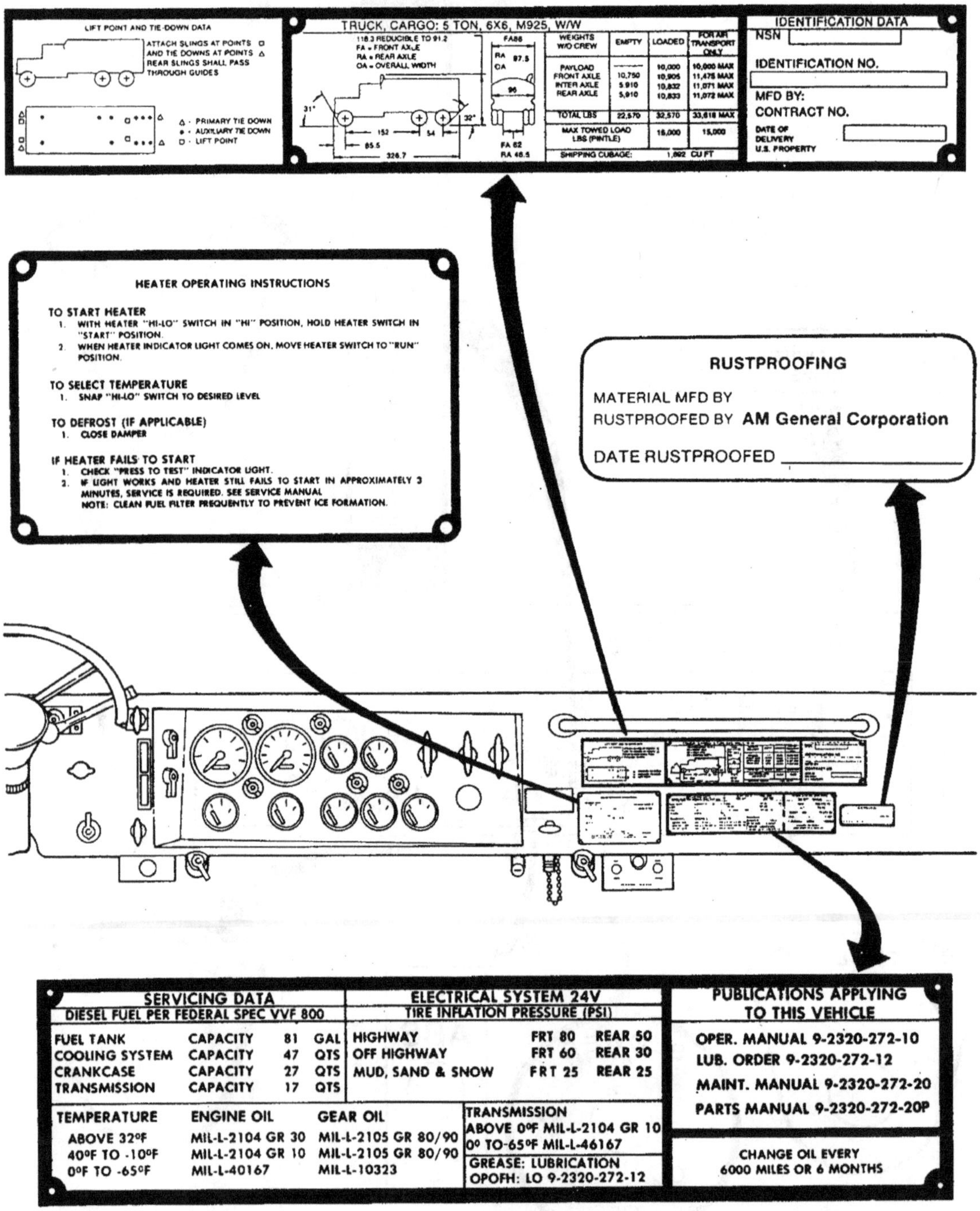
LIFT POINT AND TIE-DOWN DATA
ATTACH SLINGS AT POINTS □ AND TIE DOWNS AT POINTS △ REAR SLINGS SHALL PASS THROUGH GUIDES
△ - PRIMARY TIE DOWN
● - AUXILIARY TIE DOWN
□ - LIFT POINT
TRUCK, CARGO: 5 TON, 6X6, M925, W/W
118.3 REDUCIBLE TO 91.2
FA = FRONT AXLE
RA = REAR AXLE
OA = OVERALL WIDTH
31°
32°
152
54
85.5
326.7
FA86
RA 97.5
OA
96
FA 62
RA 48.5
WEIGHTS W/O CREW
EMPTY
LOADED
FOR AIR TRANSPORT ONLY
PAYLOAD
FRONT AXLE
INTER AXLE
REAR AXLE
10,750
5,910
5,910
10,000
10,905
10,832
10,833
10,000 MAX
11,475 MAX
11,071 MAX
11,072 MAX
TOTAL LBS
22,570
32,570
33,618 MAX
MAX TOWED LOAD LBS (PINTLE)
15,000
15,000
SHIPPING CUBAGE: 1,892 CU FT
IDENTIFICATION DATA
NSN
IDENTIFICATION NO.
MFD BY:
CONTRACT NO.
DATE OF DELIVERY
U.S. PROPERTY
HEATER OPERATING INSTRUCTIONS
TO START HEATER
1. WITH HEATER "HI-LO" SWITCH IN "HI" POSITION, HOLD HEATER SWITCH IN "START" POSITION.
2. WHEN HEATER INDICATOR LIGHT COMES ON, MOVE HEATER SWITCH TO "RUN" POSITION.
TO SELECT TEMPERATURE
1. SNAP "HI-LO" SWITCH TO DESIRED LEVEL
TO DEFROST (IF APPLICABLE)
1. CLOSE DAMPER
IF HEATER FAILS TO START
1. CHECK "PRESS TO TEST" INDICATOR LIGHT.
2. IF LIGHT WORKS AND HEATER STILL FAILS TO START IN APPROXIMATELY 3 MINUTES, SERVICE IS REQUIRED. SEE SERVICE MANUAL
NOTE: CLEAN FUEL FILTER FREQUENTLY TO PREVENT ICE FORMATION.
RUSTPROOFING
MATERIAL MFD BY
RUSTPROOFED BY AM General Corporation
DATE RUSTPROOFED
SERVICING DATA
DIESEL FUEL PER FEDERAL SPEC VVF 800
FUEL TANK CAPACITY 81 GAL
COOLING SYSTEM CAPACITY 47 QTS
CRANKCASE CAPACITY 27 QTS
TRANSMISSION CAPACITY 17 QTS
ELECTRICAL SYSTEM 24V
TIRE INFLATION PRESSURE (PSI)
HIGHWAY FRT 80 REAR 50
OFF HIGHWAY FRT 60 REAR 30
MUD, SAND & SNOW FRT 25 REAR 25
TEMPERATURE ENGINE OIL GEAR OIL
ABOVE 32°F MIL-L-2104 GR 30 MIL-L-2105 GR 80/90
40°F TO -10°F MIL-L-2104 GR 10 MIL-L-2105 GR 80/90
0°F TO -65°F MIL-L-40167 MIL-L-10323
TRANSMISSION
ABOVE 0°F MIL-L-2104 GR 10
0° TO-65°F MIL-L-46167
GREASE: LUBRICATION
OPOFH: LO 9-2320-272-12
PUBLICATIONS APPLYING TO THIS VEHICLE
OPER. MANUAL 9-2320-272-10
LUB. ORDER 9-2320-272-12
MAINT. MANUAL 9-2320-272-20
PARTS MANUAL 9-2320-272-20P
CHANGE OIL EVERY 6000 MILES OR 6 MONTHS

2-28. DECALS AND INSTRUCTION PLATES (Contd)

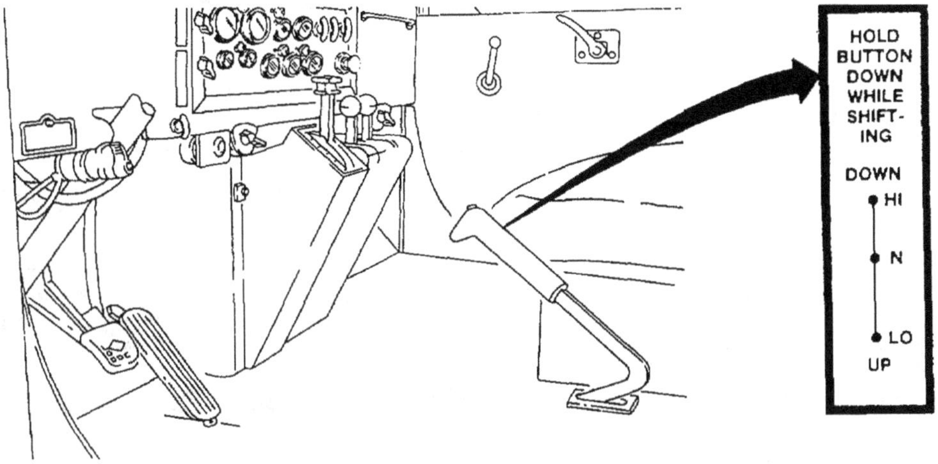
HOLD BUTTON DOWN WHILE SHIFT-ING
DOWN
HI
N
LO
UP

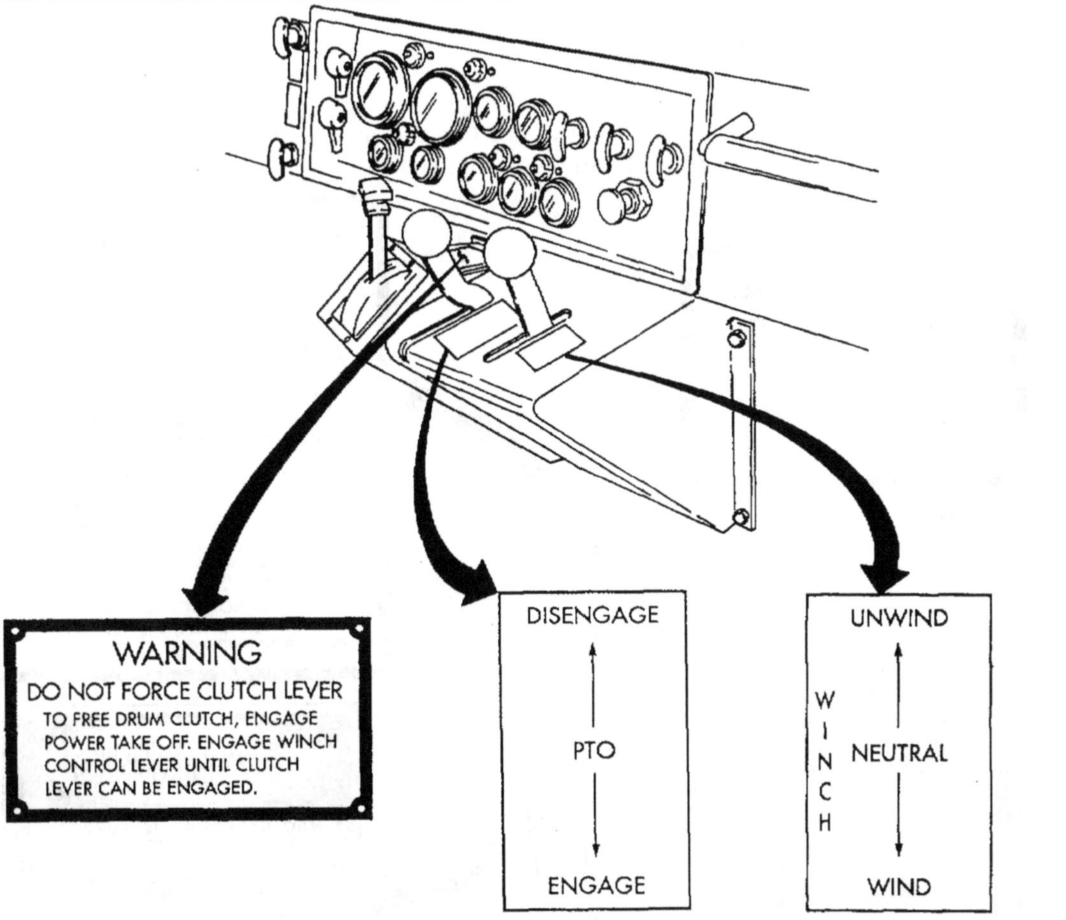
WARNING
DO NOT FORCE CLUTCH LEVER
TO FREE DRUM CLUTCH, ENGAGE POWER TAKE OFF. ENGAGE WINCH CONTROL LEVER UNTIL CLUTCH LEVER CAN BE ENGAGED.
DISENGAGE
PTO
ENGAGE
UNWIND
WINCH
NEUTRAL
WIND

2-28. DECALS AND INSTRUCTION PLATES (Contd)

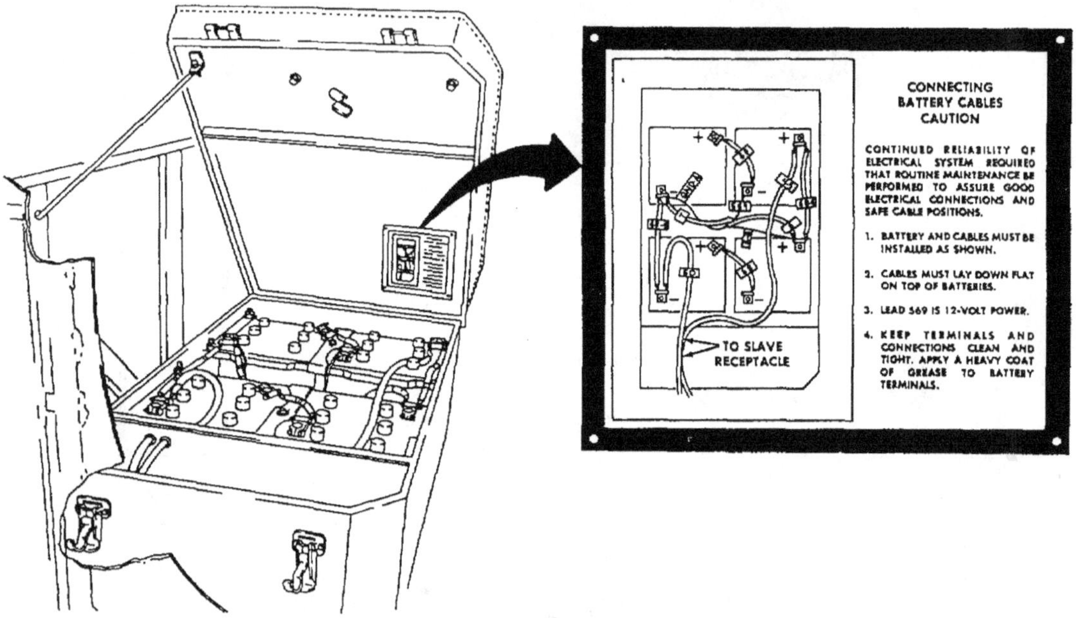

CONNECTING
BATTERY CABLES
CAUTION
CONTINUED RELIABILITY OF ELECTRICAL SYSTEM REQUIRED THAT ROUTINE MAINTENANCE BE PERFORMED TO ASSURE GOOD ELECTRICAL CONNECTIONS AND SAFE CABLE POSITIONS.
1. BATTERY AND CABLES MUST BE INSTALLED AS SHOWN.
2. CABLES MUST LAY DOWN FLAT ON TOP OF BATTERIES.
3. LEAD 569 IS 12-VOLT POWER.
4. KEEP TERMINALS AND CONNECTIONS CLEAN AND TIGHT. APPLY A HEAVY COAT OF GREASE TO BATTERY TERMINALS.
TO SLAVE RECEPTACLE

c. The plate and decals shown below are located on the air cleaner assembly and are common to all M939/A1/A2 series vehicles.

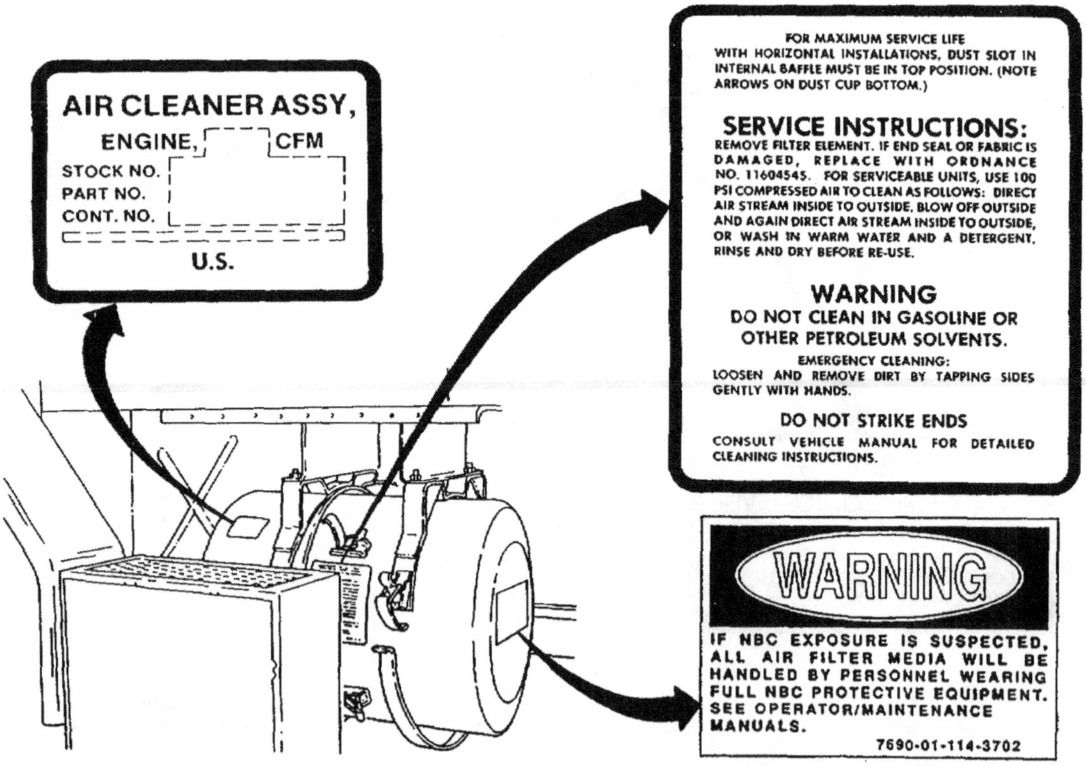

AIR CLEANER ASSY,
ENGINE, CFM
STOCK NO.
PART NO.
CONT. NO.
U.S.
FOR MAXIMUM SERVICE LIFE WITH HORIZONTAL INSTALLATIONS, DUST SLOT IN INTERNAL BAFFLE MUST BE IN TOP POSITION. (NOTE ARROWS ON DUST CUP BOTTOM.)
SERVICE INSTRUCTIONS:
REMOVE FILTER ELEMENT. IF END SEAL OR FABRIC IS DAMAGED, REPLACE WITH ORDNANCE NO. 11604545. FOR SERVICEABLE UNITS, USE 100 PSI COMPRESSED AIR TO CLEAN AS FOLLOWS: DIRECT AIR STREAM INSIDE TO OUTSIDE, BLOW OFF OUTSIDE AND AGAIN DIRECT AIR STREAM INSIDE TO OUTSIDE, OR WASH IN WARM WATER AND A DETERGENT, RINSE AND DRY BEFORE RE-USE.
WARNING
DO NOT CLEAN IN GASOLINE OR OTHER PETROLEUM SOLVENTS.
EMERGENCY CLEANING:
LOOSEN AND REMOVE DIRT BY TAPPING SIDES GENTLY WITH HANDS.
DO NOT STRIKE ENDS
CONSULT VEHICLE MANUAL FOR DETAILED CLEANING INSTRUCTIONS.
WARNING
IF NBC EXPOSURE IS SUSPECTED, ALL AIR FILTER MEDIA WILL BE HANDLED BY PERSONNEL WEARING FULL NBC PROTECTIVE EQUIPMENT. SEE OPERATOR/MAINTENANCE MANUALS.
7690-01-114-3702

2-28. DECALS AND INSTRUCTION PLATES (Contd)

d. Below are those plates located on the service and emergency couplings and are common to M939/A1/A2 series vehicles.

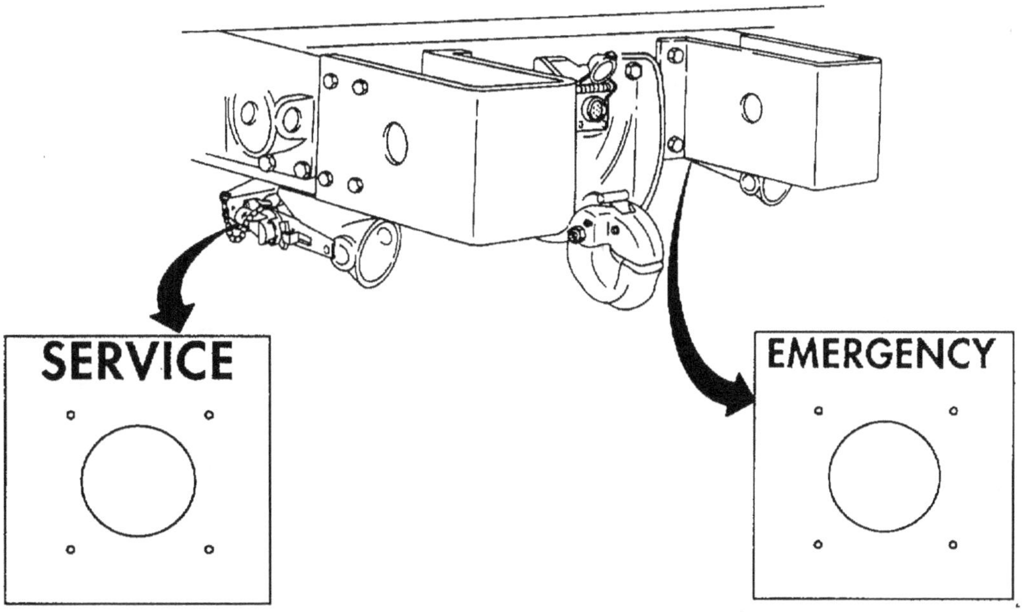
SERVICE
EMERGENCY

e. The decal shown below is located on the hydraulic reservoir tank and is common to M925/A1/A2, M928/A1/A2, M929/A1/A2, M930/A1/A2, and M932/A1/A2 models.

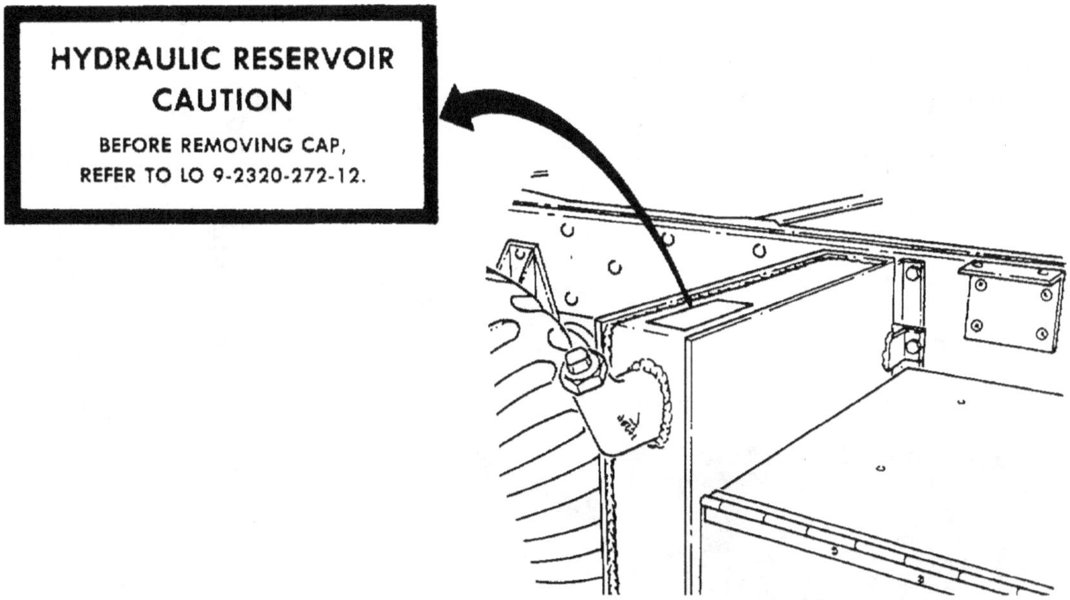
HYDRAULIC RESERVOIR
CAUTION
BEFORE REMOVING CAP,
REFER TO LO 9-2320-272-12.

2-28. DECALS AND INSTRUCTION PLATES (Contd)

f. The labels shown below are located on the central box harness and are common to all M939/A1/A2 series vehicles.

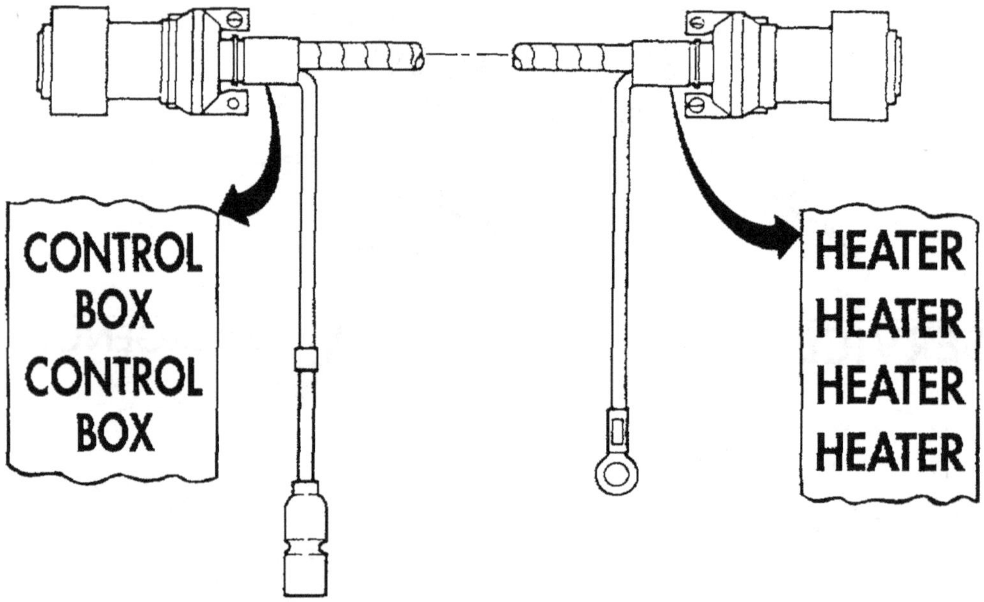
CONTROL
BOX
CONTROL
BOX
HEATER
HEATER
HEATER
HEATER

g. The decals shown below are located on the radiator and are common to all M939/A1/A2 series vehicles.

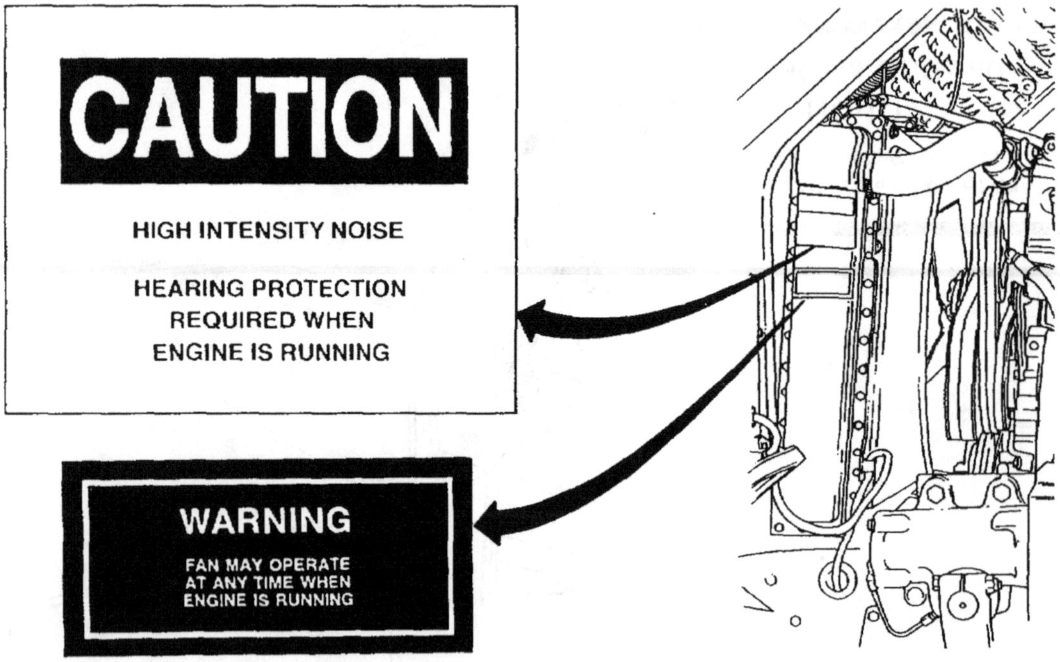
CAUTION
HIGH INTENSITY NOISE
HEARING PROTECTION REQUIRED WHEN ENGINE IS RUNNING
WARNING
FAN MAY OPERATE AT ANY TIME WHEN ENGINE IS RUNNING

2-28. DECALS AND INSTRUCTION PLATES (Contd)

h. The decal shown below is located on the surge tank and is common to all M939/A1/A2 series vehicles.

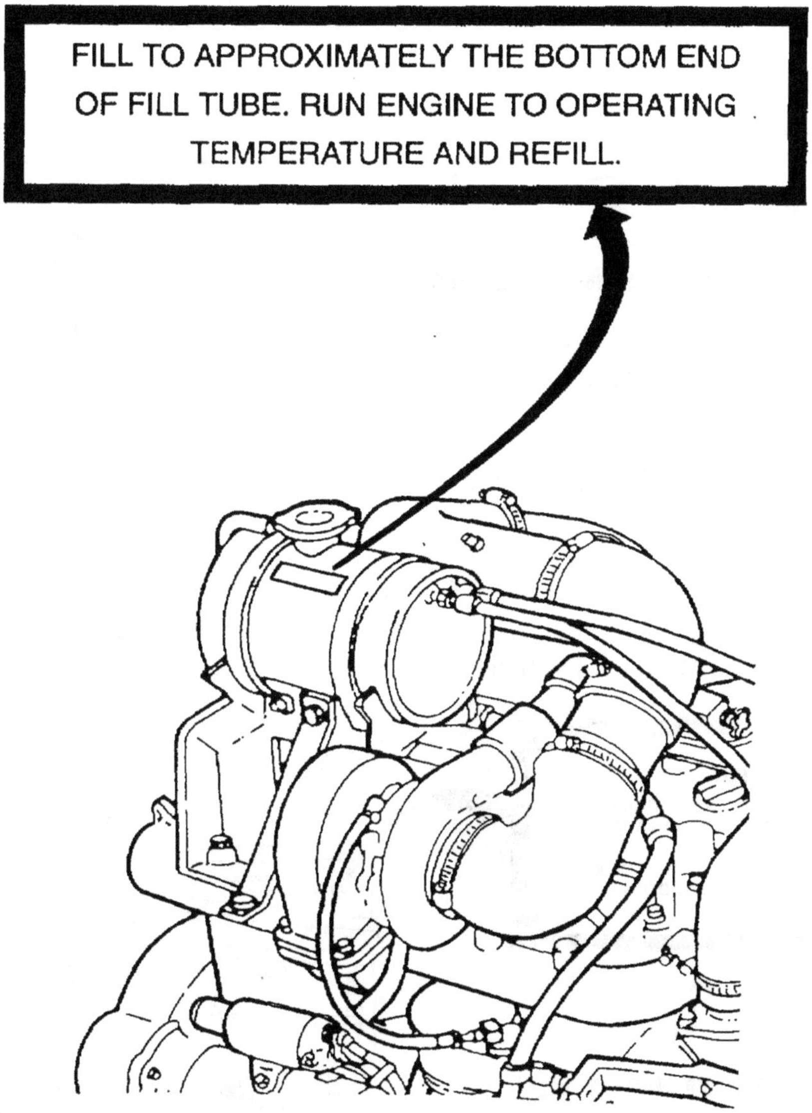
FILL TO APPROXIMATELY THE BOTTOM END
OF FILL TUBE. RUN ENGINE TO OPERATING
TEMPERATURE AND REFILL.

2-28. DECALS AND INSTRUCTION PLATES (Contd)

i. The vehicle decal shown below is common only to the dump (M929, M930), tractor (M931/A1/A2, M932/A1/A2), and wrecker (M936/A1/A2) vehicles.

WARNING
DO NOT REFUEL THIS
TANK UNTIL EXHAUST PIPE
HAS COOLED DOWN

j. The caution plates shown below are common only to the dropside cargo (M923/A1/A2, M925/A1/A2) model vehicles.

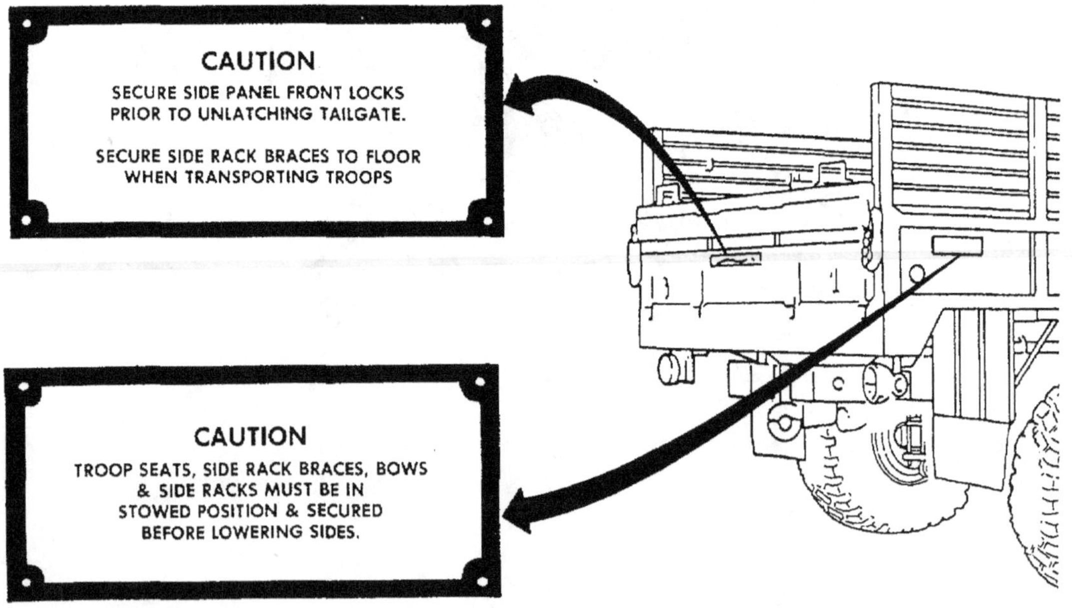
CAUTION
SECURE SIDE PANEL FRONT LOCKS
PRIOR TO UNLATCHING TAILGATE.
SECURE SIDE RACK BRACES TO FLOOR
WHEN TRANSPORTING TROOPS
CAUTION
TROOP SEATS, SIDE RACK BRACES, BOWS
& SIDE RACKS MUST BE IN
STOWED POSITION & SECURED
BEFORE LOWERING SIDES.

2-28. DECALS AND INSTRUCTION PLATES (Contd)

k. The warning plate shown below is located only on the cab protector of dump (M929/A1/A2 and M930/A1/A2) models.

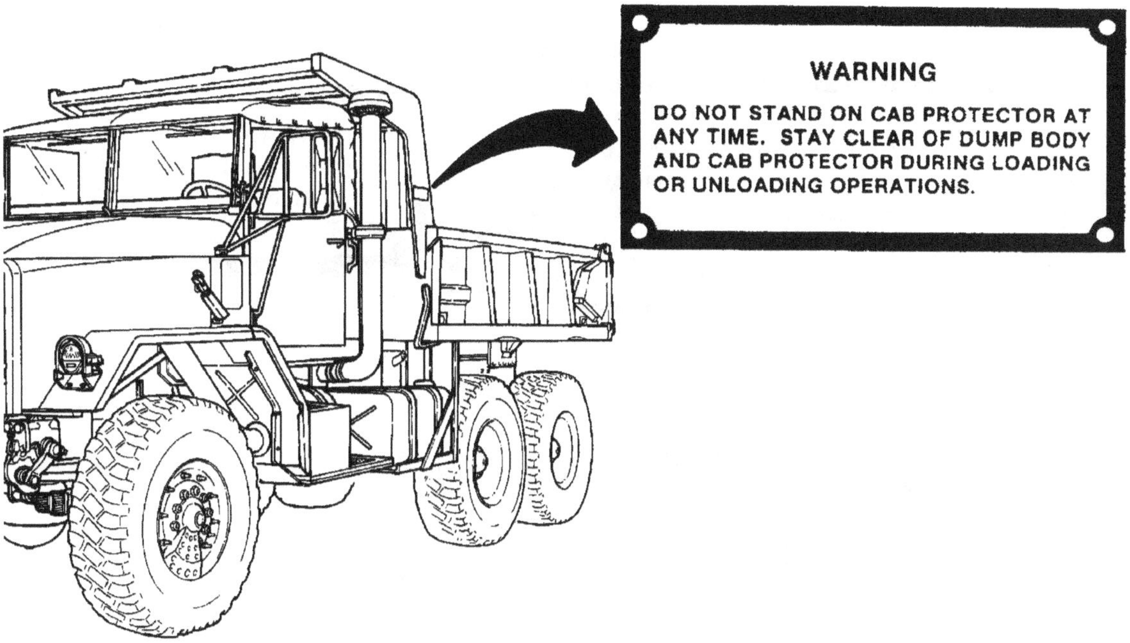
WARNING
DO NOT STAND ON CAB PROTECTOR AT ANY TIME. STAY CLEAR OF DUMP BODY AND CAB PROTECTOR DURING LOADING OR UNLOADING OPERATIONS.

l. The plate shown below is located on the hard top installed on M939/A1/A2 series vehicles.

l.1. The sticker shown below is located on driver's side front window and indicates that this vehicle is equipped with antilock brake system (ABS).

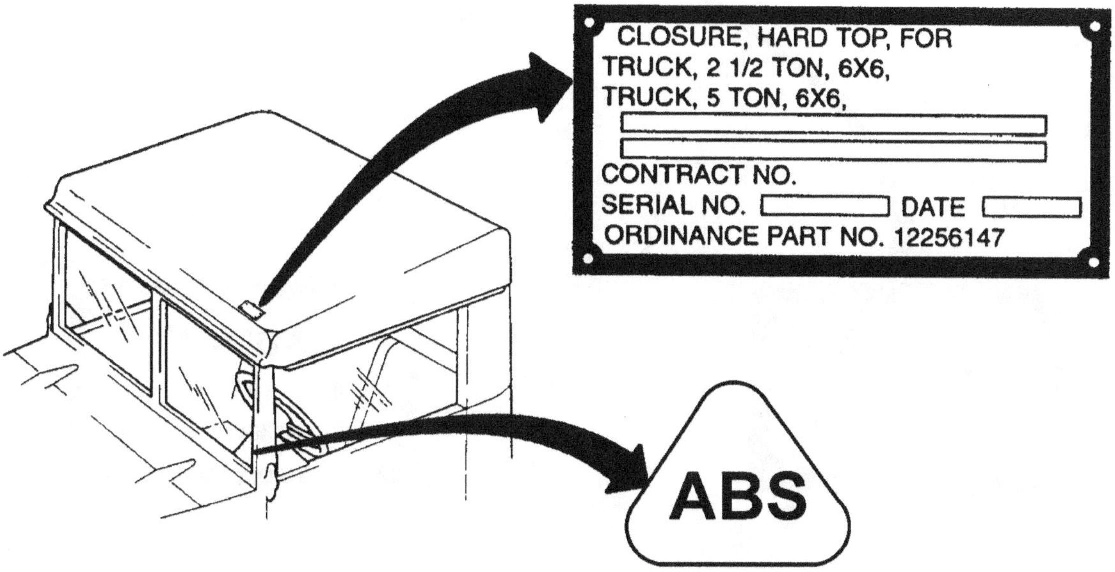
CLOSURE, HARD TOP, FOR
TRUCK, 2 1/2 TON, 6X6,
TRUCK, 5 TON, 6X6,
CONTRACT NO.
SERIAL NO.
DATE
ORDINANCE PART NO. 12256147
ABS

2-28. DECALS AND INSTRUCTION PLATES (Contd)

m. The plates shown below are common to all front winch models and wrecker rear winch (M925/A1/A2, M928/A1/A2, M930/A1/A2, M932/A1/A2, and M936/A1/A2) models.

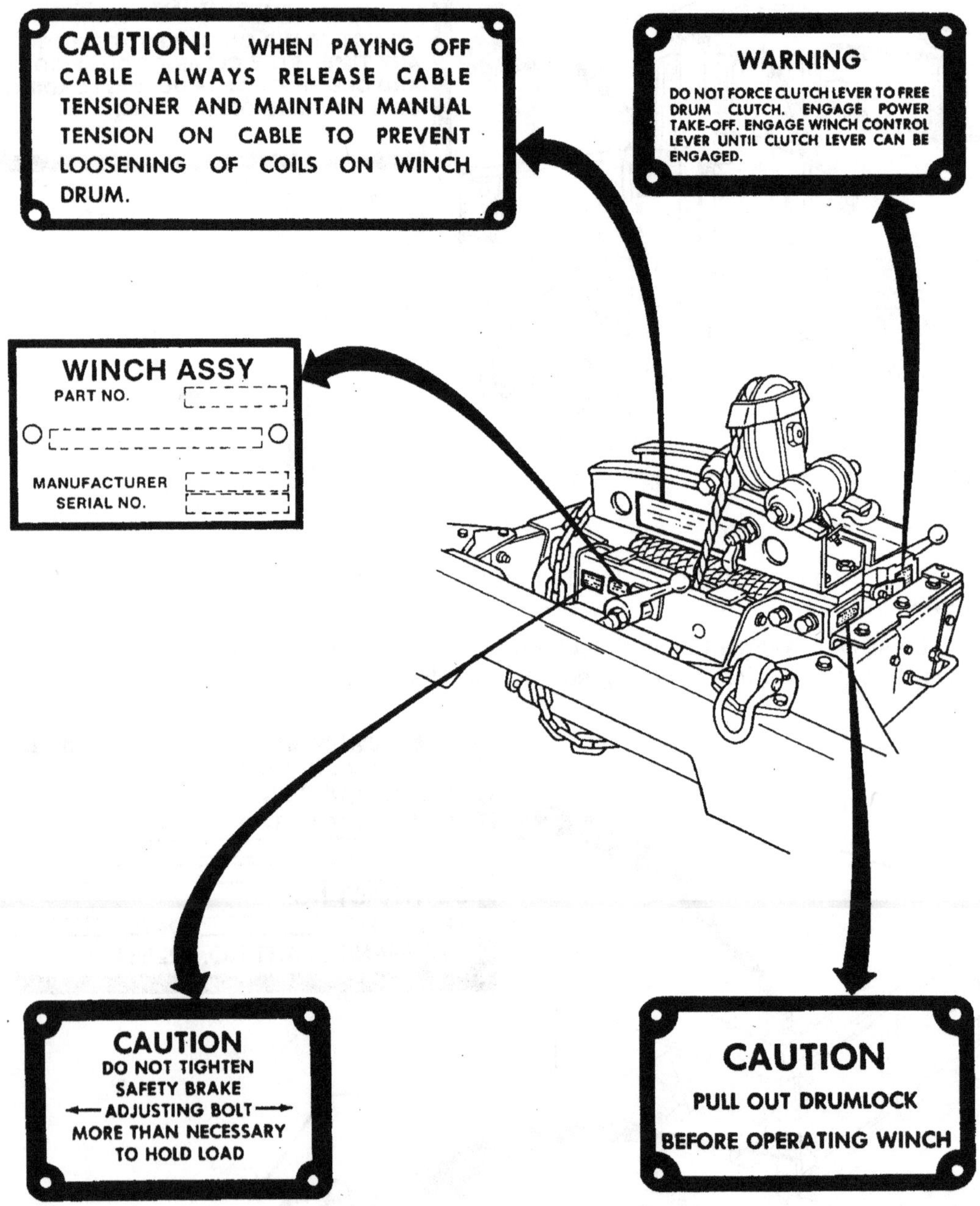
CAUTION! WHEN PAYING OFF CABLE ALWAYS RELEASE CABLE TENSIONER AND MAINTAIN MANUAL TENSION ON CABLE TO PREVENT LOOSENING OF COILS ON WINCH DRUM.
WARNING
DO NOT FORCE CLUTCH LEVER TO FREE DRUM CLUTCH. ENGAGE POWER TAKE-OFF. ENGAGE WINCH CONTROL LEVER UNTIL CLUTCH LEVER CAN BE ENGAGED.
WINCH ASSY
PART NO.
MANUFACTURER
SERIAL NO.
CAUTION
DO NOT TIGHTEN
SAFETY BRAKE
ADJUSTING BOLT
MORE THAN NECESSARY
TO HOLD LOAD
CAUTION
PULL OUT DRUMLOCK
BEFORE OPERATING WINCH

2-28. DECALS AND INSTRUCTION PLATES (Contd)

n. The plates shown below are common only to the medium wrecker (M936/A1/A2) model.

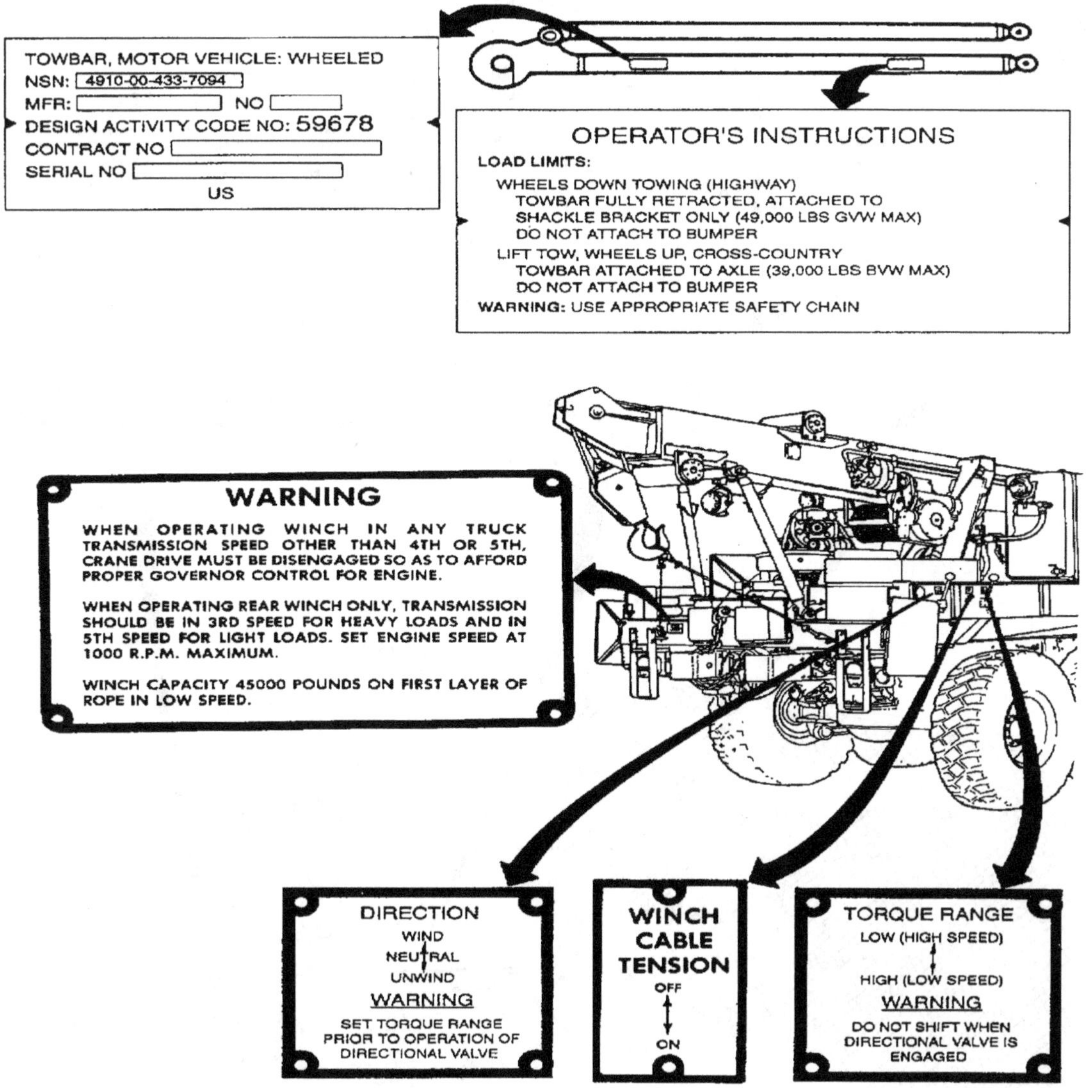
TOWBAR, MOTOR VEHICLE: WHEELED
NSN: 4910-00-433-7094
MFR: NO
DESIGN ACTIVITY CODE NO: 59678
CONTRACT NO
SERIAL NO
US
OPERATOR'S INSTRUCTIONS
LOAD LIMITS:
WHEELS DOWN TOWING (HIGHWAY)
TOWBAR FULLY RETRACTED, ATTACHED TO SHACKLE BRACKET ONLY (49,000 LBS GVW MAX)
DO NOT ATTACH TO BUMPER
LIFT TOW, WHEELS UP, CROSS-COUNTRY
TOWBAR ATTACHED TO AXLE (39,000 LBS BVW MAX)
DO NOT ATTACH TO BUMPER
WARNING: USE APPROPRIATE SAFETY CHAIN
WARNING
WHEN OPERATING WINCH IN ANY TRUCK TRANSMISSION SPEED OTHER THAN 4TH OR 5TH, CRANE DRIVE MUST BE DISENGAGED SO AS TO AFFORD PROPER GOVERNOR CONTROL FOR ENGINE.
WHEN OPERATING REAR WINCH ONLY, TRANSMISSION SHOULD BE IN 3RD SPEED FOR HEAVY LOADS AND IN 5TH SPEED FOR LIGHT LOADS. SET ENGINE SPEED AT 1000 R.P.M. MAXIMUM.
WINCH CAPACITY 45000 POUNDS ON FIRST LAYER OF ROPE IN LOW SPEED.
DIRECTION
WIND
NEUTRAL
UNWIND
WARNING
SET TORQUE RANGE PRIOR TO OPERATION OF DIRECTIONAL VALVE
WINCH CABLE TENSION
OFF
ON
TORQUE RANGE
LOW (HIGH SPEED)
HIGH (LOW SPEED)
WARNING
DO NOT SHIFT WHEN DIRECTIONAL VALVE IS ENGAGED

2-28. DECALS AND INSTRUCTION PLATES (Contd)

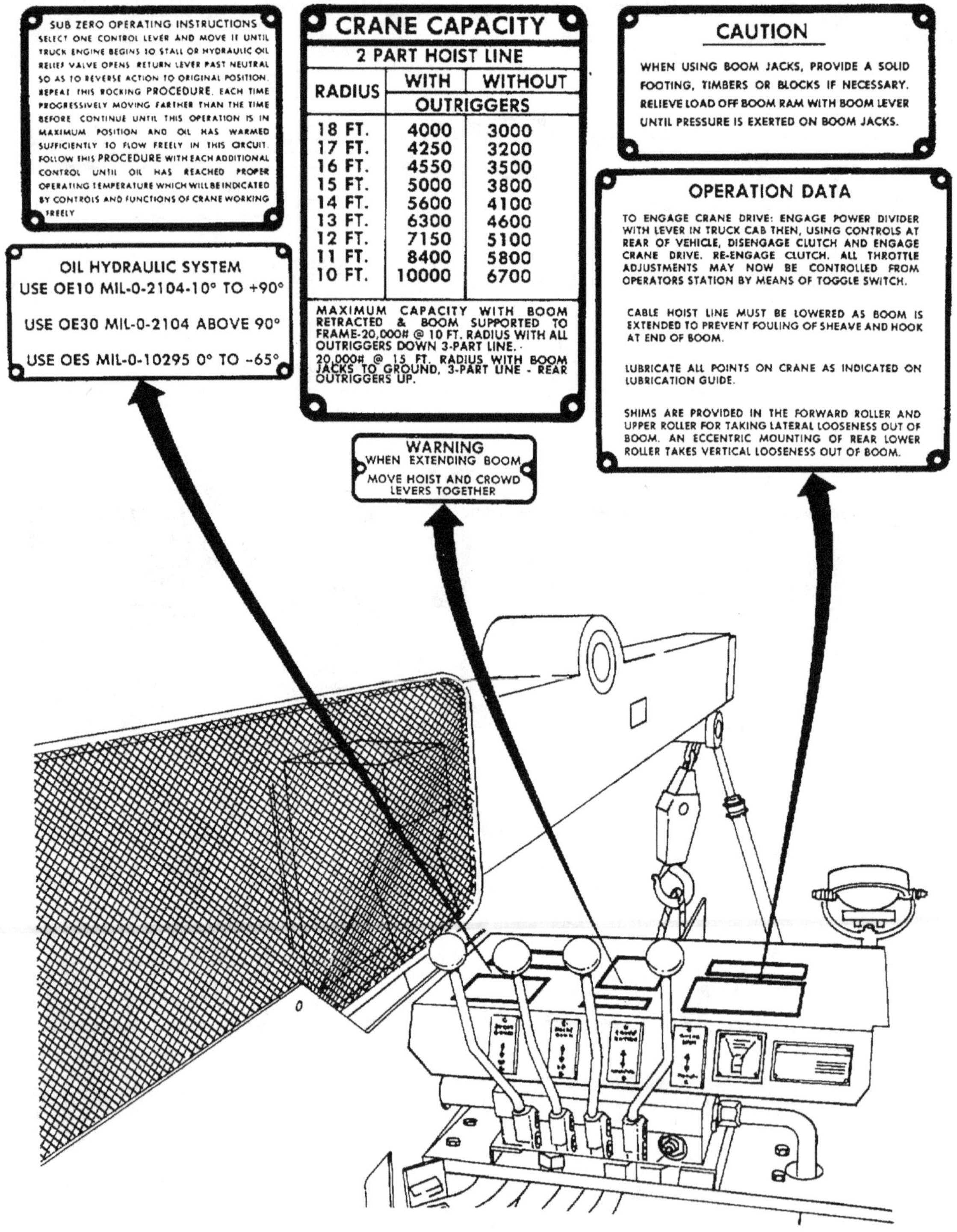
SUB ZERO OPERATING INSTRUCTIONS
SELECT ONE CONTROL LEVER AND MOVE IT UNTIL TRUCK ENGINE BEGINS TO STALL OR HYDRAULIC OIL RELIEF VALVE OPENS RETURN LEVER PAST NEUTRAL SO AS TO REVERSE ACTION TO ORIGINAL POSITION. REPEAT THIS ROCKING PROCEDURE. EACH TIME PROGRESSIVELY MOVING FARTHER THAN THE TIME BEFORE CONTINUE UNTIL THIS OPERATION IS IN MAXIMUM POSITION AND OIL HAS WARMED SUFFICIENTLY TO FLOW FREELY IN THIS CIRCUIT. FOLLOW THIS PROCEDURE WITH EACH ADDITIONAL CONTROL UNTIL OIL HAS REACHED PROPER OPERATING TEMPERATURE WHICH WILL BE INDICATED BY CONTROLS AND FUNCTIONS OF CRANE WORKING FREELY
OIL HYDRAULIC SYSTEM
USE OE10 MIL-0-2104-10° TO +90°
USE OE30 MIL-0-2104 ABOVE 90°
USE OES MIL-0-10295 0° TO -65°
CRANE CAPACITY
2 PART HOIST LINE
RADIUS
WITH
WITHOUT
OUTRIGGERS
18 FT. 4000 3000
17 FT. 4250 3200
16 FT. 4550 3500
15 FT. 5000 3800
14 FT. 5600 4100
13 FT. 6300 4600
12 FT. 7150 5100
11 FT. 8400 5800
10 FT. 10000 6700
MAXIMUM CAPACITY WITH BOOM RETRACTED & BOOM SUPPORTED TO FRAME-20,000# @ 10 FT. RADIUS WITH ALL OUTRIGGERS DOWN 3-PART LINE.
20,000# @ 15 FT. RADIUS WITH BOOM JACKS TO GROUND, 3-PART LINE - REAR OUTRIGGERS UP.
WARNING
WHEN EXTENDING BOOM
MOVE HOIST AND CROWD LEVERS TOGETHER
CAUTION
WHEN USING BOOM JACKS, PROVIDE A SOLID FOOTING, TIMBERS OR BLOCKS IF NECESSARY. RELIEVE LOAD OFF BOOM RAM WITH BOOM LEVER UNTIL PRESSURE IS EXERTED ON BOOM JACKS.
OPERATION DATA
TO ENGAGE CRANE DRIVE: ENGAGE POWER DIVIDER WITH LEVER IN TRUCK CAB THEN, USING CONTROLS AT REAR OF VEHICLE, DISENGAGE CLUTCH AND ENGAGE CRANE DRIVE. RE-ENGAGE CLUTCH. ALL THROTTLE ADJUSTMENTS MAY NOW BE CONTROLLED FROM OPERATORS STATION BY MEANS OF TOGGLE SWITCH.
CABLE HOIST LINE MUST BE LOWERED AS BOOM IS EXTENDED TO PREVENT FOULING OF SHEAVE AND HOOK AT END OF BOOM.
LUBRICATE ALL POINTS ON CRANE AS INDICATED ON LUBRICATION GUIDE.
SHIMS ARE PROVIDED IN THE FORWARD ROLLER AND UPPER ROLLER FOR TAKING LATERAL LOOSENESS OUT OF BOOM. AN ECCENTRIC MOUNTING OF REAR LOWER ROLLER TAKES VERTICAL LOOSENESS OUT OF BOOM.

2-28. DECALS AND INSTRUCTION PLATES (Contd)

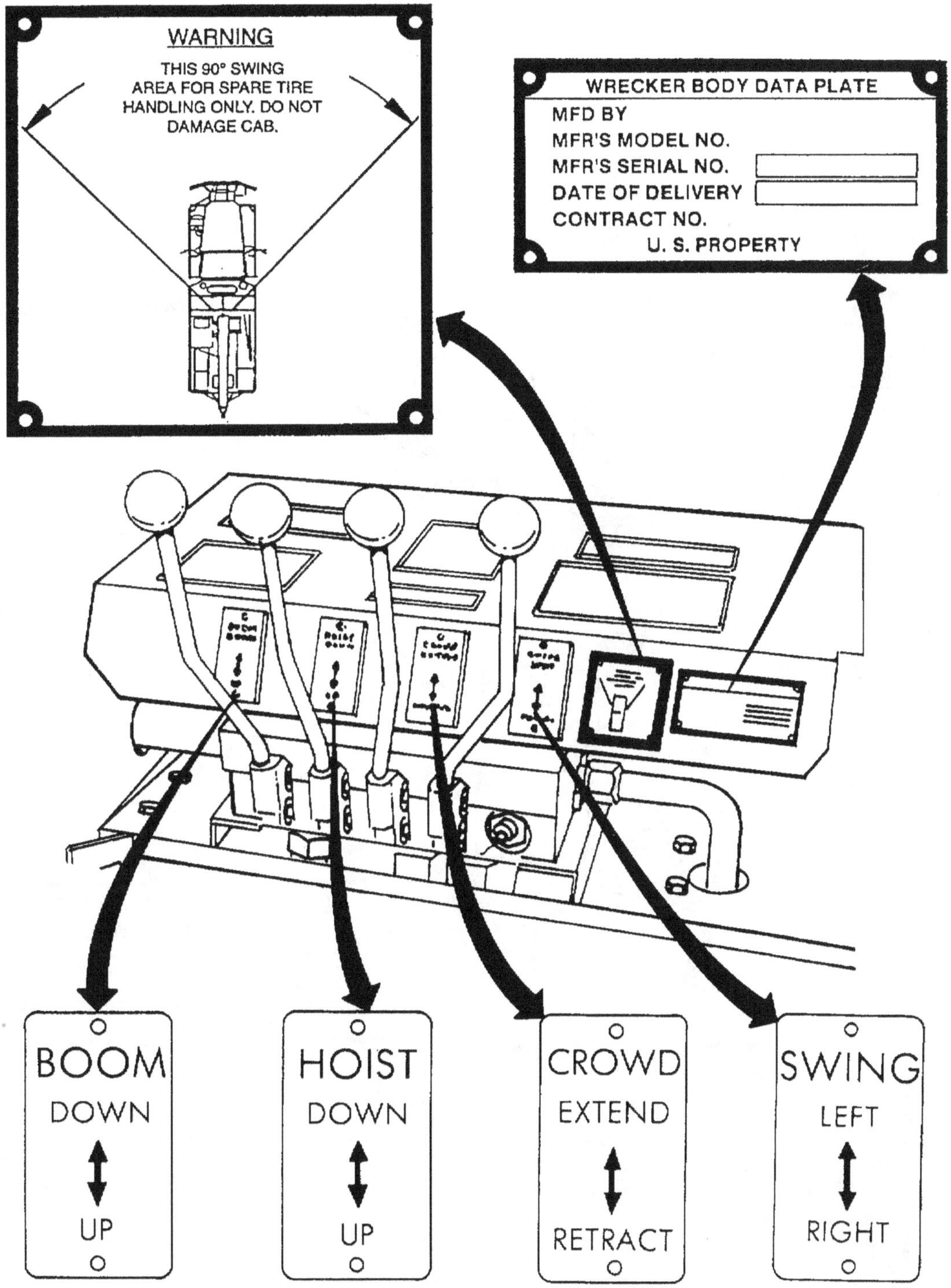
WARNING
THIS 90° SWING AREA FOR SPARE TIRE HANDLING ONLY. DO NOT DAMAGE CAB.
WRECKER BODY DATA PLATE
MFD BY
MFR'S MODEL NO.
MFR'S SERIAL NO.
DATE OF DELIVERY
CONTRACT NO.
U. S. PROPERTY
BOOM
DOWN
UP
HOIST
DOWN
UP
CROWD
EXTEND
RETRACT
SWING
LEFT
RIGHT

2-28. DECALS AND INSTRUCTION PLATES (Contd)

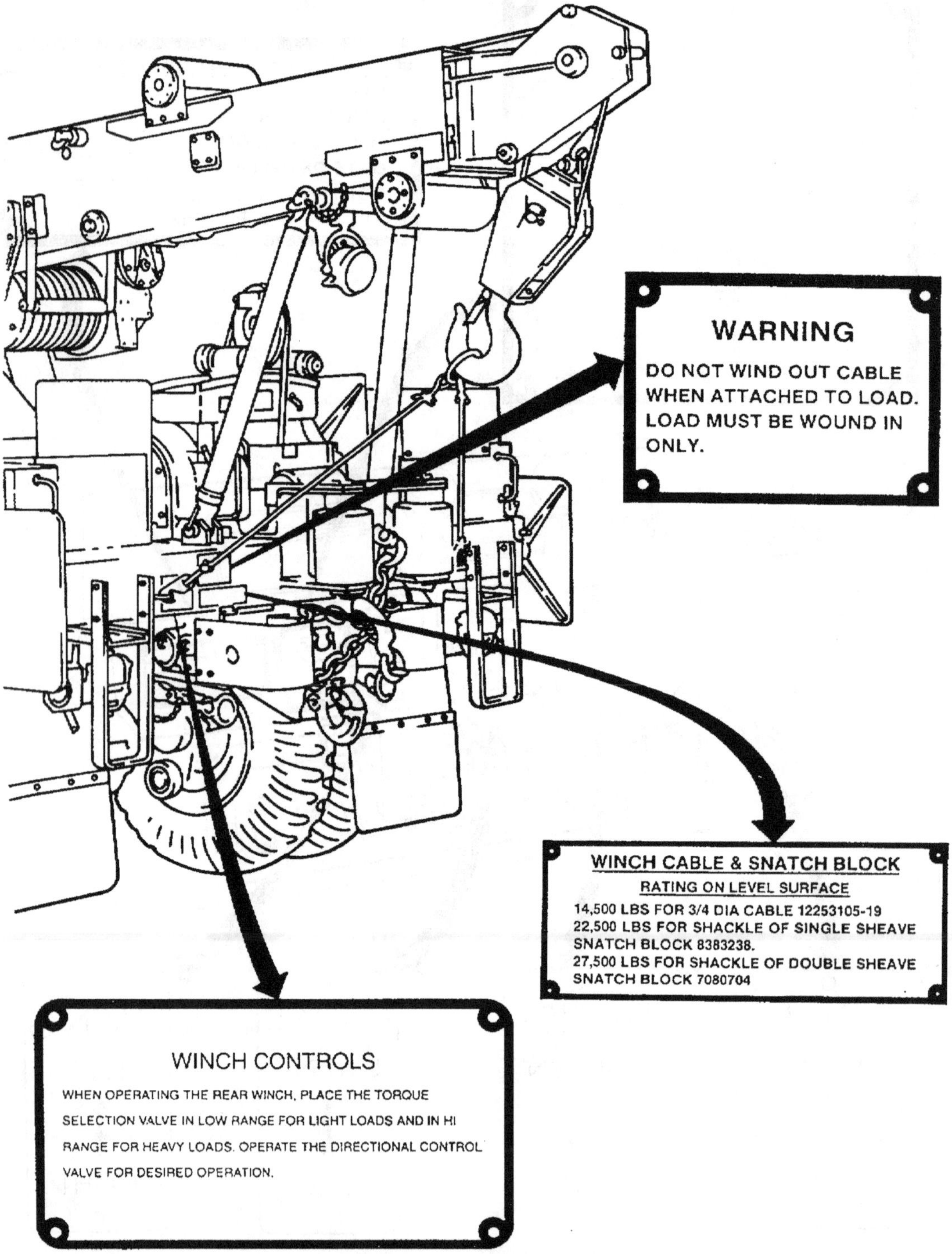
WARNING
DO NOT WIND OUT CABLE WHEN ATTACHED TO LOAD. LOAD MUST BE WOUND IN ONLY.
WINCH CABLE & SNATCH BLOCK
RATING ON LEVEL SURFACE
14,500 LBS FOR 3/4 DIA CABLE 12253105-19
22,500 LBS FOR SHACKLE OF SINGLE SHEAVE SNATCH BLOCK 8383238.
27,500 LBS FOR SHACKLE OF DOUBLE SHEAVE SNATCH BLOCK 7080704
WINCH CONTROLS
WHEN OPERATING THE REAR WINCH, PLACE THE TORQUE SELECTION VALVE IN LOW RANGE FOR LIGHT LOADS AND IN HI RANGE FOR HEAVY LOADS. OPERATE THE DIRECTIONAL CONTROL VALVE FOR DESIRED OPERATION.

2-28. DECALS AND INSTRUCTION PLATES (Contd)

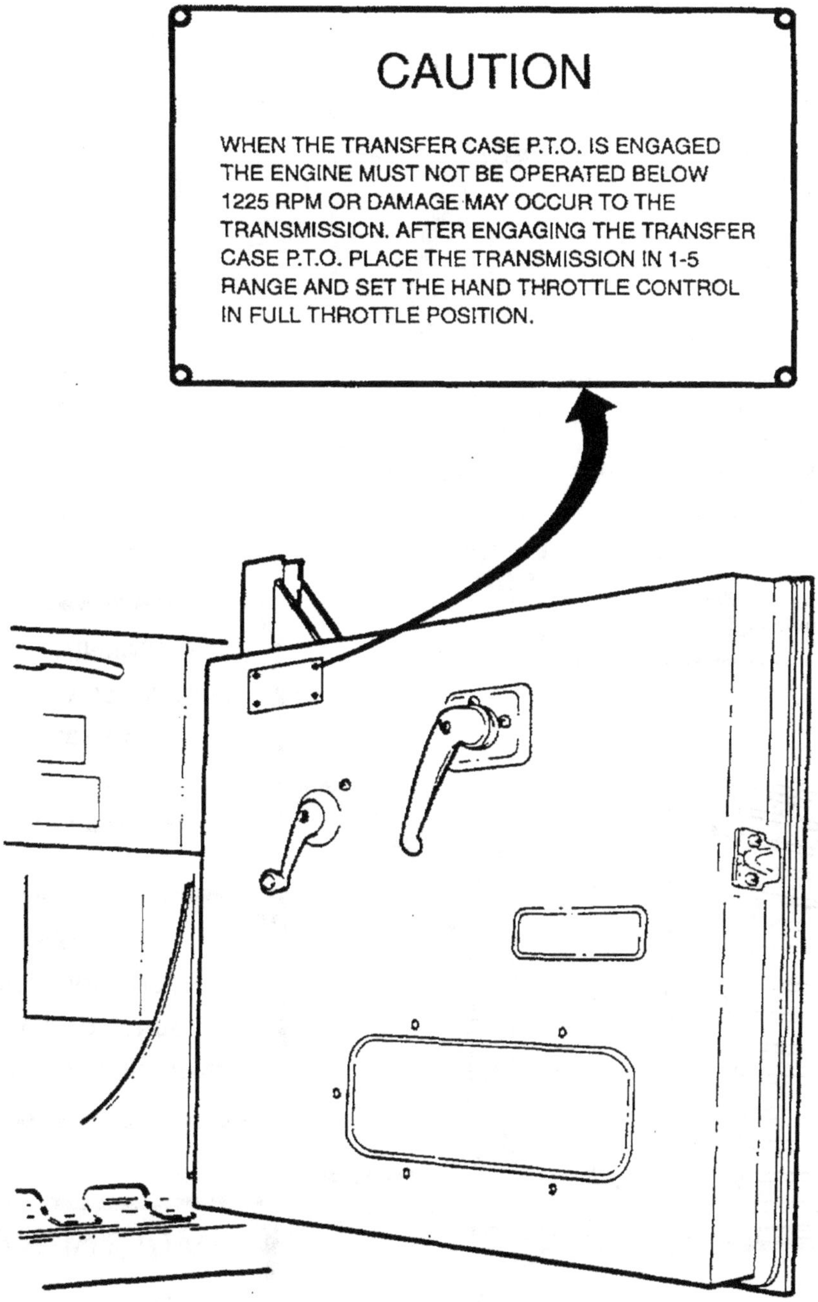
CAUTION
WHEN THE TRANSFER CASE P.T.O. IS ENGAGED THE ENGINE MUST NOT BE OPERATED BELOW 1225 RPM OR DAMAGE MAY OCCUR TO THE TRANSMISSION. AFTER ENGAGING THE TRANSFER CASE P.T.O. PLACE THE TRANSMISSION IN 1-5 RANGE AND SET THE HAND THROTTLE CONTROL IN FULL THROTTLE POSITION.

2-28. DECALS AND INSTRUCTION PLATES (Contd)

o. The vehicle plates shown below are common only to the expansible van (M934/A1/A2) model vehicles.

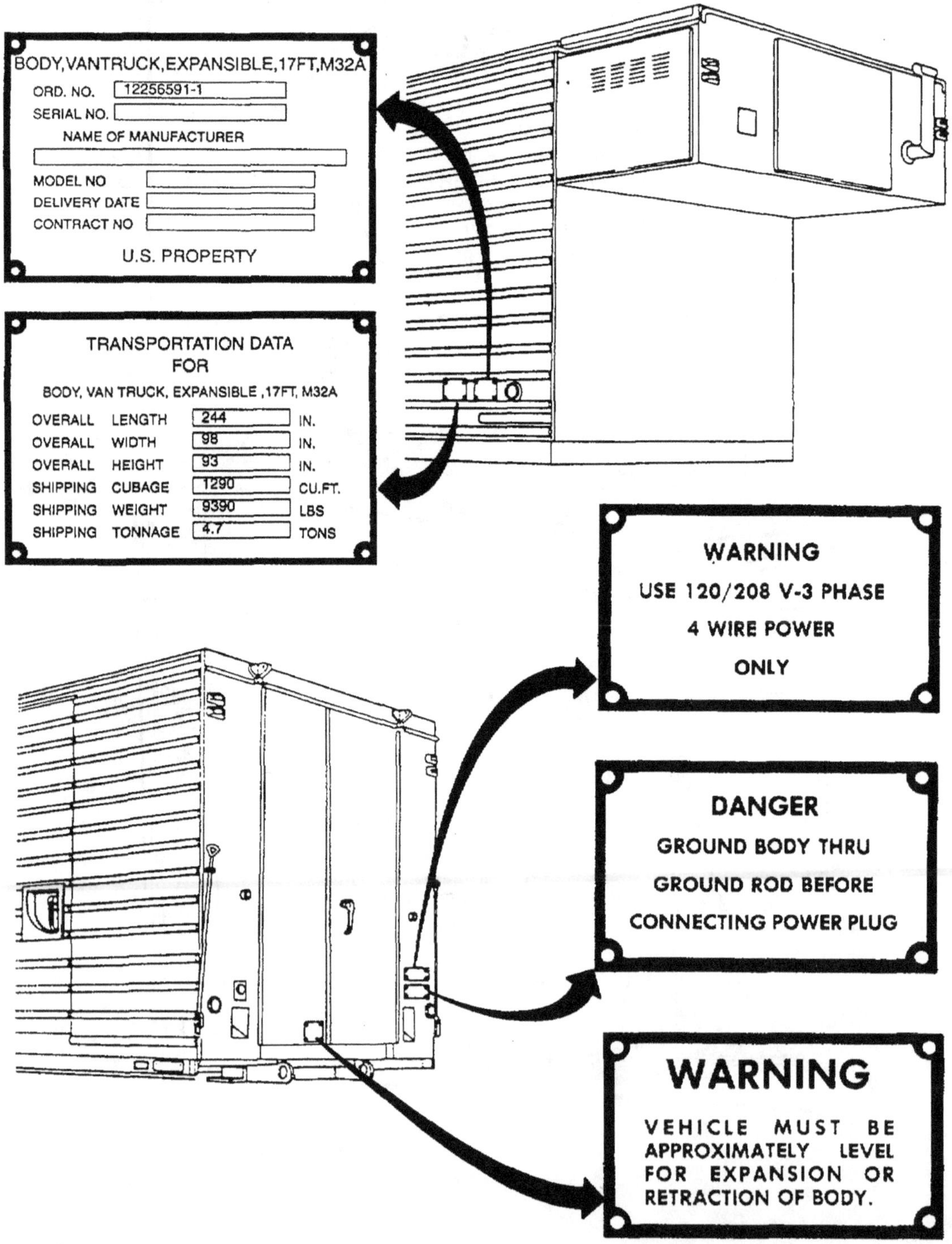
BODY,VANTRUCK,EXPANSIBLE,17FT,M32A
ORD. NO. 12256591-1
SERIAL NO.
NAME OF MANUFACTURER
MODEL NO
DELIVERY DATE
CONTRACT NO
U.S. PROPERTY
TRANSPORTATION DATA
FOR
BODY, VAN TRUCK, EXPANSIBLE ,17FT, M32A
OVERALL LENGTH 244 IN.
OVERALL WIDTH 98 IN.
OVERALL HEIGHT 93 IN.
SHIPPING CUBAGE 1290 CU.FT.
SHIPPING WEIGHT 9390 LBS
SHIPPING TONNAGE 4.7 TONS
WARNING
USE 120/208 V-3 PHASE
4 WIRE POWER
ONLY
DANGER
GROUND BODY THRU
GROUND ROD BEFORE
CONNECTING POWER PLUG
WARNING
VEHICLE MUST BE
APPROXIMATELY LEVEL
FOR EXPANSION OR
RETRACTION OF BODY.

2-28. DECALS AND INSTRUCTION PLATES (Contd)

NOTICE

TO PREVENT PERMANENT DAMAGE DO NOT ATTEMPT TO EXPAND THIS VAN WITHOUT READING THE FOLLOWING INSTRUCTIONS

VAN LEVELING INSTRUCTIONS

(A) EXPANSION SITE MUST BE LEVEL IF NECESSARY EXCAVATE AND/OR USE DUNNAGE UNDER APPROPRIATE WHEELS SO THE VAN FLOOR IS LEVEL (B) REMOVE TWO (2) LEVELING JACKS FROM JACK STORAGE BOX AND INSERT THEM INTO TWO (2) SOCKETS (MARKED "H") LOCATED UNDER REAR CORNER POSTS. SEE FIG. 1-40 OF TM 9-2320-236-14 & P
(C) RATCHET THE JACK SCREW TO RAISE OR LOWER EACH CORNER UNTIL THE VAN FLOOR IS LEVEL.
NOTE: LEVELING JACKS MUST BE USED REGARDLESS OF GROUND CONDITION TO IMPROVE STABILITY OF EXPANDED VAN.

CAUTION: DO NOT ATTEMPT TO LIFT TIRES OFF THE GROUND WITH LEVELING JACKS.

(D) USE THE TWO (2) FRONT LANDING LEGS TO LEVEL THE FRONT OF THE VAN IN THE SAME MANNER.

VAN EXPANDING INSTRUCTIONS

(A) RELEASE FOUR (4) SIDE WALL CORNER LOCKS ONE (1) LOCATED AT EACH LOWER EXTERIOR CORNER OF VAN
(B) USING SQUARE SOCKET WRENCH HANDLE (STORED ON INTERIOR SIDE OF REAR DOOR) RELEASE FOUR (4) LOCKS (TWO (2) ON EACH SIDE) BY ROTATING LOCK SHAFT (MARKED "A") COUNTERCLOCKWISE ONE QUARTER TURN
(C) USING RATCHET WRENCH WITH ATTACHED HEXAGON SOCKET (STORED ON INTERIOR SIDE OF REAR DOOR) EXPAND BOTH SIDE WALLS AS FOLLOWS. ATTACH HEXAGON SOCKET OF RATCHET WRENCH TO DRIVE SHAFT OF EITHER THE RIGHT OR LEFT EXPANDING MECHANISM (MARKED "B") LOCATED BELOW BOTH REAR DOORS. ROTATE LOCKING PLUNGER TO RELEASE LOCKING PAWL FROM TEETH ON RATCHET CAM AND TURN RATCHET WRENCH TO EXPAND EITHER SIDE WALL.
(D) OPEN FOUR (4) END PANEL DOORS, USING HOLDING RODS (MARKED "C") TO TEMPORARILY HOLD END PANEL DOORS OPEN AND OUT OF THE WAY. (E) RELEASE HINGED ROOF BY OPERATING HINGED ROOF LOCK HANDLE AND PULLING ROOF PARTIALLY OUTWARD. STANDING CLEAR (ON THE GROUND) USE THE GRAB HANDLE ATTACHED TO THE HINGED FLOOR AND PULL HINGED FLOOR DOWN OR ENTER VAN AND PUSH ON HINGED FLOOR AND HINGED ROOF. HINGED FLOOR AND HINGED ROOF ARE COUNTERBALANCED WITH EACH OTHER. REPEAT THE PROCEDURE FOR THE OPPOSITE SIDE OF THE VAN.
(F) ROTATE SIDE WALL SWIVEL HOOK (MARKED "E") TO ALIGN WITH HINGED ROOF CLAMP ASSEMBLY (MARKED "G") TYPICAL SIX (6) PLACES. (NOTE: TO RELEASE HINGED ROOF, INSERT BOARDING LADDER IN HOLES PROVIDED IN BEAM).

CLOSURE INSTRUCTIONS FOR OPERATIONAL MODE

(A) RELEASE AND STORE FOUR (4) END PANEL DOOR HOLDING RODS (MARKED "C") AND SWING ALL FOUR (4) END PANEL DOORS INWARD UNTIL CONTACT IS MADE WITH HINGED FLOOR AND ROOF SEALS. INSERT END PANEL DOOR SLIDING BOLTS INTO BATCH (MARKED "F") ON ALL FOUR (4) CORNER POSTS. NOTE: IT MAY BE NECESSARY TO PUSH UP ON FOLDING ROOF TO OBTAIN PROPER ALIGNMENT OF SEALS. (B) ENGAGE LOCKING PAWL (MARKED "B") AND RETRACT SIDE WALLS WITH RATCHET WRENCH. (C) ENGAGE EYE BOLT OF CLAMP ASSEMBLY (MARKED "G") WITH SWIVEL HOOK (MARKED "E") AND PUSH CLAMP ASSEMBLY TO THE CLOSED AND LOCKED POSITION SIX (6) PLACES.
(D) EXTEND SLIDING BOLT OF END PANEL DOORS TO THE FULLY EXTENDED POSITION INTO LATCH (MARKED "F")
(E) UNCLIP THE EXTENSION RODS LOCATED ON EACH OF THE EXTERIOR CORNER POSTS. SWING RODS DOWN AND ENGAGE IN THE FOUR (4) CORNER LOCKS. PIN LOCKS IN THE CLOSED POSITION TO TIGHTEN SEALS.

VAN RETRACTING INSTRUCTIONS

(A) TO RETRACT VAN, REVERSE THE ABOVE PROCEDURES (B) APPLY ENOUGH PRESSURE TO THE RATCHET WRENCH TO FORCE THE SIDE WALLS AGAINST THE SEALS BEFORE ATTEMPTING TO CLOSE THE FOUR (4) SIDE LOCKS (MARKED "A")

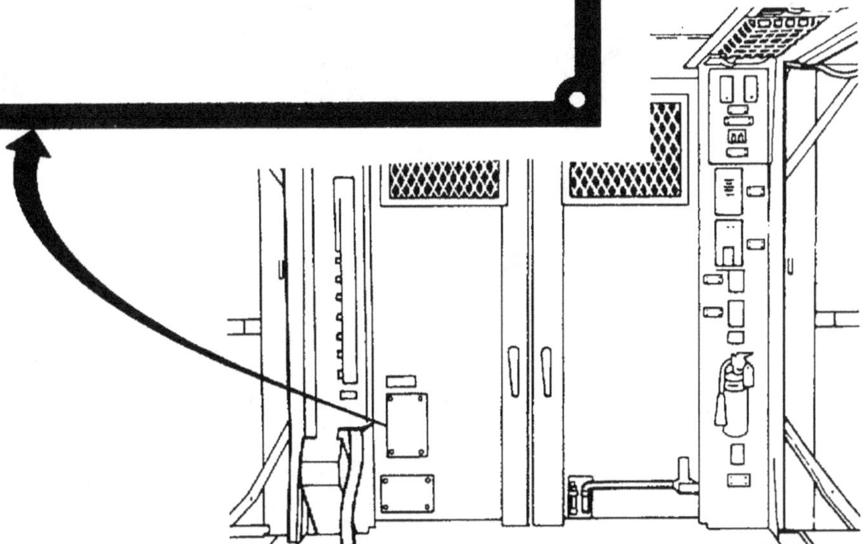

2-28. DECALS AND INSTRUCTION PLATES (Contd)

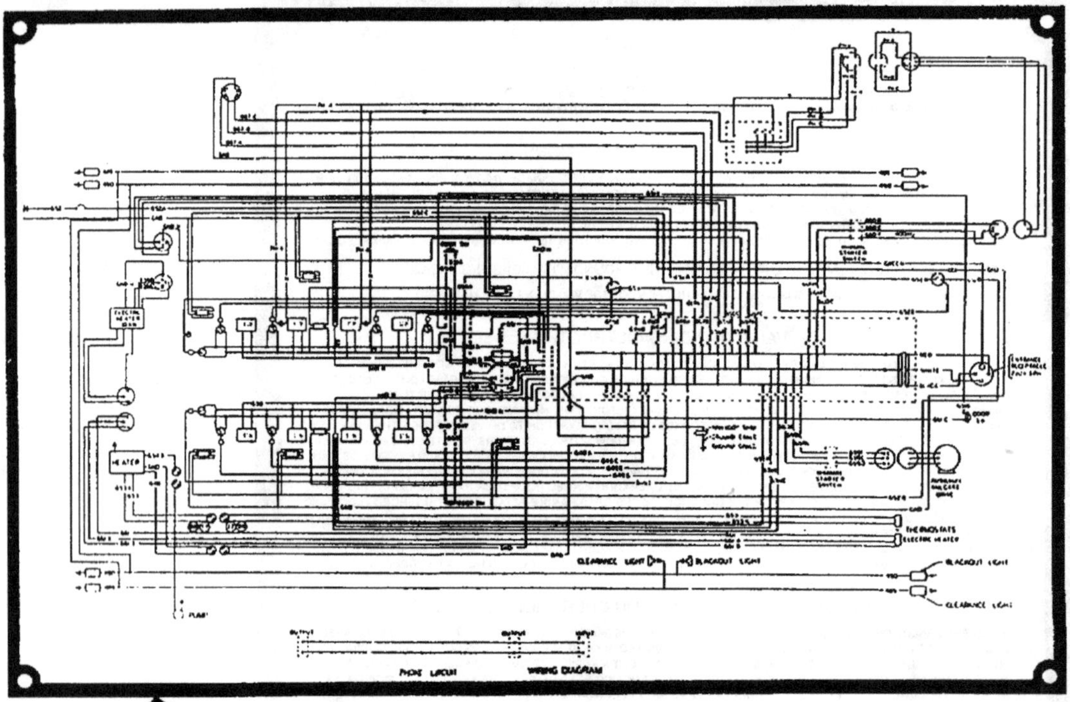

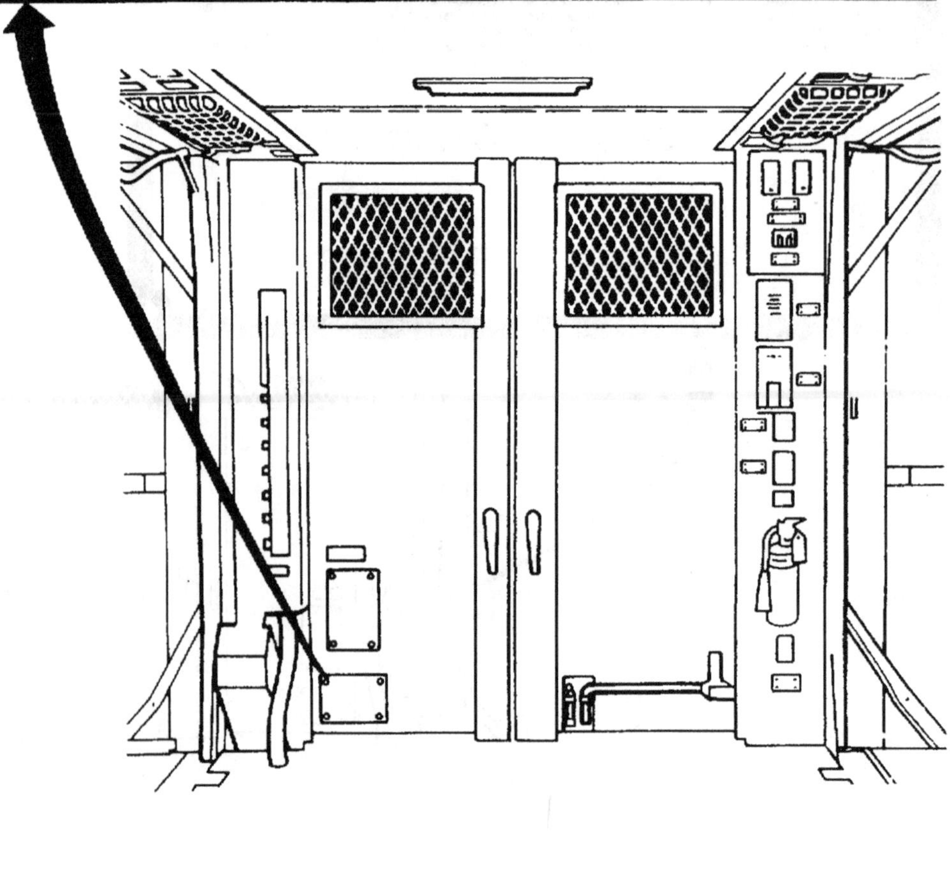

2-28. DECALS AND INSTRUCTION PLATES (Contd)

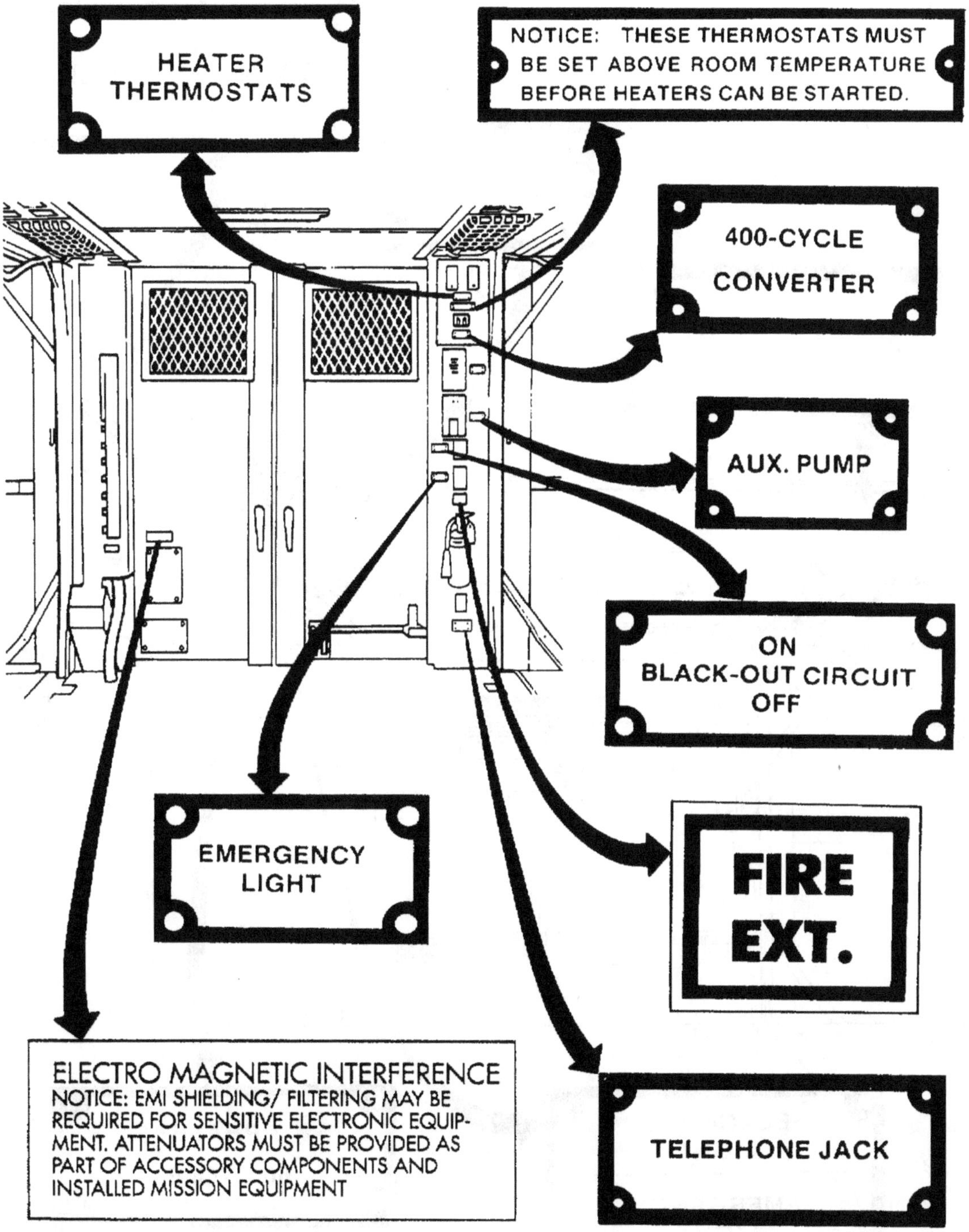
HEATER THERMOSTATS
NOTICE: THESE THERMOSTATS MUST BE SET ABOVE ROOM TEMPERATURE BEFORE HEATERS CAN BE STARTED.
400-CYCLE CONVERTER
AUX. PUMP
ON BLACK-OUT CIRCUIT OFF
EMERGENCY LIGHT
FIRE EXT.
ELECTRO MAGNETIC INTERFERENCE
NOTICE: EMI SHIELDING/ FILTERING MAY BE REQUIRED FOR SENSITIVE ELECTRONIC EQUIPMENT. ATTENUATORS MUST BE PROVIDED AS PART OF ACCESSORY COMPONENTS AND INSTALLED MISSION EQUIPMENT
TELEPHONE JACK

2-28. DECALS AND INSTRUCTION PLATES (Contd)

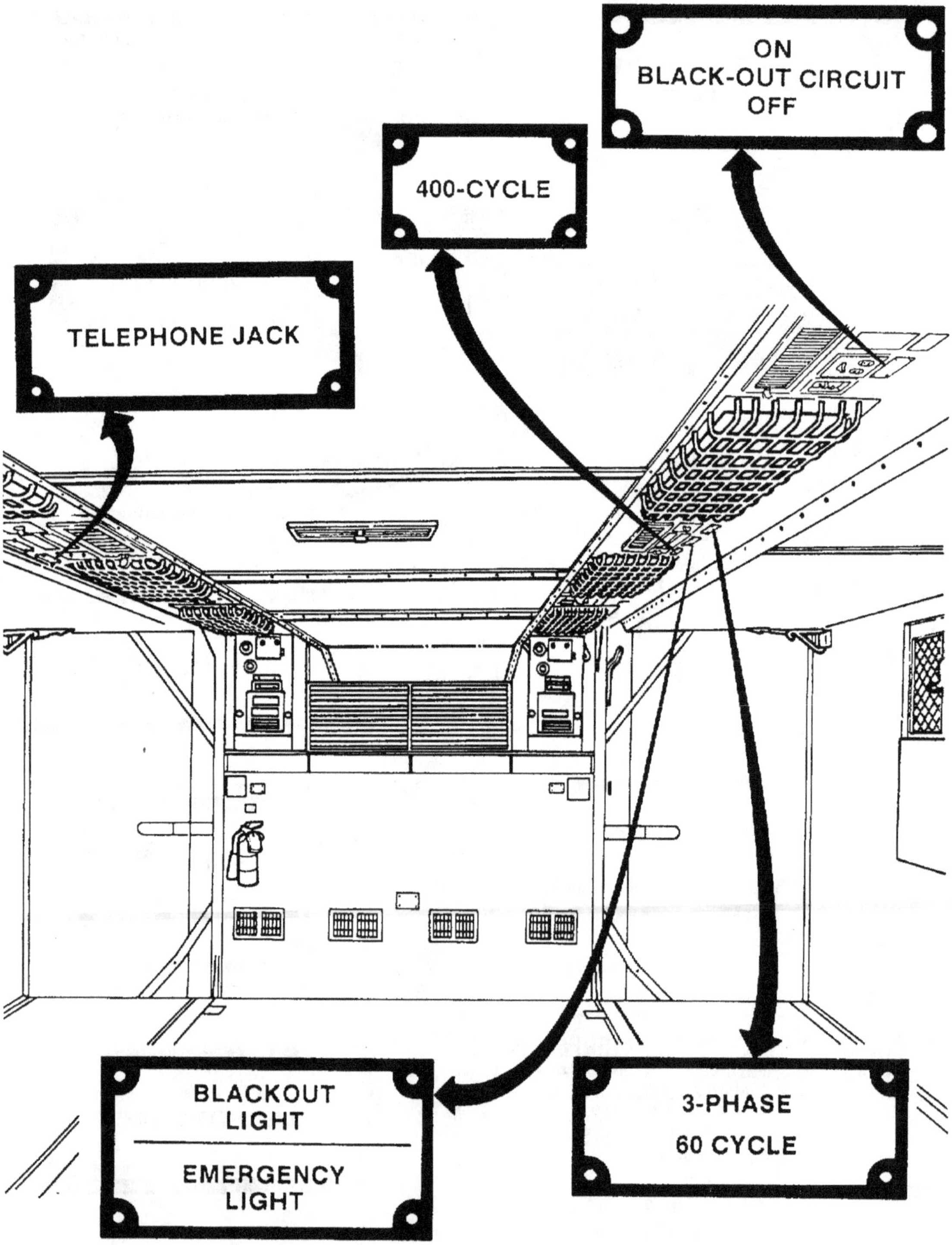
ON
BLACK-OUT CIRCUIT
OFF
400-CYCLE
TELEPHONE JACK
BLACKOUT
LIGHT
EMERGENCY
LIGHT
3-PHASE
60 CYCLE

2-28. DECALS AND INSTRUCTION PLATES (Contd)

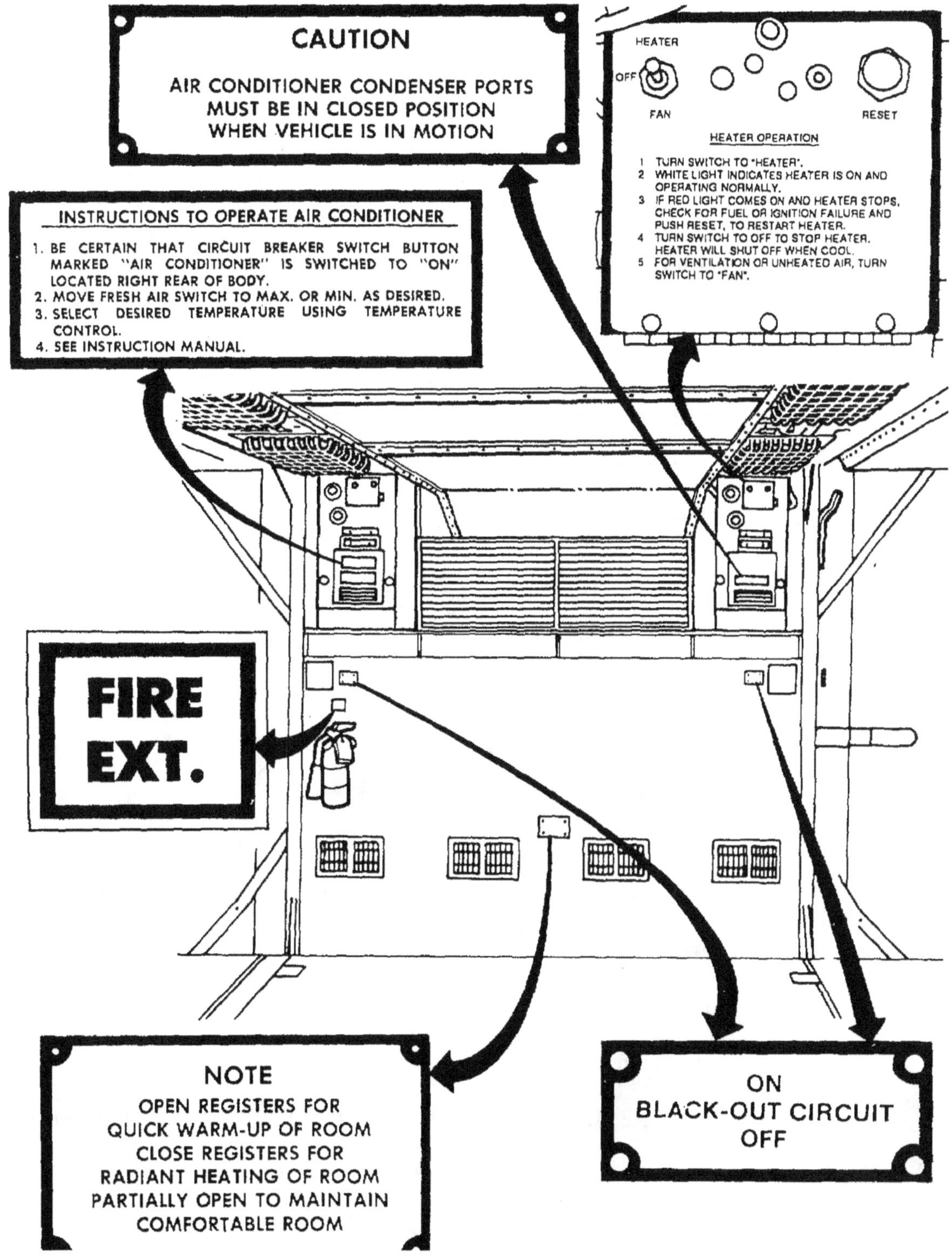
CAUTION
AIR CONDITIONER CONDENSER PORTS
MUST BE IN CLOSED POSITION
WHEN VEHICLE IS IN MOTION
HEATER
OFF
FAN
RESET
HEATER OPERATION
1 TURN SWITCH TO "HEATER".
2 WHITE LIGHT INDICATES HEATER IS ON AND OPERATING NORMALLY.
3 IF RED LIGHT COMES ON AND HEATER STOPS, CHECK FOR FUEL OR IGNITION FAILURE AND PUSH RESET, TO RESTART HEATER.
4 TURN SWITCH TO OFF TO STOP HEATER. HEATER WILL SHUT OFF WHEN COOL.
5 FOR VENTILATION OR UNHEATED AIR, TURN SWITCH TO "FAN".
INSTRUCTIONS TO OPERATE AIR CONDITIONER
1. BE CERTAIN THAT CIRCUIT BREAKER SWITCH BUTTON MARKED "AIR CONDITIONER" IS SWITCHED TO "ON" LOCATED RIGHT REAR OF BODY.
2. MOVE FRESH AIR SWITCH TO MAX. OR MIN. AS DESIRED.
3. SELECT DESIRED TEMPERATURE USING TEMPERATURE CONTROL.
4. SEE INSTRUCTION MANUAL.
FIRE
EXT.
NOTE
OPEN REGISTERS FOR
QUICK WARM-UP OF ROOM
CLOSE REGISTERS FOR
RADIANT HEATING OF ROOM
PARTIALLY OPEN TO MAINTAIN
COMFORTABLE ROOM
ON
BLACK-OUT CIRCUIT
OFF

2-29. PREPARATION FOR SHIPMENT

a. General. This paragraph provides information for personnel processing vehicles for shipment. Most tasks associated with shipment may have been performed by unit maintenance prior to delivery. If this vehicle is to be transported to a loading site, the following preparation may be necessary

b. Tools and Special Requirements. Except for spare tire removal (para. 3-11), the organization responsible for processing will provide tools and containers for stowage. Stowage of parts removed will be on vehicles or in suitable containers as specified by contracting agent. Exposed exhaust and intake pipes must be covered when vehicles are not running.

c. Minimum Reducible Height.

(1) On the M923/A1/A2, M925/A1/A2, M927/A1/A2, and M928/A1/A2 cargo trucks, the minimum reducible height is referenced on data plate located on right side of instrument panel (para. 2-28). All portions of the vehicle above this point must be removed or lowered. These may include any or all of the following:

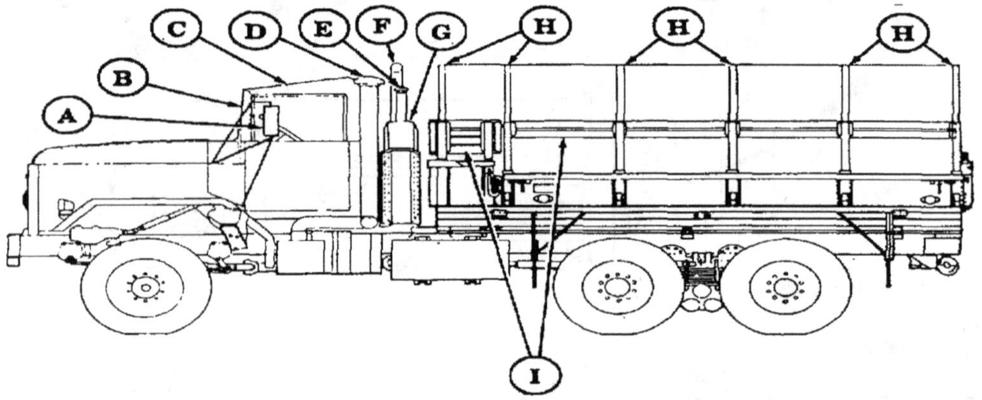
C
D
E
F
G
H
H
H
B
A
I

(A) Rear view mirrors (task d.).

(B) Windshield (para. 2-21).

(C) Cab top (para. 2-21). For hard cab tops, refer to unit maintenance.

(D) Air intake extension (task f.).

(E) Davit and boom (task g.).

(F) Exhaust pipe extension (task e.).

(G) Spare tire (M923A1/A2, M925A1/A2, M927A1/A2, M928A1/A2) (para. 3-11).

(H) Tarpaulin and bow kit (para. 2-42). Stow flat on bed of truck.

(I) Side and front racks (M927/A1/A2 and M928/A1/A2) (para. 2-23).

2-29. PREPARATION FOR SHIPMENT (Contd)

(2) On the M931/A1/A2 and M932/A1/A2 tractors, the minimum reducible height is referenced on data plate located on the side of the instrument panel (para. 2-28). All portions of the vehicle above this point must be removed or lowered. These may include any or all of the following:

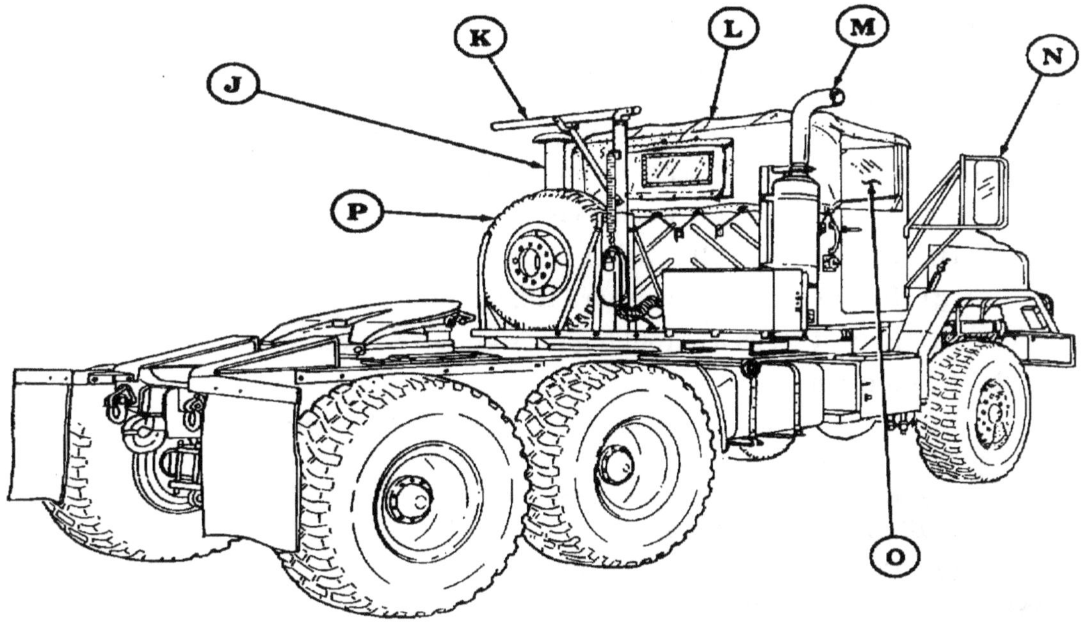
K
L
M
N
J
P
O

(J) Air intake extension (task f.).

(K) Davit and boom (task g.).

(L) Cab top (para. 2-21). For hard cab tops, refer to unit maintenance.

(M) Exhaust pipe extension (task e.).

(N) Rear view mirrors (task d.).

(O) Windshield (para. 2-21).

(P) Spare tire (M931A1/A2 and M932A1/A2) (para. 3-11).

2-29. PREPARATION FOR SHIPMENT (Contd)

d. Mirror Removal and Installation.

(1) Removal

(a) On left side of vehicle, remove locknut (11) and washer (10) from screw (8), lower brace (5), and lower door hinge (9).

(b) Remove locknut (13) and washer (14) from screw (18) support braces (16) and (17), and bracket (15).

(c) Remove locknut (12) and washer (7) from screw (2), support braces (1) and (3), and door hinge (6).

(d) Remove screws (2), (8), and (18) from mirror assembly (4), bracket (15), and door hinges (6) and (9). Remove mirror assembly (4).

(e) Install screw (18), washer (14), and locknut (13) on support bracket (15).

(f) Install screw (2), washer (7), and locknut (12) on door hinge (6).

(g) Install screw (8), washer (10), and locknut (11) on door hinge (9).

(h) Repeat steps a through g for right side of vehicle.

(2) Installation

(a) On left side of vehicle, remove screw (2), washer (7), and locknut (12) from door hinge (6).

(b) Remove screw (18), washer (14), and locknut (13) from support bracket (15).

(c) Remove screw (8), washer (10), and locknut (11) from door hinge (9).

(d) Install mirror assembly (4) on top door hinge (6) with screw (2), washer (7), and locknut (12), and on lower door hinge (9) with screw (8), washer (10), and locknut (11).

(e) Install support braces (16) and (17) on bracket (15) with screw (18), washer (14), and locknut (13).

(f) Adjust mirror as necessary to provide unobstructed rear view.

(g) Repeat steps a through f for right side of vehicle.

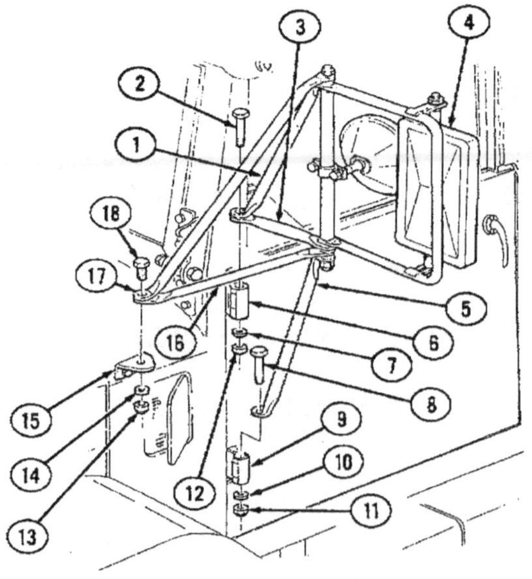

1
2
3
4
5
6
7
8
9
10
11
12
13
14
15
16
17
18

2-29. PREPARATION FOR SHIPMENT (Contd)

e. Exhaust Pipe Extension Removal and Installation.

WARNING

Do not touch hot exhaust system components with bare hands. Injury to personnel may result.

(1) Removal

(a) Remove locknut (25) and screw (22) from muffler coupling clamp (21).

(b) Remove muffler coupling clamp (21), gasket (24), and muffler extension (20) from muffler (23).

(c) Install screw (22), locknut (25), muffler coupling clamp (21), and gasket (24) on muffler extension (20).

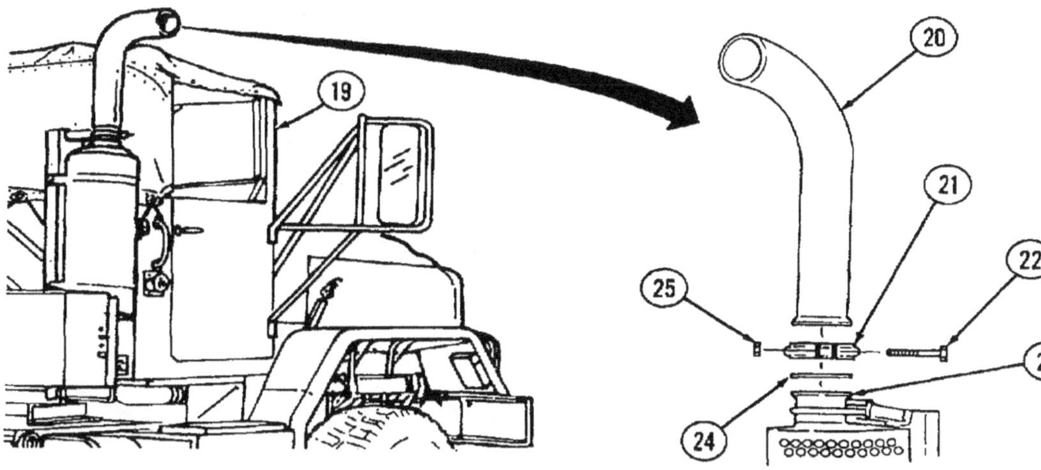
19
20
21
22
23
24
25

(2) Installation

(a) Remove locknut (25) and screw (22) from muffler coupling clamp (21).

(b) Remove muffler coupling clamp (21) and gasket (24) from muffler extension (20).

(c) Install muffler extension (20) and gasket (24) to muffler (23) with screw (22), muffler coupling clamp (21), and locknut (25).

(d) Rotate muffler exhaust extension (20) so that top points away from cab (19). Tighten screw (22) and locknut (25) to seal muffler extension (20) to muffler (23).

(e) Start engine (para. 2-12) and check for exhaust leaks.

NOTE

Report removal and installation to unit maintenance to replace

2-29. PREPARATION FOR SHIPMENT (Contd)

gasket and locknut.

CAUTION

Cover exposed intake tube when vehicle is not running. Failure to do so will result in damage to internal components.

f. Air Intake Extension Removal and Installation.

(1) Removal

(a) Remove screw (2) and locknut (3) from clamp (5) on air intake tube (4).

(b) Remove air intake extension (1) from air intake tube (4) and

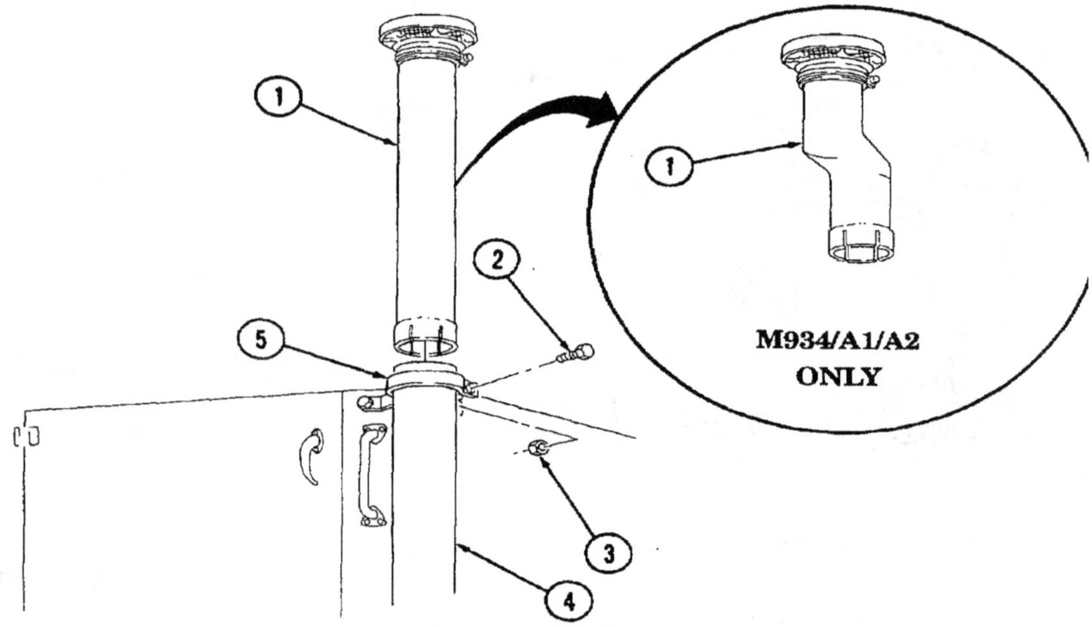
1
2
5
3
4
1
M934/A1/A2
ONLY

clamp (5).

(c) Install screw (2) and locknut (3) in clamp (5) on air intake tube (4).

(2) Installation

(a) Remove screw (2) and locknut (3) from clamp (5) on air intake tube (4).

(b) Install air intake extension (1) on air intake tube (4) with clamp (5), screw (2), and locknut (3).

2-29. PREPARATION FOR SHIPMENT (Contd)

(c) Report removal and installation of air intake extension (1) to unit maintenance as soon as possible.

g. Davit and Boom Removal and Installation.

(1) Removal (M939 series vehicles)

(a) Remove lock pin (10) and retaining pin (11) from boom (7).

(b) Remove boom extension (6) from boom (7).

(c) Loosen setscrew (8) in boom (7).

(d) Remove boom (7) from boom base (9).

(2) Installation (M939 series vehicles)

(a) Install boom (7) in boom base (9).

(b) Tighten setscrew (8) in boom (7).

(c) Install boom extension (6) in boom (7).

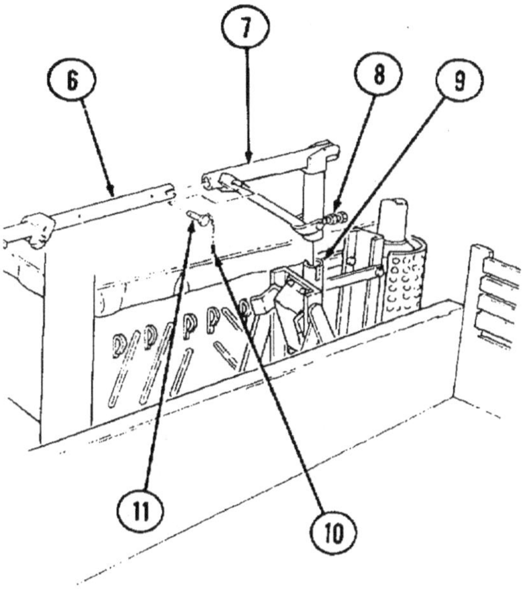
6
7
8
9
11
10

M939 SERIES VEHICLES

(d) Install retaining pin (11) in boom (8) and secure with lock pin (10).

(3) Removal (M939A1/A2 series vehicles)

(a) Remove lock pin (17) and retaining pin (18) from boom extension (12) and boom (13).

(b) Remove boom extension (12) from boom (13).

(c) Remove four locknuts (15), screws (14), and boom (13) from boom base (16).

(4) Installation (M939A1/A2)

(a) Install boom (13) on boom base (16) with four locknuts (15) and screws (14).

(b) Install boom extension (12) in boom (13).

(c) Install retaining pin (18) in

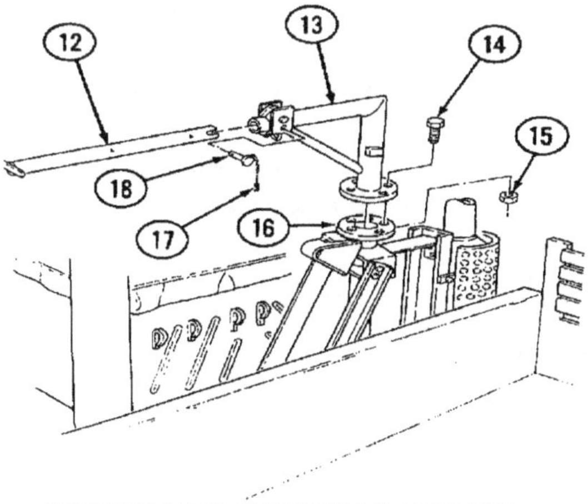
12
13
14
15
18
17
16

M939/A1/A2 SERIES VEHICLES

2-29. PREPARATION FOR SHIPMENT (Contd)

boom (13) and boom extension (12), and secure with lock pin (17).

h. Minimum Reducible Width.

(1) Shipping width is measured at the widest point on the vehicle excluding mirrors. On the M939, M939A1, and M939A2 series vehicles, it may be necessary to retract mirrors to obtain necessary clearance.

(2) Retracting mirrors.

(a) On left side of vehicle, loosen locknut (3) from screw (4).

(b) Rotate outer brace (2) inward toward door (1) until outer brace (2) no longer extends beyond fender (5).

(c) Tighten locknut (3) on screw (4) to secure outer brace (2) in this position.

(d) Repeat steps a through c for right side of vehicle.

CAUTION

With rear view mirrors retracted, use additional spotters when moving vehicle. Failure to do so may result in damage to equipment.

(3) Extending mirrors.

(a) On left side of vehicle, loosen locknut (3) from screw (4).

(b) Rotate outer brace (2) outward from the door (1) until extends it beyond fender.

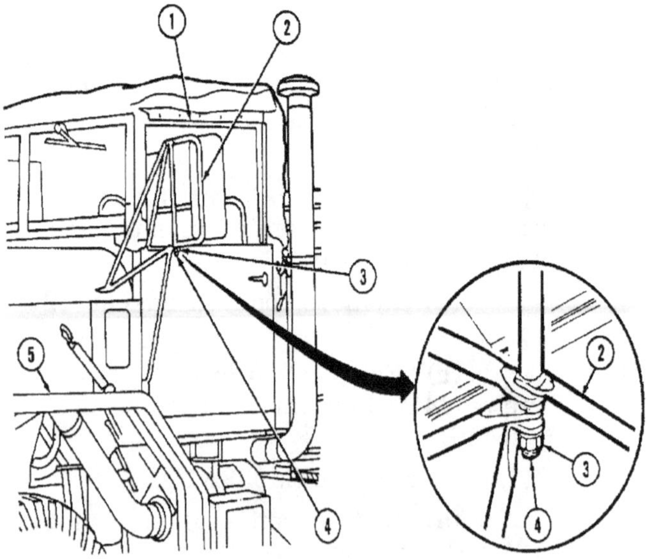
1
2
3
5
4
2
3
4

(c) Adjust as necessary to obtain unobstructed rear view.

(d) Tighten locknut (3) on screw (4) while holding outer brace (2) in place.

(e) Repeat steps a through d for right side of vehicle.

Section IV. OPERATION UNDER UNUSUAL CONDITIONS

2-30. SPECIAL INSTRUCTIONS

a. General. Special instructions for operating and maintaining vehicles under unusual conditions are included in this section. Unusual conditions are extreme high or low temperatures, humidity, and/or terrain. Special care in cleaning and lubrication must be taken to keep vehicles operating under unusual conditions.

b. Cleaning. Refer to paragraph 2-8 for cleaning instructions and precautions.

c. Lubrication.

(1) Refer to LO 9-2320-272-12 for proper lubricating instructions.

(2) Service intervals in LO 9-2320-272-12 are for normal operating conditions. Reduce service intervals when unusual conditions exist.

d. Driving Instructions.

(1) FM 21-305 contains special driving instructions for operating wheeled vehicles.

(2) FM 9-207 contains instructions on vehicle operation in extreme cold of 0° to -65°F (-18° to -54°C) or below. Other documents with information on cold weather vehicle operation are:

(a) FM 31-70 Basic Cold Weather Manual.

(b) FM 31-71 Northern Operations.

(c) FM 90-6 (HTF) Mountain Operations.

e. Reporting Materiel Failure. Report failure of vehicle, body equipment, or kits on DA form SF 368 (Quality Deficiency Report-Equipment Improvement Recommendations) as prescribed by DA Pam 738-750 and as stated in paragraph 1-5 of this manual.

f. Special Purpose Kits. Paragraphs describing special purpose kits for operation under unusual conditions are:

(1) Fuel burning personnel heater kit (para. 2-44) and engine coolant heater kit (para. 2-45).

(2) Deepwater fording kit (para. 2-40).

(3) Hardtop kit and radiator and hood cover kit (para. 2-43).

2-31. CTIS OPERATION UNDER UNUSUAL CONDITIONS

a. General. This paragraph provides instructions for M939A2 CTIS operation under emergency conditions and operation with a punctured tire.

NOTE

Except where specifically noted, the controls and indicator in this section are applicable to all M939A2 series vehicles.

2-31. CTIS OPERATION UNDER UNUSUAL CONDITIONS (Contd)

b. The selector panel (1) is part of the Electronic Control Unit (ECU) (4) and contains selectors for the four preset tire pressure modes and a run flat selector. Each selector has its own light. A steady selector light shows that the tire pressure selected has been achieved. A flashing selector light means that the system is working to change tire pressures (para. 3-5).

c. HWY Mode. The highway tire pressure selector is the normal operating mode of CTIS. The HWY mode (8) is 60 psi (414 kPa) (80 psi (552kPa) for M936A2 wrecker). If a lower tire pressure mode had been selected the last time the vehicle was operated, CTIS will automatically begin to inflate to the highway setting.

d. X-C Mode. The cross-country tire pressure selector X/C (7) is used for operating the vehicle on non-paved secondary roads and unimproved surfaces. It allows operation up to 35 mph (56 km/h) (25 mph (40 km/h) on M936A2 wrecker). When 35 mph (56 km/h) is exceeded for more than one minute, the amber overspeed warning light (2) will flash. If 35 mph (56 km/h) is exceeded for more than two minutes, CTIS will automatically begin to inflate to HWY (8) mode.

e. Sand Mode. When the mission requires maximum traction in sand, snow, or mud select SAND (6) on selector panel (1). It allows operation up to 20 mph (32 km/h) (15 mph (24 km/h) on M936A2 wrecker). When 20 mph (32 km/h) is exceeded for more than one minute, the amber overspeed warning light (2) will flash. If 20 mph (32 km/h) is exceeded for more than two minutes, CTIS will automatically begin to inflate to X/C (7) pressure.

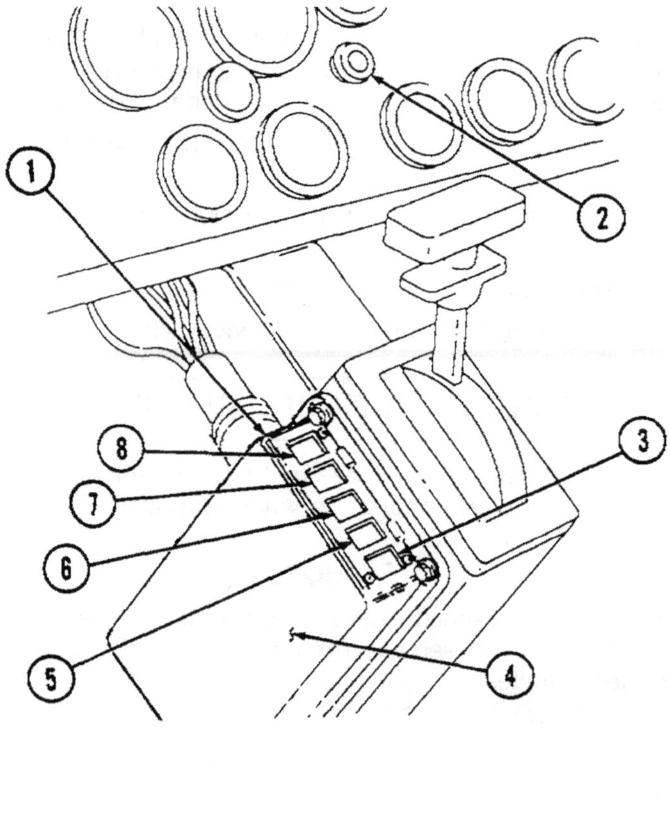
1
2
8
3
7
6
5
4

2-31. CTIS OPERATION UNDER UNUSUAL CONDITIONS (Contd)

CAUTION

Speed must be limited to 5 mph (8 km/h) in the emergency mode to prevent damage to tires.

f. Emer Mode. When the mission requires maximum traction on extremely adverse terrain, select emergency mode by depressing EMER (5) on the selector panel (1). The dash-mounted amber warning light will illuminate. Operation in emergency mode is limited to ten minutes. Then the system automatically inflates to SAND (6); if the mission demands extended emergency mode use, select EMER (5) as needed.

g. Run Flat Mode. When the mission requires operation with a punctured tire, select run flat mode by depressing RUN FLAT (3) on the selector panel (1). Run flat mode causes the CTIS to check tire pressure every fifteen seconds. Normally, checks occur every fifteen minutes. Repeated damage detection results in repeated inflation attempts. The punctured tire receives a new air supply each fifteen seconds. Operation in the run flat mode is limited to ten minutes unless reselected. If no longer required, press the RUN FLAT (3) selector a second time.

2-32. OPERATING IN EXTREME COLD

a. General. The operator must always be alert to changes in weather. The operator must take care of assigned vehicle in order to prevent damage to vehicle because of sudden changes in weather. The operator should be cautious when starting or driving a vehicle that has not been operated for a long period. Lubricants may thicken and cause parts failure. Tires may freeze to the ground, or may freeze flat on the bottom, if underinflated. The operator should be alert to such possibilities to prevent great damage to the vehicle.

b. Before Operation.

(1) Perform all before operation services listed in table 2-3, Preventive Maintenance Checks and Services (PMCS).

(2) Start engine coolant heater, if equipped, to warm vehicle coolant, engine, and batteries before attempting to start engine. Refer to paragraph 2-45 for engine coolant heater operating instructions.

c. Starting Engine.

NOTE

Shut down engine coolant heater before starting vehicle engine.

2-32. OPERATING IN EXTREME COLD (Contd)

(1) Start engine when engine coolant temperature reads 120°F (49°C) or higher as indicated by engine coolant temperature gauge (1). Refer to paragraph 2-13 for cold weather starting instructions.

(2) Check instrument readings. If any reading is not normal, stop engine. Report condition(s) to organizational maintenance if operator troubleshooting (table 3-1) cannot correct malfunction. Normal instrument readings are:

(a) Engine oil pressure gauge (7) should read 15 psi (103 kPa) on M939/A1 series vehicles and 10 psi (69 kPa) on M939A2 series vehicles, or higher with engine idling.

(b) Air pressure gauges (2) should read 90-130 psi (621-896 kPa).

(c) Voltmeter (3) should read in green area.

(d) Engine coolant temperature gauge (1) should read 175°-200°F (79°-93°C).

d. Driving Vehicle.

CAUTION

Do not allow the M939 series vehicles to exceed 5 mph (8 km/h) or M939A1/A2 series vehicles to exceed 6 mph (10 km/h) when transfer case is in low and the transmission is in 1 (first). Failure to do so will result in damage to internal engine components.

(1) Drive slowly with transfer case shift lever (5) in low range and transmission selector lever (4) in 1 (first) for 100 yd (91 m). This should be enough time to warm up gearcases and tires.

(2) Check instruments during operation. Refer to step 2c for normal readings. Transmission oil temperature gauge (6) under should read 120-220°F (49°-104°C) for normal operation.

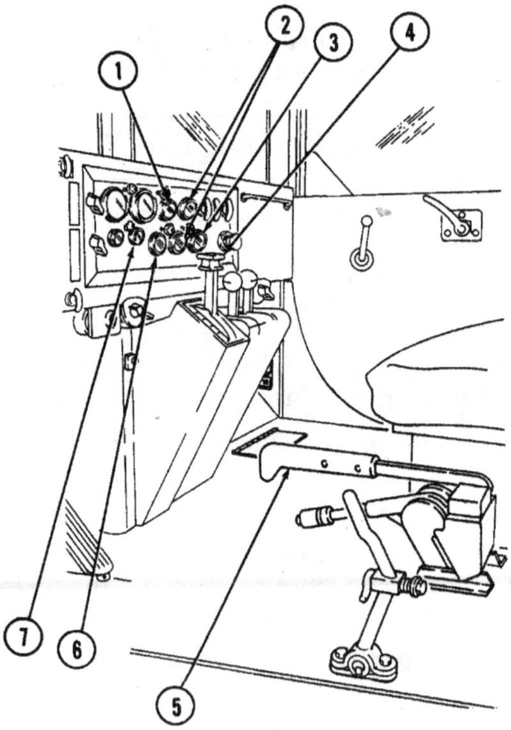
1
2
3
4
5
6
7

2-32. OPERATING IN EXTREME COLD (Contd)

e. Stopping or Parking.

CAUTION

- Operator must take every precaution to prevent snow from blowing into engine compartment when parked. Snow will melt and later form ice to jam engine controls.
- Do not apply parking brake. Brakeshoes may freeze to drum.

NOTE

- Do not idle engine for more than 15 minutes.
- Park in a sheltered area out of the wind, if possible, or park so that vehicle does not face into the wind. Park vehicle with wood planks, brush, mats, or canvas under the wheels if a long shutdown period in open area is expected.

(1) Place transmission selector lever (4) in N (neutral).

(2) Place chocks (8) in front of or behind vehicle wheels.

(3) Drain water from compressed air reservoirs by turning four drainvalves (9) counterclockwise. Close drainvalves (9) immediately after purging water.

WARNING

Alcohol used in alcohol evaporator is flammable, poisonous, and explosive. Do not smoke when adding fluid and do not drink fluid. Doing so will result in injury or death.

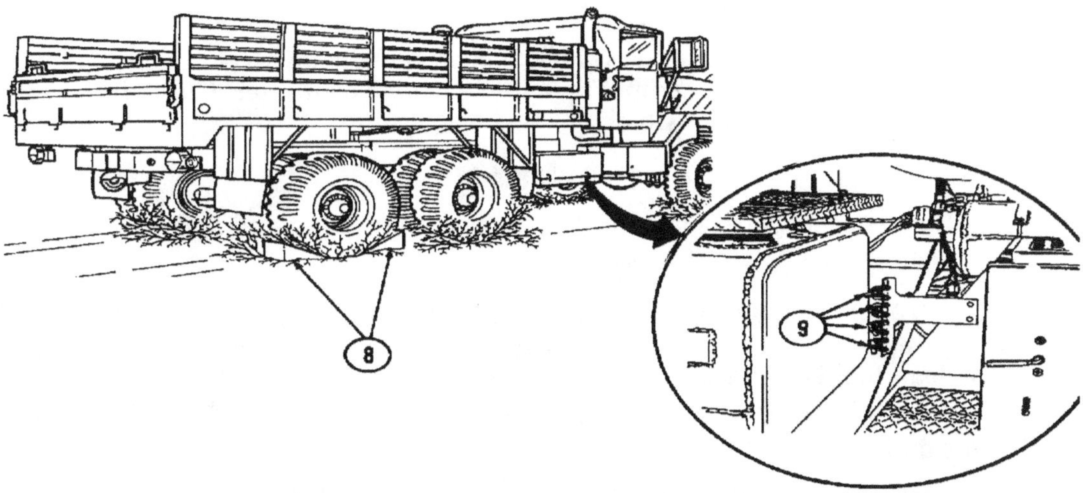
8
9

2-32. OPERATING IN EXTREME COLD (Contd)

(4) On M939/A1 series vehicles, check fluid level in alcohol evaporator (1) located on left side of engine. Add alcohol as required.

(5) Start engine coolant heater if required. Refer to paragraph 2-45 for engine coolant heater operating instructions.

(6) Perform after-operation services in Preventive Maintenance Checks and Services (PMCS table 2-3).

(7) Drain off any accumulated water in fuel filter/water separator (2). To drain water from fuel system:

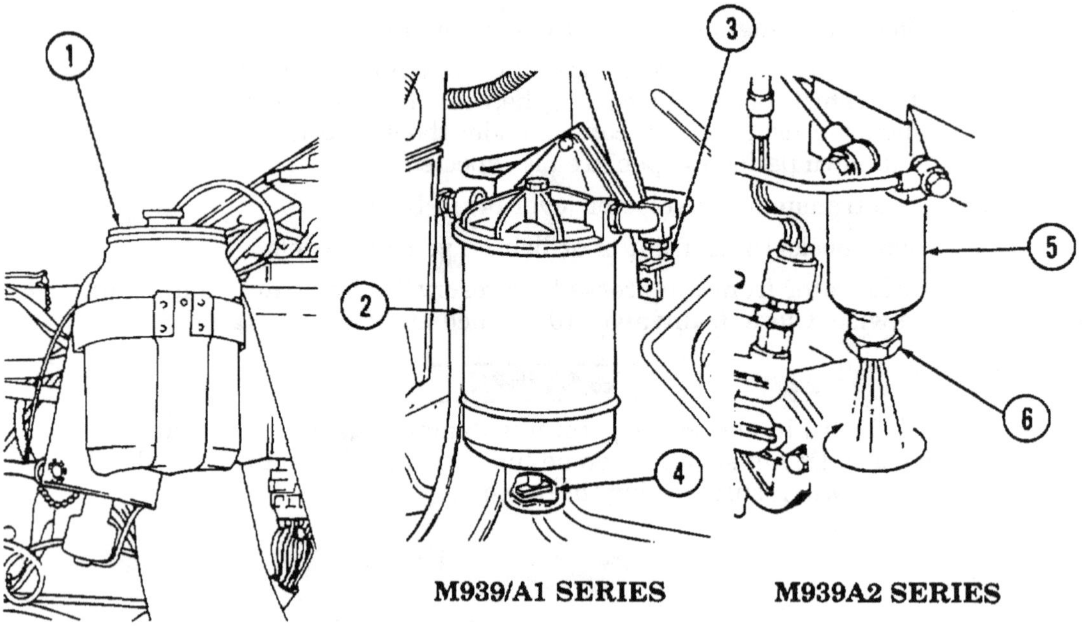
1
2
3
4
5
6

M939/A1 SERIES M939A2 SERIES

(a) On M939/A1 series vehicles, open fuel drainvalves (3) and (4) on fuel filter/water separator (2) and drain water into suitable container. Close drainvalves (3) and (4) when clear fuel is visible.

CAUTION

Do not overtighten plastic valve; damage may result and fuel may leak.

(b) On M939A2 series vehicles, loosen drainvalve (6) on bottom of fuel filter/water separator (5) and drain water into suitable container. Close drainvalve (6) when clean fuel is visible.

(c) If fuel is not clear before approximately one qt (0.946 L) has drained, notify unit maintenance.

(d) After draining has been completed, prime the fuel system (para. 3-8).

2-33. OPERATING IN SNOW

a. General.

(1) Refer to paragraph 2-13 for cold weather starting instructions.

(2) If vehicle is equipped with arctic winterization kits, refer to paragraphs 2-44 and 2-45 for description and operating instructions.

(3) Operating on snow or ice requires use of tire chains on forward-rear axle tires. Refer to FM 21-305 for installation of tire chains.

CAUTION

Attempting operation with only one driving wheel equipped with tire chain may result in damage to tire and/or power train.

NOTE

Use tire chains on forward-rear-axle-tires. For M939 series vehicles place chains on outside tires. Remove as soon as mission allows.

(4) If tire chains are not available, deflate tire pressure to 25 psi (172 kPa). Reinflate to normal pressures after operating in snow (table 1-10).

b. Driving Vehicle.

(1) Remove chocks (7) from vehicle wheels if used.

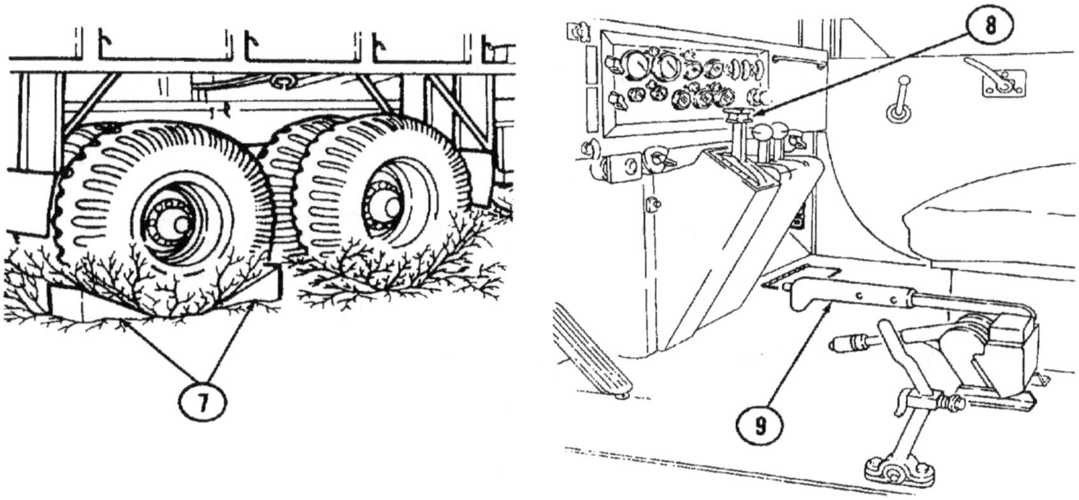
7
8
9

(2) Place transmission selector lever (8) in 1-5 (drive) and transfer case shift lever (9) in low range. Slowly accelerate without causing wheels to spin or engine to race. Place transfer case shift lever (9) in high range when under way.

2-33. OPERATING IN SNOW (Contd)

WARNING

Operators must drive at reduced speeds and be prepared to meet sudden changes in road conditions and traffic speeds. Maintain safe stopping distances. Pump brakes gradually when stopping vehicle on ice or snow. Sudden stops will cause vehicle wheels to lock, engine to stall, and loss of power steering. Failure to do this may result in injury or death.

NOTE

This warning applies to vehicles not equipped with antilock brake system (ABS). To stop a vehicle equipped with ABS, apply firm steady pressure to brake pedal to bring vehicle to a gradual stop. Do not pump brakes on a vehicle equipped with ABS when stopping. ABS will automatically release wheels that are locking and apply pressure to the other wheels.

CAUTION

If vehicle gets stuck on ice or in snow, do not rock vehicle by shifting rapidly between reverse and forward gears. This can cause power train damage.

NOTE

Use tire chains on forward-rear-axle-tires. For M939 series vehicles place chains on outside tires. Remove as soon as mission allows.

(3) If rear end skidding occurs:

(a) Turn steering wheel (1) in direction of the skid.

(b) Let up on accelerator pedal (2) and apply brake pedal (3).

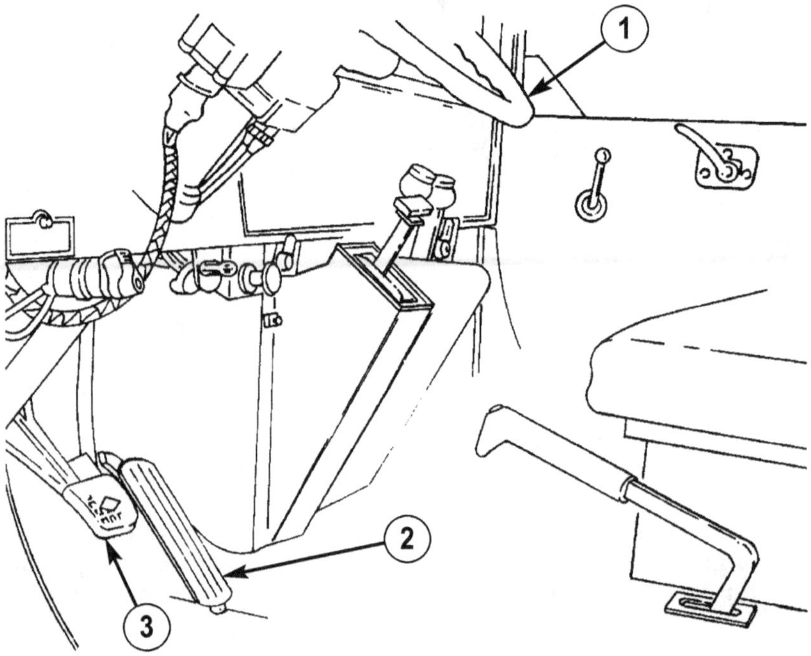
1
2
3

2-33. OPERATING IN SNOW (Contd)

c. After Operation.

(1) Remove all ice and snow build-up on vehicle.

(2) Refuel fuel tank(s) (4) as soon as possible.

CAUTION

Drain moisture from tanks in the following sequence. Failure to do so may result in reintroduction of moisture from tanks that have not been purged, resulting in equipment failures due to ice forming in air lines.

(3) Purge moisture from air tanks in the following order:

(a) Open wet tank drainvalve (7).

(b) Open primary and secondary drainvalves (5) and (6).

(c) Open spring brake drainvalve (8).

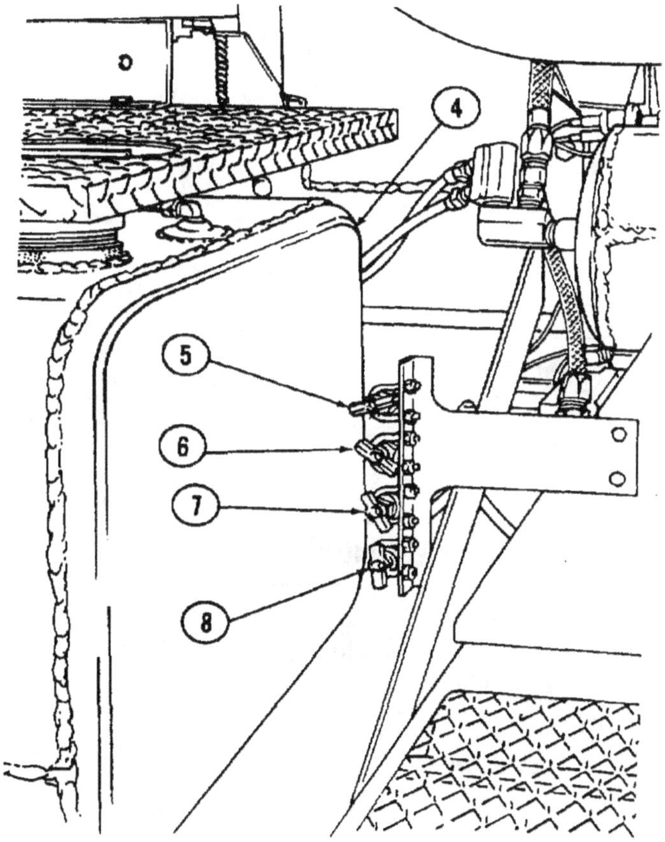
4
5
6
7
8

2-33. OPERATING IN SNOW (Contd)

(4) Drain off any accumulated water in fuel filter/water separator (1). To drain water from fuel system:

(a) On M939/A1 series vehicles, open drainvalves (2) and (3) on fuel filter/water separator (1), and drain water into suitable container. Close drainvalves (2) and (3) when clear fuel is visible.

CAUTION

Do not overtighten plastic valve; damage may result and fuel may leak.

(b) On M939A2 series vehicles, loosen drainvalve (5) on bottom of fuel filter/water separator (4) and drain water into a suitable container. Close drainvalve (5) when clean fuel is visible.

(c) If fuel is not clear before approximately one qt (0.946 L) has drained, notify unit maintenance.

(5) After draining has been completed, prime fuel system (para. 3-8).

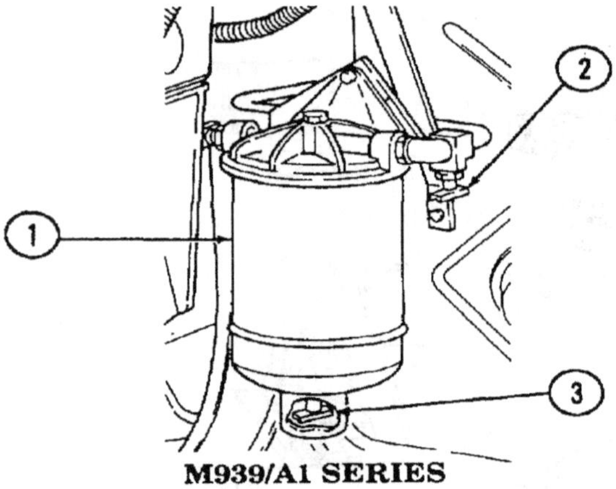
1
2
3

M939/A1 SERIES

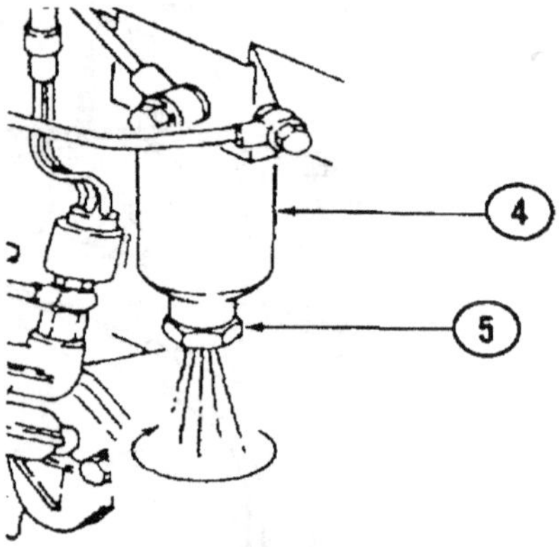
4
5

M939A2 SERIES

2-34. OPERATING IN EXTREME HEAT

a. General. Extreme heat exists when outside temperature reaches 95°F (35°C) or more. The effect of extreme heat on vehicle engine is a decrease in efficiency. Operators must adjust driving to conditions.

b. Before Operation.

(1) Perform before-operation services in table 2-3.

(2) Check for sand or insects in front of radiator. Blow out obstructions with low compressed air.

2-34. OPERATING IN EXTREME HEAT (Contd)

(3) Check tension adjustment of belts.

(4) Check coolant hoses and lines for cracks, leaks, and security of connections.

(5) Add corrosion inhibitor compound to cooling liquid.

(6) Check for correct tire inflation pressure. Do not reduce pressure if tires are hot from driving.

(7) Reduce lubrication intervals as specified in applicable LOs.

c. Driving Vehicle.

(1) Avoid continuous vehicle operation at high speeds. Avoid long, hard pulls on steep grades with transfer case shift lever (6) in low.

(2) Frequently check air cleaner indicator (7). If indicator shows red, perform emergency air cleaner servicing (para. 3-8).

(3) Frequently check engine coolant temperature gauge (8), engine oil pressure gauge (10), and transmission oil temperature gauge (9). Engine or transmission is overheating if any of the following conditions exist:

(a) Coolant temperature gauge (8) indicates more than 210°F (99°C).

(b) Oil pressure gauge (10) drops below 15 psi (103 kPa) on M939/A1 series vehicles and 10 psi (69 kPa) on M939A2 series vehicles, with engine at idle.

(c) Transmission oil temperature exceeds 300°F (149°C) on oil temperature gauge (9).

(4) If engine overheating occurs:

NOTE

Do not raise vehicle hood. Engine will cool faster at idle with hood closed.

(a) Park vehicle, allowing engine to idle.

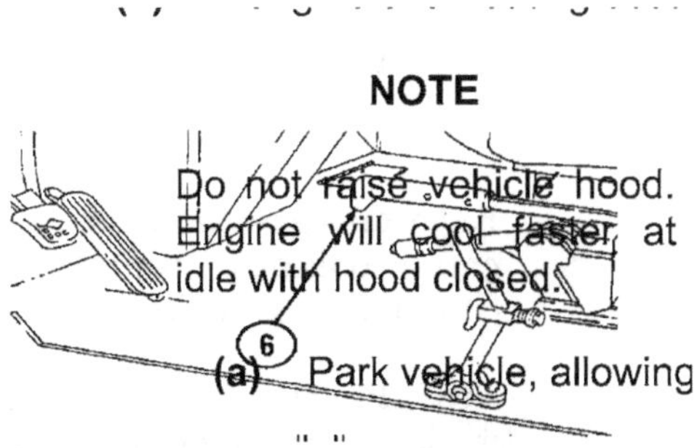
NOTE
Do not raise vehicle hood.
Engine will cool faster, at
idle with hood closed.
6
(a) Park vehicle, allowing

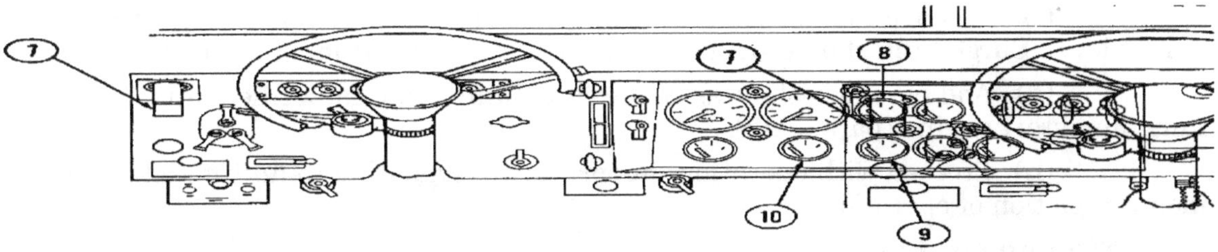
7
7
8
10
9

2-34. OPERATING IN EXTREME HEAT (Contd)

CAUTION

If engine temperature continues to rise or does not drop after two minutes of idling, shut down engine and refer to troubleshooting table 3-1.

(b) Observe coolant temperature gauge (1), engine oil pressure gauge (3), and transmission oil temperature gauge (2) for signs that engine or transmission is steadily cooling.

(c) Shut off when engine coolant temperature gauge (1) reaches normal operating temperature of 175°-200°F (79°-93 °C).

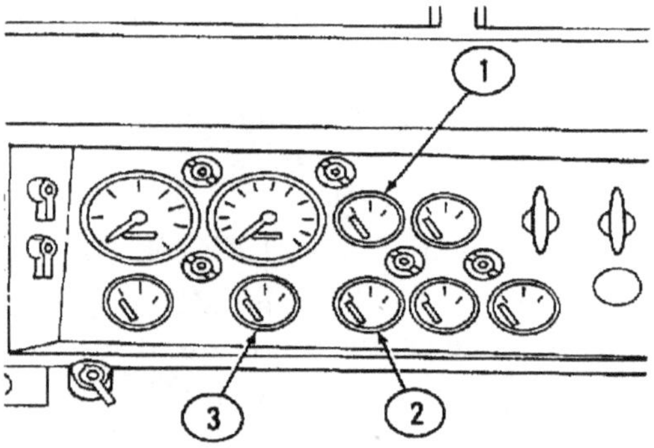
1
3
2

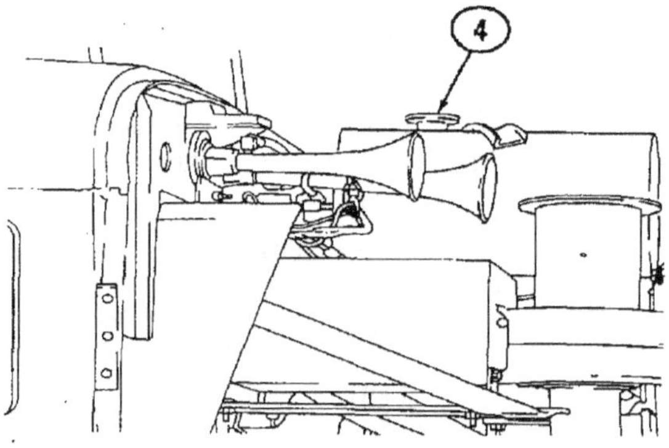
4

(d) Perform troubleshooting procedures as listed in table 3-1.

WARNING

Extreme care should be taken when removing surge tank filler cap if coolant temperature gauge reads above 175°F (79°C). Steam or hot coolant under pressure will cause injury.

(e) Place a thick cloth over surge tank filler cap (4). Carefully turn cap (4) counterclockwise to first stop to allow pressure to escape.

(f) Remove cap (4) when cooling system pressure is vented and check coolant level. Surge tank should be filled approximately to bottom of filler neck.

(g) Add engine coolant as required. Install surge tank filler cap (4) after filling and start engine (para. 2-12).

(h) Proceed with operation. Report any overheating to unit maintenance upon completion of operation.

d. Stopping or Parking.

(1) Park vehicle in a sheltered area, out of sun, if possible.

(2) Check batteries after operation and service, as required.

2-35. OPERATING IN DUSTY OR SANDY AREAS

a. General. Be aware of vehicle overheating when operating in dusty or sandy areas. Air cleaner, cooling system, and lubrication points will require frequent servicing.

NOTE

- Do not use tire chains in soft sand.
- Use a second vehicle with winch to recover vehicles stuck or sunk in sand.

b. Driving Vehicle.

(1) When starting in sand or soft ground, place transfer case shift lever (6) in low range and transmission selector lever (5) in 1-2 (second) or 1-3 (third).

(2) Deflate tires to 25 psi (172 kPa) only when operating off the road in heavy rain. Inflate tires immediately to correct pressure when operation changes to paved roads. Refer to table 1-10 for tire inflation data. Refer to paragraph 3-11 for inflating tires using vehicle air system.

(3) On M939A2 series vehicles, select SAND (8) on control panel (7). This will automatically deflate tires unless 20 mph (32 km/h) is exceeded (15 mph (24 km/h) on M936A2 wrecker) for more than two minutes. CTIS will then automatically inflate to cross-country pressure.

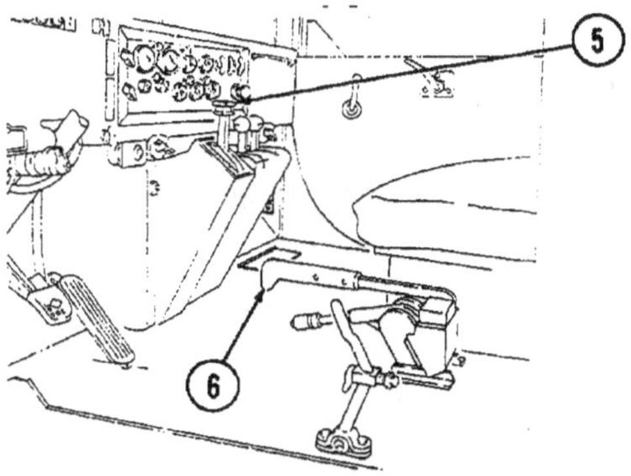
5
6

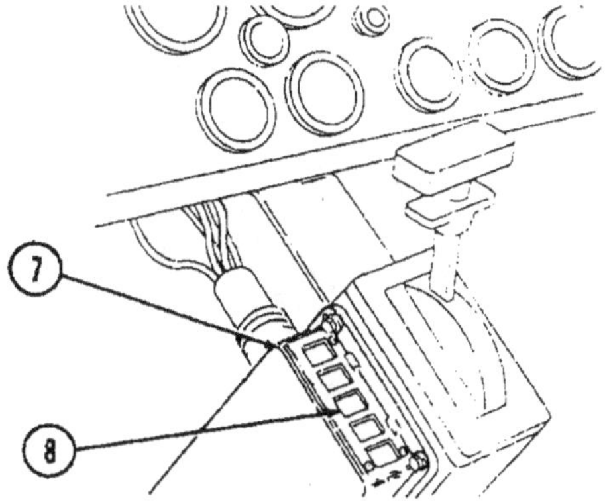
7
8

(4) Accelerate slowly so wheels will not spin and dig into sand.

(5) Inflate tires to normal pressures after vehicle has cleared deep sand (table 1-10). On M939A2 series vehicles, press the desired mode select button on the control panel (7).

(6) When moving across a slope, choose the least angle possible. Keep moving and avoid turning quickly.

(7) Keep throttle steady after reaching desired speed.

(8) Do not rock vehicle out of deep sand.

2-35. OPERATING IN DUSTY OR SANDY AREAS (Contd)

c. Stopping or Parking.

(1) Park vehicle in a sheltered area, out of blowing dust or sand whenever possible. If sheltered area is not available, park so vehicle does not face into wind and cover vehicle with tarpaulins. When entire vehicle cannot be covered, protect windows, cab, and engine compartment with tarpaulins to prevent entry of sand or dust.

(2) Use low air pressure to remove all sand from vehicle engine compartment, areas around brakes, drums, and spring seats after daily operation.

(3) Use caution while refueling to prevent dust or sand from entering fuel tank(s). Tighten filler cap securely after refueling.

2-36. OPERATING UNDER RAINY OR HUMID CONDITIONS

a. General.

(1) Vehicles inactive for long periods in hot, humid weather can rust rapidly. Fungus may grow in the fuel tank(s), on canvas tarpaulin, seats, and other components. Frequent inspection, cleaning (para. 2-8), and lubrication are necessary to maintain the readiness of vehicles.

(2) Drain off any accumulated water in fuel filter/water separator (1). To drain water from fuel system:

(a) On M939/A1 series vehicles, open drainvalves (2) and (3) on fuel filter/water separator (1), drain water into suitable container, close drainvalves (2) and (3) when clean fuel is visible.

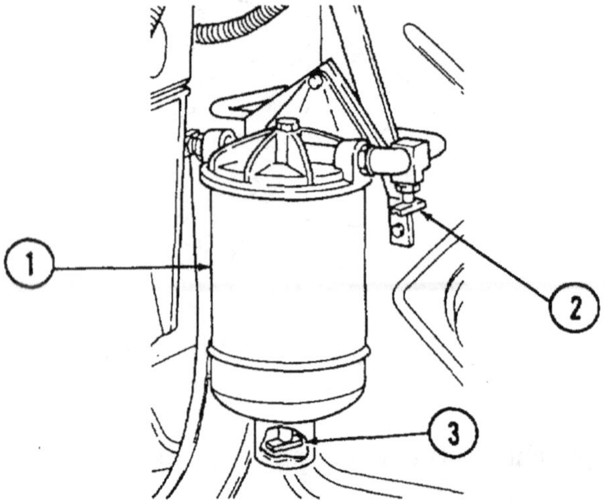
1
2
3

M939/A1 SERIES

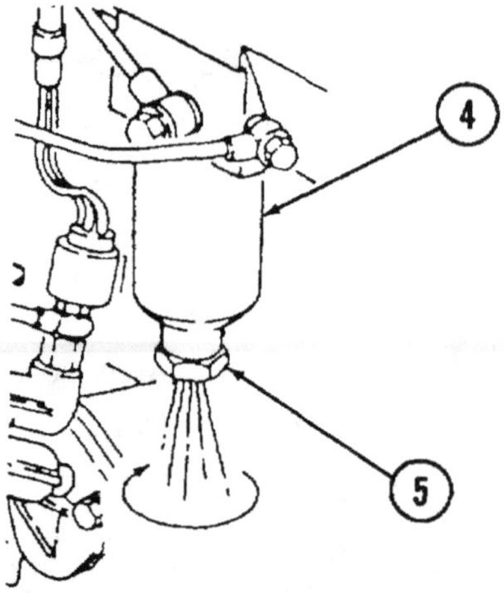
4
5

M939A2 SERIES

2-36. OPERATING UNDER RAINY OR HUMID CONDITIONS (Contd)

CAUTION

Do not overtighten plastic valve; damage may result and fuel may leak.

(b) On M939A2 series vehicles, loosen drainvalve (5) on bottom of fuel filter/water separator (4), drain water into suitable container, and close drainvalve (5) when clean fuel is visible.

(c) If fuel is not clear before approximately one qt (0.946 L) has drained, notify unit maintenance.

(d) After draining has been completed, prime the fuel system (para. 3-8).

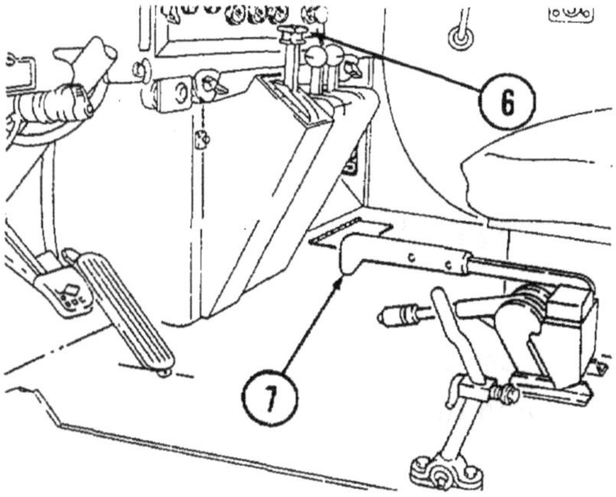
6
7

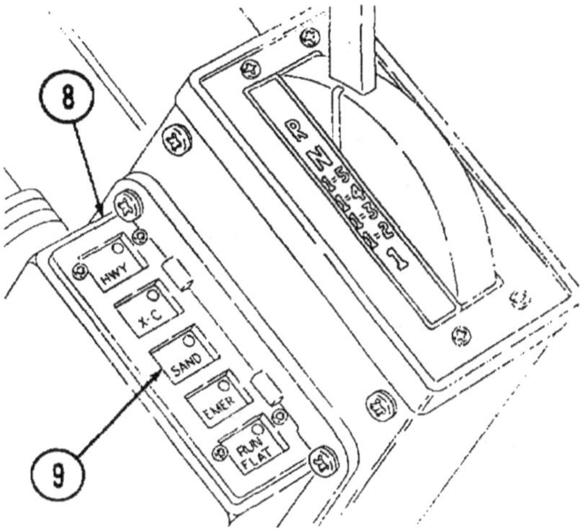
8
9
HWY
X-C
SAND
EMER
RUN FLAT

b. Driving Vehicle.

(1) Do not spin wheels when placing vehicle in motion in heavy rain or muddy conditions. If necessary, place transfer case shift lever (7) in low range and transmission selector lever (6) in 1-3 (third) to obtain a slow, firm start.

WARNING

Pump brakes gradually when slowing or stopping vehicle on wet pavement. Sudden stops will cause vehicle wheels to lock, engine to stall, and loss of power steering. Failure to do this may result in injury or death.

(2) Deflate tires to 25 psi (172 kPa) only when operating off-the-road in heavy rain. Inflate tires immediately to correct pressure when operation changes to paved roads. Refer to table 1-10 for tire inflation data. Refer to paragraph 3-11 for inflating tires using vehicle air system.

(3) On M939A2 series vehicles, select SAND (9) on control panel (8). This will automatically deflate tires unless 20 mph (32 km/h) is exceeded (15 mph (24 km/h) on M936A2 wrecker)) for more than two minutes. CTIS will automatically inflate to cross-country pressure.

2-37. OPERATING IN DEEP MUD

a. Driving Vehicle.

NOTE

- Six-wheel drive is achieved automatically when transfer case shift lever is placed in low range. In high range, the front-wheel drive lock-in switch must be engaged to achieve six-wheel drive.
- Use a second vehicle with winch to recover vehicles sunk in deep mud. Do not spin wheels. Refer to paragraph 2-22 for front winch operation.

(1) Place transmission selector lever (2) in 1-5 (drive) and transfer case shift lever (3) in low range. Place vehicle in motion slowly without causing wheels to spin or engine to race. Place transfer case shift lever (3) in high range when vehicle is under way.

(2) If rear end skidding occurs:

(a) Turn steering wheel (1) in direction of skid.

(b) Let up on accelerator pedal (4) and apply brake pedal (5) in a gradual pumping manner.

b. After Operation.

(1) Wash all mud from vehicle as soon as possible.

(2) If vehicle front winch was used, clean and lubricate. Refer to LO 9-2320-272-12.

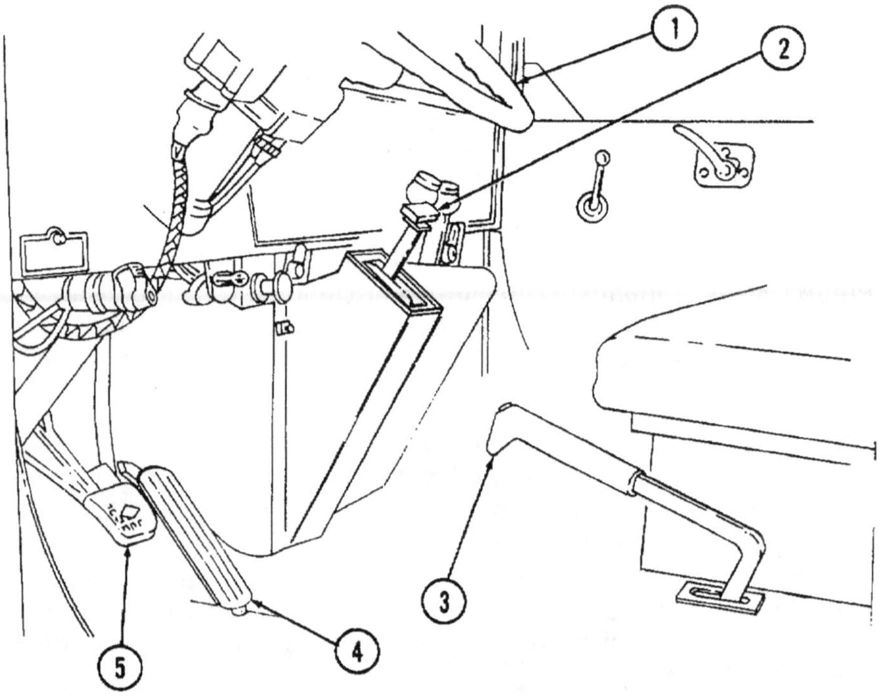
1
2
3
4
5

Section V. OPERATION OF SPECIAL PURPOSE KITS

2-38. GENERAL

This section provides information and instructions for operation of special purpose kits for M939/A1/A2 series vehicles.

Table 2-6. M939/A1/A2 Series Special Purpose Kits.

KITS	M923/A1/A2	M925/A1/A2	M927/A1/A2	M928/A1/A2	M929/A1/A2	M930/A1/A2	M931/A1/A2	M932/A1/A2	M934/A1/A2	M936/A1/A2
A-frame Kit		X		X		X		X		
Airbrake Control Kit	X	X	X	X					X	X
Bow and Tarpaulin Kit	X	X	X	X	X	X				
Chemical Agent Alarm Kit*	X	X	X	X	X	X	X	X	X	X
Deepwater Fording Kit	X	X	X	X	X	X	X	X	X	X
Fuel Burning Personnel Heater and Engine Coolant Heater Kits	X	X	X	X	X	X	X	X	X	X
Hardtop Kit	X	X	X	X	X	X	X	X	X	X
Machine Gun Mount Kit*	X	X	X	X			X	X		X
Mud Flap Kit							X	X		
Radiator and Hood Cover Kit	X	X	X	X	X	X	X	X	X	X
Rifle Mount Kit*	X	X	X	X	X	X	X	X	X	X
Troop Seat Kit					X	X				

*Information found in other section of this manual.

2-39. OPERATION OF AUXILIARY EQUIPMENT (SPECIAL PURPOSE KITS)

Operating Instructions. Operating instructions for the following special purpose kits are covered herein.

(1) Deepwater Fording Kit (page 2-218)

(2) Troop Seat Kit (page 2-223)

(3) Bow and Tarpaulin Kit (page 2-224)

(4) Hardtop Kit and Radiator and Hood Cover Kit (page 2-231)

(5) Fuel Burning Personnel Heater Kit (page 2-232)

(6) Engine Coolant Heater Kit (page 2-234)

(7) A-frame Kit (page 2-238)

(8) Airbrake Control Kit (page 2-239)

(9) Mud Flap Kit (page 2-240)

2-40. DEEPWATER FORDING KIT

a. General. Salt water causes damage to vehicle components. For this reason, do not drive needlessly in or through salt water. Vehicle components that are exposed to salt water must be washed with fresh water as soon as possible. The vehicle will ford water up to 30 in. (76 cm) in depth without a fording kit and 78 in. (198 cm) with kit installed.

b. Operator Preparation for Fording.

(1) Tighten cap(s) (1) on fuel tank(s) (2).

(a) Location of fuel tank (2) on single tank model vehicles:

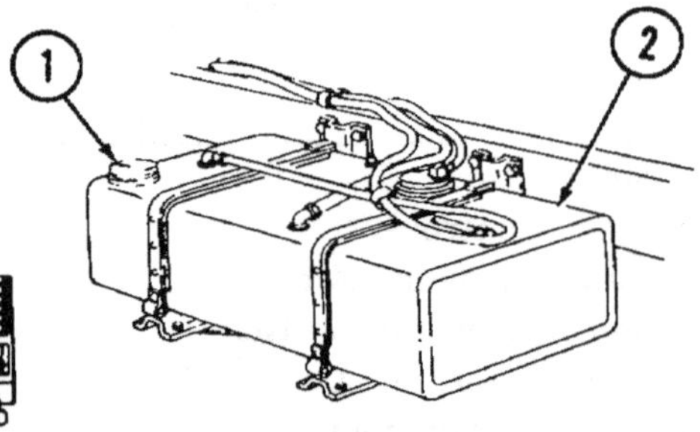
1
2

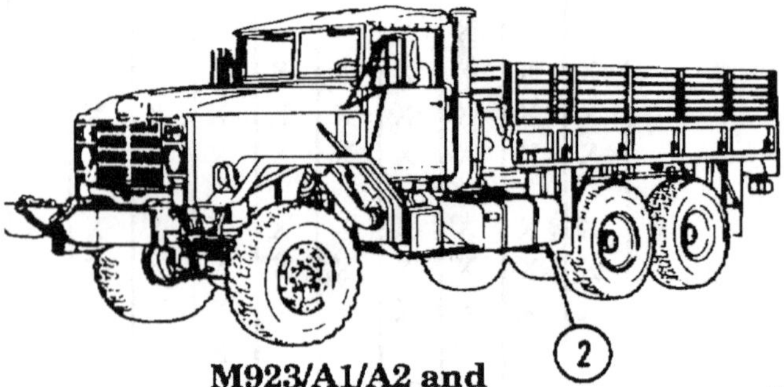
2

M923/A1/A2 and M925/A1/A2 models

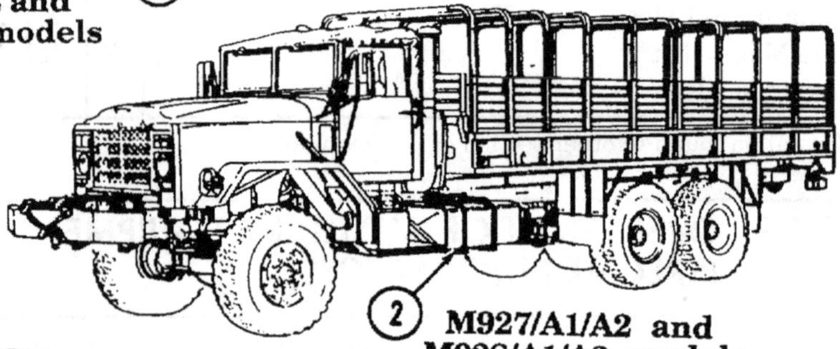
2

M927/A1/A2 and M928/A1/A2 models

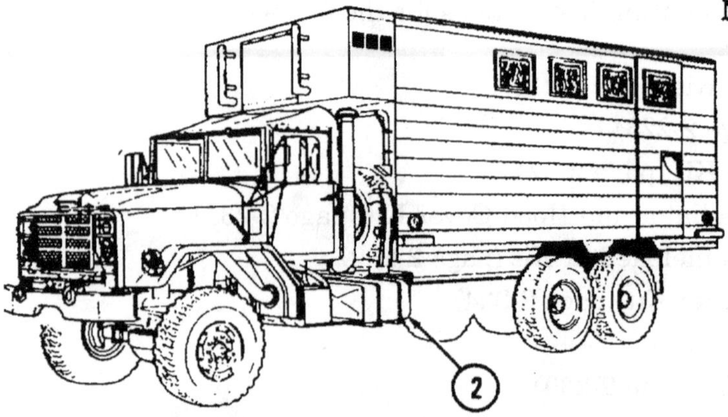
2

M934/A1/A2 models

2-40. DEEPWATER FORDING KIT (Contd)

(b) Location of fuel tanks (2) on dual tank model vehicles:

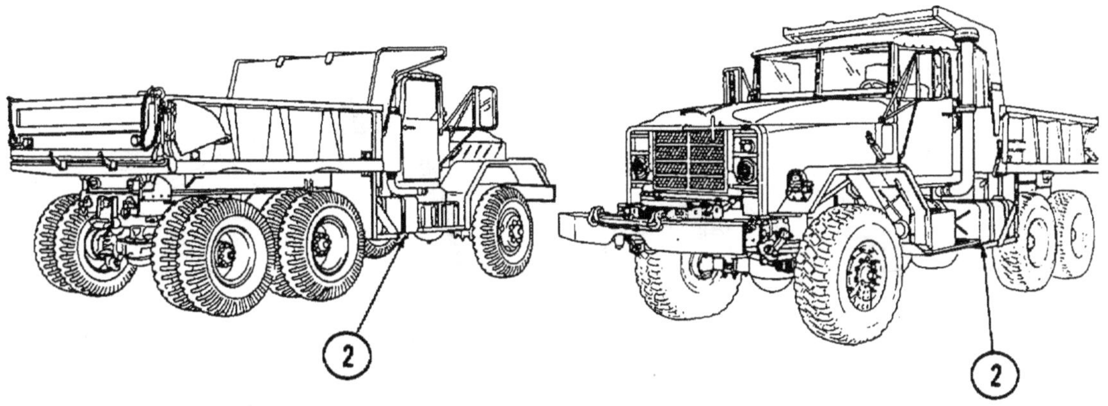
2
2

M929/A1/A2 and M930/A1/A2 models

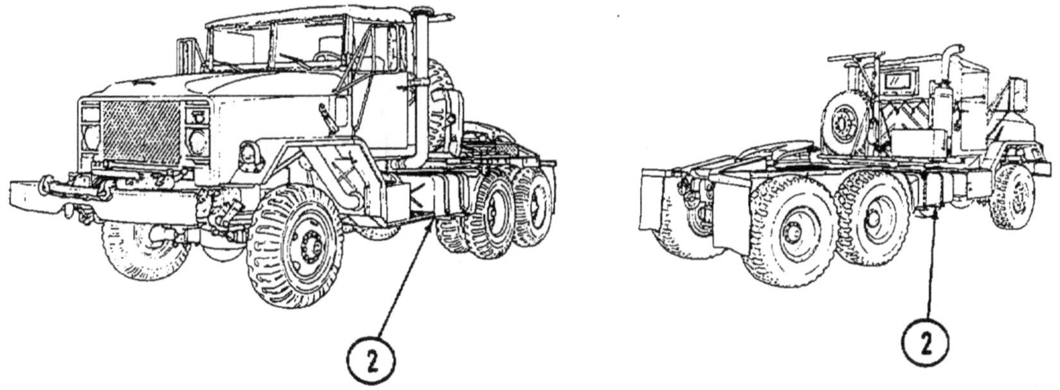
2
2

M931/A1/A2 and M932/A1/A2 models

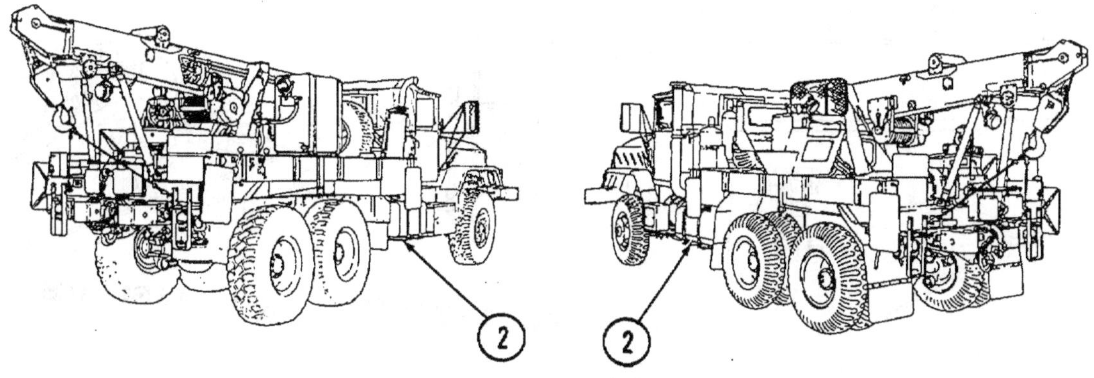
2
2

M936/A1/A2 models

2-40. DEEPWATER FORDING KIT (Contd)

(2) Secure all loose objects on vehicle.

(3) Remove flywheel housing drainplug (3) from storage boss (2). Install drainplug (3) in flywheel drain port (1).

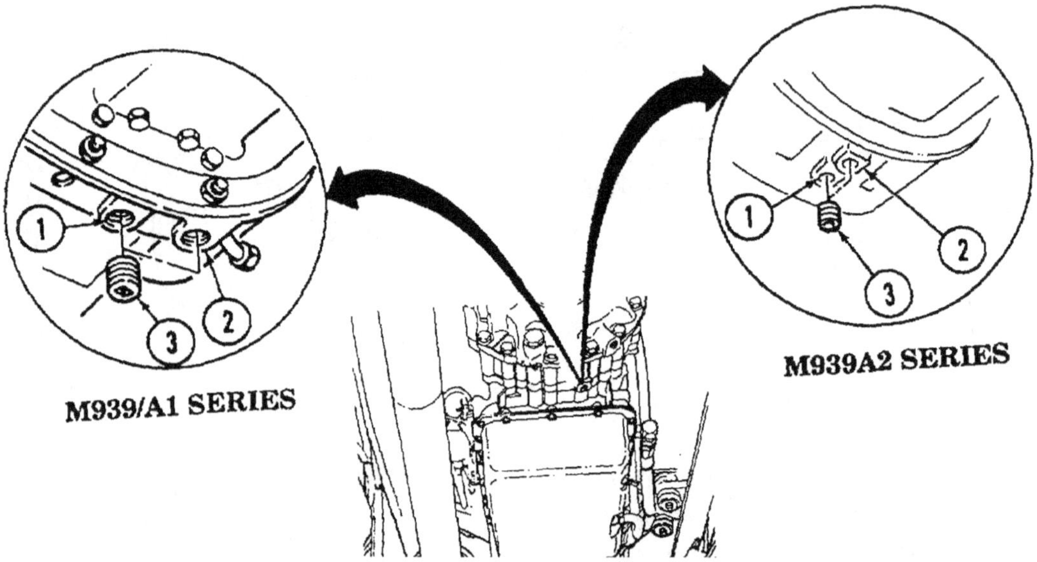
1
2
3
M939/A1 SERIES
1
2
3
M939A2 SERIES

(4) Make sure battery caps (4) are all installed and tight. Make sure transmission dipstick (5) is secured tightly.

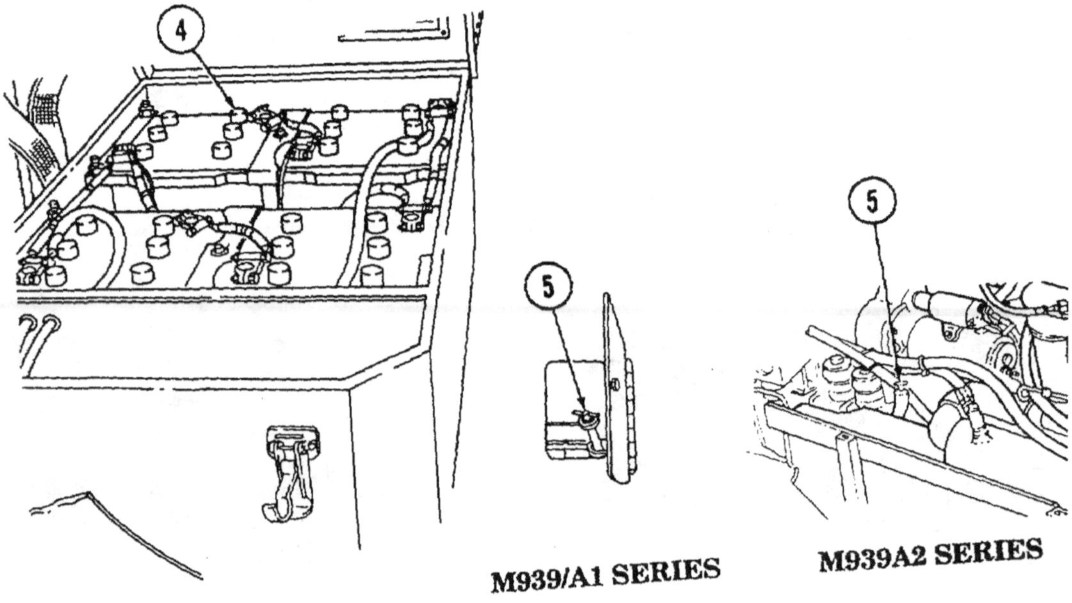
4
5
M939/A1 SERIES
5
M939A2 SERIES

2-40. DEEPWATER FORDING KIT (Contd)

c. Fording Operation.

(1) Start engine (para. 2-12). Make sure engine is running properly.

(2) Pull transfer case shift lever (7) up to low range and place transmission selector lever (6) in 1 (first).

WARNING

Do not attempt to cross water deeper than 78 in. (198 cm). Limit vehicle speed while fording to 3 or 4 mph (5 or 6 km/h). Failure to do this may result in damage to vehicle, injury, or death.

(3) Enter water slowly. Pull fording control handle (8) out immediately upon entering water.

(4) Maintain constant vehicle speed while fording, and exit water in area with gentle slope.

(5) Push fording control handle (8) in immediately upon leaving water.

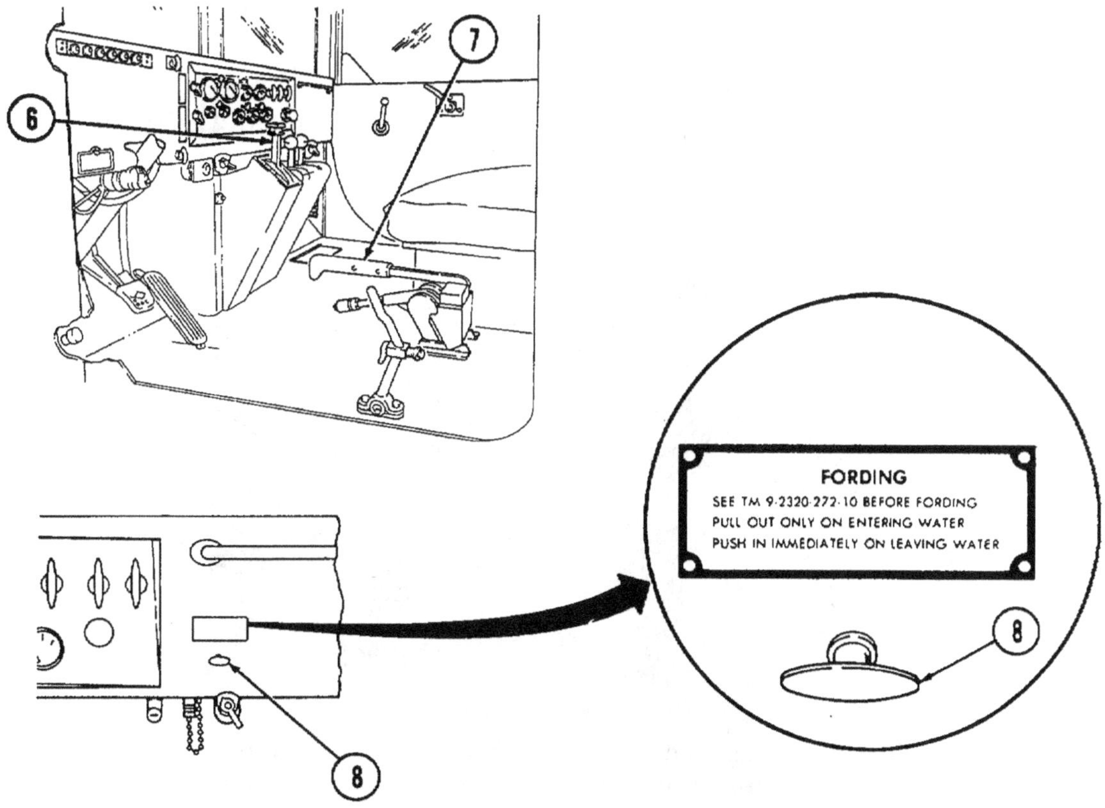
7
6
FORDING
SEE TM 9-2320-272-10 BEFORE FORDING
PULL OUT ONLY ON ENTERING WATER
PUSH IN IMMEDIATELY ON LEAVING WATER
8
8

2-40. DEEPWATER FORDING KIT (Contd)

WARNING

Do not rely on service brakes until they dry out. Keep applying brakes until uneven braking ceases. Failure to do this may result in injury or death.

d. After Fording Operation.

(1) Remove flywheel housing drainplug (3) from drain port (1). Install drainplug (3) in storage boss (2).

(2) All parts of vehicle that were in contact with salt water are to be washed with fresh water as soon as possible.

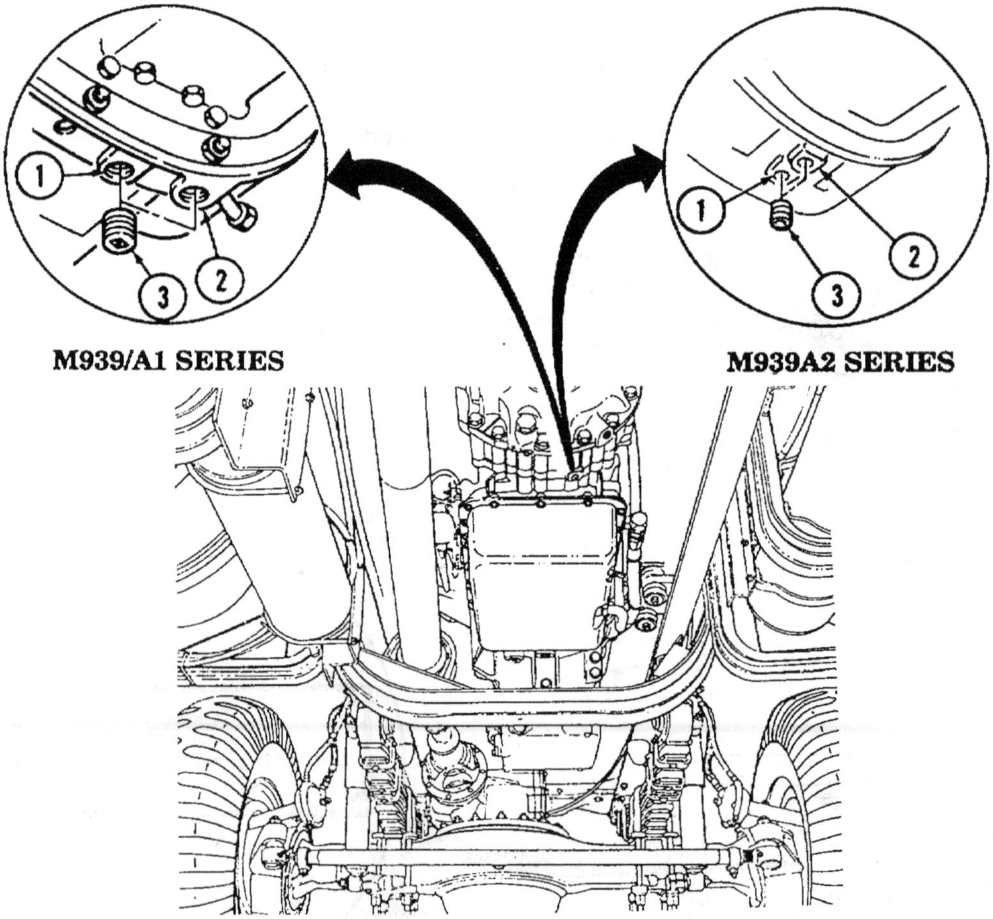
1
2
3
M939/A1 SERIES
1
2
3
M939A2 SERIES

NOTE

Vehicles completing a deepwater fording operation must be serviced by unit maintenance as soon as possible. Refer to LO 9-2320-272-12.

2-41. TROOP SEAT KIT

a. General. Troop seat kit is used to convert M929/A1/A2 and M930/A1/A2 dump trucks into troop carriers. Troop seat kit also enables dump trucks to transport bulk cargo that would otherwise extend above dump body.

b. Troop Seat Kit Installation.

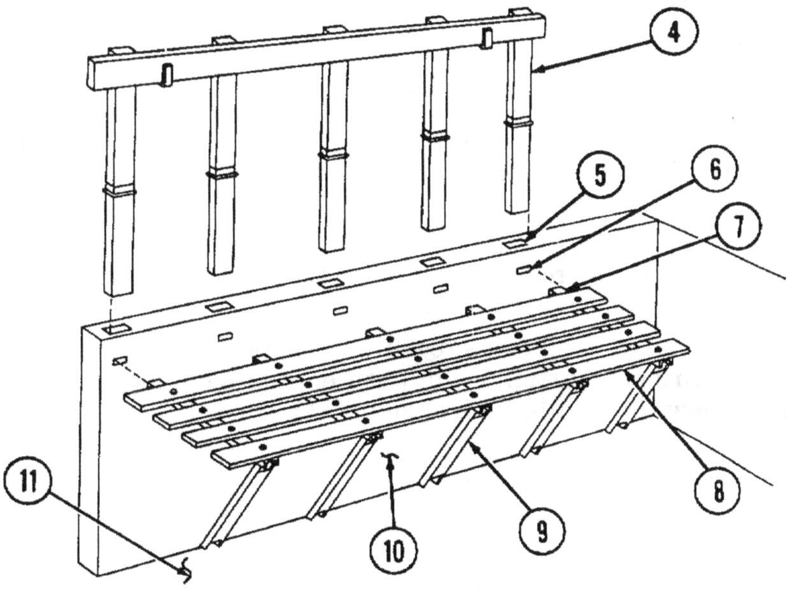
4
5
6
7
8
9
10
11

(1) Insert side racks (4) into slots (5) on side walls (10).

(2) Fold out troop seat support legs (9). Lay troop seat (8) flat on floor (11) of vehicle.

(3) Raise troop seat (8) level with slots (6) on dump body side walls (10). Insert troop seat engaging hooks (7) into slots (6), fold support legs (9) inward, and lower troop seat (8) into position.

(4) Adjust each troop seat support leg (9) until all supports evenly contact side walls (10) and floor (11) of the vehicle.

(5) Secure safety strap (13) to eyelets (12).

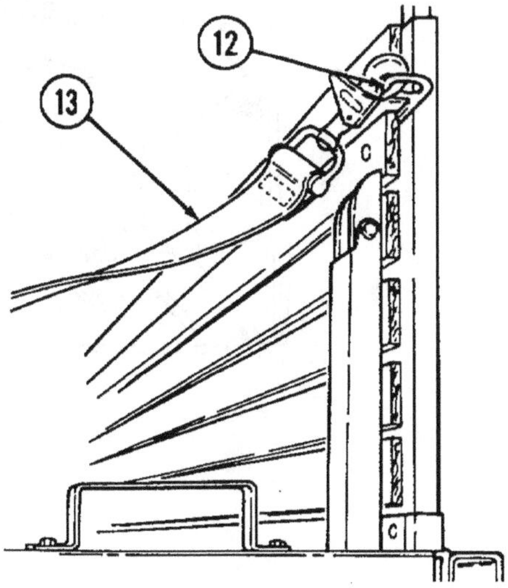
12
13

CAUTION

Troop seat kit for dump trucks must be removed and stowed off vehicle when dump truck is used for dumping operations. Failure to do this will result in damage to troop seat kit.

2-42. BOW AND TARPAULIN KIT

a. General. Kits are available for all M939/A1/A2 cargo and dump trucks, and are installed in a similar fashion. The following procedures and illustrations are installation and removal of bow and tarpaulin kits for M923/A1/A2 and M925/A1/A2 series vehicles.

b. Bow and Tarpaulin Kit Installation.

(1) Insert staves (1) into side rack sockets (4).

NOTE

Some overhead cross bows are secured in place with screws and washers instead of latches.

(2) Insert overhead cross bows (3) into staves (1). Secure each end of overhead cross bows (3) in place with stave latches (2).

(3) Thread two lashing ropes (5) into center eyelets (11) of forward end curtain (7). Place forward end curtain (7) in position and wind ropes (5) alternately around overhead bow (6) and through eyelets (11).

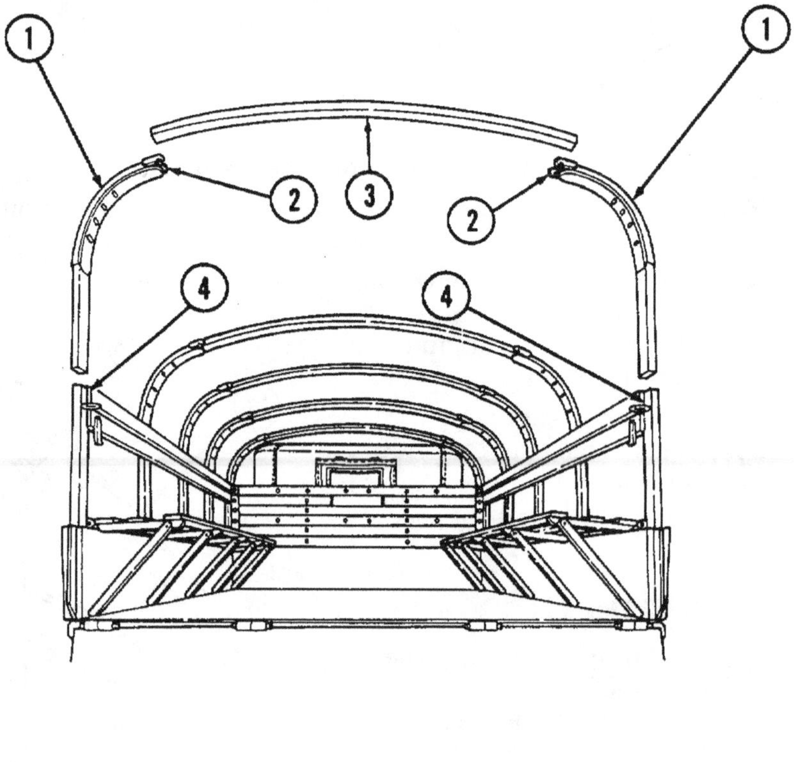
1
1
2
3
2
4
4

2-42. BOW AND TARPAULIN KIT (Contd)

(4) Secure rope (5) end on each side of vehicle to lashing hooks (8).

(5) Secure personnel safety strap (9) to eyelets (10) on side rails nearest tailgate.

(6) Repeat procedure with rear end curtain. Do not tie down bottom of rear end curtain until bow and tarp installation is completed.

NOTE

Do not tie down bottom of rear end curtain when transporting troops.

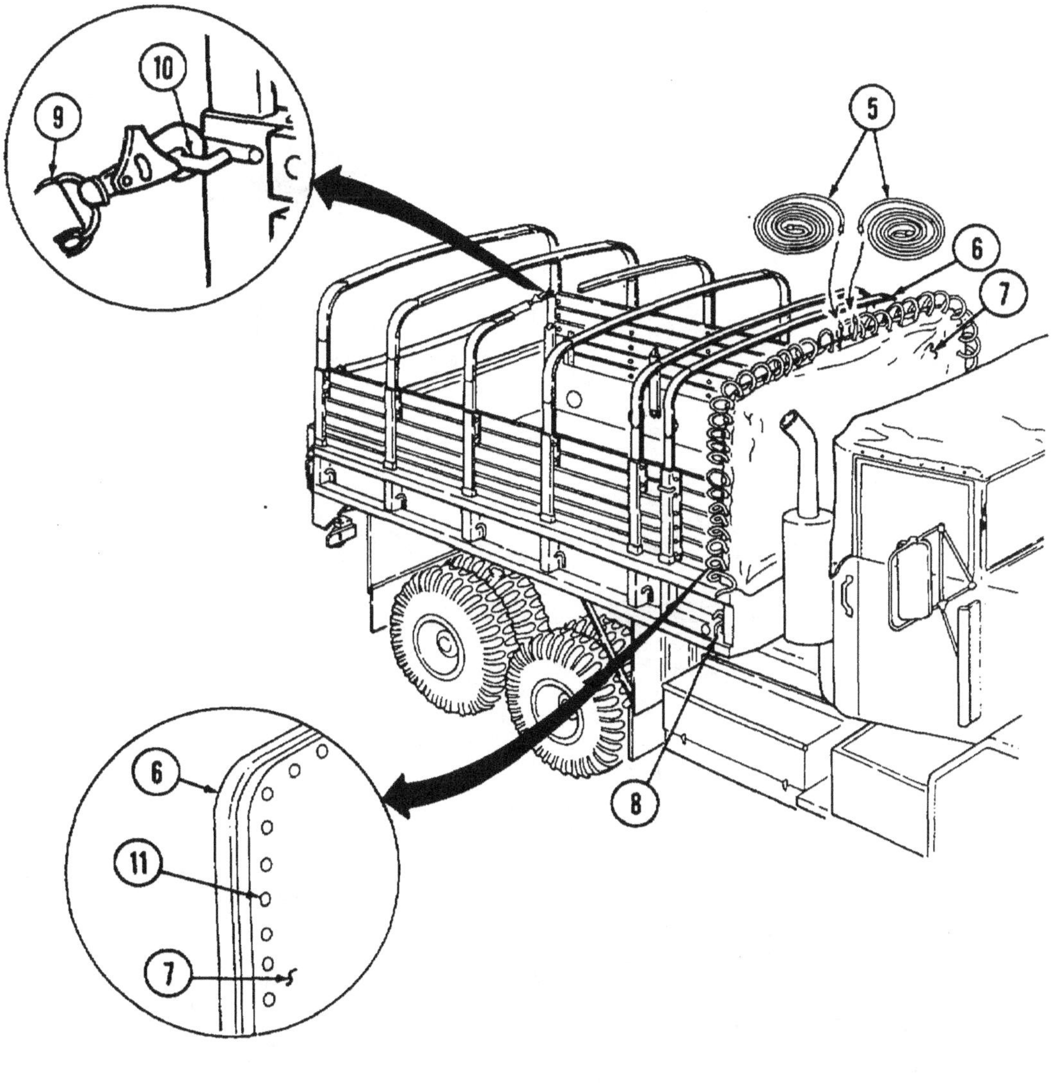
10
9
5
6
7
8
6
11
7

2-42. BOW AND TARPAULIN KIT (Contd)

(7) Place folded tarpaulin (1) across top center bow with half marked FRONT facing front of vehicle.

(8) Unfold front of tarpaulin (1) over bows (2) all the way to front of vehicle. Unfold other end of tarpaulin (1) toward rear of vehicle.

(9) Unfold one side of tarpaulin (1), then unfold other side toward sides of vehicle. Allow loose tarpaulin (1) sides to drape over side of vehicle.

NOTE

All ropes should be snug, but not too tight.

(10) Tie lashing ropes (3) to lashing hooks (4) on each side of vehicle.

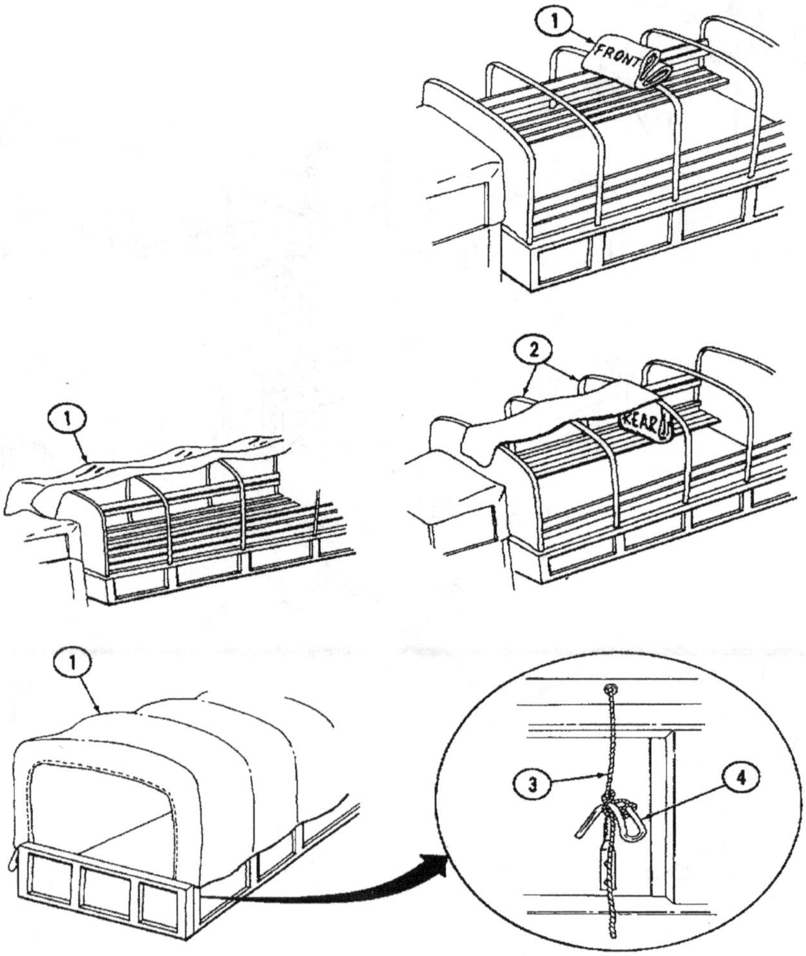
1
FRONT
2
REAR
1
1
3
4

2-42. BOW AND TARPAULIN KIT (Contd)

c. Raising Tarpaulin for Ventilation.

NOTE

This operation requires two crewmembers.

(1) Remove rear tarpaulin end curtain (5), if installed.

(2) Untie all tarpaulin lashing ropes (3).

(3) Fold up tarpaulin (1) into three to five folds until straps (6) can be attached to staves (8).

(4) Fasten folded tarpaulin (1) in place using straps (6) and buckles (7).

(5) Tie front and rear lashing ropes (3) to end staves (8).

d. Bow and Tarpaulin Kit Removal.

CAUTION

Do not fold or stow tarpaulin when wet. To do so will damage tarpaulin.

(1) Remove tarpaulin top (1) from vehicle and lay tarpaulin (1) flat on ground, with buckles (9) facing up.

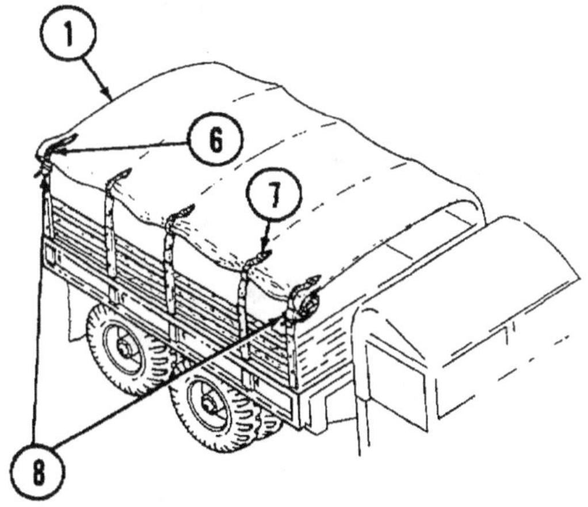
1
6
7
8

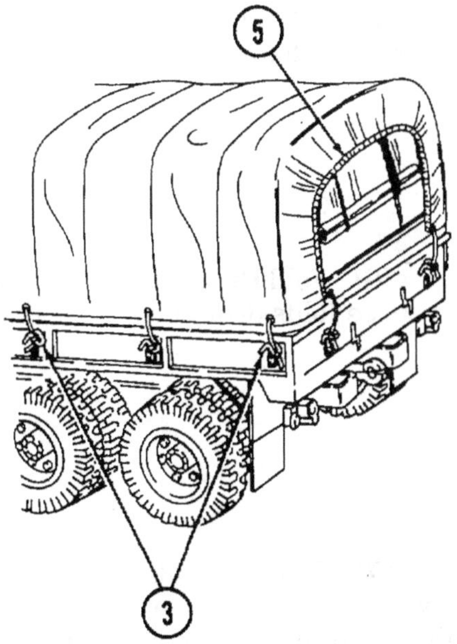
5
3

2-42. BOW AND TARPAULIN KIT (Contd)

(2) Fold eyelet side of tarpaulin (1) to first row of buckles (2).

2
1

(3) Fold tarpaulin (1) over again, and then one more time.

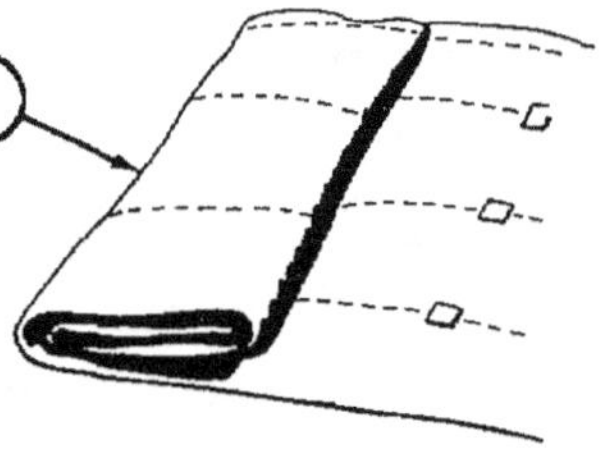
1

(4) Fold other side of tarpaulin (1) once, to the row of buckles (2).

1
2

(5) Fold tarpaulin (1) again, until the two folds meet.

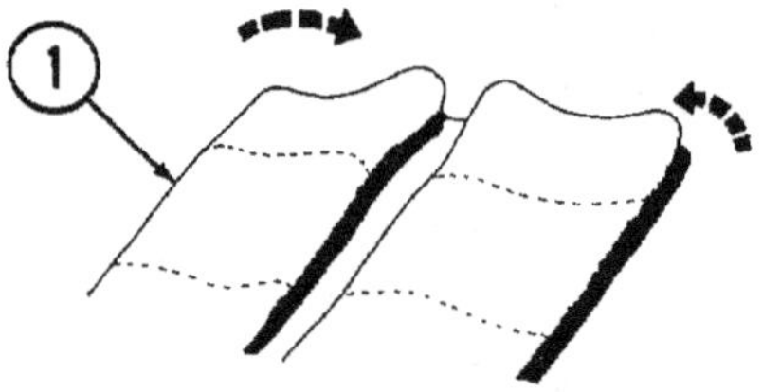
1

(6) Fold the side of tarpaulin (1) with three folds over the side with four folds.

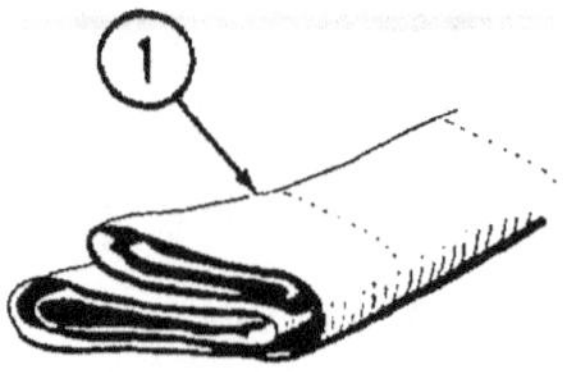
1

(7) Fold tarpaulin (1) end halfway to the first seam, and then over again, until inner edge of tarpaulin (1) is at middle.

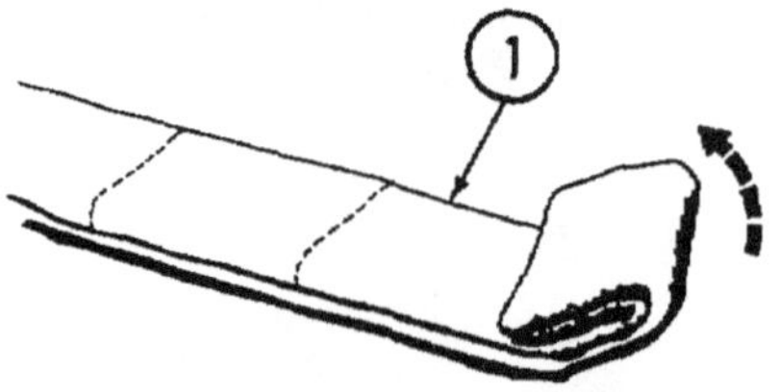
1

2-42. BOW AND TARPAULIN KIT (Contd)

(8) Repeat folding on opposite end of tarpaulin (1) until both folded ends meet.

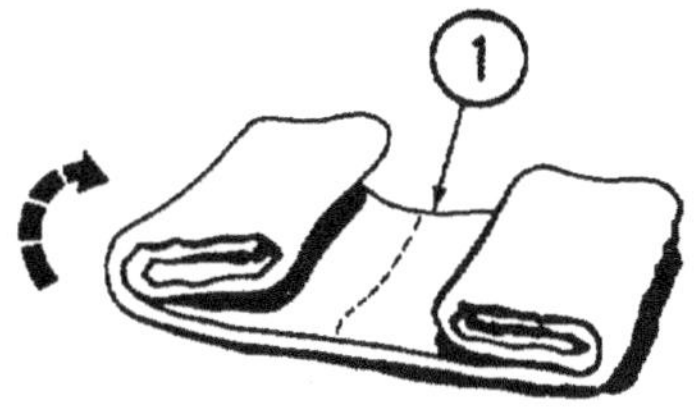
1

(9) Place folded tarpaulin (1) front end up and with chalk, mark FRONT. Make sure that letters are big enough to see.

1
FRONT

(10) Turn folded tarpaulin (1) over and mark REAR.

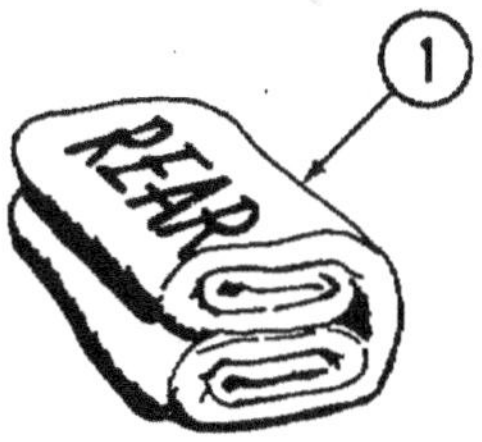
1
REAR

(11) Remove end curtains (2) and fold them to approximately the same dimensions as the tarpaulin (1).

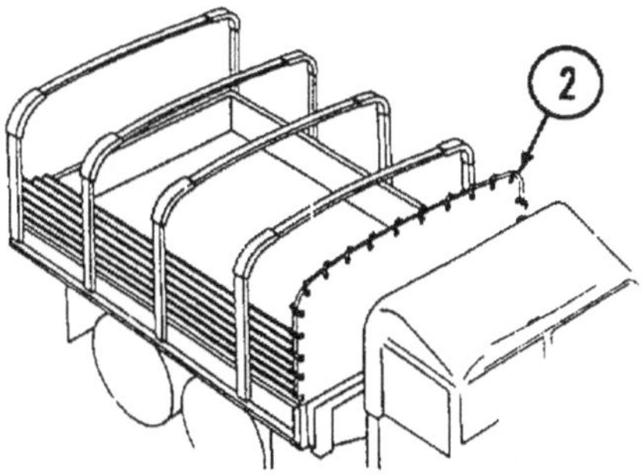
2

(12) Place tarpaulin (1) and end curtains (2) on a pallet for storage.

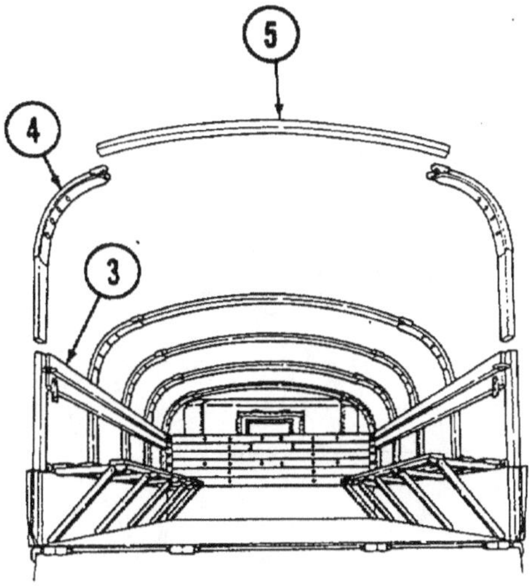
5
4
3

(13) Unlatch and remove overhead cross bows (5) from staves (4).

(14) Remove staves (4) from side rack sockets (3).

2-42. BOW AND TARPAULIN KIT (Contd)

(15) Stow staves (2) in pockets (3) on forward end of vehicle sides. On M923/A1/A2 and M925/A1/A2 model vehicles, staves are stored in pockets on cargo body directly behind vehicle cab (1).

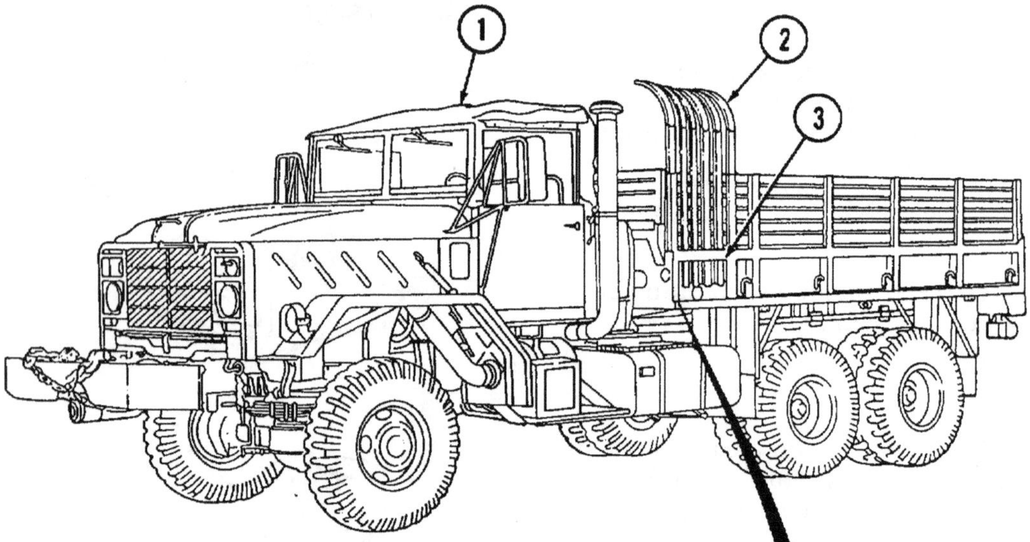
1
2
3

(16) Strap overhead cross bows (4) together and stow in storage area under cargo body.

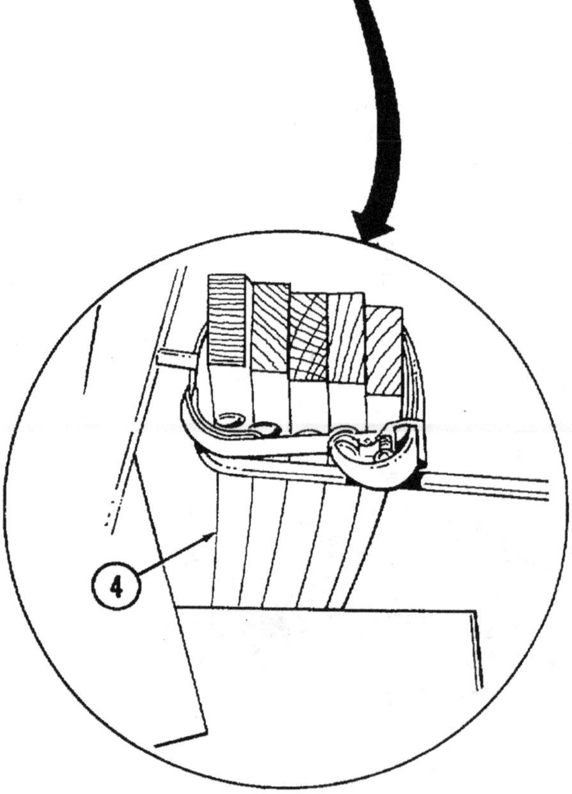
4

2-43. HARDTOP KIT AND RADIATOR AND HOOD COVER KIT

a. General. The hardtop kit and the radiator and hood cover kit are installed by direct support maintenance on vehicles operating in -25°F (-32°C) temperatures or below.

5
6

b. Operating with Engine Compartment Covers Installed.

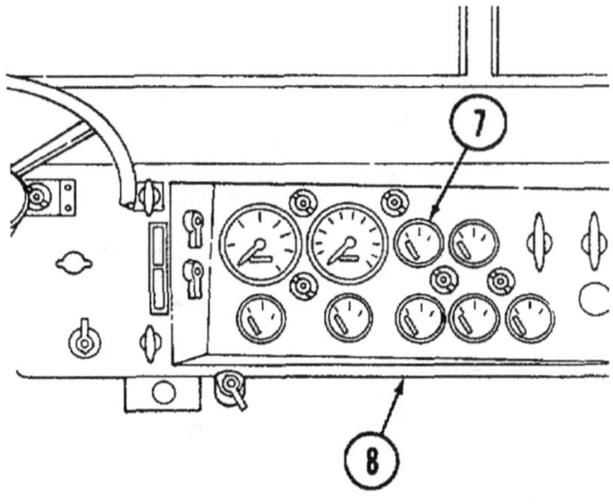
7
8

(1) Start engine with radiator cover flap (6) closed. Refer to paragraph 2-13 for cold weather starting instructions.

(2) Roll up and secure radiator cover flap (6) in open position when engine temperature rises above 175°F (79°C) as indicated by engine coolant temperature gauge (7) on instrument panel (8).

(3) If engine coolant temperature should exceed 200°F (93°C), completely remove engine compartment cover (5) to avoid overheating.

(4) Open and close cover flap (6) as required during arctic operations to maintain engine coolant temperature within normal operating range of 175°-200°F (79°-93°C).

2-44. FUEL BURNING PERSONNEL HEATER KIT

a. General. Fuel burning personnel heater provides heat and defrost to vehicle cab when the engine is operating.

b. Fuel Burning Personnel Heater Operation.

(1) Shut off engine coolant heater if operating. Refer to paragraph 2-45 for engine coolant heater shutdown instructions.

CAUTION

Do not operate the engine coolant heater control box and personnel heater control box at the same time. Electric fuel pump will not maintain fuel pressure for both heaters at same time.

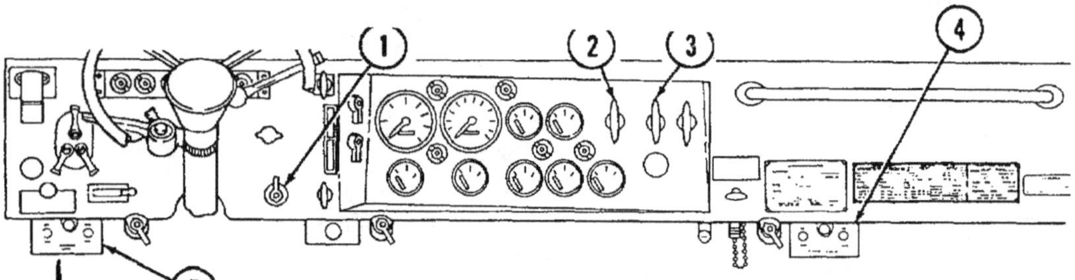
1
2
3
4
5

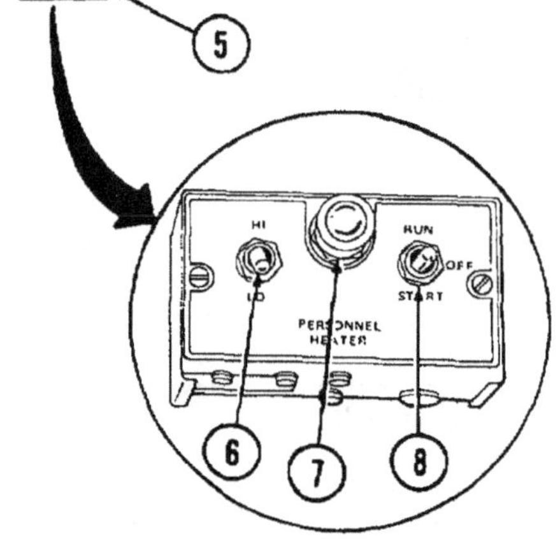
HI
LO
RUN
OFF
START
PERSONNEL HEATER
6
7
8

(2) Open electric fuel pump shutoff valve by turning valve lever (9) one-quarter turn counterclockwise. Open two valves (10) for M939A2 series vehicles.

(3) Start engine. Refer to paragraph 2-13 for cold weather starting instructions if necessary.

(4) Press PRESS-TO-TEST button (7) on personnel heater control box (5) to check operation of circuit. Observe that indicator lamp (7) illuminates.

(5) Set HI-LO switch (6) on per-

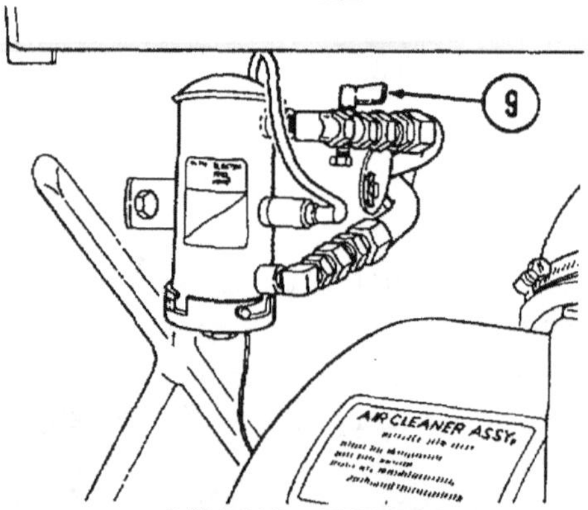
9
AIR CLEANER ASSY.

M939/A1 SERIES

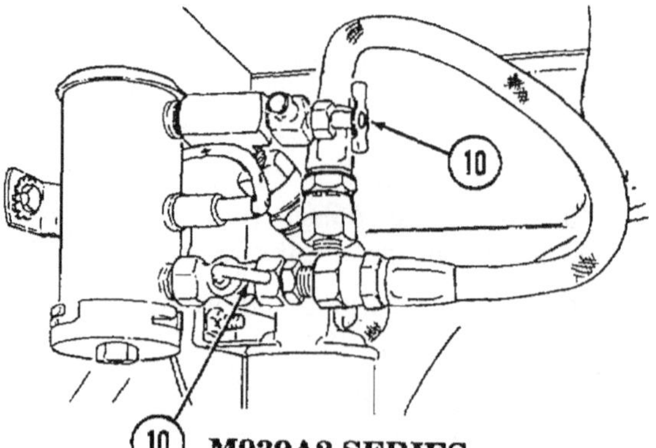
10
10

M939A2 SERIES

2-44. FUEL BURNING PERSONNEL HEATER KIT (Contd)

sonnel heater control box (5) to HI.

NOTE

Heater will not operate if switch is released from START position before indicator lamp illuminates.

(6) Hold RUN-OFF-START switch (8) on personnel heater control box (5) in START position until indicator lamp (7) illuminates.

(7) Move switch (8) from START to RUN, without hesitating at OFF, as soon as indicator lamp (7) illuminates.

NOTE

- If heater fails to start, turn RUN-OFF-START switch to OFF position and repeat steps 6 and 7. Notify your supervisor if heater fails to start after two attempts.
- Heater blower motor switch on vehicle instrument panel is not used for blower operation. HI-LO switch and RUN-OFF-START switch on personnel heater control box are used in place of blower motor switch.

(8) Adjust hot air flow with heat vent control (3).

CAUTION

Heat cab before defrosting windshield. Glass damage may result from sudden temperature changes.

(9) Adjust defroster control (2) as required to defrost windshield. All heated air is directed at windshield when defroster control (2) is pulled all the way out.

(10) To shut down personnel heater:

(a) Turn RUN-OFF-START switch (8) to OFF.

(b) Remain in vehicle cab to make sure indicator lamp (7) goes out and blower motor stops.

(c) Close defroster control (2) and heat vent control (3).

(d) Shut off engine. Refer to paragraph 2-16 for engine shutoff instructions.

(e) Close electric fuel pump shutoff valve by turning valve lever (9) one-quarter turn clockwise. Close two valves (10) for M939A2 series vehicles.

2-45. ENGINE COOLANT HEATER KIT

a. General. Engine coolant heater is not designed for use while vehicle engine is operating. This heater preheats engine coolant in preparation for starting at extremely low temperatures or to maintain engine in standby readiness.

b. Engine Coolant Heater Operation.

WARNING

Exhaust gases will kill. Do not operate engine coolant heater in closed area occupied by personnel. Failure to do this will result in injury or death.

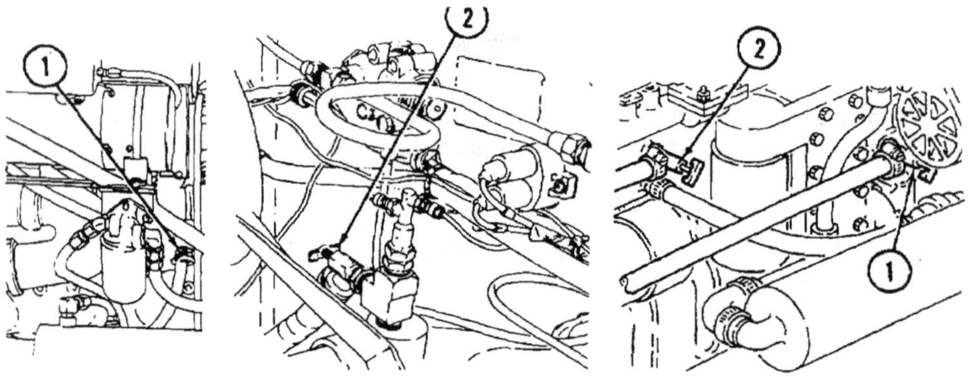
1
2
2
1

M939/A1 SERIES **M939A2 SERIES**

CAUTION

Coolant shutoff valves on engine must remain open at all times when operating heater. Failure to do this will result in damage to heater.

(1) Open coolant shutoff valves (1) and (2).

(2) Open electric fuel pump shutoff valve located near air cleaner assembly by turning valve lever (3) one-quarter turn counterclockwise. Open two valves (4) for M939A2 series vehicles.

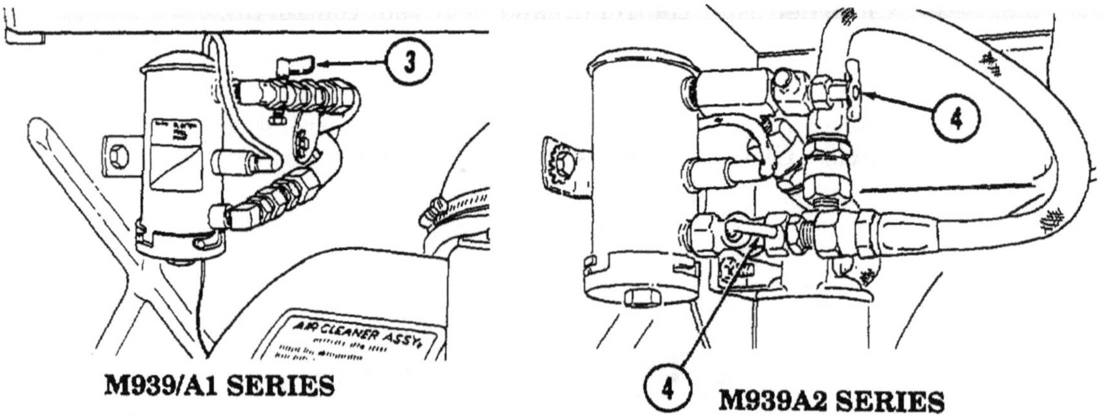
3
AIR CLEANER ASSY.
4
4

M939/A1 SERIES **M939A2 SERIES**

2-45. ENGINE COOLANT HEATER KIT (Contd)

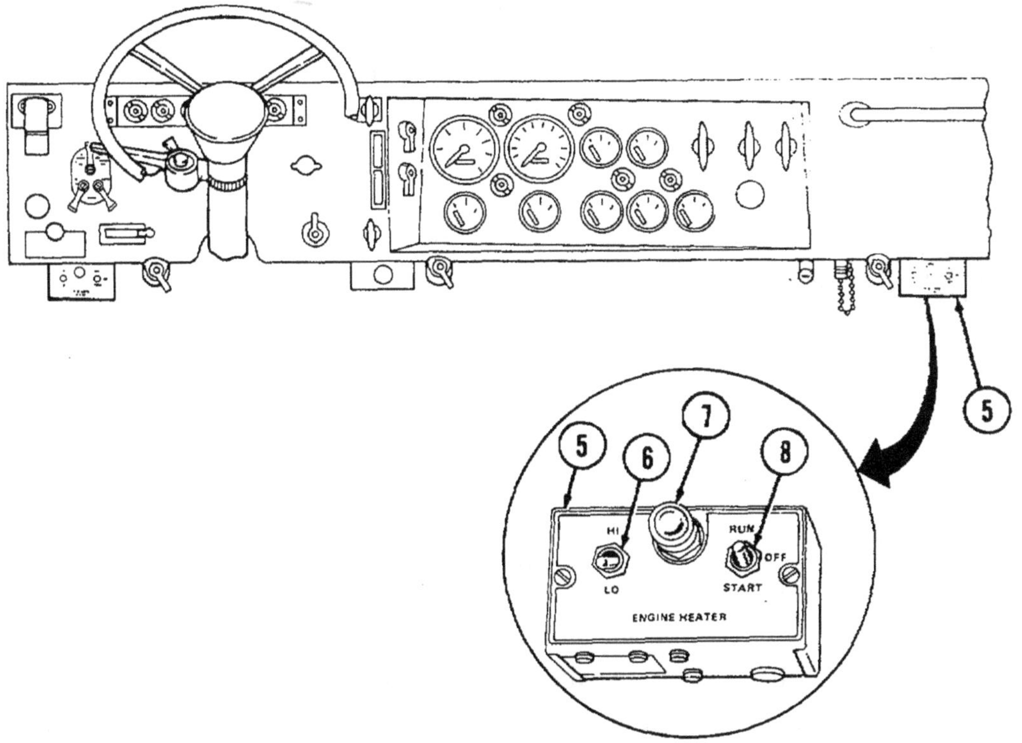
5
7
5
6
8
HI
RUN
OFF
LO
START
ENGINE HEATER

(3) Press PRESS-TO-TEST button (7) on engine coolant heater control box (5) to check operation of circuit. Observe that indicator lamp (7) illuminates.

(4) Set HI-LO switch (6) on engine coolant heater control box (5) to HI or LO.

NOTE

Select HI position if engine is cold. Switch will automatically change to LO position when coolant temperature exceeds 195°F (91°C). Switch will automatically change to HI position when coolant temperature drops below 120°F (49°C).

(5) Hold RUN-OFF-START switch (8) on engine coolant heater control box (5) in START position until indicator lamp (7) illuminates.

NOTE

Heater will not operate if switch is moved to RUN position before indicator lamp illuminates.

(6) Move switch (8) to RUN, without hesitating at OFF, as soon as indicator lamp (7) illuminates.

2-45. ENGINE COOLANT HEATER KIT (Contd)

NOTE

If heater fails to start, turn RUN-OFF-START switch to OFF position and repeat steps 5 and 6. Notify your supervisor if heater fails to start after two attempts.

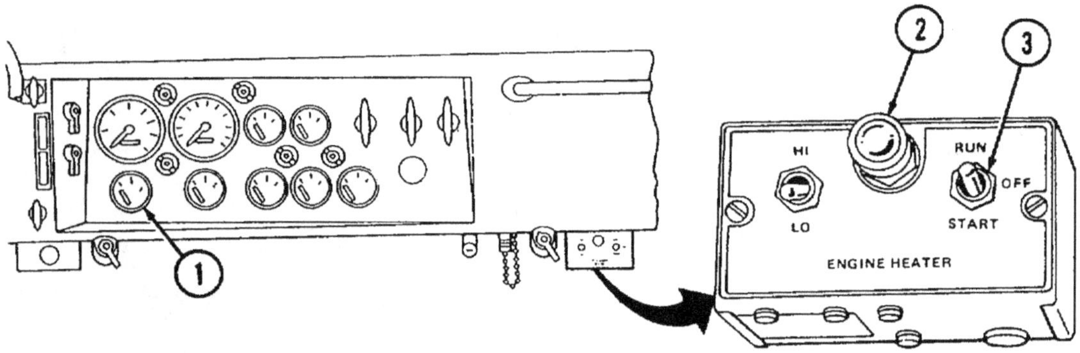
2
3
HI
RUN
OFF
LO
START
ENGINE HEATER
1

(7) Check fuel gauge (1). Make sure fuel tank(s) are full if engine coolant heater is to operate for an extended period.

(8) To shut down engine coolant heater:

(a) Turn RUN-OFF-START switch (3) to OFF.

(b) Remain in vehicle cab to make sure indicator lamp (2) goes out and blower motor shuts down (approximately 1-3 minutes).

NOTE

Omit step c if engine is to be started immediately.

(c) Close electric fuel pump shutoff valve by turning valve lever (4) one-quarter turn clockwise. Close two levers (4) on M939A2 series vehicles.

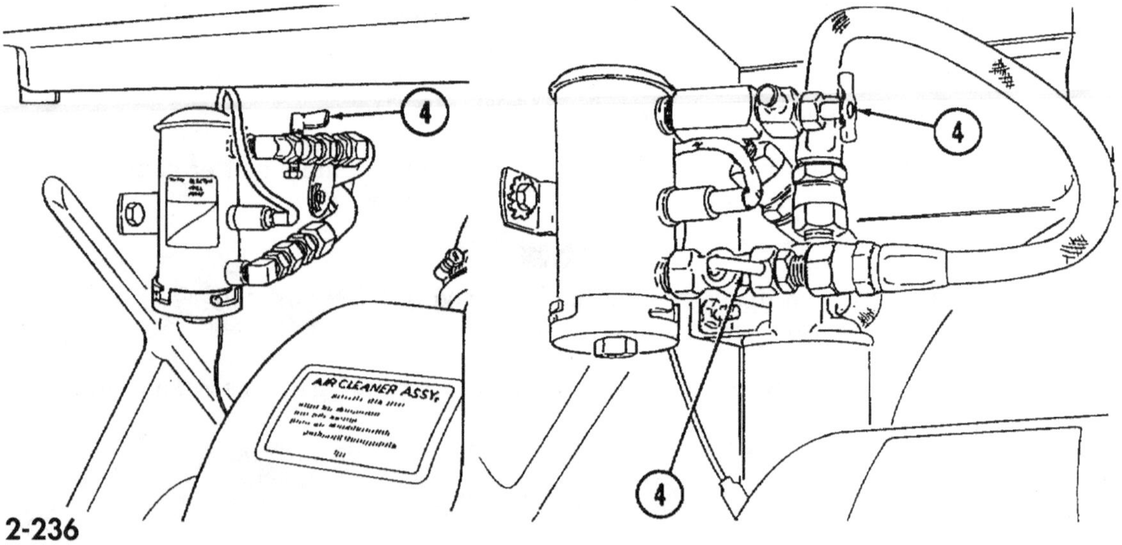
4
AIR CLEANER ASSY.
4
4

2-45. ENGINE COOLANT HEATER KIT (Contd)

CAUTION

Take care not to accidentally close coolant shutoff valves. During arctic operations, all coolant shutoff valves must remain open at all times.

(d) Do not close coolant shutoff valves (5) and (6) during any arctic operation.

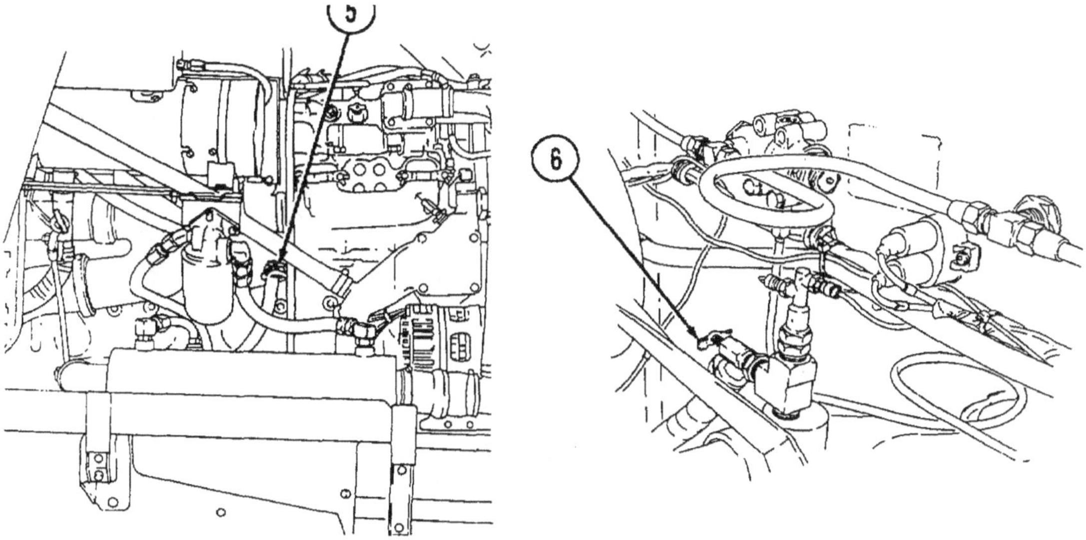
5
6

M939/A1 SERIES

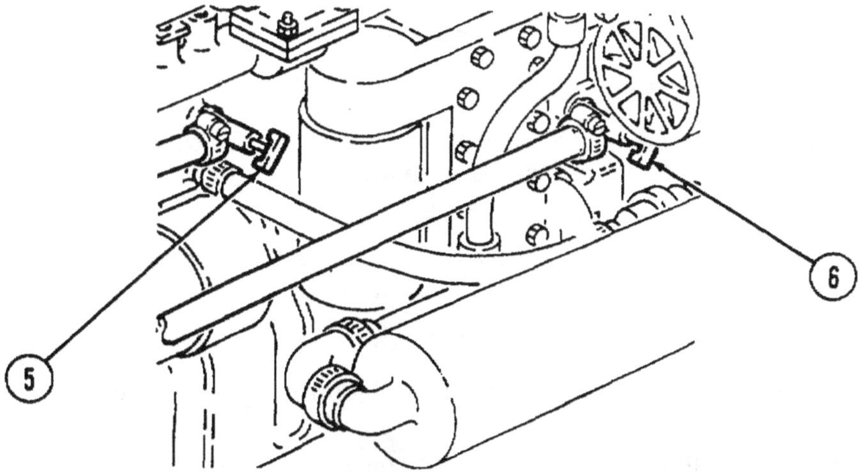
6
5

M939A2 SERIES

2-46. A-FRAME KIT

a. General. The A-frame kit is installed on cargo and tractor vehicles equipped with a front winch to provide a means for lifting, moving, loading, and unloading material and equipment. A-frame load capacity is 3,000 lb (1,362 kg).

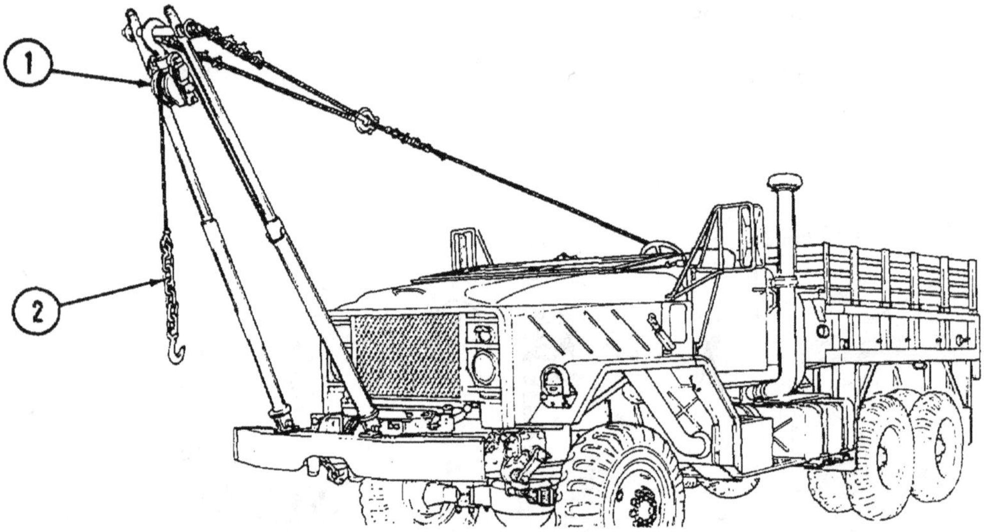
1
2

b. Preparation for Use.

WARNING

Vehicle will become charged with electricity if A-frame contacts or breaks high-voltage wire. Do not leave vehicle while high voltage line is in contact with A-frame or vehicle. Notify nearby personnel to have electrical power turned off. Failure to do this may result in injury or death.

NOTE

A-frame kit is installed and rigged by unit maintenance.

(1) Maneuver vehicle into position for operation. Be careful that A-frame does not come into contact with wires, cables, tree limbs, or other overhead obstructions.

(2) Park vehicle and apply parking brake.

c. Operating A-frame. Operate front winch to raise, lower, or hold load. Refer to paragraph 2-22 for front winch operating instructions.

CAUTION

- Do not attempt to lift more than 3,000 lb (1,362 kg) with A-frame kit.
- Do not allow cable chain (2) to contact snatch block (1).

2-47. AIRBRAKE CONTROL KIT

a. General. Airbrake control kit is installed on vehicles hauling trailers or artillery equipped with airbrakes. Airbrake kit is installed by direct support maintenance.

b. Airbrake Kit Operation.

NOTE

Inserting yoke of equipment requires two or more crewmembers, depending on size and weight of load.

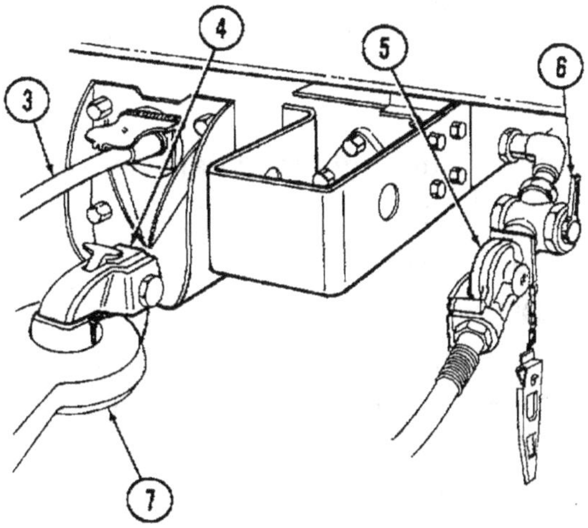

4
5
6
3
7

(1) Insert yoke (7) of equipment to be towed into pintle hook (4) of vehicle.

(2) Connect air lines from towed equipment to half couplings (5) of towing vehicle.

(3) Pull up handles (6) to open airbrake hose lines.

(4) Connect trailer brakelight cable (3) to electric receptacle above pintle hook (4).

(5) Start engine (para. 2-12).

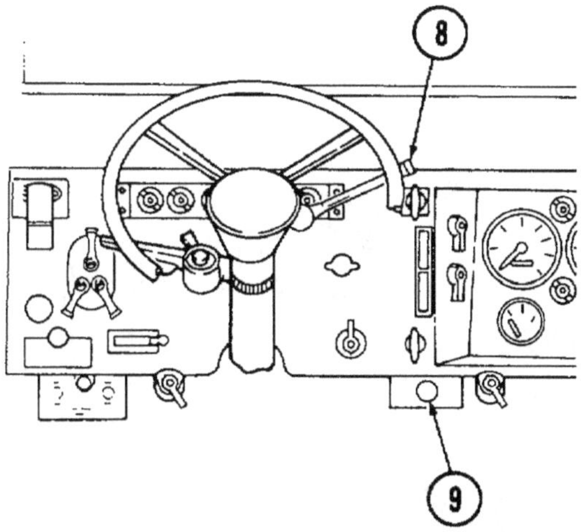

8
9

(6) Press in trailer air supply valve control knob (9) and hold in place for 15 seconds. Release valve control knob (9). Valve control knob (9) should remain in pressed-in position indicating trailer or artillery load airbrake system has proper air pressure.

NOTE

Airbrake hand control should be engaged slowly to provide steady, even braking.

(7) Pull down trailer airbrake hand control lever (8) to apply brakes of towed load.

2-48. MUD FLAP KIT

a. General. The mud flap kit is installed on the M931/A1/A2 or M932/A1/A2 tractor vehicles when a trailer is not attached.

CAUTION

Mud flaps must be removed prior to coupling to semitrailer. Failure to do this will result in damage to vehicle.

NOTE

Left and right mud flaps are replaced the same. This procedure is for the left mud flap.

b. Mud Flap Removal.

(1) Remove securing pin (3) from mud flap (1).

(2) Remove mud flap (1) from bracket (4).

(3) Unbuckle and fold back three stowage straps (5) on the backside of the tool box (6).

(4) Position mud flap (1) against tool box (6) and secure with stowage straps (5).

c. Mud Flap Installation.

(1) Unbuckle and fold back three stowage straps (5) on back of the tool box (6).

(2) Remove mud flap (1) from tool box (6) and position into bracket (4) on the frame (2).

(3) Insert securing pin (3) into mud flap (1).

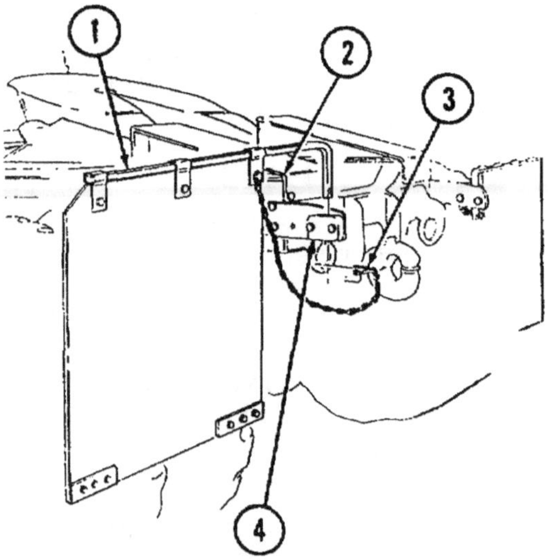
1
2
3
4

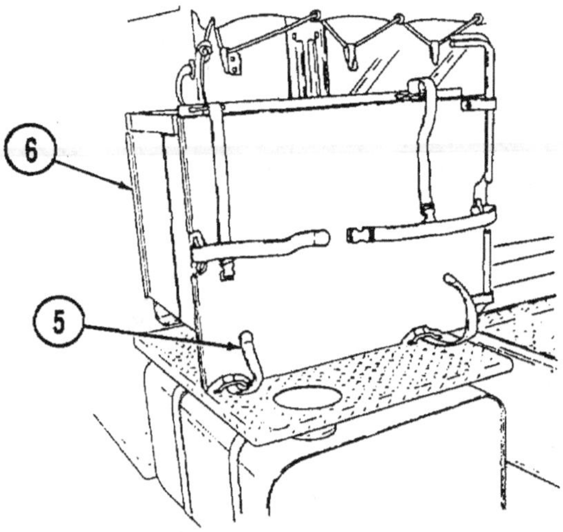
6
5

CHAPTER 3
MAINTENANCE INSTRUCTIONS

Section I. LUBRICATION INSTRUCTIONS

3-1. LUBRICATION ORDER

Lubrication instructions are contained in para. 2-10 PMCS tables. The lubrication order designates cleaning and lubricating procedures for M939/A1/A2 series vehicles. All lubrication instructions are mandatory. This document is issued with each truck and is carried in vehicle at all times. A damaged or lost lubrication order should be replaced immediately.

Proper disposal of hazardous waste material is vital to protecting the environment and providing a safe work environment. Materials such as batteries, oils, and antifreeze must be disposed of in a safe and efficient manner.

The following references are provided as a means to ensure that proper disposal methods are followed.

- Technical Guide No. 126 from the U.S. Army Environmental Hygiene Agency (USAEHA)
- National Environmental Policy Act of 1969 (NEPA)
- Clean Air Act (CAA)
- Resource Conservation and Recovery Act (RCRA)
- Comprehensive Environmental Response, Compensation, and Liability Act
- Emergency Planning and Community Right to Know Act (EPCRA)
- Toxic Substances Control Act (TSCA)
- Occupational Health and Safety Act (OHSA)

The disposal of Army Petroleum, Oils, and Lubricants (POL) products are affected by some of these regulations. State regulations may also be applicable to POL.

If you are unsure of which legislation affects you, contact state or local agencies for regulations regarding proper disposal of Army POL.

Section II. TROUBLESHOOTING PROCEDURES

3-2. SCOPE

The troubleshooting table contains instructions that will help the operator identify and correct simple vehicle malfunctions during operations. The table also helps the operator identify major mechanical difficulties that must be referred to unit maintenance.

NOTE

Operators should perform the corrective action in the order listed.

3-3. TROUBLESHOOTING PROCEDURES

This manual cannot list all malfunctions that may occur. If a malfunction occurs that is not listed in table 3-1, notify unit maintenance.

3-4. TROUBLESHOOTING SYMPTOM INDEX

3-4. TROUBLESHOOTING SYMPTOM INDEX (Contd)

3-4. TROUBLESHOOTING SYMPTOM INDEX (Contd)

MALFUNCTION NO.	MALFUNCTION	TROUBLESHOOTING PROCEDURE PAGE
39.	Ceiling lights and service receptacles fail to energize when doors are closed under blackout conditions (110 volt system)	3-15
40.	Emergency light, blackout light, and ceiling lights fail to illuminate (24 volt system)	3-15
41.	Heater will not ignite	3-15
42.	Air conditioner compressor fails to start	3-16
	SPECIAL PURPOSE KITS	
	RADIATOR AND HOOD COVER KIT	
43.	Engine fails to reach operating temperature	3-16
44.	Engine temperature exceeds 200°F (93°C)	3-16
	FUEL BURNING PERSONNEL AND ENGINE COOLANT HEATER KITS	
45.	Fuel burning personnel heater fails to start when run-off-start switch is held in start position	3-17
46.	Engine coolant heater fails to start when run-off-start switch is held in start position	3-18
47.	Heater fails to continue burning	3-18
48.	Windshield defrosters not operating	3-19
49.	Engine oil pan shroud not receiving heat (engine coolant heater only)	3-19
50.	Engine coolant system not receiving heat (engine coolant heater only)	3-19
	A-FRAME KIT	
51.	Winch inoperative	3-19
52.	A-frame misaligned	3-19
	AIRBRAKE CONTROL KIT	
53.	Trailer airbrakes do not operate when airbrake control handle is pulled down	3-20

Table 3-1. Troubleshooting

MALFUNCTION TEST OR INSPECTION CORRECTIVE ACTION

NOTE

If malfunction corrective action does not correct malfunction, notify unit maintenance.

ENGINE

1. WHEN STARTER SWITCH IS TURNED TO START, ENGINE FAILS TO CRANK.

Step 1. Check to see if battery switch is off.

If off, turn switch on.

Step 2. Check to see if transmission selector lever is in N (neutral).

If not, place in N (neutral).

WARNING

Do not smoke, have open flames, or make sparks around battery, especially if cap is off. Battery can explode and cause injury or death to personnel.

Step 3. Visually check to see if battery cables, terminals, and connections are loose, broken, or corroded. Check battery for proper water level.

If loose, tighten.

Notify unit maintenance of any damage to batteries, cables, and terminals.

2. ENGINE CRANKS BUT DOES NOT START.

NOTE

- Do not completely fill fuel tank(s) before checking visually for leaks in fuel system.
- Whenever fuel tank(s) are completely drained and then refilled, the fuel system must be primed (para. 3-8).

Step 1. Check to see if fuel gauge indicates empty.

If empty, fill fuel tank(s).

Step 2. Check to see if emergency engine stop control on instrument panel is pulled out.

Notify unit maintenance to reset fuel cutoff valve (M939/A1 series vehicles only).

Table 3-1. Troubleshooting (Contd).

MALFUNCTION TEST OR INSPECTION CORRECTIVE ACTION

Step 3. Check to see if throttle control solenoid is functioning properly (M939A2 series vehicles only).

If solenoid is malfunctioning, tie up with a strap or rope and finish mission. Report to unit maintenance as soon as possible.

3. ENGINE CRANKS BUT FAILS TO START AT OUTSIDE TEMPERATURES BELOW +32°F (0°C).

NOTE

Refer to Cold Weather Starting (para. 2-13).

Step 1. Check to see if fuel gauge indicates empty.

If empty, fill fuel tank(s).

Step 2. Check to see if emergency engine stop control on instrument panel is pulled out.

Notify unit maintenance to reset fuel cutoff valve (M939/A1 series vehicles only).

Step 3. Check ether starting system.

Operate ether starting system.

4. ENGINE STARTS BUT MISFIRES, RUNS ROUGH, OR LACKS POWER.

Step 1. Check to see if emergency engine stop control on instrument panel is pulled out.

Notify unit maintenance to reset fuel cutoff valve (M939/A1 series vehicles only).

Step 2. Check for restricted air cleaner.

If restricted, clean air cleaner element (para. 3-8).

Step 3. Check fuel supply system for water and impurities.

Perform service operation (para. 3-8).

Step 4. Check for air in fuel system.

Prime fuel system (para. 3-8).

Table 3-1. Troubleshooting (Contd).

MALFUNCTION TEST OR INSPECTION CORRECTIVE ACTION

5. ENGINE OVERHEATS AS INDICATED BY ENGINE COOLANT TEMPERATURE GAUGE

WARNING

Extreme care should be taken when removing surge tank filler cap if temperature gauge reads above 175°F (79°C). Steam or hot coolant under pressure will cause injury.

Step 1. Check radiator core for obstructions.

If clogged, remove debris (refer to table 2-1).

Step 2. Check coolant level in surge tank.

If low, add coolant to surge tank until at bottom of fill neck.

Step 3. Check for leakage from radiator, surge tank, hoses, and hose connections.

If loose, tighten. If still leaking, notify unit maintenance.

Step 4. Check engine oil level.

If low, add oil (para. 2-10).

Step 5. Check radiator fan clutch operation.

If fan blade is not turning, install override lockup bolts (para. 3-14).

6. LOW ENGINE OIL PRESSURE.

Check engine oil level.

If low, add oil (para. 2-10).

NOTE

- If oil pressure is still low, notify unit maintenance.

7. ENGINE FAILURE DURING OPERATION.

Step 1. Check to see if emergency stop cable is pulled out.

On M939A2 series vehicles, manually reset emergency stop. On M939/A1 series vehicles, notify unit maintenance.

Step 2. Check to see if throttle control solenoid is malfunctioning (M939/A2 series vehicles).

If solenoid is malfunctioning, tie up with strap or rope and finish mission. Report to unit maintenance as soon as possible.

Table 3-1. Troubleshooting (Contd).

MALFUNCTION TEST OR INSPECTION CORRECTIVE ACTION

8. EXCESSIVE EXHAUST SMOKE AFTER ENGINE REACHES NORMAL OPERATING TEMPERATURE 175°F TO 200°F (79°C TO 93°C).

Check for restricted air cleaner.

If restricted, clean air cleaner element (para. 3-8).

HEATING SYSTEM

9. HOT WATER PERSONNEL HEATER FAILS TO PRODUCE HEAT AFTER ENGINE REACHES NORMAL OPERATING TEMPERATURE.

Step 1. Check to see if blower motor switch is in OFF position.

If in OFF position, put blower motor switch in HI or LOW position.

Step 2. Check to see if coolant shutoff valves are closed (para. 2-44).

If closed, open coolant shutoff valves.

Step 3. Check for air in heater.

With engine running, open air bleed drainvalve on engine side of heater and allow air to escape (para. 2-44). Close drainvalve.

10. HEATER BLOWER MOTOR OPERATES, BUT HEAT FAILS TO REACH CAB, OR DEFROSTERS FAIL TO OPERATE.

Step 1. Check to see if heat vent control and/or defroster control levers are adjusted properly.

If not, adjust heat vent control or defroster control levers to direct heat flow to desired location (para. 2-44).

Step 2. Check to see if heat/defroster vent tubes are connected below instrument panel.

TRANSMISSION

11. EXCESSIVE CREEP IN FORWARD OR REVERSE RANGE.

Check hand throttle position.

If partially out, push hand throttle all the way in.

12. TRANSMISSION OVERHEATING AS INDICATED BY TRANSMISSION OIL TEMPERATURE GAUGE.

Check transmission oil level.

If low, add oil (para. 3-10).

Table 3-1. Troubleshooting (Contd).

MALFUNCTION TEST OR INSPECTION CORRECTIVE ACTION

13. OIL THROWN FROM FILLER TUBE.

Step 1. Check transmission oil level (para. 3-10).

If overfull, notify unit maintenance.

Step 2. Check for loose transmission oil dipstick.

If loose, turn dipstick handle clockwise until tight (para. 3-10).

14. SLIPPAGE IN ALL FORWARD RANGES.

Check transmission oil level.

If low, add oil (para. 3-10).

15. TRANSMISSION OIL LEAKAGE.

Check for loose hose and tube connections.

If loose, tighten.

TRANSFER CASE

16. TRANSFER CASE LUBRICANT LEAKAGE.

Check for loose drainplugs.

If loose, tighten drainplugs.

AIR AND BRAKE SYSTEMS

17. INSUFFICIENT AIR PRESSURE AS INDICATED BY LOW AIR PRESSURE WARNING BUZZER OR AIR PRESSURE GAUGE.

Step 1. Check to see if air reservoir drainvalves are open.

If open, close drainvalves securely (para. 3-9).

Step 2. Check all air lines for loose connections.

If loose, tighten.

Step 3. Check all air lines for damage.

If damage is not repairable, shut off affected system (para. 3-13) and notify unit maintenance.

Step 4. Check towed equipment for air leaks at drainvalves or air lines.

If leaking, tighten.

18. SERVICE BRAKES DO NOT OPERATE.

Step 1. Check to see if air reservoir drainvalves are open.

If open, close drainvalves securely (para. 3-9).

Step 2. Check all air lines for loose connections.

If loose, tighten.

Table 3-1. Troubleshooting (Contd).

MALFUNCTION TEST OR INSPECTION CORRECTIVE ACTION

19. PARKING BRAKE DOES NOT HOLD VEHICLE.

Step 1. Check parking brake handle position.

If partially applied, pull parking brake handle all the way up.

Step 2. Check handle adjustment.

Turn knob on end of lever clockwise to increase braking action (para. 3-17).

20. PARKING BRAKE DRAGS OR OVERHEATS.

Step 1. Check parking brake handle position.

If partially applied, push parking brake handle all the way down.

Step 2. Check handle adjustment.

Turn knob on end of lever counterclockwise to decrease braking action (para. 3-17).

WHEELS, TIRES, AND HUBS

21. WHEEL WOBBLES OR SHIMMIES.

Step 1. Check for loose wheel stud nuts.

If loose, tighten.

Notify unit maintenance to retighten to proper torque.

Step 2. Check for cupping or missing rubber on tire.

Replace tire (para. 3-11).

22. EXCESSIVE OR UNEVEN TIRE WEAR.

Check air pressure in tires.

Inflate or deflate tires to correct air pressure (refer to table 1-10).

23. VEHICLE WANDERS OR PULLS TO ONE SIDE ON LEVEL SURFACE OR HIGHWAY.

Check air pressure in tires.

Inflate or deflate tires to correct air pressure (refer to table 1-10).

STEERING

24. HARD STEERING.

Step 1. Check air pressure in tires.

Inflate or deflate tires to correct air pressure (refer to table 1-10).

Step 2. Check power steering reservoir oil level.

If low, add oil to FULL mark on dipstick (para. 2-10).

Table 3-1. Troubleshooting (Contd).

MALFUNCTION
TEST OR INSPECTION
CORRECTIVE ACTION

25. OIL LEAKS.

Check for loose connections.

If loose, tighten.

SPECIAL BODY EQUIPMENT:

WARNING

Wear hand protection when handling winch cable. Do not handle cable with bare hands. Broken wires will cause injury.

FRONT WINCH

26. WINCH DRUM DOES NOT TURN OR PAYOUT CABLE.

Step 1. Check to see if drum lock knob is engaged.

If engaged, pull out drum lock knob, rotate 90 degrees, and release (para. 2-22).

Step 2. On M936/A1/A2 model vehicles with level wind device, check to see if level wind lock knob and cable tensioner lock knob and lever are engaged.

If engaged, release level wind lock knob and cable tensioner lock knob and lever.

Step 3. Check if cable is binding.

If binding, free cable from drum.

27. WINCH DOES NOT WIND.

Step 1. Check to see if power takeoff is engaged.

If not, engage power takeoff.

Step 2. Check to see if winch clutch lever is engaged.

If not, engage clutch lever.

Step 3. On vehicles with level wind device, check to see if tensioner lever is positioned all the way toward right (crew side) of vehicle.

If not, pull tensioner lever all the way toward right side of vehicle.

Step 4. Check level of hydraulic oil in reservoir (para. 2-22).

If low, add oil to proper level (para. 2-10).

Table 3-1. Troubleshooting (Contd).

MALFUNCTION TEST OR INSPECTION CORRECTIVE ACTION

DUMP BODY HOIST ASSEMBLY

28. HOIST DOES NOT LIFT DUMP BODY.

Step 1. Check to see if power takeoff is engaged.

If not, engage power takeoff.

Step 2. Check to see if dump body control lever is pushed back to raise position.

If not, push lever back to raise position.

Step 3. Check level of hydraulic oil in reservoir (para 2-25).

If low, add oil to proper level (para. 2-10).

Step 4. Check for hydraulic oil leaks.

Tighten loose connections. If leaks continue, notify unit maintenance.

29. BODY RAISES TO FULL DUMP BUT DOES NOT POWER DOWN.

Step 1. Check to see if support braces are in place.

If in place, lower.

Step 2. Check to see if dump body control lever is pulled full forward to lower position.

If not, pull lever full forward to lower position.

30. HYDRAULIC PUMP NOISY.

Check level of hydraulic oil in reservoir (para. 2-25).

If low, add oil to proper level (para. 2-10).

31. TAILGATE DOES NOT OPEN.

Step 1. Check to see if tailgate control rod is pulled forward and down to unlock tailgate.

If not, pull tailgate control rod forward and down to unlock tailgate.

Step 2. Check to see if tailgate chains are restricting tailgate from opening.

If restricting opening of tailgate, reposition tailgate chains.

Table 3-1. Troubleshooting (Contd).

MALFUNCTION TEST OR INSPECTION CORRECTIVE ACTION

HYDRAULIC CRANE

32. CRANE NOT OPERATING OR LACKS POWER.

Step 1. Check to see if tachometer indicates 1,275 rpm.

If not, pull hand throttle control all the way out.

Step 2. Check to see if transmission selector lever is in 1-5 (drive).

If not, place transmission selector lever in 1-5 (drive).

Step 3. Check to see if transfer case power takeoff lever is pushed back to engaged position.

If not, push transfer case power takeoff lever back to engaged position.

Step 4. Check level of hydraulic oil in reservoir (para. 2-24).

If low, add oil to proper level (para. 2-10).

Step 5. Check for hydraulic oil leaks.

Tighten loose connections. If leaks continue, notify unit maintenance.

33. CRANE DOES NOT LIFT.

Step 1. Check level of hydraulic oil in reservoir (para. 2-24).

If low, add oil to proper level (para. 2-10).

Step 2. Check for hydraulic oil leaks.

Tighten loose connections. If leaks continue, notify unit maintenance.

34. HYDRAULIC PUMP NOISY.

Check level of hydraulic oil in reservoir (para. 2-24).

If low, add oil to proper level (para. 2-10).

If still noisy, notify unit maintenance.

35. VEHICLE ROLLS WHILE OPERATING CRANE.

Step 1. Check to see if parking brake is applied.

If not, apply parking brake.

Step 2. Check to see if chock blocks are in place.

If not, place chock blocks at wheels and notify unit maintenance.

Table 3-1. Troubleshooting (Contd).

MALFUNCTION TEST OR INSPECTION CORRECTIVE ACTION

REAR WINCH

36. WINCH NOT OPERATING OR LACKS POWER.

Step 1. Check to see if tachometer indicates 1,275 rpm.

If not, pull hand throttle control all the way out.

Step 2. Check to see if transmission selector lever is in 1-5 (drive).

If not, place transmission selector lever in 1-5 (drive).

Step 3. Check to see if transfer case power takeoff shift lever is pushed back to engaged position.

If not, push transfer case power takeoff shift lever back to engaged position.

Step 4. Check to see if level wind lock knob is released.

If not, release level wind lock knob.

Step 5. Check level of hydraulic oil in reservoir (para. 2-24).

If low, add oil to proper level (para. 2-10).

Step 6. Check for hydraulic oil leaks.

Tighten loose connections. If leaks continue, notify unit maintenance.

37. VEHICLE ROLLS WHILE OPERATING REAR WINCH.

Step 1. Check to see if parking brake is applied.

If not, apply parking brake.

Step 2. Check to see if chock blocks are in place.

If not, place chock blocks at wheels and notify unit maintenance.

EXPANSIBLE VANS

38. LIGHT SHINES THROUGH GAPS AT SIDE PANEL OF VAN BODY.

Step 1. Check toggle clamps at side panels (para. 2-27).

If toggle clamp does not draw top of side panel tight enough, loosen locknut on toggle clamp eyebolt. Turn eyebolt inward to close the gap. Tighten locknut.

Step 2. Check to see if roof is properly seated.

If not, loosen toggle clamp, push up on hinged roof, and push out on end panels, then reclose toggle clamps to ensure seal alignment.

Step 3. Check to see if blackout panels are closed properly.

If not, slide up blackout panels on van sides and rear doors until they latch in closed position.

Table 3-1. Troubleshooting (Contd).

MALFUNCTION TEST OR INSPECTION CORRECTIVE ACTION

39. CEILING LIGHTS AND SERVICE RECEPTACLES FAIL TO ENERGIZE WHEN DOORS ARE CLOSED UNDER BLACKOUT CONDITIONS (110 VOLT SYSTEM).

Step 1. Check to see if blackout circuit switch and/or main circuit breaker switches are turned off (para. 2-27).

If off, turn on switches.

Step 2. Check to see if blackout switch is turned off.

If off, turn on blackout switch.

Step 3. Check outside power cable for secure connections if electrical power is supplied from outside source.

If not, connect power cable securely to power entrance receptacle and power source.

40. EMERGENCY LIGHT, BLACKOUT LIGHT, AND CEILING LIGHTS FAIL TO ILLUMINATE (24 VOLT SYSTEM).

Step 1. Check to see if main circuit breaker and/or light switches are turned off.

If not, turn on main circuit breaker or light switches.

Step 2. Check outside power cable for secure connections if electrical power is supplied from outside source.

If not, connect power cable securely to power entrance receptacle and power source.

41. HEATER WILL NOT IGNITE.

Step 1. Check to see if main circuit breaker and/or heater switches are turned off (para. 2-27).

If off, turn on main circuit breaker or heater switches.

Step 2. Check to see if thermostat is set to desired temperature.

If not, set thermostat properly.

Step 3. Check fuel level on fuel gauge.

Fill fuel tank(s) as necessary.

Table 3-1. Troubleshooting (Contd).

MALFUNCTION TEST OR INSPECTION CORRECTIVE ACTION

42. AIR CONDITIONER COMPRESSOR FAILS TO START.

Step 1. Check to see if bonnet door is closed (para. 2-27).

If closed, push bonnet door control rod forward to open bonnet door.

Step 2. If closed, check to see if main circuit breaker and/or air conditioner switches in circuit breaker box are turned off.

If off, turn on main circuit breaker or air conditioner switches.

Step 3. Check to see if power input switch and/or compressor circuit breaker are turned off.

If off, turn on power input switch or compressor circuit breaker.

Step 4. Check to see if compressor switch is turned to HIGH when starting air conditioner.

If not, turn compressor switch to HIGH when starting air conditioner.

SPECIAL PURPOSE KITS:

RADIATOR AND HOOD COVER KIT

43. ENGINE FAILS TO REACH OPERATING TEMPERATURE.

Check to see if radiator cover flap is opened (para. 2-43).

If open, roll cover flap down.

44. ENGINE TEMPERATURE EXCEEDS 200°F (93°C).

Step 1. Check to see if radiator cover flap is closed (para. 2-43).

If closed, roll up cover flap and secure.

WARNING

Extreme care should be taken when removing surge tank filler cap if temperature gauge reads above 175°F (79°C). Steam or hot coolant under pressure will cause injury.

Step 2. Check radiator core for obstructions.

If clogged, remove debris (refer to table 2-1).

Step 3. Check coolant level in surge tank.

If low, add coolant to surge tank until at bottom of filler neck.

Step 4. Check for leakage from tank, hoses and hose connections.

If loose, tighten. If still leaking, notify unit maintenance.

Table 3-1. Troubleshooting (Contd).

MALFUNCTION TEST OR INSPECTION CORRECTIVE ACTION

Step 5. Check engine oil level.

If low, add oil (para. 2-10).

Step 6. Check radiator fan clutch operation.

If fan blade is not turning, install override lockup bolts (para. 3-14).

FUEL BURNING PERSONNEL AND ENGINE COOLANT HEATER KITS

45. FUEL BURNING PERSONNEL HEATER FAILS TO START WHEN RUN-OFF-START SWITCH IS HELD IN START POSITION.

WARNING

Exhaust gases can kill. Do not operate engine coolant heater in closed area occupied by personnel. Such action will result in injury or death.

NOTE

Heater will not operate if RUN-OFF-START switch is moved to RUN position before indicator lamp illuminates.

Step 1. Press PRESS-TO-TEST button on heater control box to check operation of circuit.

If indicator lamp does not illuminate, notify unit maintenance.

Step 2. Check to see if HI-LO switch on heater control box is set to HI.

If not, set HI-LO switch to HI.

Step 3. Check fuel level on fuel gauge.

Fill fuel tank(s) if necessary.

Step 4. Check to see if electric fuel pump shutoff valve is closed (para. 2-44).

If closed, open fuel pump shutoff valve.

Table 3-1. Troubleshooting (Contd).

MALFUNCTION TEST OR INSPECTION CORRECTIVE ACTION

46. ENGINE COOLANT HEATER FAILS TO START WHEN RUN-OFF-START SWITCH IS HELD IN START POSITION.

WARNING

Exhaust gases can kill. Do not operate engine coolant heater in closed area occupied by personnel. Such action will result in injury or death.

CAUTION

Do not operate engine coolant heater and personnel heater at the same time. Damage to equipment may result.

NOTE

- Select HI position if engine is cold. Select LOW position if engine is already well heated. Switch will automatically change to LOW position when coolant temperature exceeds 195°F (91°C). Switch will automatically change to HI position when coolant temperature drops below 120°F (49°C).
- Heater will not operate if RUN-OFF-START switch is moved to RUN position before indicator lamp illuminates.

Step 1. Press PRESS-TO-TEST button on heater control box to check operation of circuit.

If indicator lamp does not illuminate, notify unit maintenance.

Step 2. Check fuel level on fuel gauge.

Fill fuel tank(s) if necessary.

Step 3. Check to see if electric fuel pump shutoff valve located near air cleaner assembly is closed (para. 2-45).

If closed, open fuel shutoff valve.

47. HEATER FAILS TO CONTINUE BURNING.

Check fuel level on fuel gauge.

Fill fuel tank(s) if necessary.

Table 3-1. Troubleshooting (Contd).

MALFUNCTION TEST OR INSPECTION CORRECTIVE ACTION

48. WINDSHIELD DEFROSTERS NOT OPERATING.

Step 1. Check adjustment of defroster control handle.

Adjust defroster control handle.

Step 2. Check for restrictions in defroster deflectors.

If restricted, clear restriction.

49. ENGINE OIL PAN SHROUD NOT RECEIVING HEAT (ENGINE COOLANT HEATER ONLY).

Step 1. Check to see if coolant heater is operating (para. 2-45).

Start heater. If inoperative, notify unit maintenance.

Step 2. Check to see if coolant heater exhaust tube is disconnected from oil pan shroud.

If disconnected, reconnect heater exhaust tube.

50. ENGINE COOLANT SYSTEM NOT RECEIVING HEAT (ENGINE COOLANT HEATER ONLY).

Step 1. Check to see if coolant heater is operating (para. 2-45).

Start heater. If inoperative, notify unit maintenance.

Step 2. Check to see if one or more coolant shutoff valves are closed at engine.

If closed, open coolant shutoff valve(s).

A-FRAME KIT

51. WINCH INOPERATIVE.

Check winch.

See malfunctions 26 and 27.

52. A-FRAME MISALIGNED.

Check to see if cable is secured in towing pintle.

If not, secure cable in towing pintle and lock pintle in closed position.

Table 3-1. Troubleshooting (Contd).

MALFUNCTION TEST OR INSPECTION CORRECTIVE ACTION

AIRBRAKE CONTROL KIT

53. TRAILER AIRBRAKES DO NOT OPERATE WHEN AIRBRAKE CONTROL HANDLE IS PULLED DOWN.

Step 1. Check to see if trailer air supply valve control knob is not in pressed-in position (para. 2-47).

If not, press in trailer air supply valve control knob and hold in place for 15 seconds. Release valve. If valve does not stay in, notify unit maintenance.

Step 2. Check to see if trailer air lines are securely connected to air couplings of towing truck.

If not, securely connect trailer air lines of air coupling, and open air coupling valves.

3-5. CTIS STATUS INDICATION AND MALFUNCTIONS

a. Central Tire Inflation System (CTIS). CTIS is designed to work automatically in case of tire leakage. Additionally, CTIS will adjust tire pressure when a road surface selection is made, to preset valves. Refer to table 3-2, CTIS Indications, for explanation of indications or malfunction.

b. CTIS Integration with the Airbrake System. The Central Tire Inflation System uses the same air compressor that supplies air to the vehicle brakes. The vehicle brakes are always given priority over the CTIS. When brake operation causes the air pressure in the brake reservoir to fall below a preset limit, inflation will stop until the air compressor has refilled the brake reservoir. If the CTIS was deflating the tires, it would continue to do so.

CAUTION

To prevent damage to transmission, refer to paragraph 2-15 for proper transmission shifting procedures.

c. Rapid Inflation. The most rapid tire pressure increase is achieved during vehicle operation. Keep the transmission downshifted to a lower gear. Keep engine operation at 2,000 rpm. Engaging the turbocharger supplies extra air pressure directly to the air compressor.

Table 3-2. CTIS Indications

CTIS STATUS INDICATION SYSTEM MALFUNCTION ACTION REQUIRED

1. SINGLE MODE LIGHT: STEADY.

None. Air pressure achieved, wheel valves closed, and system stable.

None.

2. SINGLE MODE LIGHT: FLASHING.

None. System working to achieve new air pressure.

None.

3. TWO MODE LIGHTS ON: STEADY.

System has shut off with tire pressure between two settings, but vehicle and CTIS are still operational.

Monitor and, if indication is repeated frequently, notify supervisor.

4. FOUR MODE LIGHTS FLASHING.

System has shut off due to air leak or possible tire damage and is waiting for operator instruction.

a. Select RUN FLAT, if tire damage is minimal (vehicle is still operational).

b. Change tire if tire damage is not minimal (para. 3-11).

c. Check for air leaks (para. 3-3).

5. FIVE LIGHTS FLASHING.

NOTE

If CTIS is not operational, disconnect electrical connector from Electronic Control Unit (ECU) and complete mission (para. 3-16).

System has shut off due to fault detection of CTIS component or system has major air leak.

a. Check for air leaks (para. 3-3) and reset CTIS (para. 3-16).

b. Disable CTIS (para. 3-16) and notify unit maintenance.

6. RUN FLAT SELECTOR FLASHING (WITH A STEADY OR FLASHING MODE LIGHT).

None. RUN FLAT has been selected, and tire pressure is being checked at frequent intervals (every 15 seconds).

May be turned off by depressing run flat mode.

Table 3-2. CTIS Indications (Cont'd).

CTIS STATUS INDICATION SYSTEM MALFUNCTION ACTION REQUIRED

7. **SYSTEM REPEATEDLY RESUMES CYCLING 30 SECONDS AFTER MODE LIGHT STOPS FLASHING.**

   Undetermined.

   Notify unit maintenance.

8. **SYSTEM SHUTS OFF DURING INFLATION, SINGLE MODE LIGHT CONTINUES TO FLASH.**

   Undetermined.

   Notify unit maintenance.

9. **SYSTEM FAILS TO DEFLATE, PARTIALLY DEFLATES, OR TIRE PRESSURES ARE IMBALANCED.**

   Undetermined.

   Disable CTIS (para. 3-16) and notify unit maintenance.

10. **SELECTOR PANEL LIGHTS WORK, SYSTEM FAILS TO INFLATE OR DEFLATE.**

    Undetermined.

    Notify unit maintenance.

11. **LOSS OF OVERSPEED WARNING LIGHT AND/OR OVERSPEED PRESSURE CHANGE.**

    Undetermined.

    Disable CTIS (para. 3-16), maintain vehicle speed within limits of tire pressure setting (table 1-10), and notify unit maintenance.

12. **SYSTEM IS OVERINFLATING THE TIRES.**

    Undetermined.

    Disable CTIS (para. 3-16), readjust tire pressure for road conditions (table 1-10), and notify unit maintenance.

13. **SLOW AIR RECOVERY OR OCCASIONAL LOW AIR WARNING DURING BRAKE OPERATION.**

    Step 1. Minor leak in air system.

    Troubleshoot air system (para. 3-3), and notify unit maintenance.

    Step 2. Major leak in air system.

    Disable CTIS (para. 3-16) and notify unit maintenance.

Section III. MAINTENANCE PROCEDURES

3-6. GENERAL

The operator/crew is responsible for daily, weekly, and monthly checks listed in the Preventive Maintenance Checks and Services, table 2-3. Certain other maintenance services, also the responsibility of the operator/crew, are listed in this section.

3-7. BREAK-IN OPERATION

a. Road Test.

CAUTION

Do not go faster than the maximum allowable speeds shown on the maximum road speed data plate. Do not drive continuously at maximum allowable speeds. Be alert for signs of equipment failure. Failure to do this may result in equipment damage.

All vehicles received by the using organization must be road-tested to check operation and condition of all reconditioned vehicles, except those previously driven 50 mi (80 km). The operator will check the instrument panel and gauges as often as possible for signs of unsatisfactory performance. Stops will be made at least every 10 mi (16 km) to give the operator a chance to check the vehicle for possible coolant, oil, fuel, or exhaust leakage and any signs that may show the engine, transmission, wheel hubs, brake drums, axles, differentials, or transfer case assemblies are overheated. The vehicle must be checked thoroughly for any control that is hard to operate and any instrument not operating properly. Unusual noises and vibration will be noted. All unusual conditions will be reported to unit level maintenance.

b. After Road Test. After the road test, correct any faulty condition that can be done at operator's maintenance level. Notify unit maintenance about any other faulty condition.

3-8. ENGINE SERVICE

a. General. To perform engine service, the hood must be unlatched and secured in the opened position. After completing engine service, release hood, lower it to fixed position, and latch it (para. 2-19).

3-8. ENGINE SERVICE (Contd)

b. Engine Crankcase Oil Level.

CAUTION

- Never operate engine in M939/A1 series vehicles with oil level below L (low) level mark or above H (high) level mark.
- Never operate engine in M939A2 series vehicles below ADD 2 QTS (low) mark or above FULL (high) mark on dipstick.

(1) Engine oil level dipstick (4) is located on right side of engine below coolant surge tank (1) on M939/A1 series vehicles, and behind alternator (2) on M939A2 series vehicles.

(2) On M939/A1 series vehicles, turn dipstick (4) handle counterclockwise to release from dipstick tube (3).

(3) Pull dipstick (4) from dipstick tube (3).

(4) Wipe dipstick (4) clean and return to dipstick tube (3).

(5) Slowly pull dipstick (4) from dipstick tube (3) and read level.

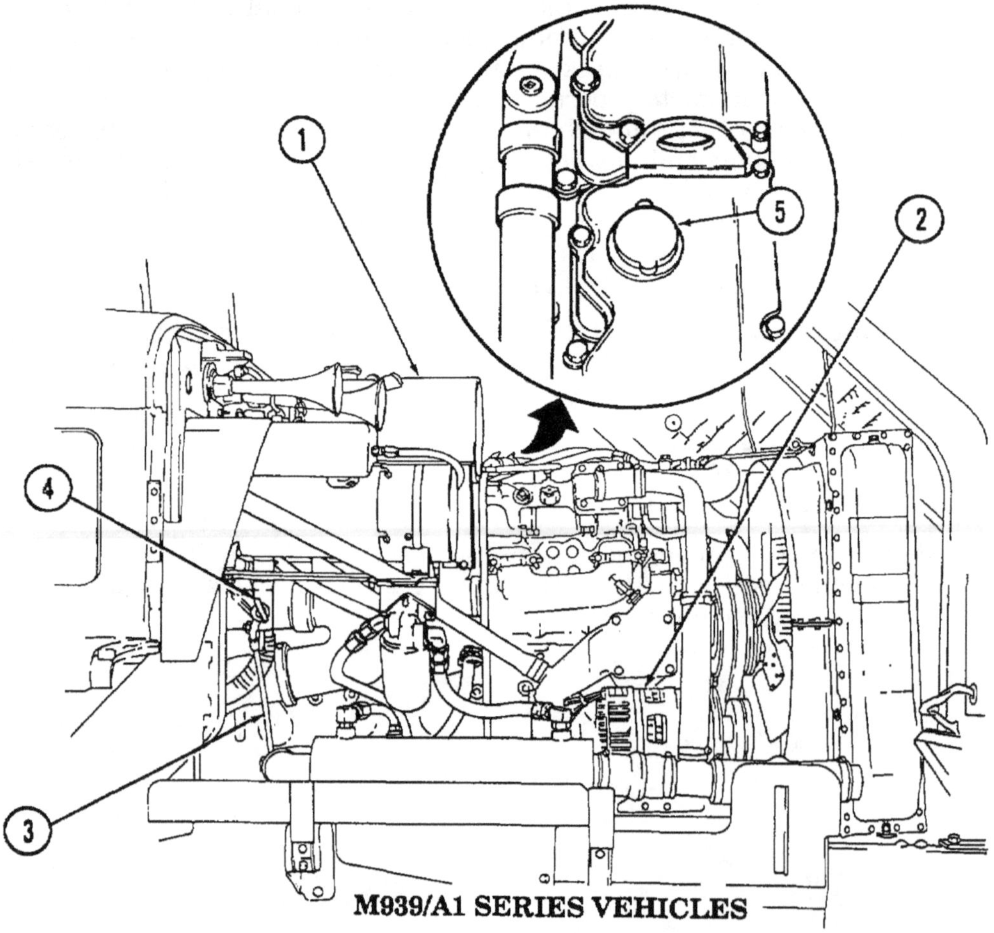
1
2
3
4
5

M939/A1 SERIES VEHICLES

3-8. ENGINE SERVICE (Contd)

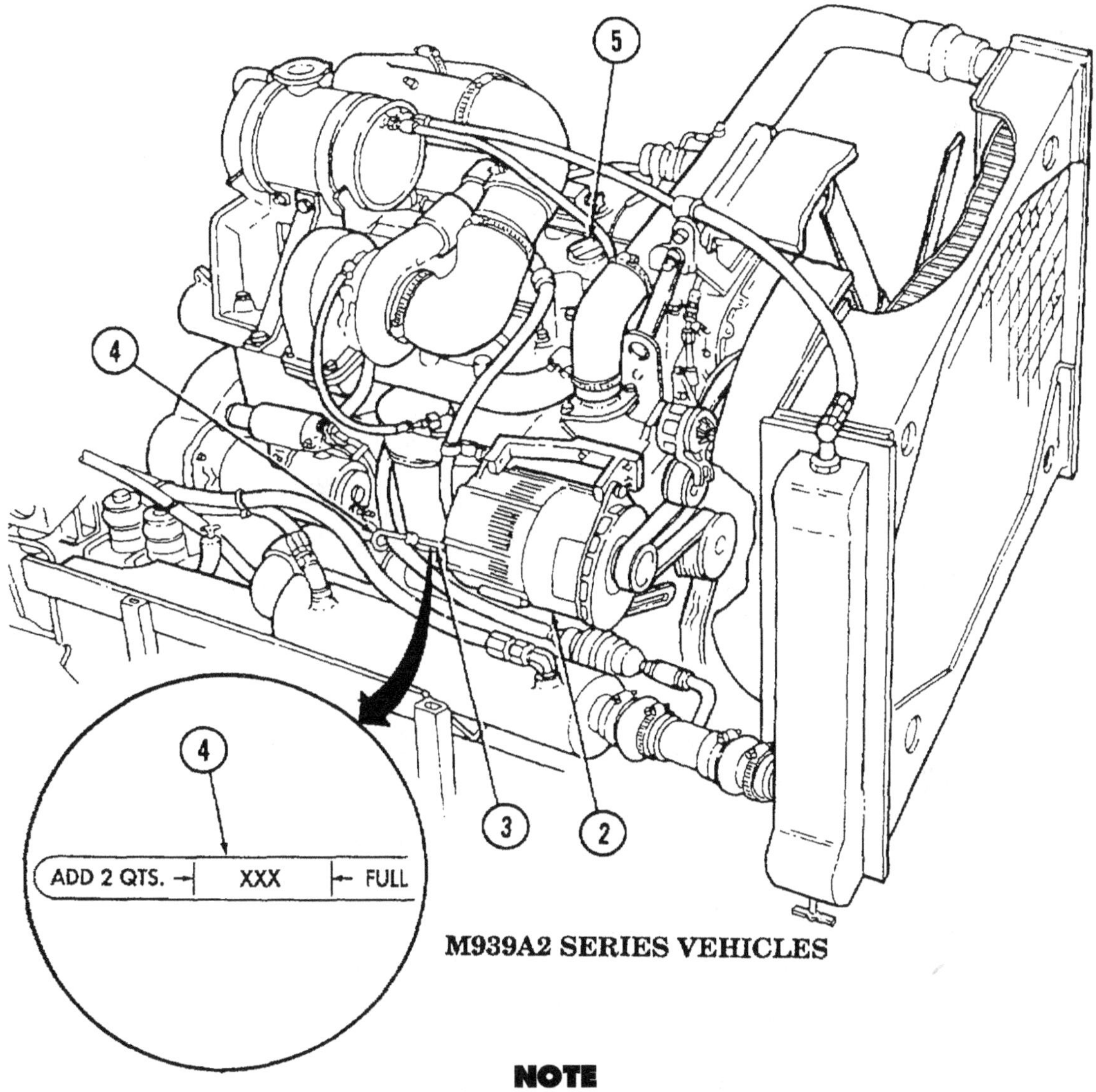
5
4
3
2
4
ADD 2 QTS.
XXX
FULL

M939A2 SERIES VEHICLES

NOTE

On M939/A1 series vehicles, 7 qts (6.6 L) of oil are required to raise oil level from L mark to H mark on dipstick. On M939A2 series vehicles, 2 qts (1.9 L) are required to raise oil level from ADD 2 QTS. mark to FULL mark. Refer to para. 2-10 for oil specifications.

(6) If engine oil level is low, remove engine oil filler cap (5), add engine oil, and replace engine oil filler cap (5). Tighten cap securely wipe away any spilled oil.

(7) After checking or adjusting oil level, wipe dipstick (4) clean and reinstall dipstick (4) in dipstick tube (3). Make sure dipstick (4) is seated securely.

(8) On M939/A1 series vehicles, turn dipstick (4) handle clockwise to tighten in dipstick tube (3).

3-8. ENGINE SERVICE (Contd)

c. Coolant Surge Tank.

WARNING

Extreme care should be taken when removing surge tank filler cap if temperature gauge reads above 175°F (79°C). Steam or hot coolant under pressure may cause injury.

(1) Remove coolant surge tank filler cap (1). Visually check coolant level. Surge tank (2) should be filled to approximately bottom end of filler tube (3) with engine cold, and slightly above with engine at normal operating temperature.

NOTE

- Have suitable container ready to catch liquid contaminants.
- If surge tank on M939A2 series vehicles is found to be empty, open drainvalve on aftercooler and fill surge tank. Close drain-valve when coolant is observed flowing from drain, and continue filling to approximately bottom end of fill tube.

(2) If coolant level is low, add coolant as necessary.

(3) Run engine until temperature gauge (4) reads 175°-200°F (79°-93°C). Check and refill as necessary.

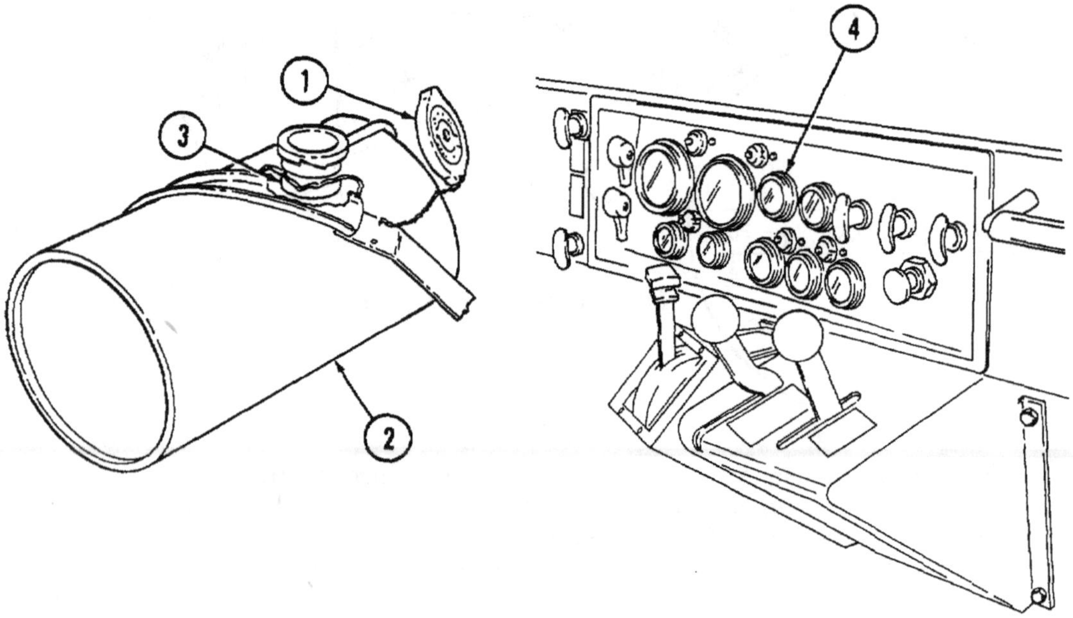
1
3
2
4

d. Power Steering Reservoir.

CAUTION

Do not overfill power steering reservoir. Oil will overflow into vent system.

3-8. ENGINE SERVICE (Contd)

NOTE

- Power steering reservoir oil level is checked with engine stopped.
- With engine cold, use COLD FULL mark on dipstick. If engine is at normal operating temperature, 175°-200°F (79°-93°C), use HOT FULL mark on dipstick.

(1) The power steering reservoir (7) is located on left side of engine near the radiator. Remove oil filler cap (6) and wipe off dipstick (5). Reinstall and remove dipstick (5) to check reservoir (7) oil level.

(2) If oil level is low, add oil (para. 2-10). Replace oil filler cap (6), tighten securely, and wipe up any oil spilled.

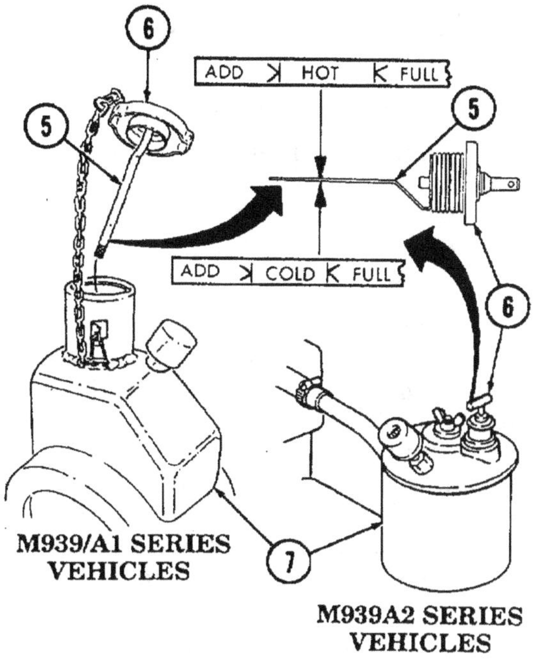
6
ADD HOT FULL
5
5
ADD COLD FULL
6
7
M939/A1 SERIES VEHICLES
M939A2 SERIES VEHICLES

e. Fuel Filter/Water Separator.

(1) Service Operation.

(a) The fuel filter/water separator (9) is located under left-front fender on M939/A1 series vehicles, and attached to left side of engine on M939A2 series vehicles, and requires daily maintenance.

(b) On M939/A1 series vehicles, open inlet drainvalve (8) located near top of fuel filter/water separator (9).

(c) On M939A2 series vehicles, open drainvalve (10) at bottom of fuel filter/water separator (9) and drain off 1 pt. (0.47 L) of liquid into a container.

(d) If larger amounts of water and impurities are detected, drain until fuel is clear. Notify unit maintenance.

(e) After service has been completed, close drainvalves (8) or (10), and prime fuel system.

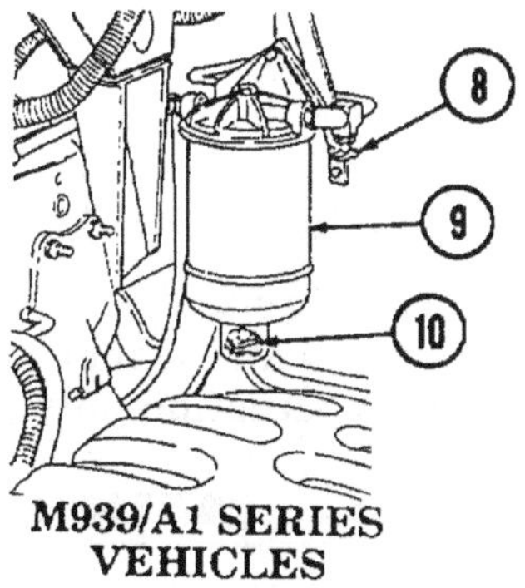
8
9
10
M939/A1 SERIES VEHICLES

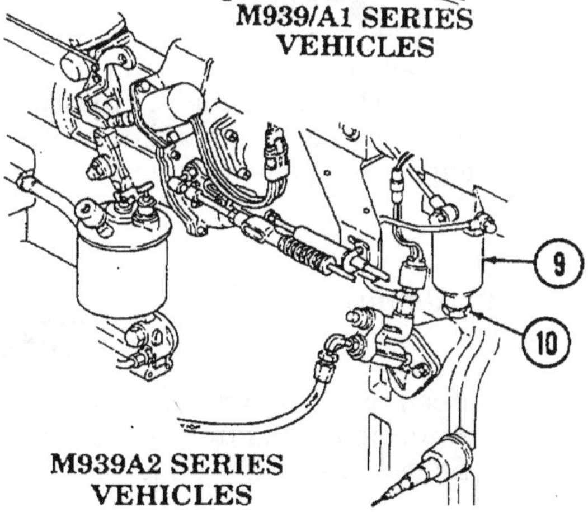
9
10
M939A2 SERIES VEHICLES

3-8. ENGINE SERVICE (Contd)

e. Fuel Filter/Water Separator (Contd).

(2) Priming Fuel System.

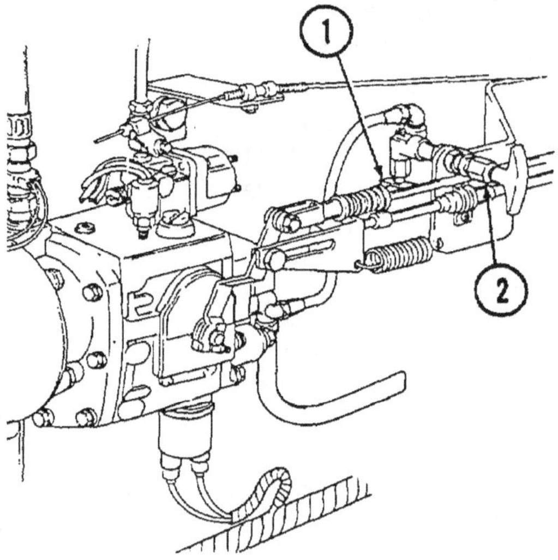
1
2

(a) The fuel system must be primed whenever the fuel filter/water separator element is replaced and after any draining of the fuel system.

(b) Open air purge drainvalve (1) at hand primer pump (2).

(c) Place a container under air purge drainvalve (1). Operate hand primer pump (2) to discharge a combination of fuel and air from the fuel system. Continue pumping until all air is expelled and a steady flow of fuel is observed. Stop hand primer pump (2) operation and close air purge drainvalve (1). Dispose of waste fuel properly.

(3) Final Inspection. Start engine (para. 2-12) and check for unusual noises and fuel system leaks.

f. Air Cleaner.

WARNING

If NBC exposure is suspected, all air filter media should be handled by personnel wearing protective equipment. Consult your unit NBC officer or NBC NCO for appropriate handling or disposal instructions.

(1) General. Air cleaner service is required whenever the red band is visible in window of air cleaner indicator (3). The operator will service the air cleaner in an emergency situation only. Notify unit maintenance as soon as possible.

CAUTION

Do not operate engine without an air cleaner element. Failure to do so may result in internal engine component damage.

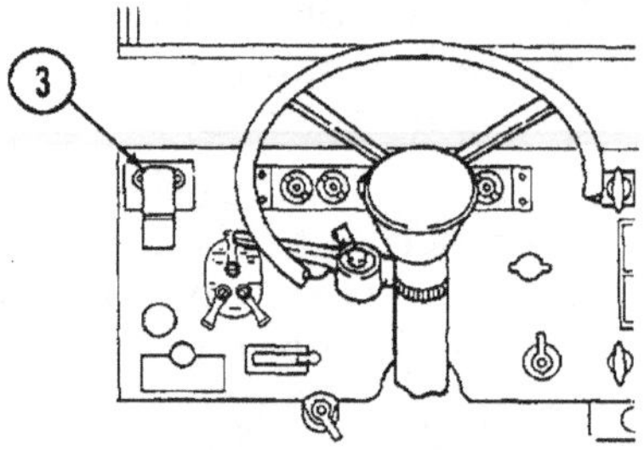
3

(2) Removal.

WARNING

Keep fingers out from under or directly above the locking end of securing latches during removal or installation. Injury will result if fingers are caught under latches and/or if fingers are struck by latch when unsnapped.

3-8. ENGINE SERVICE (Contd)

(2) Removal (Contd).

(a) Release latch (6) securing rear retaining strap (5) to hanger (11) and spread apart retaining strap (5).

(b) Release five latches (8) securing air cleaner body (7) to air cleaner manifold (4).

(c) Remove air cleaner body (7), gasket (9), and element (10).

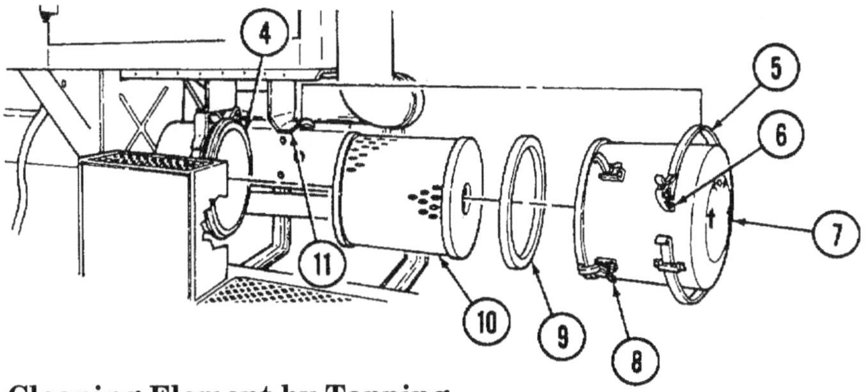
4
5
6
7
11
10
9
8

(3) Cleaning Element by Tapping.

CAUTION

Do not strike ends of element on hard surface. Damage will result.

(a) Hold element (10) so open end faces ground.

(b) Gently tap completely around element (10) with hand to free trapped dirt.

10

(4) Installation.

(a) Position air cleaner element (10) in air cleaner manifold (4) with closed end of element (10) facing outward.

WARNING

Keep fingers out from under or directly above the locking end of securing latches during removal or installation. Injury will result if fingers are caught under latches and/or if fingers are struck by latch when unsnapped.

(b) Position gasket (9) on air cleaner body (7) and install air cleaner body (7) over element (10) with arrows on end of air cleaner body (7) pointing up.

(c) Secure air cleaner body (7) to manifold (4) with five latches (8).

(d) Secure rear retaining strap (5) to hanger (11).

(5) Final Inspection. Start engine (para. 2-12) and press top of air cleaner indicator (3) to release red band. If green band does not appear, report condition to unit maintenance.

3-9. AIR RESERVOIRS

a. General. Four drainvalves, located on right side of vehicle next to toolbox, are used to drain moisture from air reservoirs.

b. Service.

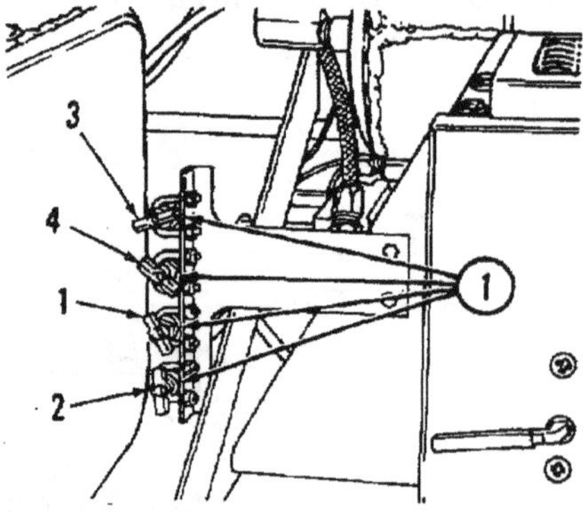
3
4
1
2
1

(1) Turn drainvalves (1) counterclockwise to drain moisture from:

(a) Airbrake system wet tank reservoir drainvalve 1.

(b) Spring brake air reservoir drainvalve 2.

(c) Primary airbrake system air reservoir drainvalve 3.

(d) Secondary airbrake system air reservoir drainvalve 4.

(2) After all moisture has been drained and only air is coming out, turn drainvalves (1) clockwise to close.

c. Final Inspection. Make sure drainvalves (1) are closed tight to prevent air from escaping. If air escapes after drainvalves are closed tightly, notify unit maintenance immediately.

3-10. TRANSMISSION OIL LEVEL

CAUTION

When checking transmission oil level, do not permit dirt, dust or grit to enter transmission filler tube. Make sure dipstick handle and end of filler tube are clean to prevent internal transmission damage.

a. General. The transmission oil level is checked weekly with engine running at idle, transmission in neutral, and parking brake applied. In M939/A1 series vehicles, transmission dipstick (3) is located under access door (4) in center of cab floor. In M939A2 series vehicles, transmission dipstick (3) is on right side of engine compartment.

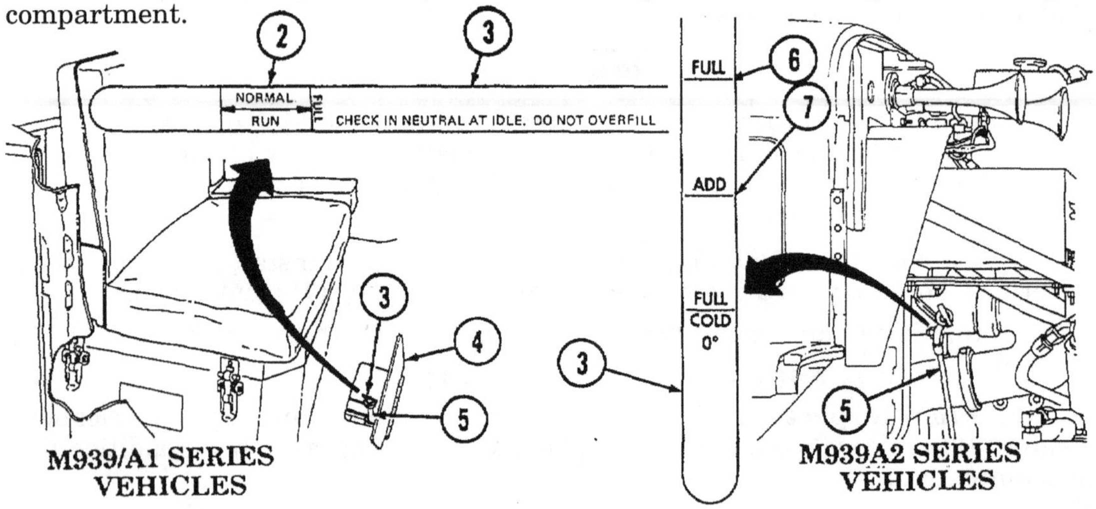
2
3
NORMAL
RUN
FULL
CHECK IN NEUTRAL AT IDLE. DO NOT OVERFILL
3
4
5
M939/A1 SERIES VEHICLES
FULL
6
7
ADD
FULL
COLD
0°
3
5
M939A2 SERIES VEHICLES

3-10. TRANSMISSION OIL LEVEL (Contd)

b. Check Oil Level.

WARNING

Do not check transmission oil level if transmission oil temperature gauge indicates temperature over 220°F (104°C). Stop engine and allow transmission to cool. Hot oil may result in injury to personnel.

(1) Open access door (4) on cab floor in M939/A1 series vehicles or open hood in M939A2 series vehicles to access dipstick (3). Clean around end of filler tube (5) before removing dipstick (3).

WARNING

Hot turbocharger and exhaust manifold can cause injury to personnel.

(2) Turn dipstick (3) handle counterclockwise and pull out dipstick (3) on M939/A1 series vehicles. On M939A2 series vehicles, pull dipstick (3) straight out.

(3) Wipe clean and insert dipstick (3) in filler tube (5).

(4) On M939/A1 series vehicles, withdraw dipstick (3) slowly to prevent a false reading. If transmission oil temperature gauge (8) on instrument panel reads 180°F (82°C) or below, level on dipstick (3) should show between marks designated for normal run (2).

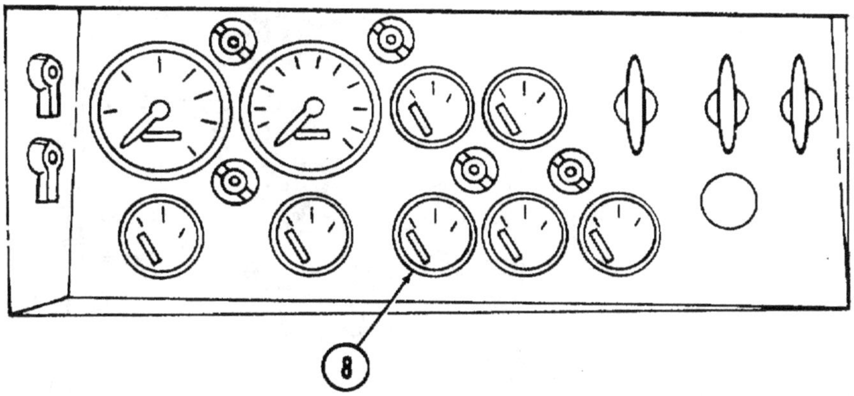
8

(5) On M939A2 series vehicles, slowly pull dipstick (3) out of filler tube (5). Level on dipstick (3) should be between ADD mark (7) and FULL mark (6).

CAUTION

Overfilling transmission will result in internal transmission damage.

(6) If transmission oil level is low, add oil through filler tube (5) (LO 9-2320-272-12). Return dipstick (3) to filler tube (5), tighten dipstick handle (M939/A1 series vehicles), and wipe up any oil spilled.

3-11. WHEELS AND TIRES

a. General. Tires are checked as part of Preventive Maintenance Checks and Services (PMCS) (table 2-3). If tire becomes flat while operating, stop vehicle immediately, if tactical situation permits. All tires on M939/A1/A2 series vehicles are bi-directional and do not require any special mounting.

b. Spare Tire. Models M923/A1/A2, M925/A1/A2, M927/A1/A2, M928/A1/A2, M931/A1/A2, and M932/A1/A2 (cargo trucks, tractors) are equipped with expendable spare tire davit boom. The M929/A1/A2, M930/A1/A2 (dump trucks), and the M934 (expansible van) are equipped with an eyebolt for attaching a chain fall. The M934A1 and M93A2 (expansible vans) are equipped with a davit-expendable boom and a built-in hand operated winch, while the M936/A1/A2 (wrecker) models use the vehicle boom and chain to lift and lower spare tire.

(1) Removal (M923/A1/A2, M925/A1/A2, M927/A1/A2, M928/A1/A2, M931/A1/A2, and M932/A1/A2).

NOTE

- This operation requires personnel.
- Procedures and illustrations are for M931A2 and M932A2 other models listed are similar.

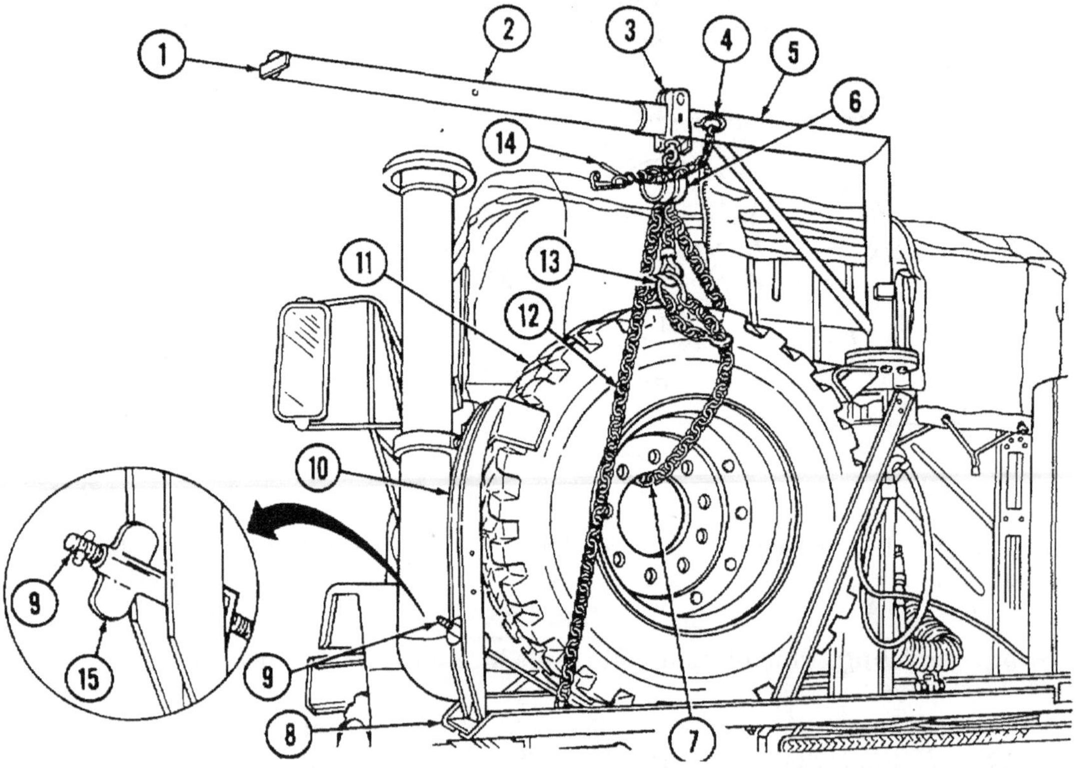
1
2
3
4
5
6
14
11
13
12
10
9
15
9
8
7

(a) Obtain utility chain (7) and chain fall (6) from tool compartment.

(b) Secure utility chain (7) around spare tire (11) and take up slack.

3-11. WHEELS AND TIRES (Contd)

(c) Hook chain fall (6) to davit boom roller (3) and to utility chain (7) securing spare tire (11) and take up slack.

(d) Remove lock pin (14) from davit boom (2), roller (3) and retaining pin (4) from davit boom (2) and davit support (5).

(e) Slide out davit boom (2) until holes in davit boom (2) and davit support (5) are aligned. Secure in place with retaining pin (4) and lock pin (14).

(f) Hook chain fall (6) to ring (13) on utility chain (7). Pull hand chain (12) to take up slack.

(g) Loosen wingnut (15) securing wheel brace (10) to threaded bar (9).

(h) Guide threaded bar (9) from notch in davit boom base (8). Lift wheel brace (10) and threaded bar (9) from davit boom base (8).

WARNING

- Stand clear during hoisting operations. Injury may result if personnel are struck by a swinging spare tire.
- Use caution when operating chain hoist. Injury may result if fingers are caught in chain hoist pulley sheave.

(i) Pull hand chain (12) to slightly raise spare tire (11). Slide spare tire (11) along davit boom (2) until roller (3) hits stop (1).

NOTE

Spare tire must be turned flat to vehicle body during lowering operation.

(j) Pull hand chain (12) to lower spare tire (11) to ground.

(k) Remove utility chain (7) from spare tire (11) and chain fall (6).

(2) Installation (M923/A1/A2, M925/A1/A2, M927/A1/A2, M928/A1/A2, M931/A1/A2, and M932/A1/A2).

(a) Place utility chain (7) through rim of spare tire (11) and ring (13) on utility chain (7). Center ring (13) at top of spare tire. Take up slack and hook utility chain (7) back onto utility chain (7).

(b) Lower chain fall (6) hook and attach to ring (13).

(c) Raise spare tire (11) until it clears davit boom base (8).

(d) Using roller (3), place spare tire (11) in davit boom base (8).

(e) Lower wheel brace (10), guide threaded bar (9) into notch in davit boom base (8) and tighten wingnut (15) until snug.

(f) Remove utility chain (7) from spare tire (11) and chain fall (6).

(g) Slide davit boom (2) into davit support (5) until holes in davit boom (2) and davit support (5) are aligned. Secure in place with retaining pin (4) and lock pin (14).

(h) Return utility chain (7) and chain fall (6) to tool compartment.

3-11. WHEELS AND TIRES (Contd)

(3) **Removal** (M929/A1/A2, M930/A1/A2, and M934).

NOTE

This procedure requires two personnel.

(a) Obtain utility chain (1) and chain fall (4) from tool compartment.

(b) Secure utility chain (1) around spare tire (6) and take up slack.

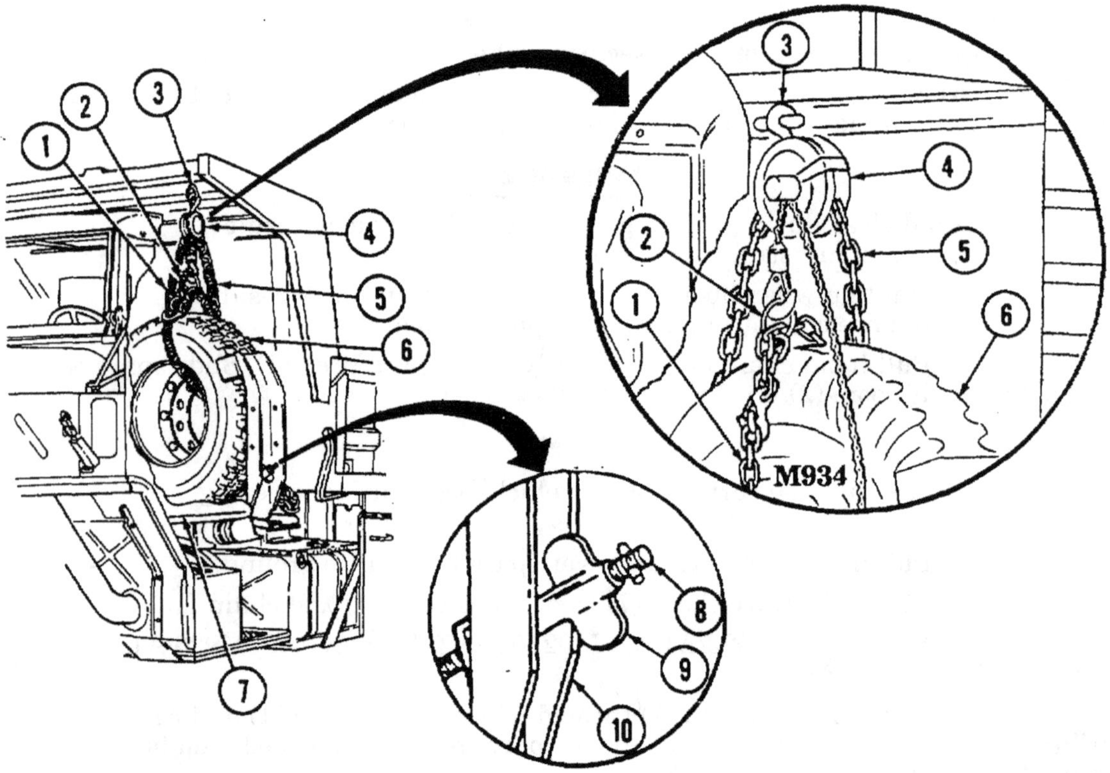
1
2
3
4
5
6
7
8
9
10
M934

(c) Hook chain fall (4) through support loop (3) and to utility chain (1), securing spare tire (6) with ring (2) and take up slack.

(d) Loosen wingnut (9) securing wheel brace (10) to threaded bar (8).

(e) Pull wheel brace (10) and threaded bar (8) from spare tire (6).

WARNING

- Keep spare tire from swinging. Injury may occur if personnel are struck by a swinging spare tire.
- Use caution when operating chain fall. Fingers may be caught in chain fall pulley sheave causing injury.

3-11. WHEELS AND TIRES (Contd)

(f) Pull hand chain (5) to slightly raise spare tire (6) from spare tire carrier base (7). Move spare tire (6) out of bracket area towards back of cab. Pull spare tire (6) towards side of vehicle until sufficient clearance is obtained to lower spare tire (6) to ground.

(g) Pull on hand chain (5) to lower spare tire (6) to ground.

(h) Remove utility chain (1) securing spare tire (6) to chain fall (4).

(4) Installation (M929/A1/A2, M930/A1/A2, and M934).

(a) Place utility chain (1) through rim of spare tire (6) and ring (2) on utility chain (1). Center ring (2) at top of spare tire (6). Take up slack and hook utility chain (1) back onto utility chain (1).

(b) Lower chain fall (4) hook and attach to utility chain ring (2).

(c) Raise spare tire (6) until it clears tire carrier base (7). Place tire on tire carrier base (7).

(d) Lower wheel brace (10), guide threaded bar (8) to notch in spare tire carrier base (10), and tighten wingnut (9) until snug.

(e) Remove utility chain (1) from spare tire (6) and chain fall (4) from support loop (3). Return utility chain (1) and chain fall (4) to tool compartment.

(5) Removal (M934A1/A2).

NOTE

- This procedure requires two personnel.
- Cable and hook will be attached to spare tire during normal vehicle operation.

(a) Push and hold button (14) and remove retaining pin (13) and lift brace (11) up. Reinstall pin (13) with brace (11) in up position.

(b) Turn winch handle (12) counterclockwise to remove tension.

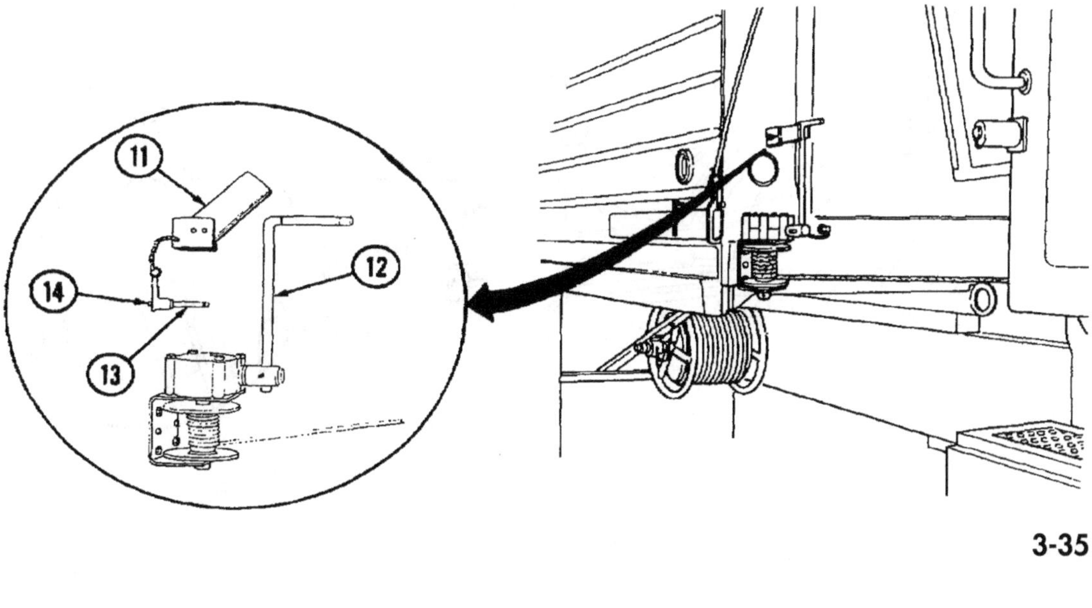
11
14
12
13

3-11. WHEELS AND TIRES (Contd)

(c) Loosen wingnut (7) attaching wheel brace (8) to threaded bar (6).

(d) Guide threaded bar (6) from notch in spare tire carrier (4). Lift wheel brace (3) and threaded bar (6) from spare tire carrier (4).

WARNING

Stand clear during hoisting operations. Injury may result if personnel are struck by a swinging spare tire.

(e) Turn handle (9) clockwise to slightly raise spare tire (5). Swing spare tire (5) and davit boom (2) toward side of vehicle until sufficient clearance is obtained to lower spare tire (5) to ground.

NOTE

Spare tire must be turned flat to vehicle body during lowering operation.

(f) Turn winch handle (9) counterclockwise to lower spare tire (5) to ground.

(g) Remove davit chain (1) from spare tire (5).

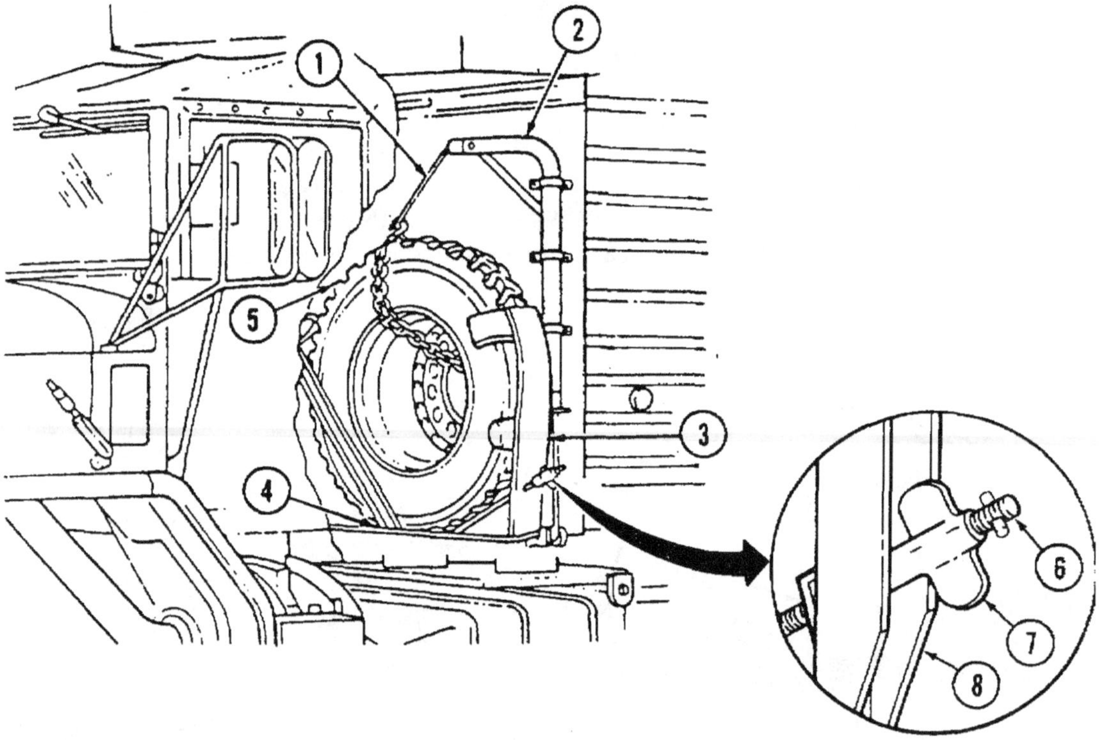
1
2
5
3
4
6
7
8

3-11. WHEELS AND TIRES (Contd)

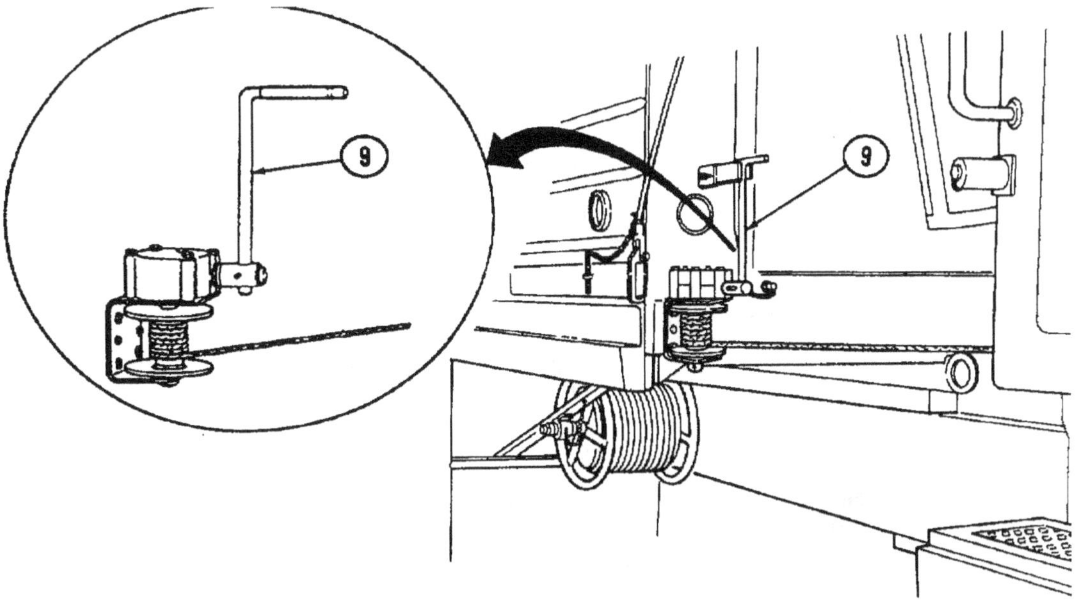
9
9

(h) Reverse steps (a) through (g) as necessary to install spare tire (5).

(6) Removal (M936/A1/A2).

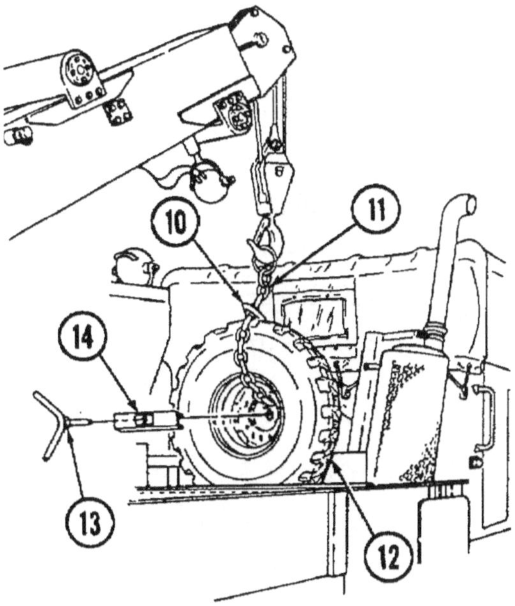
10
11
14
13
12

WARNING

Do not touch hot exhaust system components. Injury to personnel may result.

NOTE

- Utility chain hook must be centered on top of tire to keep tire from lifting.
- This operation requires an assistant to guide spare tire.

(a) Obtain utility chain (11) from tool compartment, install around spare tire (12) with utility chain hook (10) centered on top of spare tire (12), and take up slack.

(b) Remove wingnut (13) and brace (14) securing spare tire (12).

3-11. WHEELS AND TIRES (Contd)

(c) Prepare wrecker for boom (1) operation (para. 2-24).

(d) Position boom (1) with boom hook (2) centered over spare tire (5). Attach boom hook (2) to chain ring (3).

(e) Raise spare tire (5) from spare tire support (6).

WARNING

Keep spare tire from swinging. Injury may occur if personnel are struck by swinging spare tire.

(f) Position boom (1) to right side of vehicle until sufficient clearance is obtained to lower spare tire (5) to ground, and lower spare tire (5).

(g) Remove chain (4) from spare tire (5) and boom hook (2).

(h) Secure wrecker from boom (1) operation (para. 2-24).

(7) Installation (M936/A1/A2).

(a) Install utility chain (4) around spare tire (5) with utility chain hook (7) centered on top of spare tire (5), and take up slack.

(b) Prepare wrecker for boom (1) operation (para. 2-24).

WARNING

Keep spare tire from swinging. Injury may occur if personnel are struck by swinging spare tire.

NOTE

Assistant will guide spare tire to tire support.

(c) Position boom (1) with boom hook (2) centered over spare tire (5). Attach boom hook (2) to chain ring (3).

(d) Raise spare tire (5) and place in spare tire support (6).

(e) Remove chain (4) from spare tire (5) and boom hook (2).

(f) Install brace (9) and wingnut (8) on spare tire (5) and support (6).

(g) Return utility chain (4) to tool compartment.

(h) Secure wrecker from boom (1) operation (para. 2-24).

3-11. WHEELS AND TIRES (Contd)

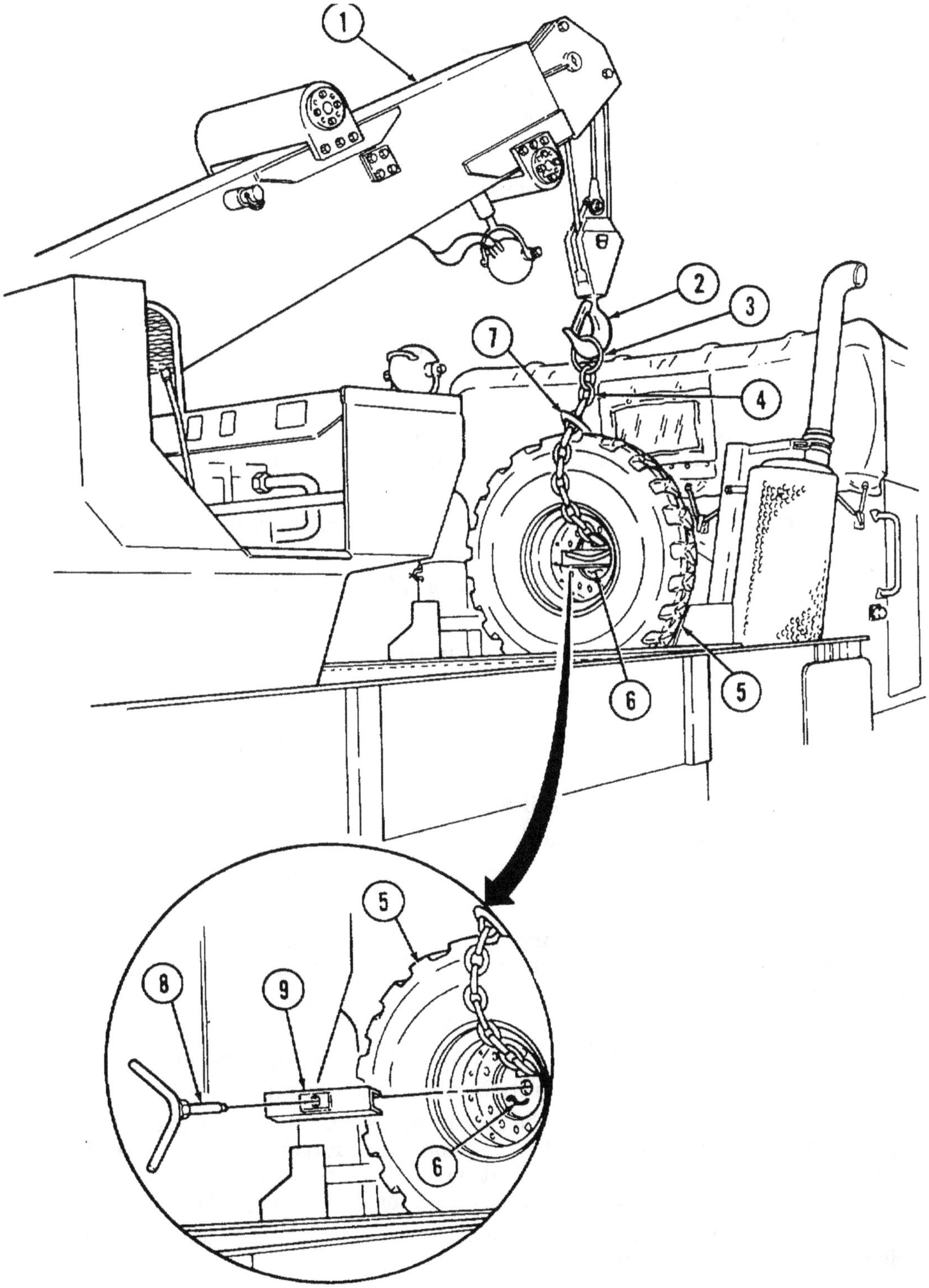
1
2
3
7
4
6
5
5
8
9
6

3-11. WHEELS AND TIRES (Contd)

c. Tire Replacement.

WARNING

Engage parking brake and chock wheels on both sides to keep vehicle from rolling. Failure to do so may result in injury to personnel.

(1) Tire Removal (M939 series vehicles).

(a) Remove hydraulic jack (10), jack handle (12), wheel stud nut wrench (1), and wrench handle (14) from tool compartment. Remove spare tire from storage location.

NOTE

Wheel stud nuts on left side have left-hand threads. Those on right side have right-hand threads. Studs and nuts are marked L and R accordingly.

(b) Install wrench handle (14) through wheel stud nut wrench (1) and position wrench (1) on wheel stud nuts (2). Loosen ten wheel stud nuts (2) but do not remove.

(c) Turn jack screw (9) of jack (10) out approximately 3 in. (7.6 cm) by hand.

(d) Turn valve (11) at base of jack (10) by turning clockwise with slotted end of jack handle (12) until closed securely.

WARNING

Do not work under vehicle supported by jack only. Jack may slip causing vehicle to fall, which will result in injury or death.

NOTE

- Place a block under jack if used on soft terrain.
- Use jack stands if available.
- Expansible van bodies must be retracted on M934 models before jacking vehicle up.

(e) Position jack (10) under axle housing (8) near tire to be removed. Insert jack handle (12) into jack (10). Move jack handle (12) up and down until tire is off the ground.

NOTE

Use special second wheel stud wrench NSN 5120-00-378-4411 to hold inner adapter spacer nuts in place when braking outer lug nut.

(f) If rear dual tire is to be replaced, remove ten wheel stud nuts (2) from adapter spacer nuts (4) and outer tire (3) from axle hub (13). If rear inner wheel (6) is to be removed, reverse wheel stud nut wrench (1), remove handle (14), and install near large end of wrench (1). Remove ten adapter spacer nuts (4) and inner tire (6).

(g) If front tire is to be replaced, remove ten wheel stud nuts (2) from studs (7) and tire (6) from hub (13).

3-11. WHEELS AND TIRES (Contd)

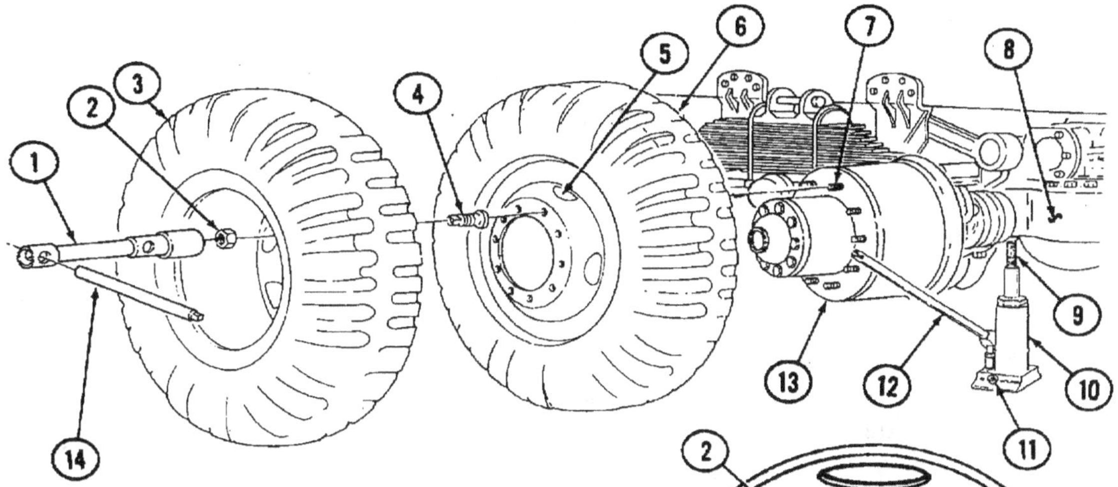
1
2
3
4
5
6
7
8
9
10
11
12
13
14

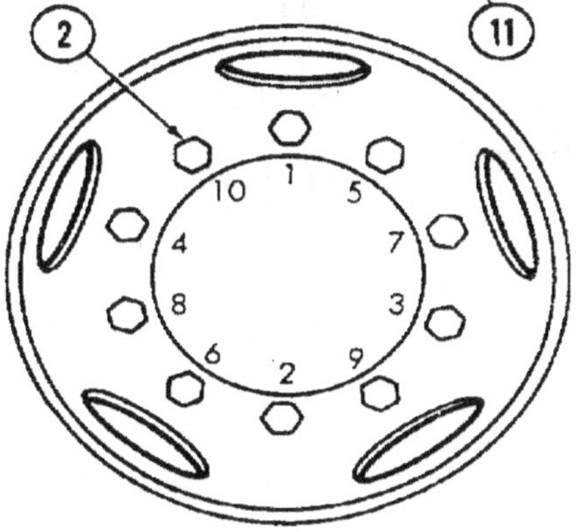
2
1
2
3
4
5
6
7
8
9
10

TIGHTENING SEQUENCE

(2) Tire Installation (M939 series vehicles).

NOTE

- Use jack handle as pry bar to raise tire over wheel studs.
- Have all nuts torqued by unit maintenance as soon as possible.
- Return unserviceable wheel and tire to unit maintenance for repair, replacement, or exchange.

(a) If installing rear inner dual tire (6), position tire (6) on axle hub (13) shallow side out, and install ten adapter spacer nuts (4) on studs (7). Tighten securely in sequence shown, and install outer dual tire (3).

(b) If installing a rear outer dual tire (3), align valve stem (5) with ventilation holes, position tire (3) on axle hub (13) with deep side out, and install ten wheel stud nuts (2) on adapter spacer nuts (4), and tighten wheel stud nuts (2) securely in sequence shown.

(c) Installation of front tire is accomplished by following step (a) except that ten wheel stud nuts (2) are connected directly to studs (7).

(d) Turn valve (11) at base of jack (10) counterclockwise with slotted end of jack handle (12) to lower vehicle tire to ground. Remove jack (10) from under axle housing (8).

(e) Return jack (10), jack handle (12), wheel stud nut wrench (1), and wrench handle (14) to tool compartment and remove wheel chocks.

(f) Secure damaged tire in spare tire carrier (para. 3-11b.).

3-11. WHEELS AND TIRES (Contd)

(3) Tire Removal (M939A1 series vehicles).

WARNING

Engage parking brake and chock wheels on both sides. Failure to do so may result in injury to personnel.

(a) Remove hydraulic jack (9), jack handle (7), wheel stud nut wrench (2), and wrench handle (1) from tool compartment. Remove spare tire from storage location.

NOTE

- This procedure requires two personnel.
- Wheel stud nuts on left side have left-hand threads. Those on right side have right-hand threads. Studs and nuts are marked L and R accordingly.

(b) Install wrench handle (1) through wheel stud nut wrench (2) and position wheel stud nut wrench (2) on wheel stud nuts (3). Loosen ten wheel stud nuts (3) but do not remove.

(c) Turn jack screw (10) of jack (9) out approximately 3 in. (7.6 cm).

(d) Turn valve (8) at base of jack (9) by turning clockwise with slotted end of jack handle (7) until closed securely.

WARNING

Do not work under vehicle supported by jack only. Jack may slip causing vehicle to fall, which will result in injury or death.

NOTE

- Place a block under jack if used on soft terrain.
- Use jack stands if available.
- Expansible van bodies must be retracted on M934A1 model trucks before jacking vehicle up.

(e) Position jack (9) under axle housing (6) near tire (4) or (12) to be removed. Insert handle (7) into jack (9). Move jack handle (7) up and down until tire (4) or (12) is off the ground.

(f) Remove ten wheel stud nuts (3) from studs (11) and tire (4) or (12) from hub (5).

3-11. WHEELS AND TIRES (Contd)

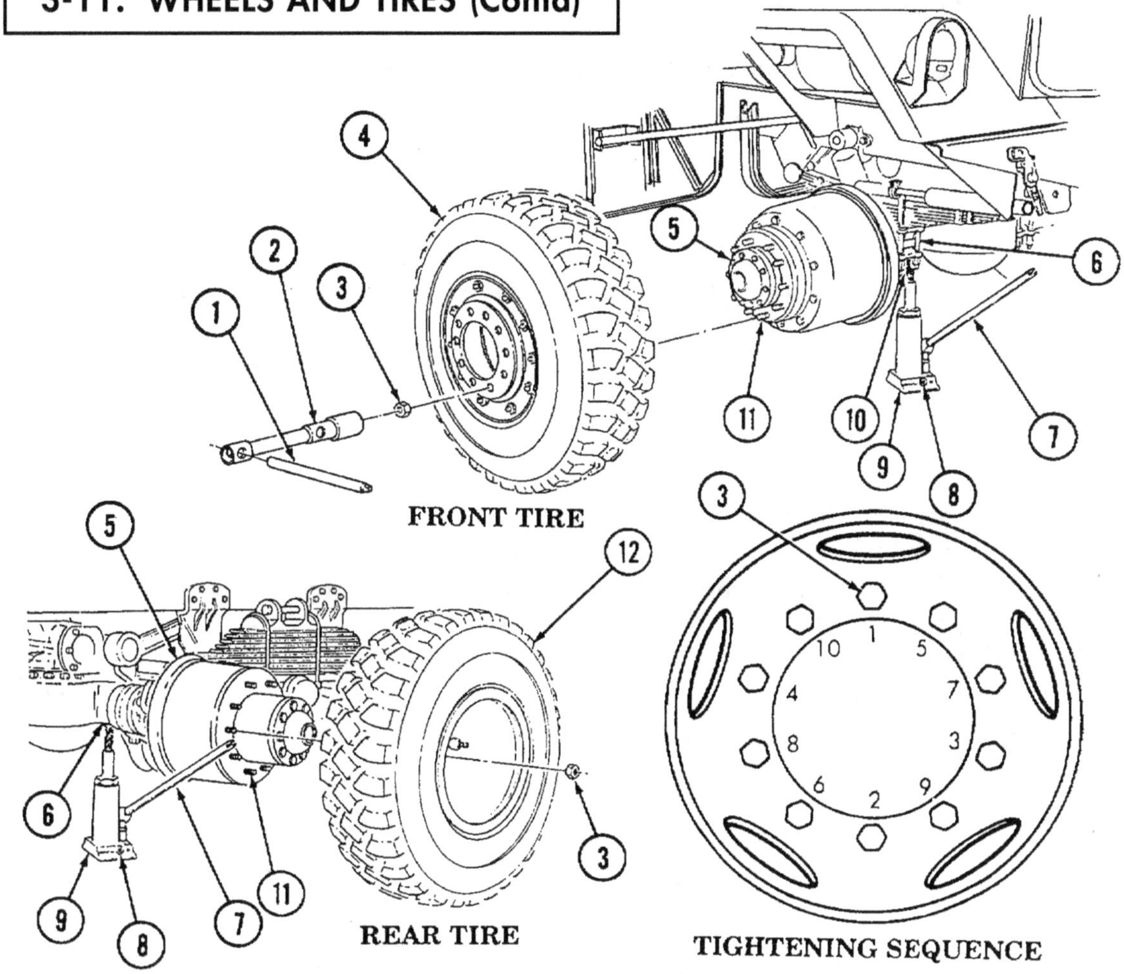
FRONT TIRE
REAR TIRE
TIGHTENING SEQUENCE

(4) Tire Installation (M939A1 series vehicles).

NOTE

- Use jack handle as pry bar to raise tire over wheel studs.
- Have all nuts torqued by unit maintenance as soon as possible.
- Return unserviceable wheel and tire to unit maintenance for repair, replacement, or exchange.

(a) Position tire (4) or (12) on axle hub (5) over wheel studs (11), front tire (4) is mounted shallow side out and rear tire (12) is mounted deep side out. Tighten wheel stud nuts (3) securely in sequence shown.

(b) Turn valve (8) at base of jack (9) counterclockwise with slotted end of jack handle (7) to lower vehicle tire to ground. Remove jack (9) from under axle housing (6).

(c) Return jack (9), jack handle (7), wheel stud nut wrench (2), and wrench handle (1) to tool compartment and remove wheel chocks.

(d) Secure damaged tire in spare tire carrier (para. 3-11b.).

3-11. WHEELS AND TIRES (Contd)

(5) Front Tire Removal (M939A2 series vehicles).

(a) Remove CTIS tools from tool compartment.

(b) Remove spare tire from vehicle.

WARNING

- Engage parking brake and chock wheels on both sides to keep vehicle from rolling. Failure to do so may result in injury to personnel.
- Air in system is under pressure. Make sure engine is shut down and air reservoirs are drained before disconnecting CTIS components to prevent serious injury to personnel.

NOTE

Temporarily store CTIS assembly removed during this operation on spare tire while removing damaged tire to prevent loss of critical parts.

(c) Remove valve core cap (2) and valve core (3) from tank valve (7) to exhaust air pressure from tire (19) and install valve core (3) securely back in tank valve (7). Install valve core cap (2) on valve core (3).

(d) Remove two nuts (26) and washers (25) from rim studs (8) and (12).

(e) Remove two screws (28), washers (27), shield (1), and spacer (24) from hub (14).

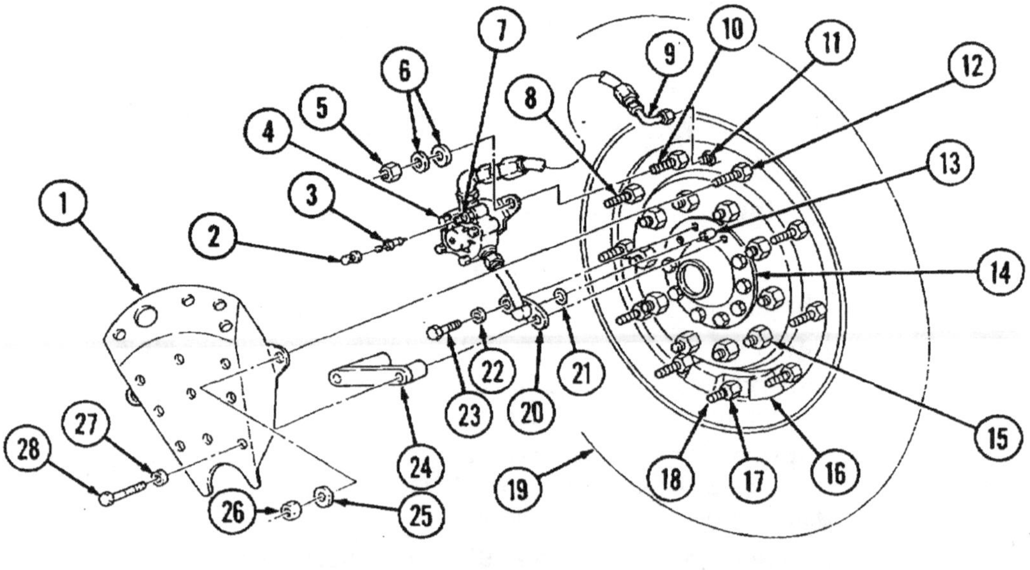
1
2
3
4
5
6
7
8
9
10
11
12
13
14
15
16
17
18
19
20
21
22
23
24
25
26
27
28

(f) Remove nut (5) and two washers (6) from rim stud (10) and wheel valve (4).

(g) Disconnect hose assembly (9) from turret valve (11).

(h) On spare tire (30), remove valve core cap (32) and valve core (31) from turret valve (29).

3-11. WHEELS AND TIRES (Contd)

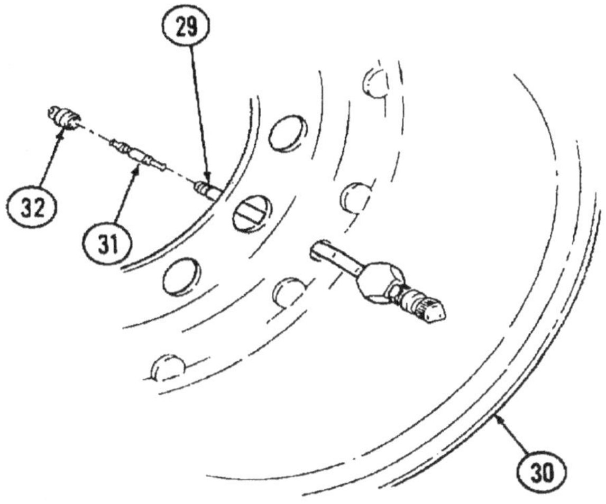
29
32
31
30

CAUTION

Do not damage or lose O-ring when removing manifold. Damage or loss of O-ring will result in loss of CTIS pressure and damage to equipment.

(i) Remove screw (23), washer (22), and O-ring (21) from manifold (20).

(j) Remove hose assembly (9), wheel valve (4), and manifold (20) from manifold tube (13) as a complete assembly.

(k) Remove hydraulic jack (36), handle (34), wheel stud nut wrench (39), and wrench handle (40) from tool compartment.

WARNING

Do not work under vehicle supported by jack only. Jack may slip causing vehicle to fall, which will result in injury or death.

NOTE

- This procedure requires two personnel.
- Place a block under jack if used on soft terrain.
- Use jack stands if available.
- Expansible van bodies must be retracted on M934A2 models before jacking vehicle up.
- Wheel stud nuts on left side have left-hand threads. Those on right side have right-hand threads. Stud and nuts are marked L and R accordingly.

(l) Install wrench handle (40) through wheel stud nut wrench (39), loosen ten wheel stud nuts (15) and rimnut (17), but do not remove.

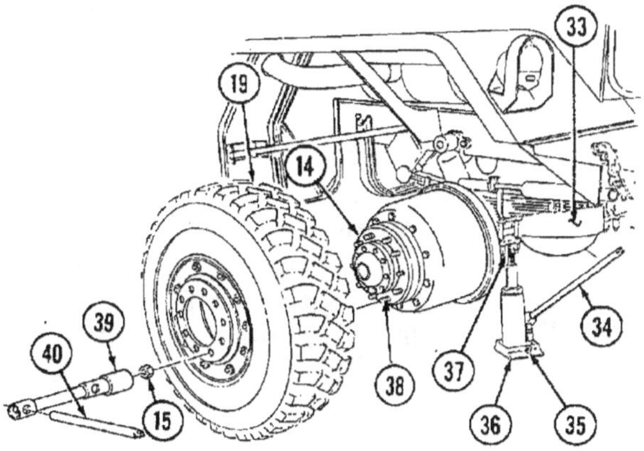
33
19
14
39
40
15
38
37
34
36
35

(m) Turn jack screw (37) out approximately 3 in. (7.6 cm). Position jack (36) under axle housing (33), close valve (35), and jack up vehicle until tire is off the ground.

(n) Remove ten wheel stud nuts (15) from wheel studs (38) and remove tire (19).

(o) Remove rimnut (17) and counterweight (16) from rim stud (18). Replace rimnut back on rim stud (18).

(p) Remove rimnut (17) from spare tire (30) and install counterweight (16) on rim stud (18) of spare tire (30) with rimnut (17).

3-11. WHEELS AND TIRES (Contd)

(6) Front Tire Installation (M939A2 series vehicles)

NOTE

Use jack handle to lift tire over hub and stud.

(a) Install spare tire (4) on hub (5) with shallow side out over hub (5) so that turret valve (22) and manifold tube (24) are aligned.

(b) Install ten wheel stud nuts (3) on wheel studs (11) and tighten until tire (4) is against face of hub (5) using the wheel stud wrench (2) and wrench handle (1).

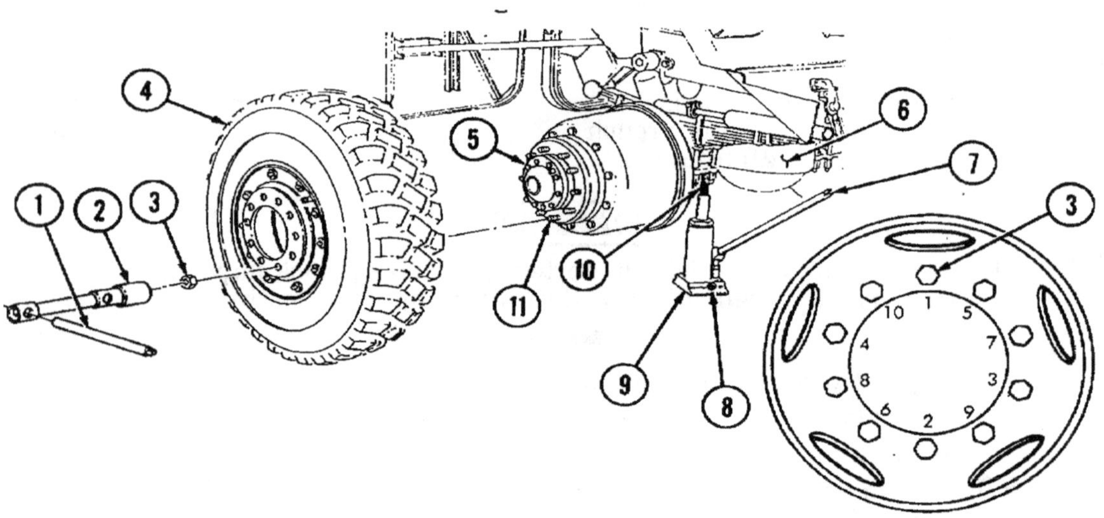
4
5
6
7
1
2
3
3
10
11
9
8
10
1
5
4
7
8
3
6
2
9

TIGHTENING SEQUENCE

(c) Turn valve (8) at the base of the jack (9) counterclockwise with slotted end of jack handle (7) to lower tire (4) to ground, and remove jack (9) from under axle (6) and turn jack screw (10) back into jack (9).

(d) Tighten ten wheel stud nuts (3) securely in sequence shown. Tighten rimnut (26) on counterweight (25) and rim stud (27).

NOTE

- Ensure O-ring seal is on manifold tube before installing manifold.
- Ensure valve core has been removed from turret valve when installing hose assembly.

(e) Install hose assembly (20), wheel valve (15), and manifold (29) as a complete assembly. With manifold (29) and O-ring (28) over manifold tube (24), install wheel valve (15) onto rim studs (19) and (21), and hose assembly (20) to turret valve (22).

3-11. WHEELS AND TIRES (Contd)

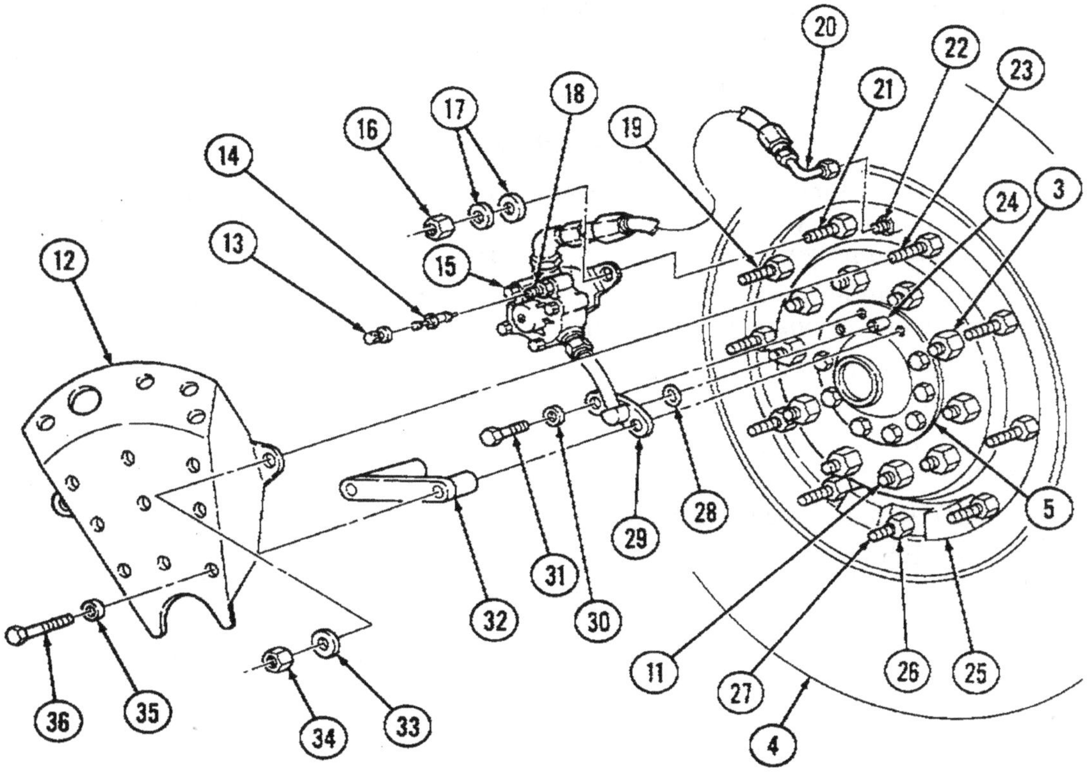

20
22
23
21
18
17
16
19
14
24
3
13
15
12
5
28
29
31
32
30
11
26
25
27
4
36
35
34
33

(f) Install nut (16) and two washers (17) on rim stud (21).

(g) Install screw (31) and washer (30) in manifold (29) and hub (5).

(h) Install one of two screws (36), washer (35), shield (12), and short end of spacer (32), on manifold (29) and hub (5), and the second of two screws (36), washer (35), shield (12), and long end of spacer (32) on hub (5).

(i) Install two nuts (34) and washers (33) on shield (12) and rim studs (19) and (23).

(j) Tighten valve core (14) in tank valve (18) and install valve cap (13).

(k) Start vehicle and select desired CTIS mission mode.

NOTE

- Check replaced tire to ensure tire inflates automatically.
- Report tire change to unit maintenance as soon as possible. Most screws will require torquing to specific limits and O-ring seal will have to be replaced.

(l) Return jack (9), jack handle (7), wheel stud nut wrench (2), wrench handle (1), and CTIS tools to tool compartment and remove chocks.

(m) Secure damaged tire in spare tire carrier (para. 3-11b.).

3-11. WHEELS AND TIRES (Contd)

(7) Rear Tire Removal (M939A2 series vehicles).

(a) Remove CTIS tools from tool compartment.

(b) Remove spare tire from vehicle.

WARNING

- Air in system is under pressure. Make sure engine is shut down and air reservoirs are drained before disconnecting CTIS components to prevent serious injury to personnel.
- Air is under pressure and creates danger to eyes. Shield eyes to prevent serious personal injury.
- Ensure brake is set and vehicle is properly chocked. Failure to do so may result in serious injury to personnel.

(c) Remove valve core cap (12) and valve core (13) from tank valve (14) to exhaust all air from tire (16). Replace valve core (13) and valve core cap (12).

(d) Remove hose assembly (11) from turret valve (15).

(e) Remove valve cap (18) and valve core (19) from turret valve (20) of spare tire (21) to exhaust air. Install valve core (19) and valve cap (18) from spare tire (21) in turret valve (15) of damaged tire (16).

(f) Remove two locknuts (8) and washers (7) holding shield (6) on wheel valve (5).

(g) Remove two screws (10), washers (9), and shield (6) from hub cap (17) and wheel valve (5). Replace two screws (10) and washers (9) in hub cap (17) and hand-tighten.

(h) Loosen adapter nut (4) until wheel valve (5) can be removed.

(i) Remove wheel valve (5) and hose assembly (11) as a complete assembly from adapter nut (4).

(j) Before removing adapter, clean surface of hub body of dirt and foreign material which could clog CTIS air passages. Failure to do so will result in improper inflation or damage to wheel valve.

CAUTION

- Temporarily seal hole in hub body by rapping electrical tape around hub body at least twice, ensuring that tape completely covers hole. Failure to do so may result in introduction of dirt or foreign material into critical CTIS components.
- Temporarily reattach adapter, O-ring, and washer to wheel valve and rap with electrical tape to seal end and hold O-ring and washer in place.

(k) Clean area around adapter.

(l) Reinstall parts and tape together.

(m) Remove adapter nut (4), O-ring (3), and washer (2) from hub body (1). Seal hole in hub body (1) with electrical tape.

3-11. WHEELS AND TIRES (Contd)

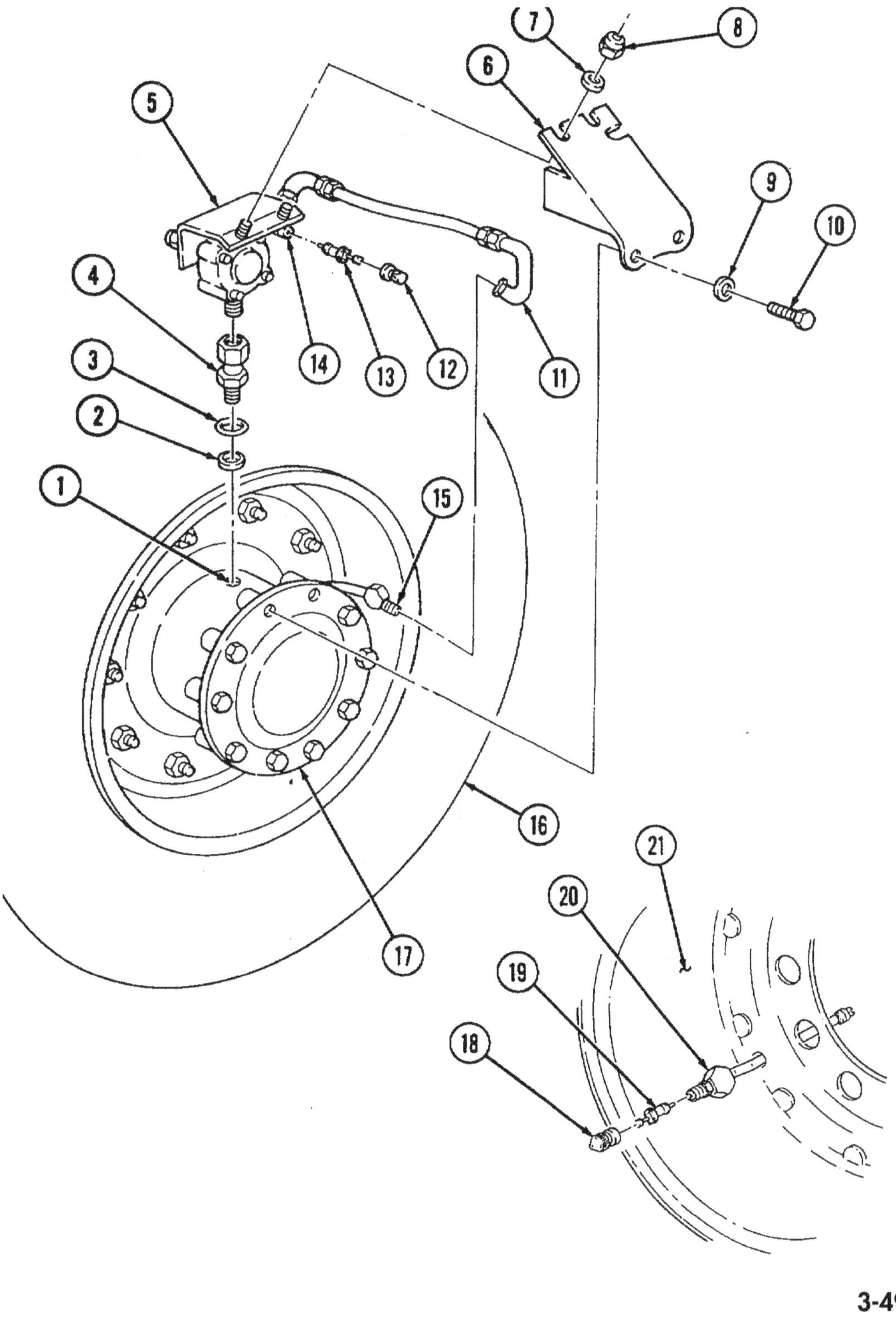
1
2
3
4
5
6
7
8
9
10
11
12
13
14
15
16
17
18
19
20
21

3-11. WHEELS AND TIRES (Contd)

(n) Remove hydraulic jack (9), jack handle (7), wheel stud nut wrench (2), and wrench handle (1) from tool compartment.

WARNING

Do not work under vehicle supported by jack only. Jack may slip causing vehicle to fall. Failure to do so will result in injury or death.

(o) Install wrench handle (1) through hole in side of small end of wheel stud nut wrench (2) and use wheel stud nut wrench (2) to loosen ten wheel stud nuts (5), but do not remove.

(p) Turn jack screw (10) of jack (9) out approximately 3 in. (7.6 cm).

(q) Turn valve (8) at base of jack (9) by turning clockwise with slotted end of jack handle (7) until closed securely.

NOTE

- This procedure requires two personnel.
- Place a block under jack if used on soft terrain.
- Use jack stands if available.
- Expansible van bodies must be retracted on M934A2 models before jacking vehicle up.
- Wheel stud nuts on left side have left-hand threads. Those on right side have right-hand threads. Stud and nuts are marked L and R accordingly.

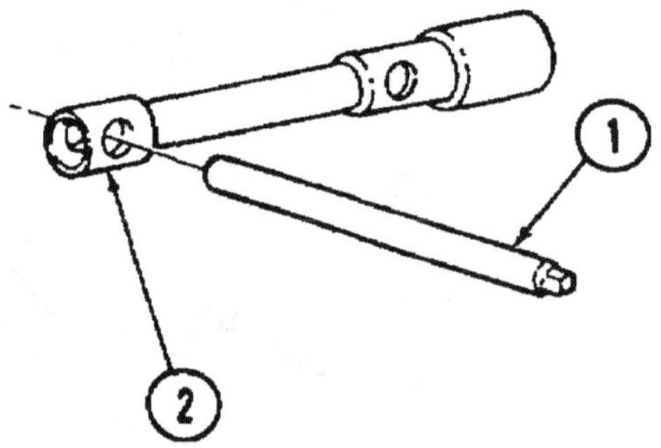
1
2

(o) Position jack (9) under axle housing (11) near tire (4) that is damaged. Insert jack handle (7) into jack (9), and move jack handle (7) up and down until tire (4) is off the ground.

(p) Remove ten wheel stud nuts (5) from wheel studs (6) and tire (4) from hub (3).

3-11. WHEELS AND TIRES (Contd)

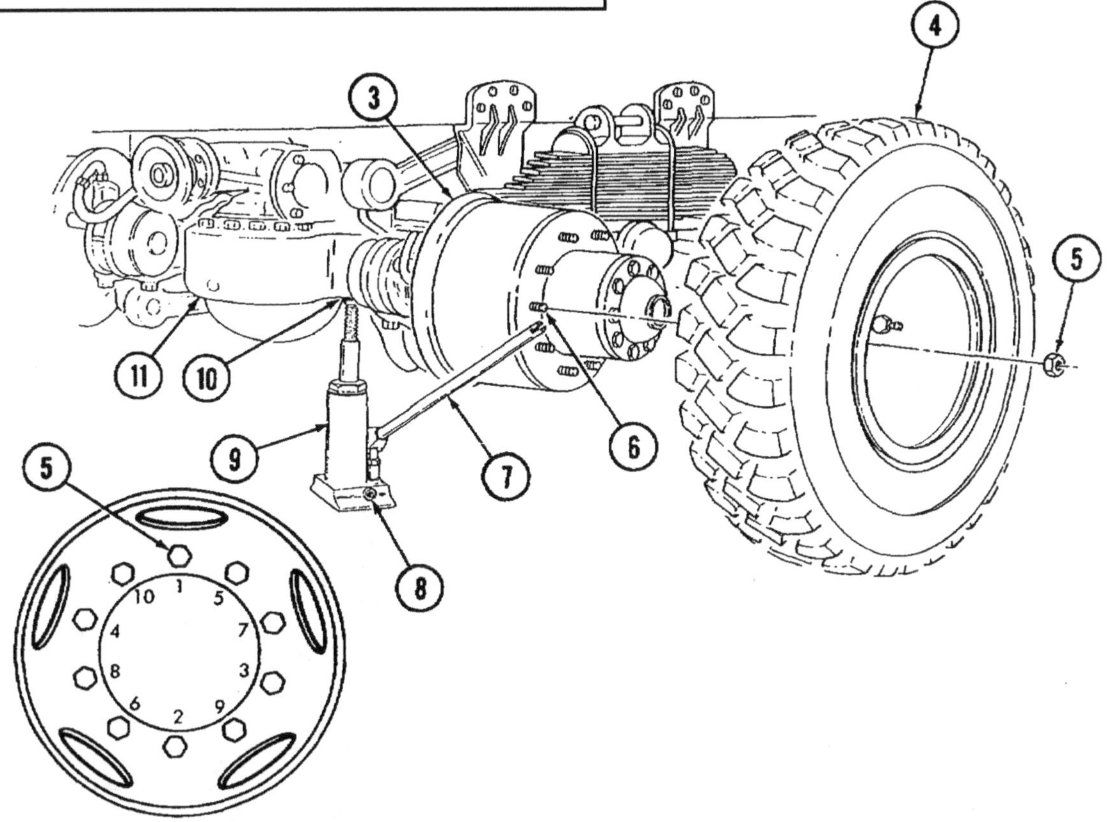

4
3
5
11
10
9
7
6
8
5
1
2
3
4
5
6
7
8
9
10

TIGHTENING SEQUENCE

(8) **Rear Tire Installation (M939A2 series vehicles).**

NOTE

- Use jack handle to lift tire over hub and studs.
- Mount tire on hub and studs with deep side out.
- Ensure hole in hub body is centered between third and fourth stud hole (counterclockwise from the turret valve) on the spare.

(a) Install tire (4), deep side out, over hub (3) and wheel studs (6).

(b) Install ten wheel stud nuts (5) on wheel studs (6) and tighten until tire (4) is against the face of the hub (3) using wheel stud wrench (2).

(c) Turn valve (8) at base of jack (9) counterclockwise with slotted end of jack handle (7) to lower tire (4) to ground and remove jack (9) from under axle housing (11).

(d) Tighten ten wheel stud nuts (5) securely in sequence shown.

(e) Return jack (9), jack handle (7), wheel stud nut wrench (2), and wrench handle (1) to tool compartment and remove chocks.

3-11. WHEELS AND TIRES (Contd)

(f) Remove tape from hub body (1). Remove adapter (4), O-ring (3), and washer (2) from temporary storage on the wheel valve (5), and install in hole in hub body (1).

(g) Install hose assembly (11) on turret valve (16) and hand-tighten.

(h) Install wheel valve (5) on adapter (4). Tighten adapter nut (15) and hose assembly (11).

(i) Install shield (6) on wheel valve (5) with two nuts (8) and washers (7), and on hub cap (18) with two screws (10) and washers (9).

(j) Start vehicle and select desired CTIS mission mode.

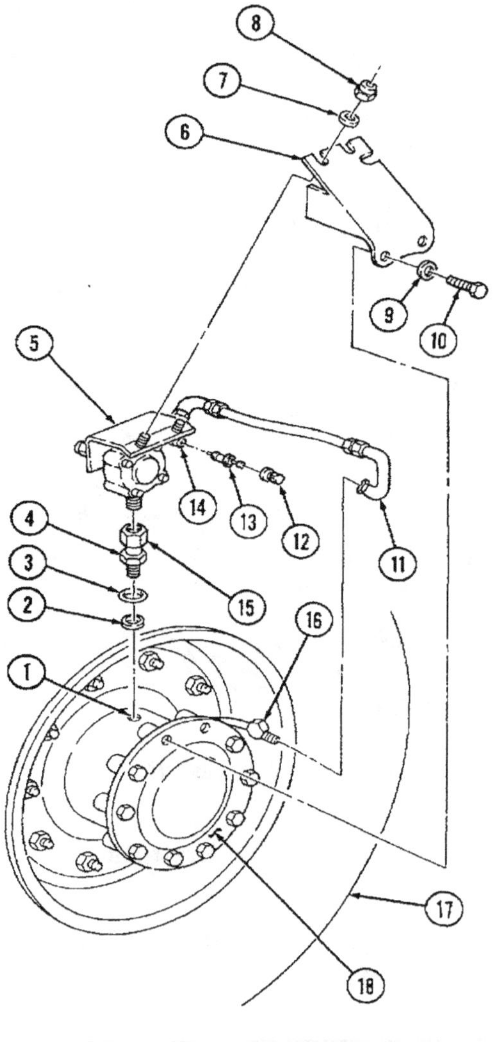

8
7
6
9
10
5
14
13
12
11
4
3
2
15
16
1
17
18

NOTE

- Check replaced tire to ensure tire inflates automatically.
- Report tire change to unit maintenance as soon possible. Most screws will require torquing to specific limits, and O-ring seal will have to be replaced.

(k) Return CTIS tools to tool compartment.

(l) Secure damaged tire in spare tire carrier.

d. Tire Inflation.

(1) General. Never decrease pressure of warm tires (17) except for operation in snow, sand, or mud (para. 2-33, 2-35, or 2-37). After operations are completed, reinflate tires (17) to recommended pressure (refer to table 1-10).

3-11. WHEELS AND TIRES (Contd)

NOTE

Chock wheels if necessary.

(2) Tire Gauging and Inflation.

(a) Remove valve cap (12) from valve stem (14) for access to valve core (13).

(b) Remove tire inflation gauge and hose assembly (19) from tool compartment.

(c) Start engine and engage parking brake. Make sure air reservoir pressure is higher than recommended tire pressure by checking primary (20) and secondary (21) air pressure gauges on instrument panel (22).

(d) Remove emergency air coupling cover (25). Install tire inflation hose assembly coupling (26) on left front emergency air coupling (23) to inflate front tires, and on right rear emergency air coupling for rear tires.

(e) Align air valve handle (24) at air coupling (23) with piping to release compressed air into tire inflation gauge and hose assembly (19).

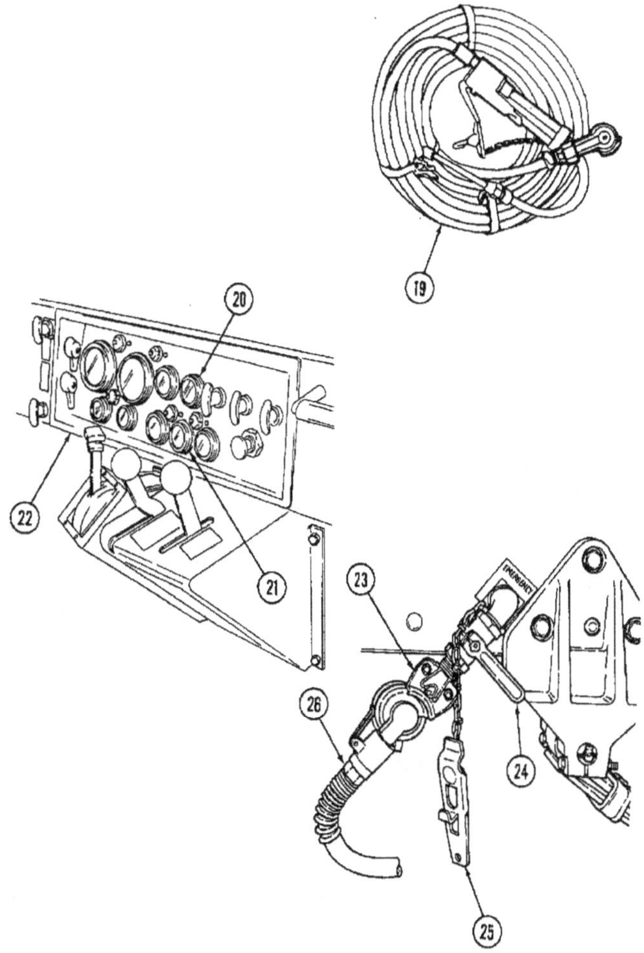
19
20
22
21
23
26
24
25

3-11. WHEELS AND TIRES (Contd)

(f) Start at one corner of vehicle to gauge and adjust pressure, as necessary, of all tires (6). Remove tire valve cap (1), apply tire gauge air chuck (7) on tire valve stem (2) and press down firmly to read tire pressure on gauge dial (5). Press air chuck lever (4) to inflate tire (6) as necessary. Release air chuck lever (4) momentarily to read pressure on gauge dial (5).

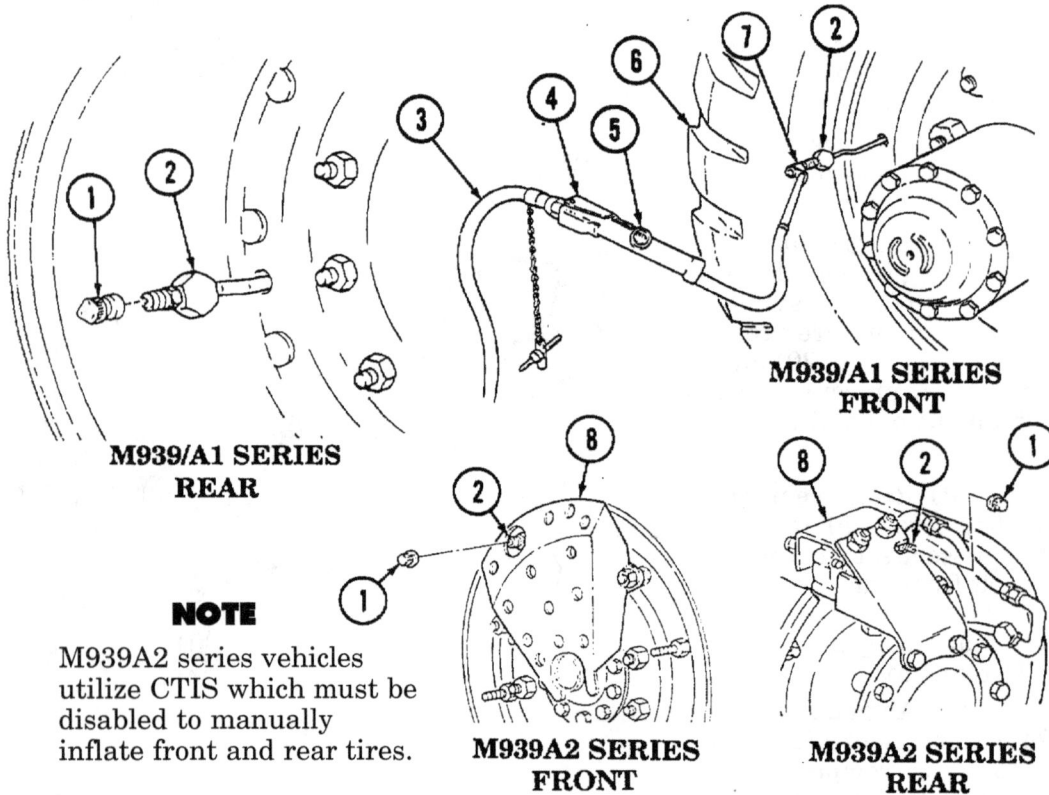

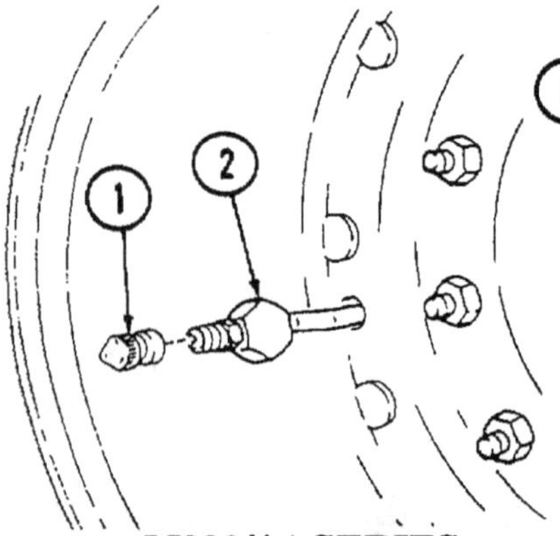
1
2

M939/A1 SERIES REAR

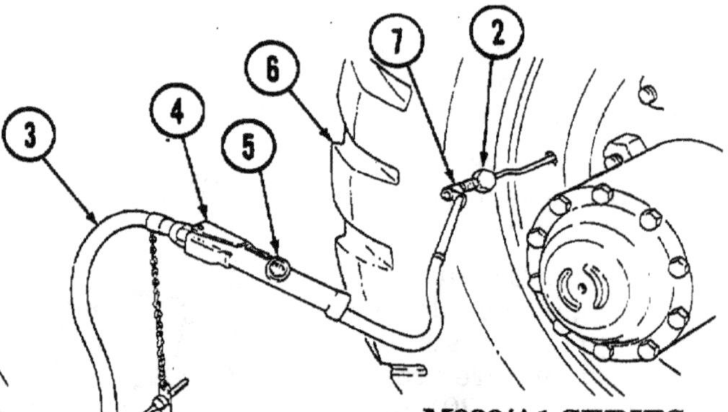
7
2
6
4
3
5

M939/A1 SERIES FRONT

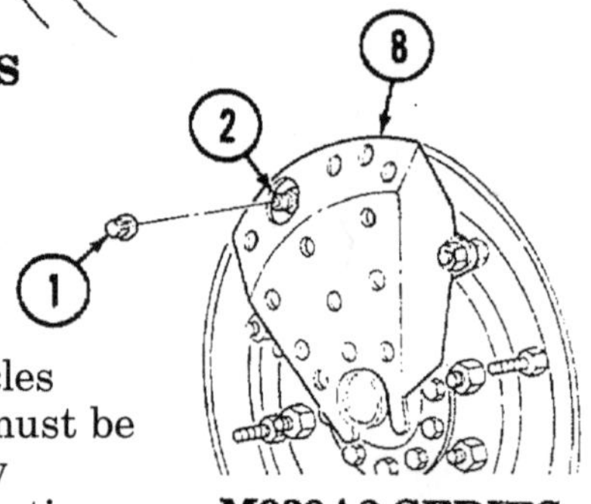
8
2
1

M939A2 SERIES FRONT

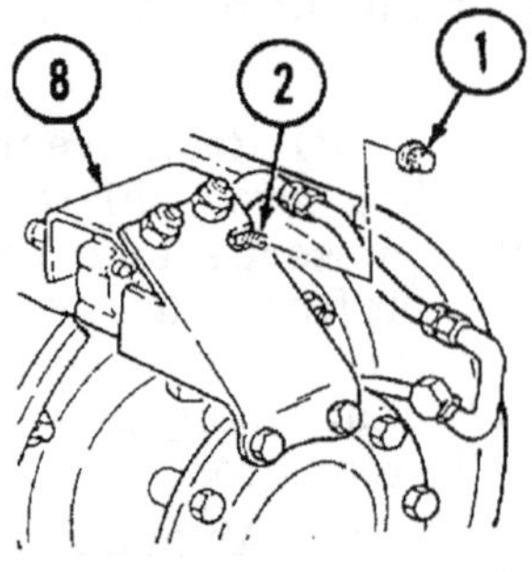
8
2
1

M939A2 SERIES REAR

NOTE

M939A2 series vehicles utilize CTIS which must be disabled to manually inflate front and rear tires.

(g) On M939A2 series vehicles, remove valve cap (1) from the valve stem (2) through CTIS cover access (8).

(3) Remove tire gauge air chuck (7) from tire valve stem (2) when tire pressure is adjusted to recommended inflation pressures in table 1-10. Install tire valve cap (1) and tighten finger-tight.

(4) When tire inflation operation is completed, turn air valve handle (10) crosswise to piping. Uncouple gauge and hose assembly (3) from air coupling (9), and install cover (11) on air coupling (9).

(5) Return tire inflation gauge and hose assembly (3) to tool compartment.

NOTE

Remove wheel chocks if used.

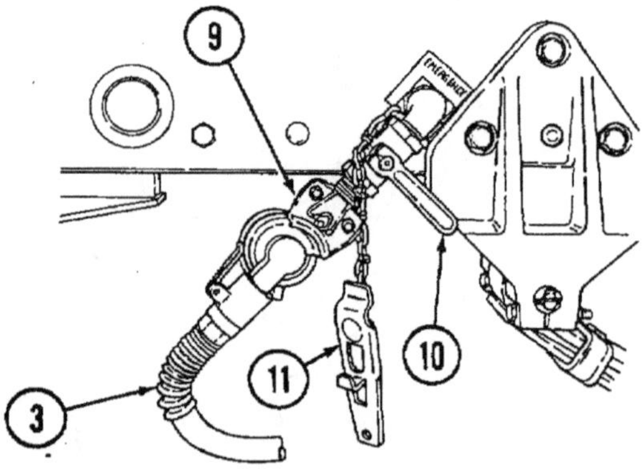
9
11
10
3

3-12. AIR SYSTEM SHUTOFF VALVES

General. Primary and secondary air systems are provided with shutoff valves to isolate them from the rest of the vehicle air system. These valves are closed by the operator only in an emergency situation.

(1) The primary system air shutoff valve (1) on the M939/A1 series vehicles is located at the forward end of the wet tank. On the M939A2 series vehicles, it is at the side of the wet tank reservoir. If air leaks in the primary system and the vehicle must be driven, close air shutoff valve (1). Closing the primary system air shutoff valve (1) will isolate the primary system and maintain air pressure to brake system components and enable operator to slow down or stop the vehicle.

(2) The secondary system air shutoff valve (2) is located on top of the wet tank reservoir. In the event of an air leak in the secondary system and the vehicle must be driven, close air shutoff valve (2). Closing the secondary system and air shutoff valve (2) will isolate the secondary system and maintain air pressure to brake system components and enable operator to slow down or stop the vehicle.

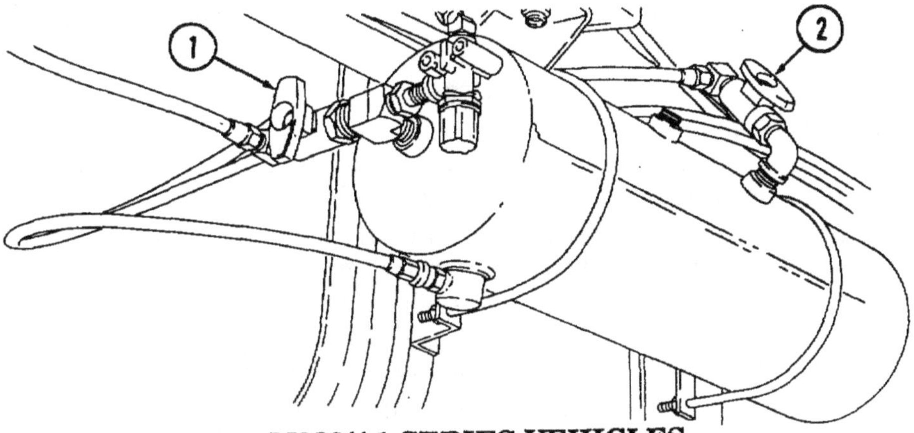
1
2

M939/A1 SERIES VEHICLES

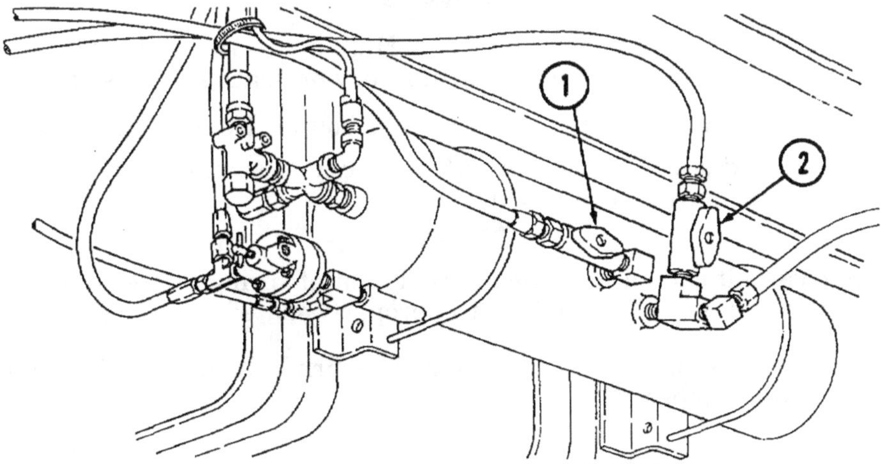
1
2

M939A2 SERIES VEHICLES

3-13. SPRING BRAKE SERVICE

a. General. Spring brakes on M939/A1/A2 series vehicles lock automatically and stop the vehicle whenever a large loss of air pressure occurs. Before the vehicle can be towed away for repairs, the spring in each spring brake must be manually released. This procedure is performed by operator only in an emergency.

b. Applying Spring Brake. Switches located on the emergency brake and PTO levers reduce the air pressure to the spring brake system, resulting in automatic activation of spring brakes. Testing of these switches can be accomplished by manually lifting switch and increasing engine speed while in drive.

c. Spring Brake Override. An air-actuation switch on the dash supplies normal air pressure to spring brakes preventing them from engaging.

d. Releasing the Spring (Caging Brakes).

WARNING

Make sure vehicle parking brake is engaged and wheels are chocked before releasing springs in spring brakes. Failure to do so will result in vehicle rolling out of control, which may cause injury or death.

NOTE

- Do not lose rubber plugs. After removal, store all four plugs in map compartment inside cab. Be sure to notify unit maintenance of where they are stored so they may be reinstalled later.
- If inside of brake chamber is clogged with mud, sand, or dirt, do not proceed with spring release operation unless the chamber can be cleared. Notify unit maintenance if chamber cannot be cleared.

(1) Remove rubber plug (1) from spring brake chamber (2).

(2) Visually inspect spring brake chamber (2) for mud, sand, or dirt.

WARNING

Do not remove rim clamp bolt or nut. High pressure inside of spring brake chamber may result in injury or death if released.

(3) Remove nut (6), washer (5), and release bolt (4) from storage housing (3).

(4) Insert T-end of release bolt (4) all the way into spring brake chamber (2) and turn bolt (4) one-quarter turn clockwise.

NOTE

If release bolt cannot be pulled directly out of spring brake chamber after it has been turned, bolt is properly seated.

(5) Pull on release bolt (4) to make sure it is firmly holding spring plate within the spring brake chamber (2).

3-13. SPRING BRAKE SERVICE (Contd)

(6) Install nut (6) and washer (5) on end of release bolt (4). Tighten nut (6) down until it is in contact with spring brake chamber (2).

NOTE

Measurement can be accomplished using the paper ruler in the back of this manual.

CAUTION

Measurement of release bolt from nut must be at least 3 in. (7.6 cm). If distance is not correct, damage to brakes will result.

(7) Turn nut (6) clockwise until at least 3 in. (7.6 cm) of threads of release bolt (4) extends out of nut (6).

(8) Reverse steps 1 through 7 to remove manual release of spring brake.

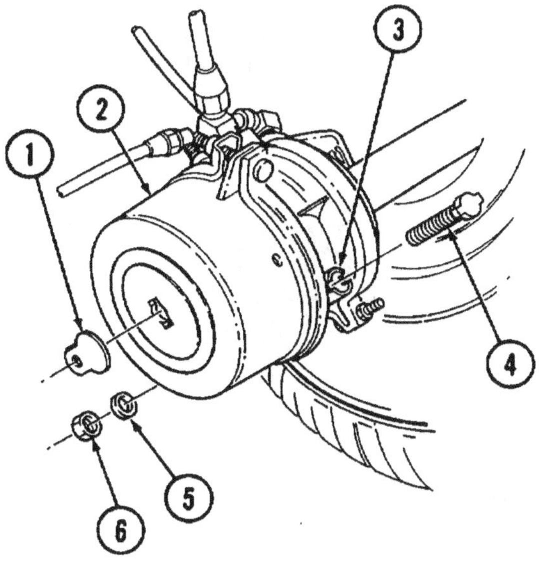
3
2
1
4
5
6

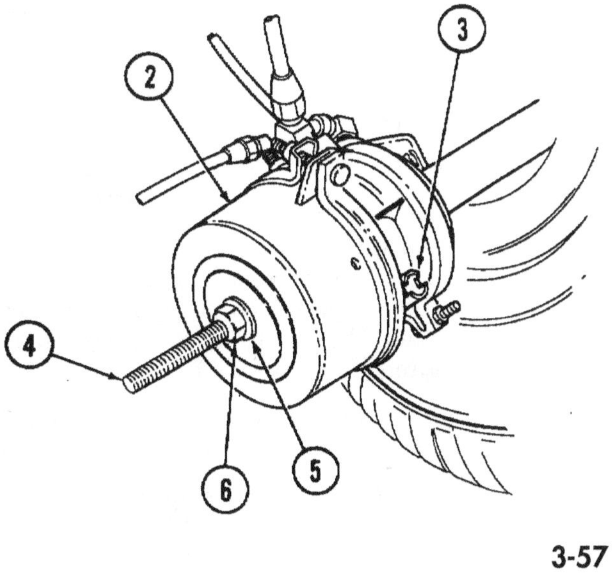
3
2
4
6
5

3-14. RADIATOR FAN CLUTCH EMERGENCY SERVICE

a. General. The radiator fan on M939/A1 series vehicles normally activates when the engine coolant temperature exceeds 185°F (85°C), which is within the normal operating range of 175°-195°F (79°-91°C). It is possible, however, for the thermostat governing fan operation to become damaged. This will result in engine overheating. In an emergency, the operator can bypass the fan thermostat by bolting the fan to the engine's fan clutch assembly. This procedure is performed by the operator only when service by unit maintenance is not available.

b. Symptoms.

(1) Engine coolant temperature exceeds 195°F (91°C) as indicated by engine coolant temperature gauge (3).

(2) Operator has stopped vehicle and allowed engine to idle as described in paragraph 2-16.

(3) Engine coolant temperature gauge (3) indicated engine is not cooling, or engine coolant temperature continues to rise after two-minute idle period.

c. Inspection. With engine continuing to idle, raise hood and inspect fan (4) for operation. If fan is not turning, fan clutch thermostat is damaged. Operator must immediately shut down engine and notify unit maintenance. If service is not available, perform step d., emergency service.

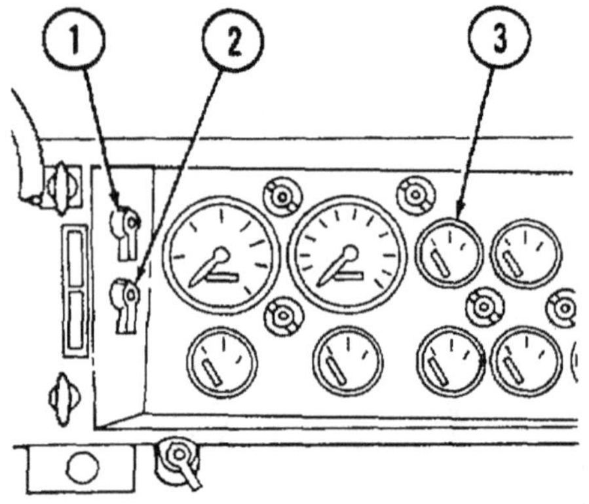
1
2
3

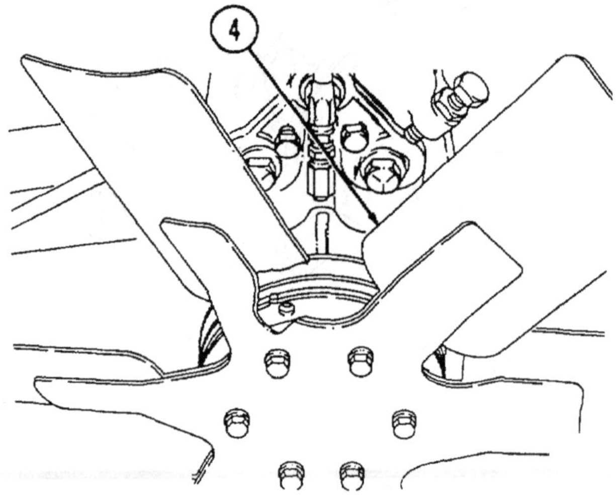
4

d. Emergency Service.

WARNING

- Make sure battery switch and ignition switch are OFF. Make sure crewmembers inside vehicle cab are aware of danger in engaging these switches while emergency service is being performed. Failure to do so may result in fan blade suddenly engaging and cause injury or death.
- Do not allow hands to contact engine during emergency service. Burns will result from contact with engine.

(1) Stop engine by shutting off ignition switch (2) and battery switch (1).

3-14. RADIATOR FAN CLUTCH EMERGENCY SERVICE (Contd)

(2) Raise and secure hood (paragraph 2-19).

(3) Remove two clutch override lockup bolts (6) from storage boss on fan clutch support bracket (5).

NOTE

Because fan clutch assembly is a moving part when engine is running, alignment mark may be located in different position from position shown in illustration. Without starting engine, tap engine ignition switch to move alignment mark to proper position.

(4) Line up alignment mark on side of fan mounting plate (8) with alignment mark on side of fan clutch assembly (7). Fan mounting plate (8) turns freely by hand.

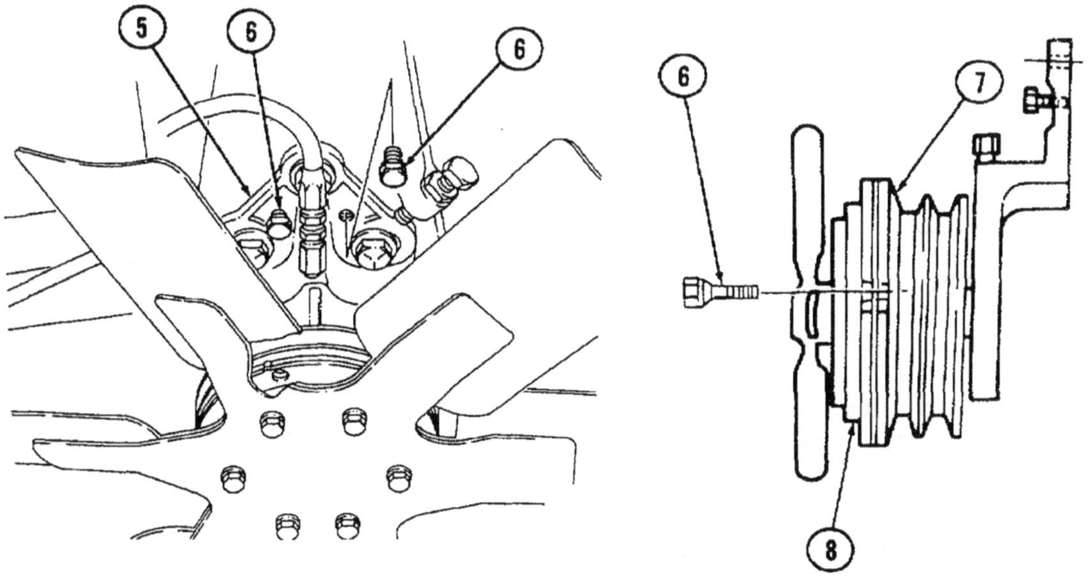
5
6
6
6
7
8

(5) With alignment marks lined up, insert two clutch override lockup bolts (6) into holes of fan mounting plate (8) and hand-tighten.

(6) Tighten override lockup bolts (6) until fully seated to secure fan mounting plate (8) to fan clutch assembly (7).

(7) Close and secure hood (para. 2-19).

(8) Start engine and allow engine to cool at idle speed until engine coolant temperature drops to normal operating temperature range of 175°-195°F (79°-91°C).

(9) Make certain unit maintenance is notified of emergency service performed on vehicle.

3-15. ENGINE SPLASH SHIELD REMOVAL

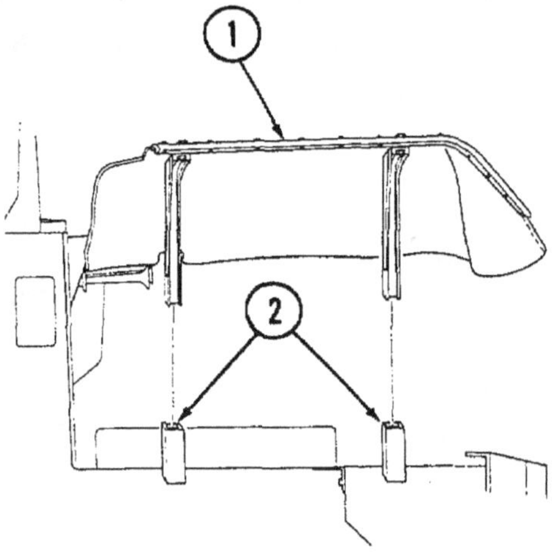
1
2

a. General. Splash shields are removed to gain access to the engine.

NOTE

Removal or installation of left and right engine splash shields is the same.

b. Removal.

(1) Raise and secure hood (para. 2-19).

(2) Lift engine splash shield (1) out of two brackets (2).

c. Installation.

(1) Guide engine splash shield (1) ends into two brackets (2) and push down splash shield (1) until it touches bottom of brackets (2).

(2) Close and secure hood (para. 2-19).

3-16. CTIS DISABLE AND RESET

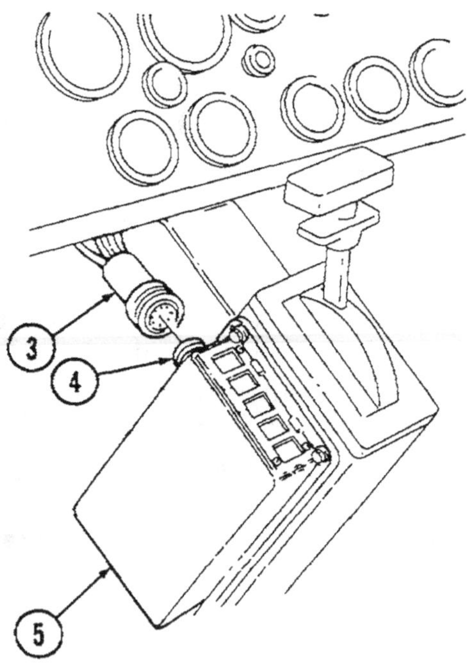
3
4
5

a. General. Central Tire Inflation System (CTIS) is a completely automatic system with the capability for operator to select several different modes of operation. When system malfunction can not be resolved by switching modes, it may be necessary to disable the Electronic Control Unit (ECU).

b. Disable ECU.

(1) Ensure engine is not running and battery switch is turned off (para. 2-16).

(2) Turn cable connector (3) clockwise one quarter turn and remove from connector (4) of ECU (5).

c. Reset ECU.

(1) Ensure engine is not running and battery switch is turned off (para. 2-16).

(2) Connect cable connector (3) to connector (4) of ECU (5) and turn counterclockwise one-quarter turn.

(3) Start engine (para. 2-12) and check system for proper operation (para. 2-14).

3-17. EMERGENCY BRAKE SERVICE

a. General. The emergency brake combines the holding strength of the spring brake system (activated by a switch on emergency brake lever), with brake shoes attached to the output shaft of the transfer case.

b. Operation. The emergency brake can be applied simply by pulling the emergency brake (3) backward and upward at the same time until it is in the full upright position.

c. Testing With Engine Running (para. 2-12).

(1) With service brake (5) applied, depress override button (2) on dash.

(2) Set emergency brake. Make sure parking brake warning light (6) illuminates.

(3) Shift transmission selector (1) to 1-5 drive.

(4) With engine at idle speed, release service brakes (5). If vehicle moves, adjust emergency brake (3).

d. Adjusting Emergency Brake.

(1) With emergency brake (3) off (forward and down position), rotate knurled knob (4) clockwise one-quarter turn.

(2) Repeat testing procedure and step (1) until emergency brake holds vehicle at idle. Rotate knurled knob (4) an additional one-quarter turn.

CAUTION

If emergency brake cannot be adjusted, or drags when released, notify unit maintenance. Failure to do so may result in damage to equipment.

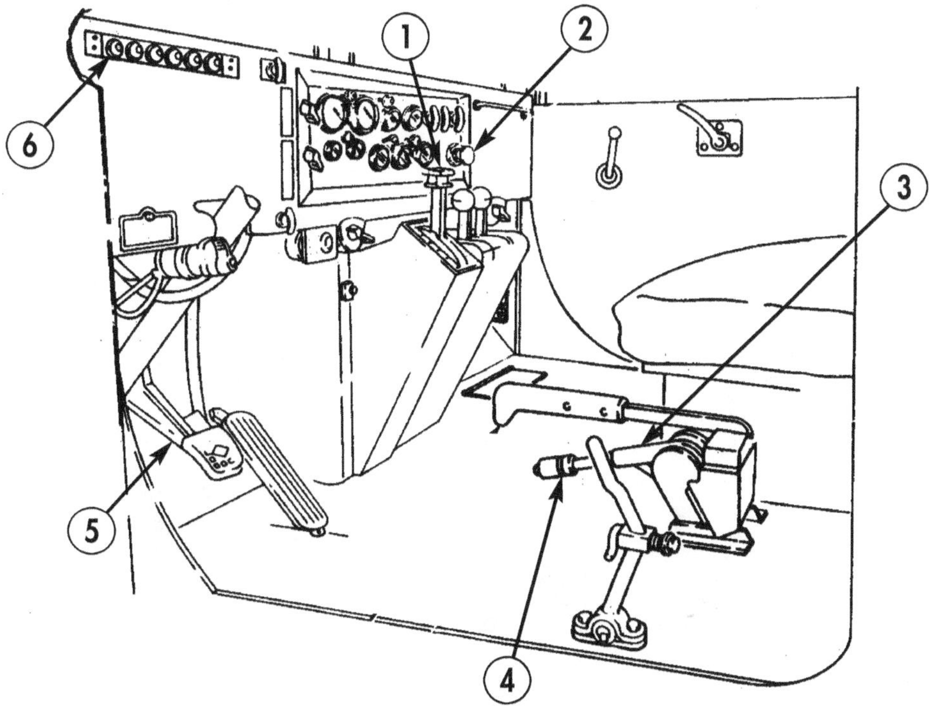
1
2
3
4
5
6

APPENDIX A
REFERENCES

A-1. PUBLICATION INDEX

The following index should be consulted frequently for latest changes or revisions and for new publications relating to material covered in this manual:

Consolidated Army Publications and Forms Index . . . . DA Pam 25-30

A-2. OTHER PUBLICATIONS

a. Technical Manuals.

Cleaning Materials . . . . TM 9-247
Hand Receipt Manual . . . . TM 9-2320-272-10-HR
Maintenance Manual for Decontaminating Apparatus . . . . TM 3-4230-204-12&P
Maintenance Manual for Machine Gun Mounts . . . . TM 9-1005-245-13&P
Procedures and Destruction of Tank–Automotive Equipment to Prevent Enemy Use . . . . TM 750-244-6
Care, Maintenance Repair and Inspection of Pneumatic Tires and Inner Tubes . . . . TM 9-2610-200-14
Load-Testing Vehicles Used to Handle Missiles and Rockets . . . . TB 9-352

b. Technical Bulletins.

Hearing Conservation Program . . . . PAM 40-501
Safety Inspection and Testing of Lifting Devices . . . . TB 43-0142
Security of Tactical Wheeled Vehicles . . . . TB 9-2300-422-20
Use of Antifreeze Solutions and Cleaning Compounds in Engine Cooling Systems . . . . TB 750-651
Tactical Wheeled Vehicles: Repair of Frames . . . . TB 9-2300-247-40

c. Field Manuals.

Basic Cold Weather Manual . . . . FM 31-70
Driver's Manual, Wheeled Vehicles . . . . FM 21-305
First Aid . . . . FM 4-25.11
Mountain Operations . . . . FM 3-97.6
Northern Operations . . . . FM 31-71
Operation and Maintenance of Ordnance Materiel in Cold Weather (0° to -65°F) . . . . FM 9-207
Recovery and Battlefield Damage Assessment and Repair . . . . FM 9-43-2

APPENDIX A (Contd)

A-2. OTHER PUBLICATIONS (Contd)

d. General Publications.

Army Acquisition Policy	AR 70-1
Army Information Management	AR 25-1
Driver Selection, Testing, and Licensing	AR 600-55
The Army Publishing Program	AR 25-30
Preventive Medicine	AR 40-5
Prevention of Motor Vehicle Accidents	AR 385-55
The Army Maintenance Management System (TAMMS)	DA Pam 738-750

e. Forms.

Equipment Control Record	DA Form 2408-9
Equipment Inspection and Maintenance Worksheet	DA Form 2404
Maintenance Request	DA Form 2407
Pre-printed Hand Receipts	DA Form 2062
Quality Deficiency Report	SF 368
Recommended Changes to DA Publications	DA Form 2028

APPENDIX B
COMPONENTS OF END ITEM (COEI) AND BASIC ISSUE ITEMS (BII) LISTS

Section I. INTRODUCTION

B-1. SCOPE

This appendix lists integral components and basic issue items for M939, M939A1, and M939A2 series vehicles. The appendix is designed to help you inventory items required for safe and efficient operation.

B-2. GENERAL

a. Section II, Components of the End Item (COEI). These items are installed in the vehicle at time of manufacture or rebuild. (None authorized for M939, M939A1, and M939A2 series.)

b. Section III, Basic Issue Items (BII). These are the minimum essential items required to place and maintain M939, M939A1, and M939A2 series vehicles in operation. BII must accompany the vehicle during operation and whenever it is transferred between accountable officers. The illustrations will assist you in identifying each basic issue item.

B-3. EXPLANATION OF COLUMNS

The following provides an explanation of columns found in the tabular listings:

a. Column (1) – Illustration Number (Illus Number). This column indicates the number of the illustration in which the item is shown.

b. Column (2) – National Stock Number. Indicates the national stock number assigned to the item and will be used for requisitioning purposes.

c. Column (3) – Description.

(1) Left side of column. Indicates the federal item name and, if required, a minimum description to identify and locate the item. The last line for each item indicates the Commercial and Government Entity Code (CAGEC) (in parentheses) followed by the part number.

B-3. EXPLANATION OF COLUMNS (Contd)

(a). Commercial and Government Entity Code (CAGEC).

CAGEC	MANUFACTURER
03306	Ampco Metal Div., Ampco
04741	White Motor Corp., Ogden Truck Plant
18876	Army Missile Command
19204	Rock Island Arsenal
19207	U.S. Army Tank-automotive and Armaments Command, AMSTA-TFP
21108	G T Price Products Inc.
21450	Army Weapons Command, ATTN: AMSWE-REE-S Ordnance Corps Engineering Standards Rock Island Arsenal
24076	Sargent Industries — Gar Wood Div.
24617	General Motors Corp.
28047	Hein – Werner Corp.
32779	Alert Stamping and Mfg. Co., Inc.
50980	Department of the Army, U.S. Army General Materiel and Petroleum Activity
55719	Snap-On Tools Corp.
57068	Stanley Works Corp.
65814	TRW
77348	Rayetter R Plumb Inc.
80063	U.S. Army Communications and Electronics Materiel Readiness
80244	General Services Administration Federal Supply Services
81337	Army Natick Research and Development Center
81348	Federal Specifications Promulgated by General Service Administration
81349	Military Specification Promulgated by Standardization Div. Directorate of Logistic Services DSA
81902	Craig Systems Corp
95879	Stewart Warner Alemite Corp.
96906	Military Standards

(b). Part Number. Indicates primary number used by manufacturer for control of design and characteristics of item, through engineering drawings, specifications, standards, and inspection requirements, used to identify an item or range of items.

(2) Right side of column. If item needed differs for different models of this equipment, the model is shown under the "Usable On Code" heading in this column.

B-3. EXPLANATION OF COLUMNS (Contd)

USABLE ON CODE	VEHICLE USED ON
A	All
DAW	M923A1 wo/w (Dropside)
DAX	M925A1 w/w (Dropside)
DAC	M927A1 wo/w (XLWB)
DAD	M928A1 w/w (XLWB)
DAE	M929A1 wo/w
DAF	M930A1 w/w
DAG	M931A1 wo/w
DAH	M932A1 w/w
DAJ	M934A1 wo/w
DAL	M936A1 w/w
V15	M923 wo/w (Dropside)
V14	M925 w/w (Dropside)
V17	M927 wo/w (XLWB)
V16	M928 w/w (XLWB)
V20	M929 wo/w
V19	M930 w/w
V22	M931 wo/w
V21	M932 w/w
V24	M934 wo/w
V18	M936 w/w
ZAA	M923A2 wo/w (Dropside)
ZAB	M925A2 w/w (Dropside)
ZAC	M927A2 wo/w (XLWB)
ZAD	M928A2 w/w (XLWB)
ZAE	M929A2 wo/w
ZAF	M930A2 w/w
ZAG	M931A2 wo/w
ZAH	M932A2 w/w
ZAJ	M934A2 wo/w
ZAL	M936A2 w/w

d. Column (4) – U/I (Unit of Issue). Indicates how the item is issued for the National Stock Number shown in column two.

e. Column (5) – Qty Rqd. Indicates the quantity required.

Section II. COMPONENTS OF END ITEM LIST (COEI)

B-4. GENERAL

These items are installed in the vehicle at the time of manufacture or rebuild. (None authorized for M939, M939A1, and M939A2 series.)

Section III. BASIC ISSUE ITEMS (BII)

B-5. GENERAL

These are the minimum essential items required to place and maintain M939, M939A1, and M939A2 vehicles in operation. Although shipped separately packaged, BII must accompany the vehicle during operation and whenever it is transferred between accountable officers. The illustrations will assist you to identify each basic issue item. Refer to para. B-3 for explanation of columns.

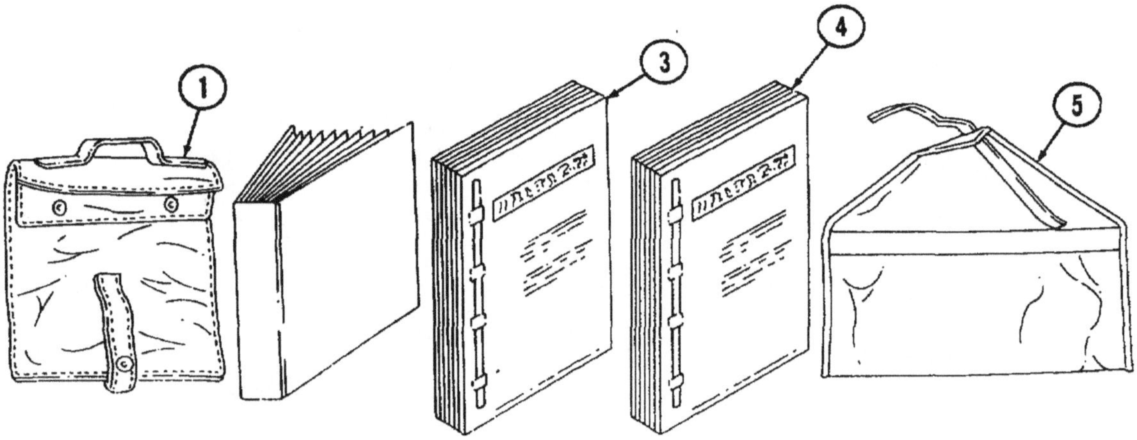
1
3
4
5

(1) Illus Number	(2) National Stock Number	(3) Description CAGEC and Part Number	Usable On Code	(4) U/I	(5) Qty Rqd
		COMMON EQUIPMENT, COMMON TOOLS			
1	2540-00-670-2459	BAG: pamphlet, cotton duck, 3 x 9-1/4 x 11-1/4-in. (in map compartment behind crew seat) (19207) 7961712	A	EA	1
Deleted					
3		TM 9-2320-272-10: Technical Manual, Operator's (in pamphlet bag)	A	EA	1
4		TM 9-2320-272-10HR: Technical Manual, Hand Receipts (in pamphlet bag)	A	EA	1
5	5140-00-315-2775	BAG: tool, cotton duck, 10 x 20-in., w/flap (81337) 5-7-1	A	EA	1

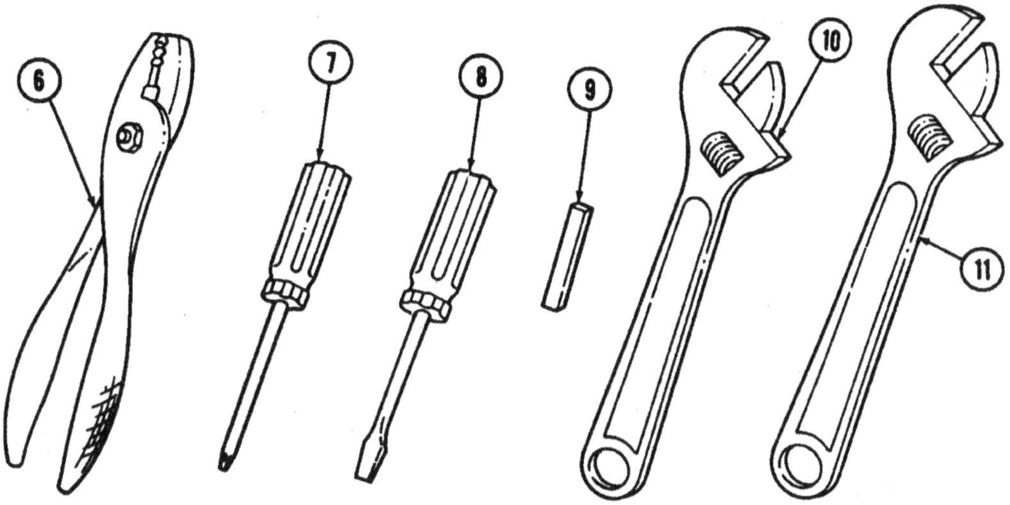
6
7
8
9
10
11

Section III. BASIC ISSUE ITEMS (Contd)

(1) Illus Number	(2) National Stock Number	(3) Description CAGEC and Part Number	Usable On Code	(4) U/I	(5) Qty Rqd
		COMMON EQUIPMENT, COMMON TOOLS (Contd)			
6	5120-00-223-7397	PLIERS: combination, slip-joint, straight nose, w/cutter, 8-in. long, phosphate finish (in toolbag) (19207) 11655775-3	A	EA	1
7	5120-00-234-8913	SCREWDRIVER: cross tip, Phillips, plastic handle, point no. 2, 4-in. blade, 7-1/2-in. overall length (in toolbag) (19207) 11655777-12	A	EA	1
8	5120-00-222-8852	SCREWDRIVER: flat tip, flared sides, plastic handle, round blade, 1/4-in. wide tip, 4-in. long blade, 7-3/4-in. overall length (in toolbag) (19207) 116755777-2	A	EA	1
9	5315-00-732-1019	WRENCH: key, oil drainplug, straight bar, 1/2-in. square x 2-1/2-in. long (in toolbag) (96906) MS20066-543	A	EA	1
10	5120-00-240-5328	WRENCH: open end, adjustable, .95-in. jaw opening, 8-in. long (in toolbag) (19207) 11655778-3	A	EA	1
11	5120-00-264-3796	WRENCH: open end, adj. 0-in. to 1.322-in. jaw opening, 12-in. long, phosphate finish, type I, class A (in toolbag) (19207) 11655778-5	A	EA	1

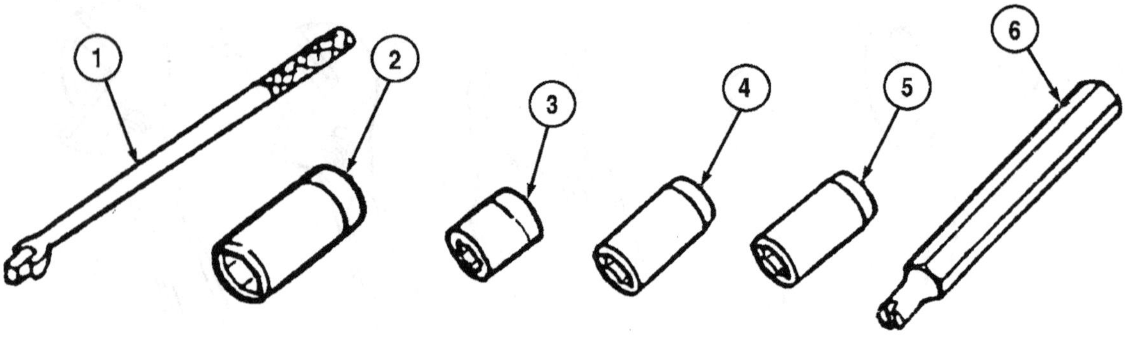
1
2
3
4
5
6

Section III. BASIC ISSUE ITEMS (Contd)

(1) Illus Number	(2) National Stock Number	(3) Description CAGEC and Part Number	Usable On Code	(4) U/I	(5) Qty Rqd
		EQUIPMENT FOR CTIS SERVICE			
1	5120-00-236-7590	HANDLE, SOCKET WRENCH: hinged, 1/2-in. drive end, 14-1/2-in. to 19-in. O/A lg (81348) GGG-W-641; Type III, Class 1	ZAA,ZAB, ZAC,ZAD, ZAE,ZAF, ZAG,ZAH, ZAJ,ZAL	EA	1
2	5130-00-714-0600	SOCKET, SOCKET WRENCH: 1/2-in. sq. drive, 15/16-in. (81348) GGG-W-660 Deep Style, Type I, Class 1, Style A (Use to remove CTIS wheel valve shield)	ZAA,ZAB, ZAC,ZAD, ZAE,ZAF, ZAG,ZAH, ZAJ,ZAL	EA	1
3	5120-00-189-7985	SOCKET, SOCKET WRENCH: 1/2-in. sq. drive, 12 point, 3/4-in., regular length (81348) GGG-W-641, Type II, Class 2, Style A (Use to remove CTIS wheel valve shield and manifold)	ZAA,ZAB, ZAC,ZAD, ZAE,ZAF, ZAG,ZAH, ZAJ,ZAL	EA	1
4	5120-00-189-7913	SOCKET, SOCKET WRENCH: 1/2-in. sq. drive, 1-1/16-in., deep style Type I, Class 1, Style A (05506) ST-1234	ZAA,ZAB, ZAC,ZAD, ZAE,ZAF, ZAG,ZAH, ZAJ,ZAL	EA	1
5	5130-00-203-6448	SOCKET, SOCKET WRENCH: 1/2-in. sq. drive, 1-1/8-in., impact type (81348) GGG-W-660	ZAA,ZAB, ZAC,ZAD, ZAE,ZAF, ZAG,ZAH, ZAJ,ZAL	EA	1
6	5120-00-541-4687	TOOL: valve core removal: 3-1/4-in. lg. (53477) 2688	ZAA,ZAB, ZAC,ZAD, ZAE,ZAF, ZAG,ZAH, ZAJ,ZAL	EA	1

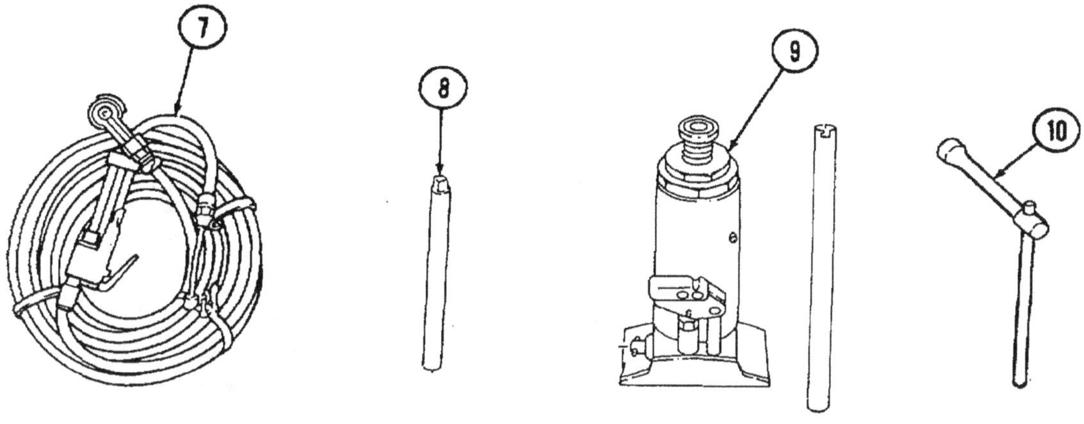
7
8
9
10

Section III. BASIC ISSUE ITEMS (Contd)

(1) Illus Number	(2) National Stock Number	(3) Description CAGEC and Part Number	Usable On Code	(4) U/I	(5) Qty Rqd
		COMMON EQUIPMENT FOR TIRE SERVICE			
7	4910-01-038-2820	GAUGE AND HOSE ASSEMBLY: tire inflation, self-contained, w/30-ft hose (in toolbox – vehicle right side) (19207) 11677140-5	A	EA	1
8	5120-00-243-2419	HANDLE: bar, wheel stud nut wrench, 3/4-in. diameter x 30-in. long, phosphate finish (in toolbox – vehicle right side) (19207) 6196547	A	EA	1
9	5120-00-595-8396	JACK: hydraulic, hand, 8-ton capacity, 11-in. closed, 23-1/8-in. open w/operating lever, type I, class 2, style A, size 8-6 (in toolbox – vehicle right side) (04741) 16W233	V14,V15, V16,V17, V18,V19, V20,V21, V22,V24	EA	1
9	5120-01-374-0532	JACK: hydraulic, hand, self-contained (in toolbox-vehicle right side) (M939A1/A2) (19207) 12375464	ALL Except V14,V15, V16,V17, V18,V19, V20,V21, V22,V24	EA	1
10	5120-00-316-9217	WRENCH: wheel stud nut, straight, double socket, 1-1/2-in. hexagon opening, 13/16-in. square opening, 17-in. to 19-in. long, type II, size 1 (in toolbox – vehicle right side) (19207) 11677000-3	A	EA	1

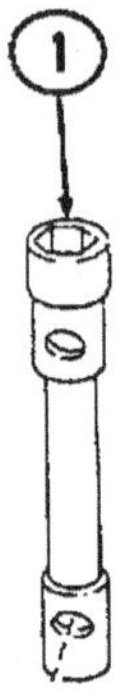
1

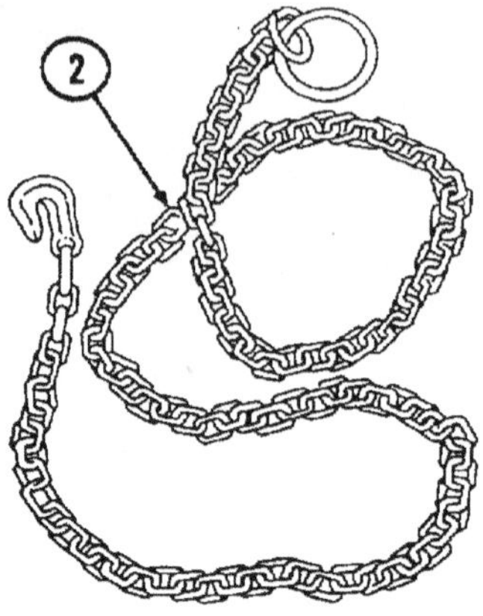
2

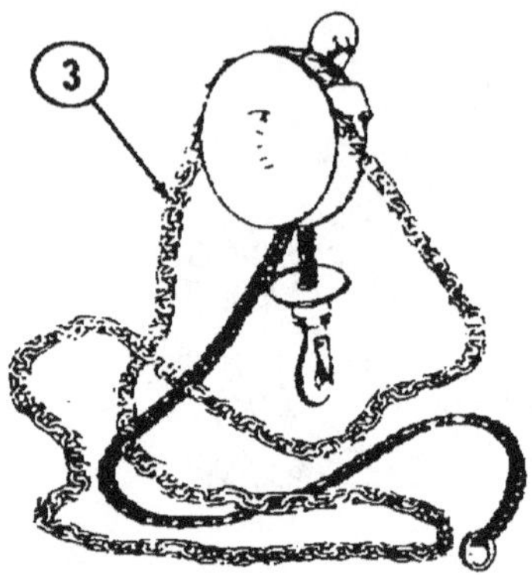
3

Section III. BASIC ISSUE ITEMS (Contd)

(1) Illus Number	(2) National Stock Number	(3) Description CAGEC and Part Number	Usable On Code	(4) U/I	(5) Qty Rqr
		SPECIALIZED EQUIPMENT FOR TIRE SERVICE			
1	5120-00-378-4411	WRENCH: Wheel stud nut, straight, double socket, 1-1/2-in. hexagon opening, socket within a socket/with handle. (87641) 151	V14,V15 V16,V17, V18,V19, V20,V21, V22,V24		
2	4010-01-114-3728	CHAIN ASSEMBLY: single leg w/grab hook, w/ring end link, 750-lb work load, zinc finish (in toolbox – vehicle right side) (19207) 12256287	V14,V15, V16,V17, V19,V20, V21,V22, V24	EA	1
2	4010-01-238-0518	CHAIN ASSEMBLY: single leg w/grab hook, w/ring end link, 1250-lb work load, zinc finish (in toolbox – vehicle right side) (19207) 12302917	DAC,DAD, DAE,DAF, DAG,DAH, DAL,DAJ, DAW,DAX, V18,ZAA, ZAB,ZAC, ZAD,ZAE, ZAF,ZAG, ZAH,ZAJ, ZAL	EA	1
3	3950-01-238-0504	HOIST ASSEMBLY: chain, hand-operated, hook suspension, 500-lb rated load (in toolbox – vehicle right side) (19207) 12301088	A	EA	1

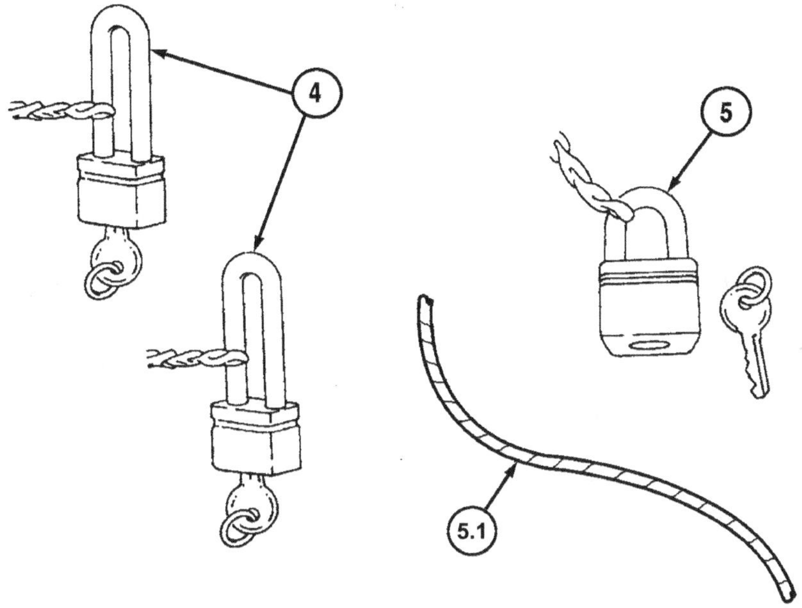
4
5
5.1

Section III. BASIC ISSUE ITEMS (Contd)

(1) Illus Number	(2) National Stock Number	(3) Description CAGEC and Part Number	Usable On Code	(4) U/I	(5) Qty Rqd
		EQUIPMENT — MISCELLANEOUS			
4	5340-01-050-7059	PADLOCK SET: keyed alike, 1-1/2-in. size, w/clevis and chain, composed of 2 padlocks, 2 keys per set (in toolbox – vehicle right side) (96906) MS21313-160	DAC,DAD, DAG,DAH, DAW,DAX, V14,V15, V16,V17, V21,V22, ZAA,ZAB, ZAC,ZAD, ZAG,ZAH,	ST	1
5	5340-00-682-1508	PADLOCK SET: keyed individually, 1-1/2-in. size, w/clevis and chain, w/2 keys (in toolbox – vehicle right side) (96906) MS35647-3	DAE,DAF, V19,V20, ZAE,ZAF	EA	1
5.1	6150-00-772-8814	HARNESS, TRAILER LIGHTS INTERVEHICLER 144 in. (19207) 7728814	DAG,DAN V21,V22 ZAG,ZAE	EA	1

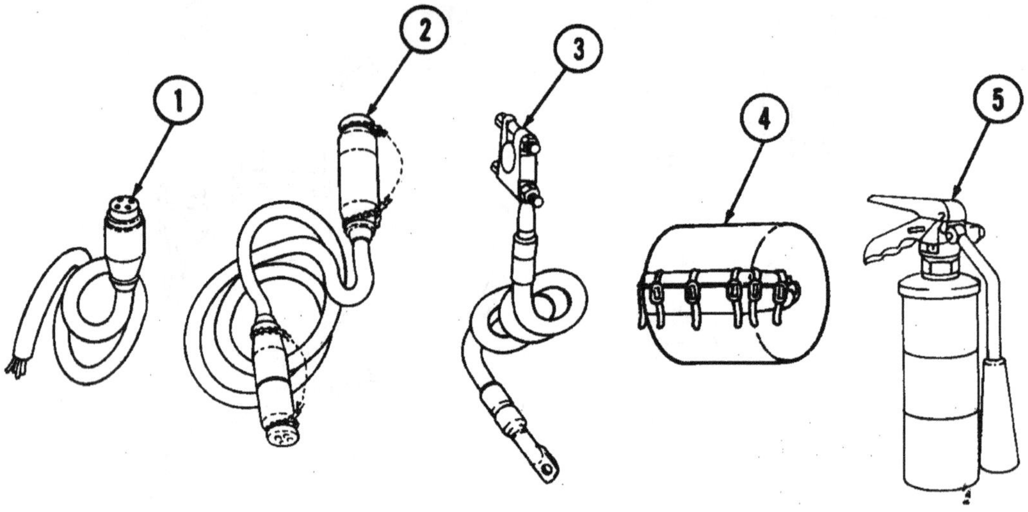
1
2
3
4
5

Section III. BASIC ISSUE ITEMS (Contd)

(1) Illus Number	(2) National Stock Number	(3) Description CAGEC and Part Number	Usable On Code	(4) U/I	(5) Qty Rqd
		TOOLS AND EQUIPMENT – EXPANSIBLE VANS (M934, M934A1, AND M934A2)			
1	6150-00-134-0848	CABLE: electric, auxiliary, 600-volt, 39-1/4-in. long (on ceiling, front of van) (19207) 11601641	DAJ,V24, ZAJ	EA	1
2	6150-00-134-0847	CABLE: electric, jumper, 600-volt, 100-ft long w/coupling ends (on power cable reel, right side of van) (19207) 11601643	DAJ,V24, ZAJ	EA	1
3	6140-00-851-4573	CABLE: ground, 48-in. long, used w/rod (8380403) (in left side box) (19207) 7017575	DAJ,V24, ZAJ	EA	1
4	3950-00-870-9939	COVER: cable reel, cotton duck (over power cable – ahead of right forward rear tire) (19207) 8735021	DAJ,V24, ZAJ	EA	1
5	4210-01-189-6452	EXTINGUISHER, FIRE: purple “K” dry chemical (located in van body) (19207) 12255633-3	DAJ,V24, ZAJ	EA	2

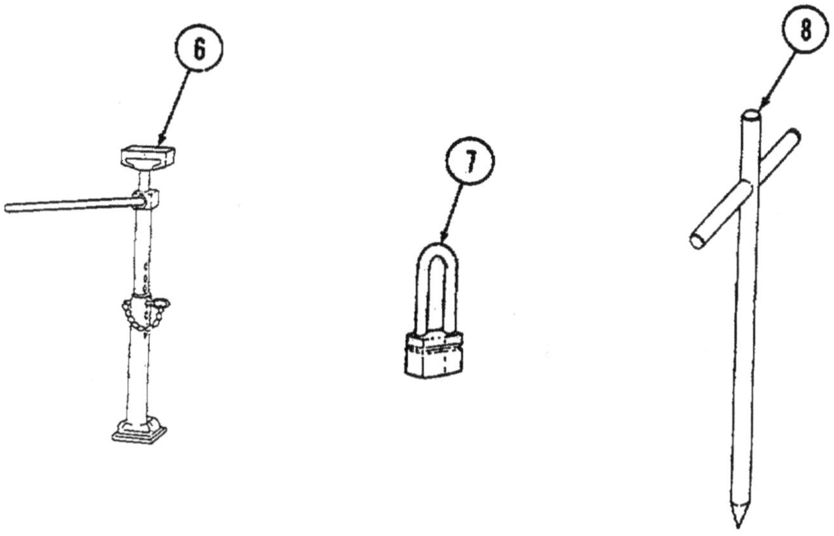
6
7
8

Section III. BASIC ISSUE ITEMS (Contd)

(1) Illus Number	(2) National Stock Number	(3) Description CAGEC and Part Number	Usable On Code	(4) U/I	(5) Qty Rqd
		TOOLS AND EQUIPMENT – EXPANSIBLE VANS (M934, M934A1, AND M934A2) (Contd)			
6	5120-00-566-0617	JACK: leveling vehicle, portable (rear compartment) (19207) 7534672	DAJ,V24, ZAJ	EA	4
7	5340-01-050-7059	PADLOCK SET: keyed alike, 1-1/2-in. size, long shackle, composed of 5 padlocks and 7 keys, class 2 (on rear door, ladders, storage compartments) (96906) MS21313-53	DAJ,V24, ZAJ	EA	1
8	2510-00-790-2296	ROD: ground, 3/4-in. diameter x 30-in. long, w/crossbar. Used with ground cable (7017575) (in left side box) (19207) 8380403	DAJ,V24, ZAJ	EA	1

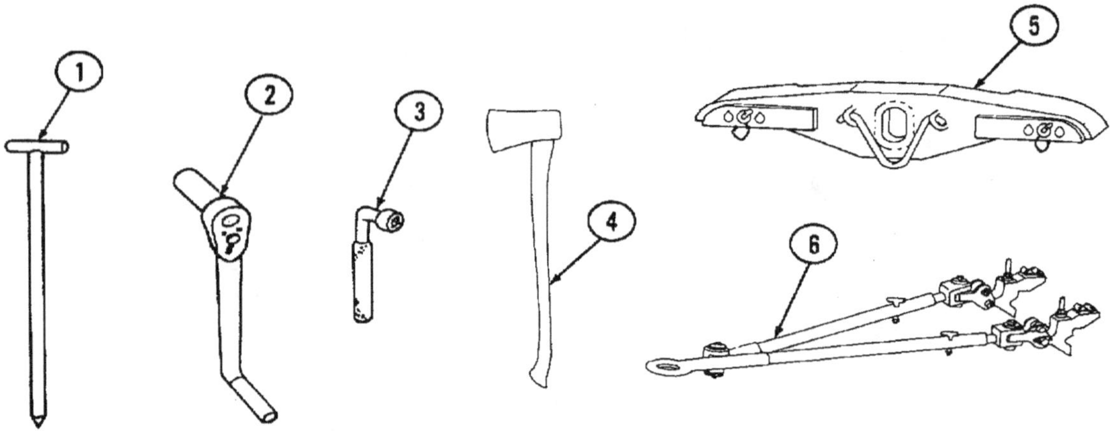

1
2
3
4
5
6

Section III. BASIC ISSUE ITEMS (Contd)

(1) Illus Number	(2) National Stock Number	(3) Description CAGEC and Part Number	Usable On Code	(4) U/I	(5) Qty Rqd
		TOOLS AND EQUIPMENT – EXPANSIBLE VANS (M934, M934A1, AND M934A2) (Contd)			
1	2590-00-870-9936	SPIKE: stabilizer anchor, welded (in rear exterior compartment) (19207) 7534689	DAJ,V24, ZAJ	EA	8
2	5120-00-650-7830	WRENCH: ratchet, reversible w/removable socket, 3/4-in. square drive (stowed on interior of rear door, left-hand side) (19207) 7759181	DAJ,V24, ZAJ	EA	1
3	5120-00-650-7829	WRENCH: socket, 90 degree offset, 1/2-in. square opening (stowed on interior of rear door, left-hand side) (19207) 8380406	DAJ,V24, ZAJ	EA	1
		TOOLS AND EQUIPMENT – MEDIUM WRECKER (M936, M936A1, AND M936A2)			
4	5110-00-293-2336	AX: single bit, 4-lb head weight, 4-3/4-in. cutting edge, 35-1/2-in. to 36-1/2-in. long, type I, class 1, design A, olive drab finish [in compartment no. 1B (pg B-24)] (19207) 6150925	DAL,V18, ZAL	EA	1
5	4910-00-347-9703	BAR: lifting, whiffletree [on right deck (pg B-24)] (19207) 8690061	DAL,V18, ZAL	EA	1
6	4910-01-365-9304	BAR: towing, V universal type w/bumper axle clamp assembly [on right deck (pg B-24)] (59678) 7551383	DAL,V18, ZAL	EA	1

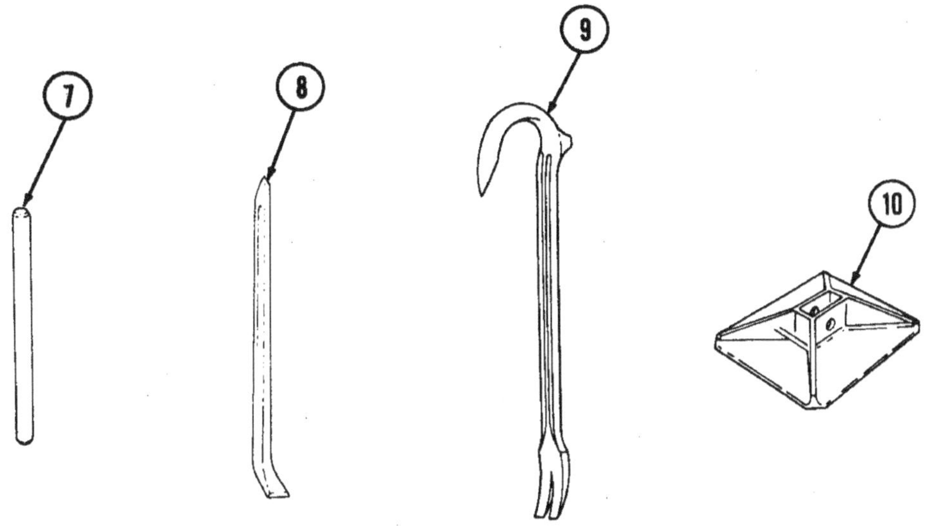
7
8
9
10

Section III. BASIC ISSUE ITEMS (Contd)

(1) Illus Number	(2) National Stock Number	(3) Description CAGEC and Part Number	Usable On Code	(4) U/I	(5) Qty Rqd
		TOOLS AND EQUIPMENT – MEDIUM WRECKER (M936, M936A1, AND M936A2) (Contd)			
7	3040-00-860-2359	BAR: cranking, outrigger, 1-in. diameter, 12-in. long, olive drab finish [one in compartment no. 3 and one in compartment no. 2 (pg B-24)] (19207) 10900233	DAL,V18, ZAL	EA	2
8	5120-00-224-1384	BAR: pinch, offset and tapered ends, 1-in. hexagon stock, 36-in. long, olive drab finish, type III [in compartment no. 1B (pg B-24)] (18348) GGG-B-101	DAL,V18, ZAL	EA	1
9	5120-00-293-0665	BAR: wrecking, gooseneck pinch point, w/claw, 3/4-in. hexagon stock, 36-in. long, olive drab finish, type V, class 2, style A [in compartment no. 1B (pg B-24)] (59068) 55-130	DAL,V18, ZAL	EA	1
10	2540-00-040-2299	BASE PLATE: boom jack [in mounting brackets, left and right side of rear winch (pg B-24)] (19207) 8330155	DAL,V18, ZAL	EA	2

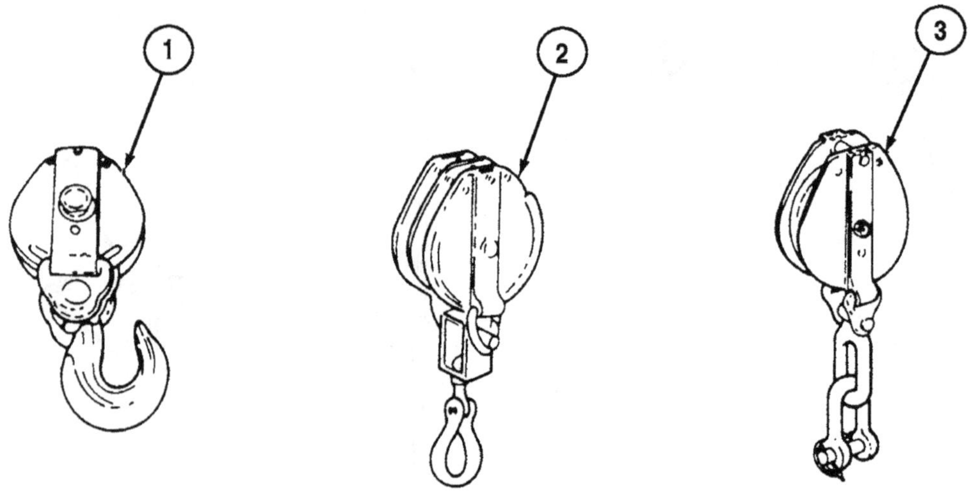
1
2
3

Section III. BASIC ISSUE ITEMS (Contd)

(1) Illus Number	(2) National Stock Number	(3) Description CAGEC and Part Number	Usable On Code	(4) U/I	(5) Qty Rqd
		TOOLS AND EQUIPMENT – MEDIUM WRECKER (M936, M936A1, AND M936A2) (Contd)			
1	3940-00-105-9933	BLOCK: rigging, steel wire rope, single 8-in. sheave, w/swivel hook, 5/8-in. diameter rope, 10-ton capacity, olive drab finish [in compartment no. 1B (pg B-24)] (19207) 11631726	DAL,V18, ZAL	EA	1
2	3940-00-708-0704	BLOCK: rigging, wire rope, double 8-in. sheave, w/swivel shackle, 7/8-in. diameter rope, 25-ton capacity, olive drab finish [in compartment no. 1B (pg B-24)] (81349) MIL B 11837	DAL,V18, ZAL	EA	2
3	3940-00-899-1352	BLOCK: rigging, single 10-in. sheave w/swivel eye and shackle, 3/4-in. diameter rope, 15-ton capacity, olive drab finish [in compartment no. 1B (pg B-24)] (19207) 8383238	DAL,V18 ZAL	EA	2

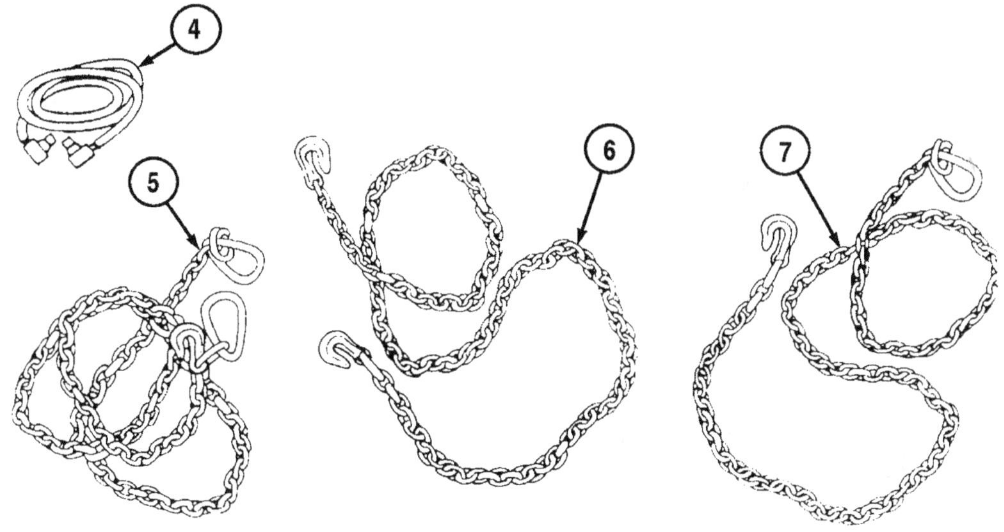
4
5
6
7

Section III. BASIC ISSUE ITEMS (Contd)

(1) Illus Number	(2) National Stock Number	(3) Description CAGEC and Part Number	Usable On Code	(4) U/I	(5) Qty Rqd
		TOOLS AND EQUIPMENT – MEDIUM WRECKER (M936, M936A1, AND M936A2) (Contd)			
4	6150-01-022-6004	CABLE: slave, electric, 24 volts, 20-ft long, (NATO), [in compartment no. 1A (pg B-24)] (19207) 11682336-1	DAL,V18, ZAL	EA	1
5	4010-00-473-6166	CHAIN: tow, single leg, 5/8-in. link, 16-ft long, w/2 pear-shaped coupling links, w/1 grab hook end [in compartment no. 1B (pg B-24)] (19207) 7077063	DAL,V18, ZAL	EA	1
6	4010-00-443-4845	CHAIN: utility, single leg, 3/8-in. link, 14-1/2-ft long w/2 grab hooks, zinc plate finish [in compartment no. 1B (pg B-24)] (19207) 10944642-2	DAL,V18, ZAL	EA	1
7	4010-01-010-2536	CHAIN: utility, single leg, 3/4-in. link, 12-ft long, w/grab hook, w/pear-shaped coupling link [in compartment no. 1B (pg B-24)] (19207) 8744250	DAL,V18, ZAL	EA	1

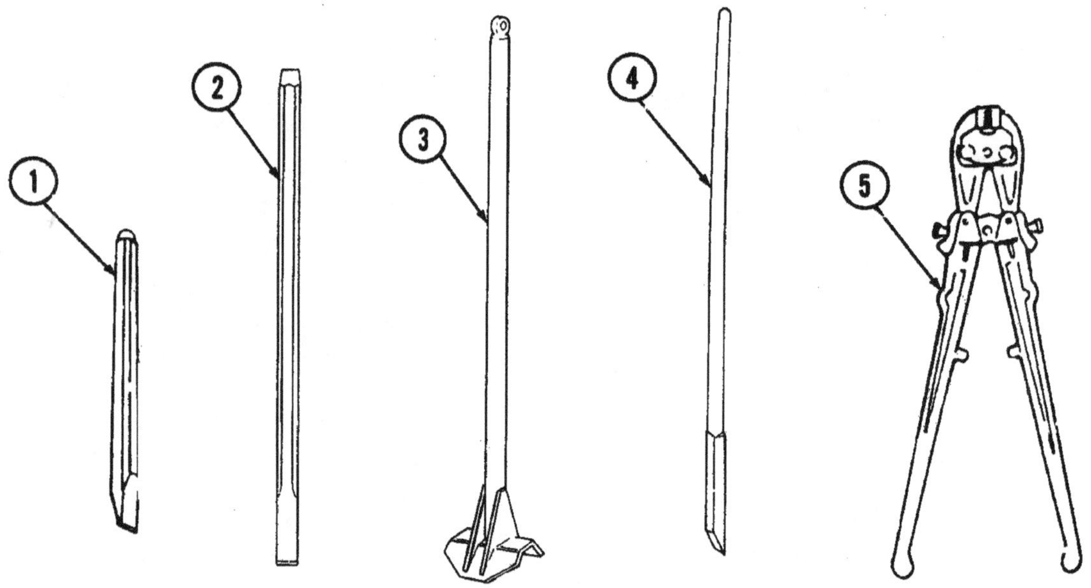
1
2
3
4
5

Section III. BASIC ISSUE ITEMS (Contd)

(1) Illus Number	(2) National Stock Number	(3) Description CAGEC and Part Number	Usable On Code	(4) U/I	(5) Qty Rqd
		TOOLS AND EQUIPMENT – MEDIUM WRECKER (M936, M936A1, AND M936A2) (Contd)			
1	5110-00-236-3272	CHISEL: cold, hand, 3/4-in. cutting edge, 6-1/2-in. long, type IV, class 1 [in compartment no. 4B (pg B-24)] (81348) GGG-C-313	DAL,V18, ZAL	EA	1
2	5110-00-238-8296	CHISEL: machinist's, cold, hand, long length, 1-in. cutting edge, 24-in. long, olive drab finish, type IV, class 2 [in compartment no. 1B (pg B-24)] (O1DJ4) C6	DAL,V18, ZAL	EA	1
3	2540-00-315-2306	CHOCK: field [on left deck (pg B-24)] (19207) 8330150	DAL,V18, ZAL	EA	2
4	5120-00-224-1390	CROWBAR: pinch-point, 1-1/4-in. diameter stock, 59-in. to 62-in. long, olive drab finish, type II, class 1, size 4 [in compartment no. 1B (pg B-24)] (18876) 9150189	DAL,V18, ZAL	EA	1
5	5110-00-188-2524	CUTTER: bolt, rigid head-type, clipper cut-type, 9/16-in. diameter, mild steel rod cutting capacity, 35-in. to 39-in. long, olive drab finish [in compartment no. 1A (pg B-24)] (81348) GGG-C-740	DAL,V18, ZAL	EA	1

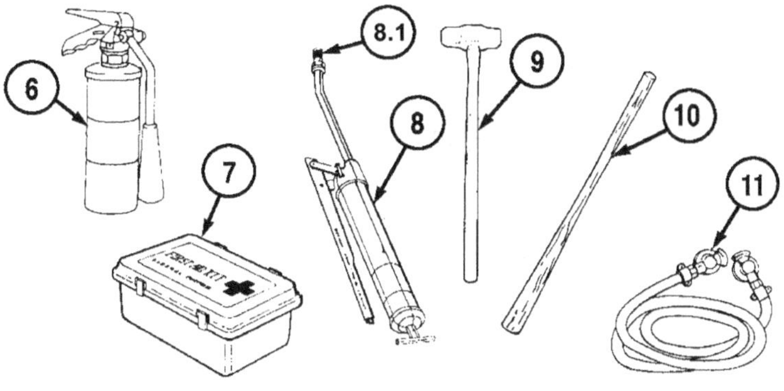

6
7
8.1
8
9
10
11

Section III. BASIC ISSUE ITEMS (Contd)

(1) Illus Number	(2) National Stock Number	(3) Description CAGEC and Part Number	Usable On Code	(4) U/I	(5) Qty Rqd
		TOOLS AND EQUIPMENT – MEDIUM WRECKER (M936, M936A1, AND M936A2) (Contd)			
6	4210-01-189-6452	EXTINGUISHER, FIRE: hand, purple "K" dry chemical [in brackets behind wrecker crane gondola (pg B-24)] (19207) 12255633-3	DAL,V18, ZAL	EA	2
7	6545-00-922-1200	FIRST-AID KIT: general purpose, 12 unit [in compartment no. 1 under crew seat (pg B-24)] (19207) 11677011	DAL,V18, ZAL	EA	1
8	4930-00-253-2478	GREASE GUN: hand lever operated, 14-oz cartridge or bulk load [in compartment no. 1B (pg B-24)] (81349) MIL-G-3859	DAL,V18, ZAL	EA	1
8.1	4930-00-288-1511	ADAPTER, GREASE GUN: coupling (19207) 6300333	DAG,DAH, DAQ,V21, V22,ZAG, ZAH	EA	1
9	5120-00-900-6098	HAMMER: hand, sledge, blacksmith's, double-face, 12-lb, 30-in. to 33-in. handle length, olive drab finish, type X, class 1 [in compartment no. 1B (pg B-24)] (58536) A-A-1293	DAL,V18, ZAL	EA	1
10	5120-00-288-6574	HANDLE: mattock, pick, railroad or clay pick, 36-in. long, olive drab finish grade AA [in compartment no. 1B (pg B-24)] (19207) 11677021	DAL,V18, ZAL	EA	1
11	4720-00-740-9662	HOSE: air connecting, inter-vehicular, 10-1/2-ft long, w/2 coupling ends [in compartment no. 1A (pg B-24)] (19207) 7061338	DAL,V18, ZAL	EA	2

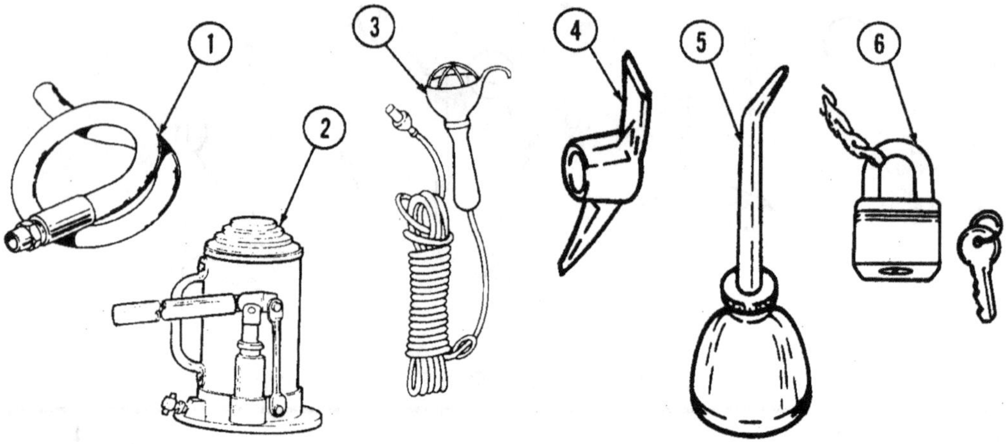
1
2
3
4
5
6

Section III. BASIC ISSUE ITEMS (Contd)

(1) Illus Number	(2) National Stock Number	(3) Description CAGEC and Part Number	Usable On Code	(4) U/I	(5) Qty Rqd
		TOOLS AND EQUIPMENT – MEDIUM WRECKER (M936, M936A1, AND M936A2) (Contd)			
1	4720-00-899-6721	HOSE: tank drain, hydraulic oil, 1-3/16-in. outside diameter x 5-ft long, olive drab finish [in compartment no. 1A (pg B-24)] (19207) 10900093	DAL,V18, ZAL	EA	1
2	5120-00-188-1790	JACK: hydraulic, hand, self-cont, 30-ton cap., w/oper lever, OD finish [in compartment no. 1A (pg B-24)] (28047) RHD160	DAL,V18, ZAL	EA	1
3	6230-00-274-4018	LIGHT: extension, w/single plug and plug socket, 24V, 25-ft, w/o lamp [in compartment no. 4 (pg B-24)] (32779) 2000 G2A	DAL,V18, ZAL	EA	1
4	5120-00-243-2395	MATTOCK: pick-type, 5-lb w/o handle, olive drab finish, type II, class F [in compartment no. 1B (pg B-24)] (19207) 11677022	DAL,V18, ZAL	EA	1
5	4930-00-266-9182	OILER: hand, push bottom, 8-oz capacity, 4-in. long spout [on left deck (pg B-24)] (96906) MS15764-1	DAL,V18, ZAL	EA	1
6	5340-00-838-5266	PADLOCK SET: low-security, keyed alike, regular (open) shackle, class 2 [in position securing compartments nos. 1, 1A, 1B, 2, 3, and 4 (pg B-24)] (96906) MS21313-124	DAL,V18, ZAL	EA	1

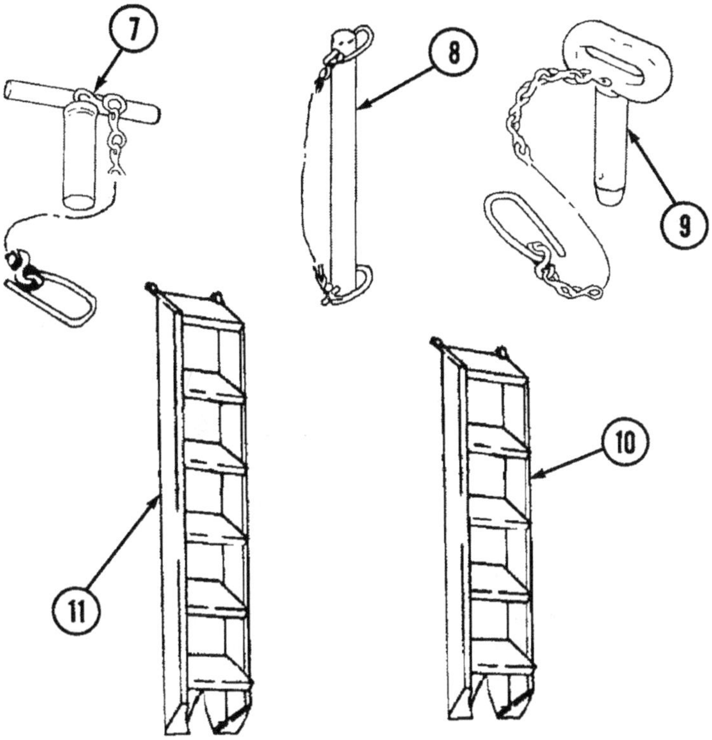
7
8
9
10
11

Section III. BASIC ISSUE ITEMS (Contd)

(1) Illus Number	(2) National Stock Number	(3) Description CAGEC and Part Number	Usable On Code	(4) U/I	(5) Qty Rqd
		TOOLS AND EQUIPMENT – MEDIUM WRECKER (M936, M936A1, AND M936A2) (Contd)			
7	5315-00-316-1008	PIN: tie-bar yoke, w/lockpin, olive drab finish [in compartment no. 3 (pg B-24)] (19207) 8327939	DAL,V18, ZAL	EA	2
8	5315-00-854-4431	PIN: inner, boom jack, w/lockpin, olive drab finish [in compartment no. 2 (pg B-24)] (19207) 10876413	DAL,V18, ZAL	EA	1
9	5315-00-740-9834	PIN: boom jack, w/lockpin, olive drab finish [in compartment no. 3 (pg B-24)] (19207) 7409834	DAL,V18, ZAL	EA	2
10	2540-01-098-5079	LADDER, BOARDING 5-STEP (19207) 8759434	V24	EA	1
11	2540-01-372-6677	LADDER, BOARDING 6-STEP (19207) 12375500	DAJ,ZAJ	EA	1

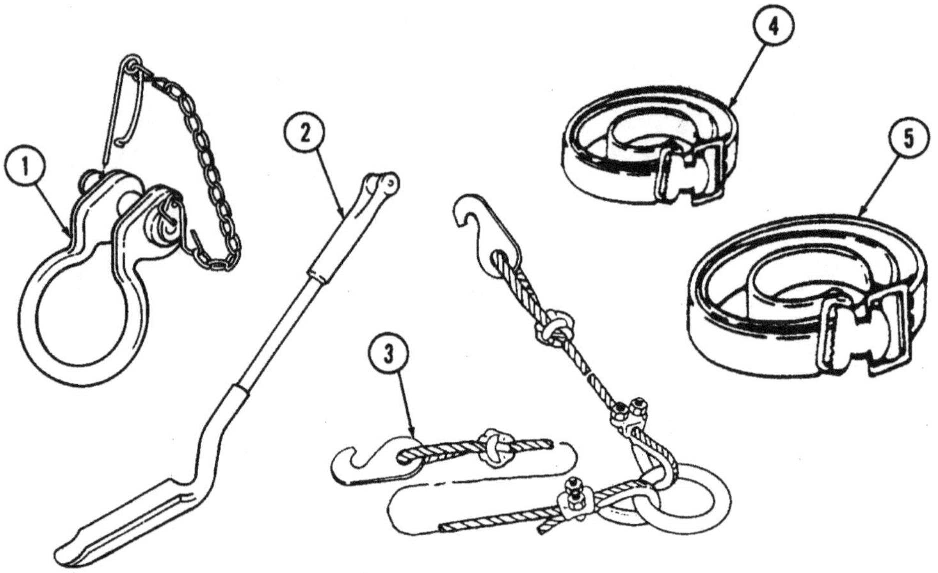
1
2
3
4
5

Section III. BASIC ISSUE ITEMS (Contd)

(1) Illus Number	(2) National Stock Number	(3) Description CAGEC and Part Number	Usable On Code	(4) U/I	(5) Qty Rqd
		TOOLS AND EQUIPMENT – MEDIUM WRECKER (M936, M936A1, AND M936A2) (Contd)			
1	2540-00-318-0326	SHACKLE: lifting, round pin, 7/8-in. diameter, olive drab finish [in compartment no. 2 (pg B-24)] (19207) 7357967	DAL,V18, ZAL	EA	2
2	5120-00-293-3336	SHOVEL: hand, round point, D-handle, short, size 2, olive drab finish, type IV, class A, style 1 [in compartment no. 1B (pg B-24)] (19207) 11655784	DAL,V18, ZAL	EA	1
3	2590-00-040-2297	SLING: wire rope, double leg w/ring, w/2 hook ends (ring ends attach to block, two hook ends attach to vehicle rear bumperettes) (19207) 8330151	DAL,V18, ZAL	EA	1
4	5340-00-543-3034	STRAP: webbing, 1-1/2-in. wide x 24-in. long, w/buckle (securing field chocks to vehicle deck) [on left deck (pg B-24)] (19207) 8690516	DAL,V18, ZAL	EA	1
5	5340-00-753-3744	STRAP: webbing, 1-1/2-in. wide x 36-in. long, w/buckle (securing boom jacks to vehicle deck) [on left deck (pg B-24)] (19207) 8690473	DAL,V18, ZAL	EA	1

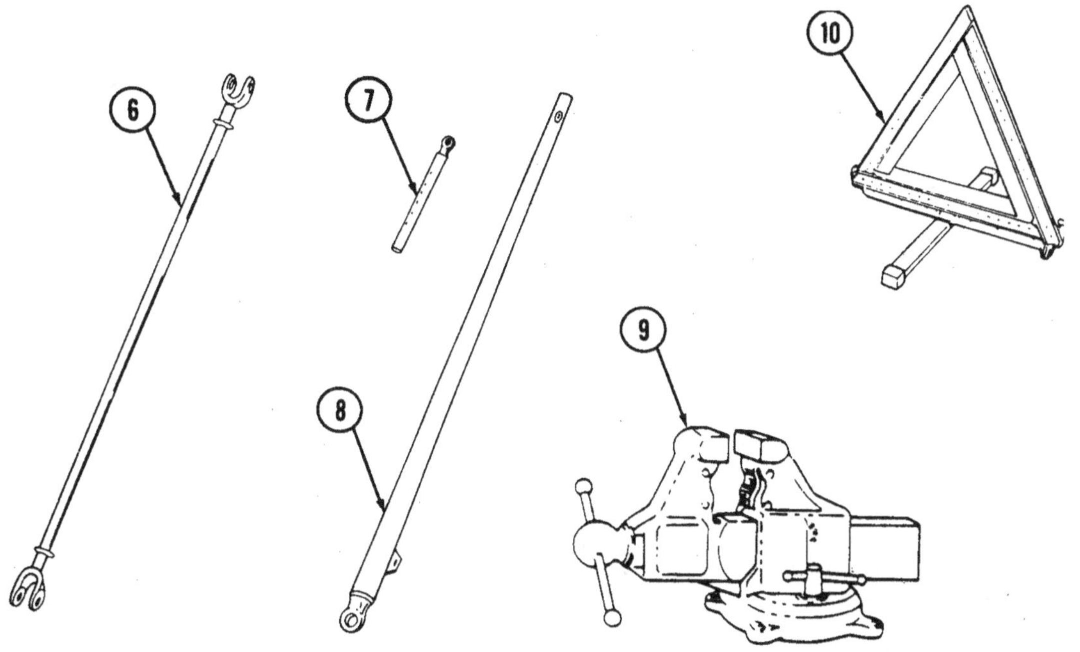
10
6
7
9
8

Section III. BASIC ISSUE ITEMS (Contd)

(1) Illus Number	(2) National Stock Number	(3) Description CAGEC and Part Number	Usable On Code	(4) U/I	(5) Qty Rqd
		TOOLS AND EQUIPMENT – MEDIUM WRECKER (M936, M936A1, AND M936A2) (Contd)			
6	2540-00-040-2298	TIE BAR: boom jack [on left deck (pg B-24)] (19207) 8330152	DAL,V18, ZAL	EA	1
7	2540-00-040-2300	TUBE: boom jack, top [assembled and stored on left deck (pg B-24)] (19207) 8330157	DAL,V18, ZAL	EA	2
8	2540-00-040-2301	TUBE: boom jack, bottom [assembled and stored on left deck (pg B-24)] (19207) 8330158	DAL,V18, ZAL	EA	2
9	5120-00-243-9072	VISE: bench and pipe, swivel base, 5-in. stationary jaw, w/1/8-in. to 4-in. pipe jaw [vehicle front bumper (pg B-24)] (81348) GGG-V-410	DAL,V18, ZAL	EA	1
10	9905-00-148-9546	WARNING DEVICE KIT, HIGHWAY REFLECTIVE TRIANGLE: [in compartment no. 2 (pg B-24)] (19207) 11669000, set no. 3	DAL,V18, ZAL	EA	1

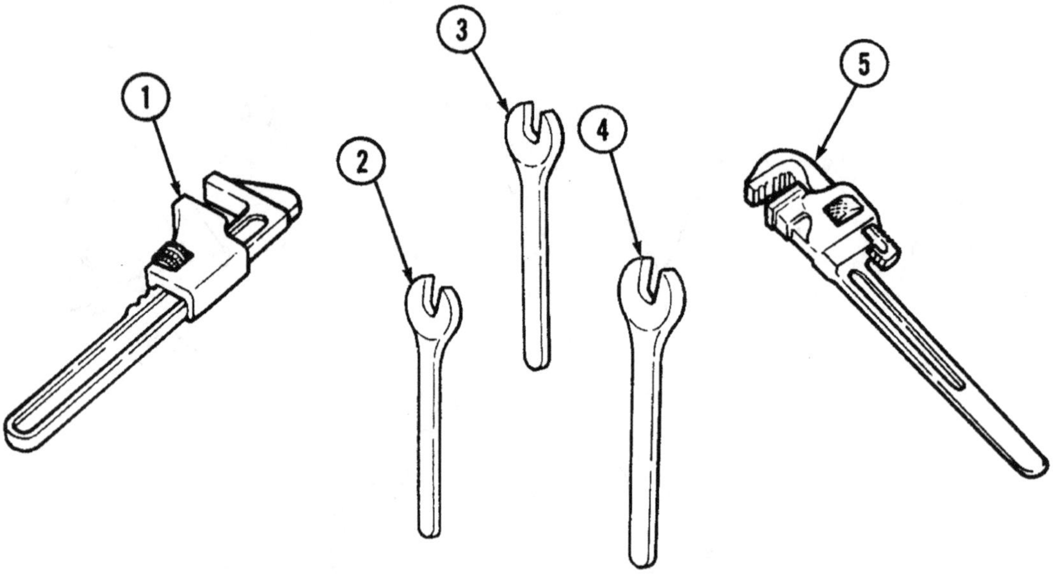
1
2
3
4
5

Section III. BASIC ISSUE ITEMS (Contd)

(1) Illus Number	(2) National Stock Number	(3) Description CAGEC and Part Number	Usable On Code	(4) U/I	(5) Qty Rqd
		TOOLS AND EQUIPMENT – MEDIUM WRECKER (M936, M936A1, AND M936A2) (Contd)			
1	5120-00-264-3793	WRENCH: auto, adjustable, 0-in. to 3-5/8-in. jaw opening, 15-in. long [stowed in compartment no. 4 (pg B-24)] (24617) 2117080	DAL,V18, ZAL	EA	1
2	5120-00-277-1244	WRENCH: open end, fixed, single head, 15-degree head angle, 1-5/8-in. opening, 15-in. long [stowed in compartment no. 4 (pg B-24)] (65814) 710	DAL,V18, ZAL	EA	1
3	5120-00-277-1245	WRENCH: open-end, fixed, single-head, 15-degree head angle, 1-11/16-in. opening, 15-in. long [stowed in compartment no. 4 (pg B-24)] (65814) 1010A	DAL,V18, ZAL	EA	1
4	5120-00-277-1242	WRENCH: open-end, fixed, single head, 15-degree head angle, 1-13/16-in. opening, 16-1/2-in. long [stowed in compartment no. 4 (pg B-24)] (19207) 6012498	DAL,V18, ZAL	EA	1
5	5120-00-277-1461	WRENCH: pipe, heavy duty, adjustable, 1-in. to 2-in. pipe capacity, 18-in. long [in compartment no. 4 (pg B-24)] (21450) 41W664	DAL,V18, ZAL	EA	1

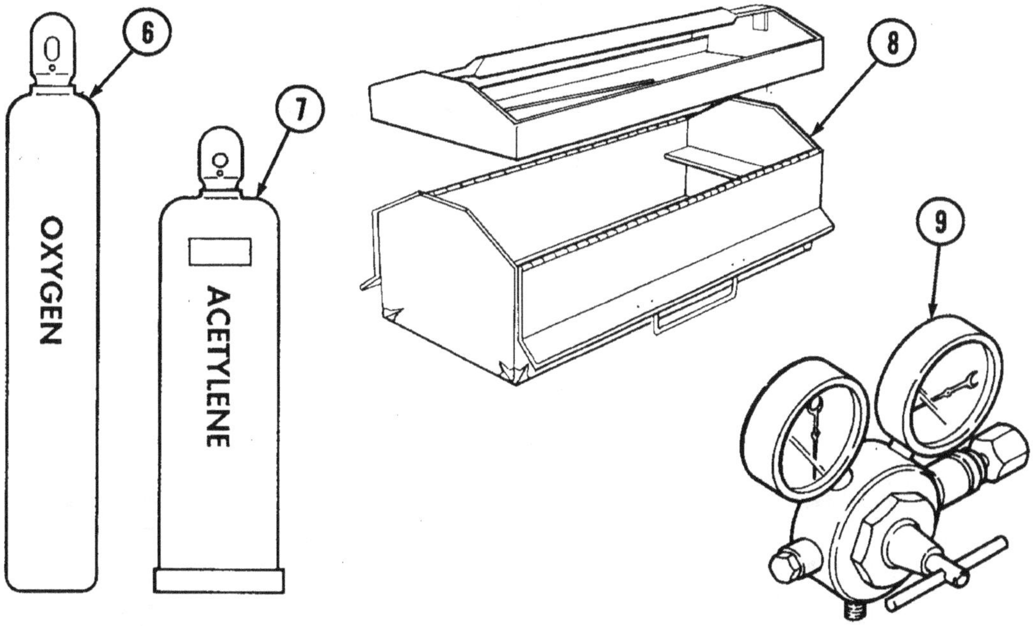
6
7
8
9
OXYGEN
ACETYLENE

Section III. BASIC ISSUE ITEMS (Contd)

(1) Illus Number	(2) National Stock Number	(3) Description CAGEC and Part Number	Usable On Code	(4) U/I	(5) Qty Rqd
		EQUIPMENT - WELDING AND CUTTING (M936, M936A1, AND M936A2)			
6	8120-00-357-7992	CYLINDER: compressed gas, oxygen, 250-cu ft capacity [behind cab, forward left side of wrecker body (pg B-24)] (81348) C-901/1-15	DAL,V18, ZAL	EA	1
7	8120-00-268-3360	CYLINDER: compressed gas, acetylene, 360-cu ft capacity [behind cab, forward left side of wrecker body (pg B-24)] (81349) MIL-C-3701	DAL,V18, ZAL	EA	1
8	5180-00-754-0661	TOOL KIT: welder's [in compartment no. 1A) (pg B-24)] (50980) SC5180-90-CL-N39	DAL,V18, ZAL	EA	1
9	4940-00-357-7778	TORCH OUTFIT: cutting and welding [in compartment no. 1B (pg B-24)] (19204) SC4940-95-CL-B23	DAL,V18, ZAL	EA	1

Section III. BASIC ISSUE ITEMS (Contd)

STOWED EQUIPMENT LOCATIONS
(M936, M936A1, AND M936A2
MEDIUM WRECKER)

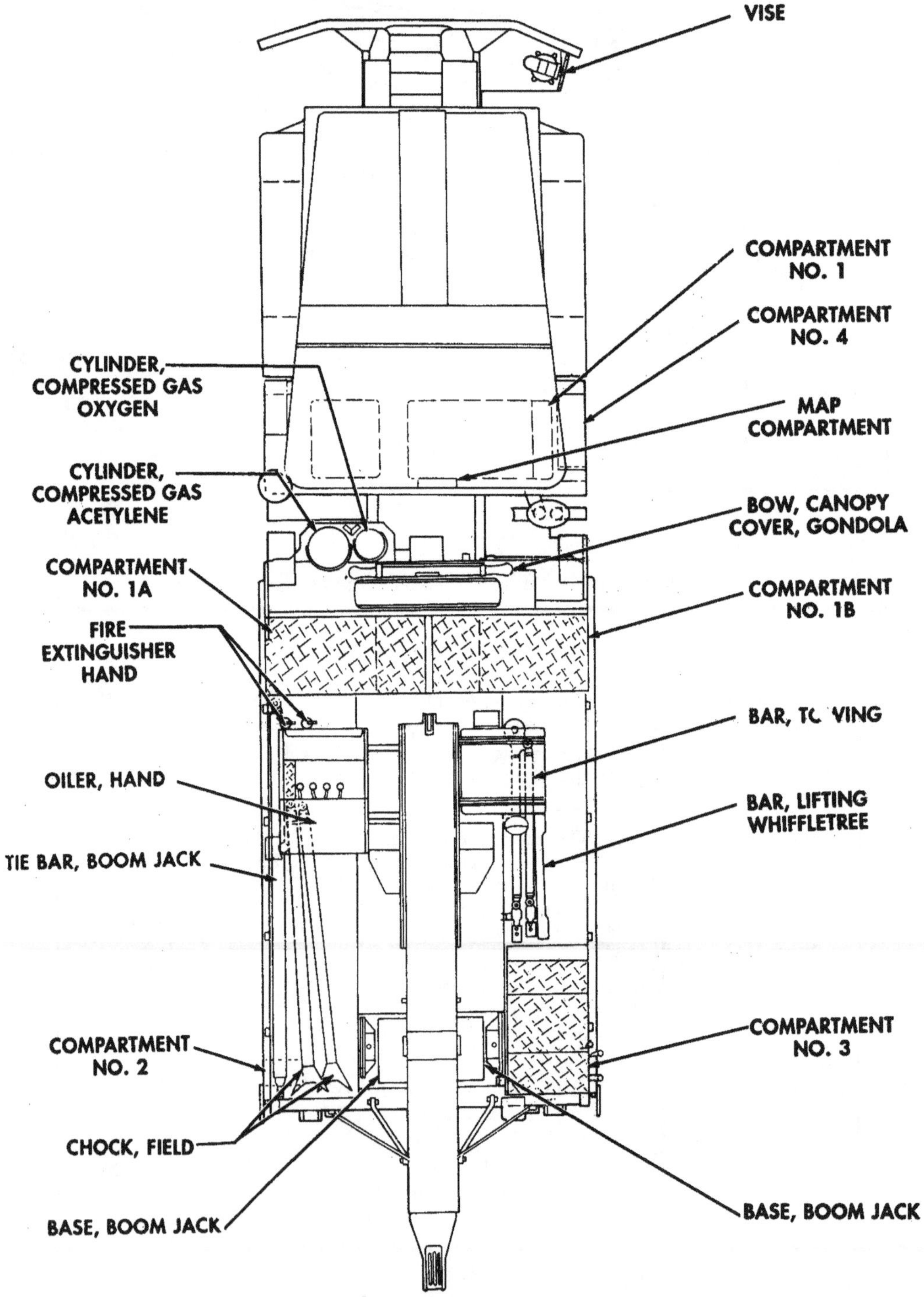
VISE
COMPARTMENT NO. 1
COMPARTMENT NO. 4
MAP COMPARTMENT
CYLINDER, COMPRESSED GAS OXYGEN
CYLINDER, COMPRESSED GAS ACETYLENE
BOW, CANOPY COVER, GONDOLA
COMPARTMENT NO. 1A
COMPARTMENT NO. 1B
FIRE EXTINGUISHER HAND
BAR, TC VING
OILER, HAND
BAR, LIFTING WHIFFLETREE
TIE BAR, BOOM JACK
COMPARTMENT NO. 2
COMPARTMENT NO. 3
CHOCK, FIELD
BASE, BOOM JACK
BASE, BOOM JACK

APPENDIX C
ADDITIONAL AUTHORIZATION LIST (AAL)

Section I. INTRODUCTION

C-1. SCOPE

This appendix lists additional items authorized for support of M939, M939A1, and M939A2 series vehicles.

C-2. GENERAL

This list identifies items that do not have to accompany your truck and do not have to be turned in with it.

C-3. EXPLANATION OF LISTING

a. Descriptions, national stock numbers, and part numbers are provided to help you identify and request additional items you require to support this equipment. If item required differs for different models, the model is shown under the "USABLE ON CODE" heading. Codes are the same as in appendix B, Basic Issue Items.

b. Column (1) – National Stock Number. Indicates the national stock number assigned to the item and will be used for requisitioning purposes.

c. Column (2) – Description. Indicates the federal item name and, if required, a minimum description to identify and locate the item. The last line for each item indicates the Commercial and Government Entity Code (CAGEC) (in parentheses) followed by the part number. If item needed differs for different models of this equipment, the model is shown under the "Usable On Code" heading in this column. Refer to appendix B for cage codes used.

d. Column (3) – U/I (Unit of Issue). Indicates how the item is issued for the national stock number shown in column two.

e. Column (4) – Qty Auth. Is the quantity of the item authorized.

Section II. ADDITIONAL AUTHORIZATION LIST

(1) NATIONAL STOCK NUMBER	(2) DESCRIPTION CAGEC & PART NUMBER	USABLE ON CODE	(3) U/I	(4) QTY AUTH
	ABS Diagnostic Info Centre (OUKB6) WHMB364317001		EA	1
5935-00-322-8959	ADAPTER, CONNECTOR SLAVE 2-PIN (19207) 11677570	DAW,DAX ZAA,ZAB	EA	1
4930-00-204-2550	ADAPTER: grease gun, rigid, thin-stem, 6-in. long, type IV, class 2 (19207) 5349744	DAL,V18, ZAL	EA	1
5110-00-293-2336	AX: single bit, 4-lb head (19207) 6150925	All Except DAL,V18, ZAL	EA	1
6135-00-835-7210	BATTERY: dry, 1.5 volt, BA 30, (81349) BA30/U	DAL,V18, ZAL	EA	4
3940-00-105-9933	BLOCK: rigging, steel, wire rope, single 8-in. sheave, w/swivel hook, 5/8-in. diameter rope, 10-ton safe work load (19207) 11631726	DAD,DAF, DAH,DAL, DAX,V14, V16,V18, V19,V21, ZAB,ZAD, ZAF,ZAH, ZAL	EA	1
3940-00-926-3719	BLOCK: tackle, manila rope, single 4-1/2-in. sheave, 1-in. diameter rope, 3-in. rope circumference, w/loose side hook, w/becket, 5,100-lb capacity, type II, class 1 [in compartment no 1B (pg B-24)] (81348) GGG-B-490	DAL,V18, ZAL	EA	1
3940-00-926-3710	BLOCK: tackle, manila rope, single 4-3/4-in. sheave, 1-in. diameter rope, 3-in. rope circumference, w/loose side hook, w/becket, 3,300-lb capacity, type II, class 1 [in compartment no 1B (pg B-24)] (81348) GGG-B-490	DAL,V18, ZAL	EA	1
2590-00-473-6331	BRACKET: gas/water can (19207) 6566675	DAL,V18, ZAL	EA	1
7240-01-337-5269	CAN: gasoline, MIL type, 5 gallon, (81349) MIL-C-53109	A	EA	1
7240-00-089-3827	CAN: water, MIL type, 5 gallon plastic (81349) MIL-C-43613	A	EA	1
5140-00-860-2354	CASE: crosscut saw, cotton duck, 63-3/4-in. long (closed) (61465) 3005187	DAL,V18, ZAL	EA	1

Section II. ADDITIONAL AUTHORIZATION LIST (Contd)

(1) NATIONAL STOCK NUMBER	(2) DESCRIPTION CAGEC & PART NUMBER	USABLE ON CODE	(3) U/I	(4) QTY AUTH
2540-00-933-9022	CHAIN: pneumatic tire, truck, single tire, type TS, 11:00 x 20 (96906) MS500055-22	V14,V15, V16,V17, V18,V19, V20,V21, V22,V24	PR	1
2540-21-911-1360	CROSS CHAIN, TIRE: pneumatic tire, truck, single tire, type TS, 11:00 x 20 (4N506) CL97	V14,V15, V16,V17, V18,V19, V20,V21, V22,V24	PR	1
2540-00-933-9033	CHAIN: pneumatic tire, truck, single tire, type TS, 14.00 x 20 (96906) MS500055-27	DAC,DAD, DAE,DAF, DAG,DAH, DAJ,DAL, DAW,DAX ZAA,ZAB, ZAC,ZAD, ZAE,ZAF, ZAG,ZAH, ZAJ,ZAL	PR	1
4010-00-473-6166	CHAIN tow, single leg, 5/8-in. link, 16-ft long, w/grab hook, w/2 pear-shaped coupling links, olive drab finish (19207) 7077063	DAD,DAF, DAH,DAL, DAX,V14, V16,V18, V19,V21, ZAB,ZAD, ZAF,ZAH, ZAL	EA	1
2540-00-912-1848	CHOCK BLOCKS: (97403) 13211E3357	A	EA	1
4030-01-477-0524	CLAMP, LINE SLIDING GRIP (098P0) NEI PR054-001-B	ZAA,ZAB	DZ	1
2540-00-860-2355	COVER: fitted, gondola (19207) 10876433	DAL,V18, ZAL	EA	1
9390-01-204-1161	EXHAUST STACK WEATHER: All wrap non-metallic special, rubber (17284) FB0037	A	EA	1
4930-00-288-1511	EXTENSION: grease gun, flex hose, 12-in. long to 14-in. long (19207) 6300333	DAG,DAH, DAL,V18, V21,V22, ZAG,ZAH	EA	1
4210-01-149-1356	EXTINGUISHER, FIRE: 5-lb purple K dry chemical, w/bracket (19207) 12255633-1	All Except DAL,V18, ZAL	EA	1

Section II. ADDITIONAL AUTHORIZATION LIST (Contd)

(1) NATIONAL STOCK NUMBER	(2) DESCRIPTION CAGEC & PART NUMBER	USABLE ON CODE	(3) U/I	(4) QTY AUTH
6230-00-264-8261	FLASHLIGHT: electric, hand, 2-cell, w/lamp and lens filter, w/o batteries, type I, class A (81349) MIL-F-3747	DAL,V18, ZAL	EA	2
6125-01-020-7268	FREQUENCY GENERATOR: motor generator, AC input, continuous cycle, three phase input (60 Hz), three phase output (420 Hz) (91723) 30-154	DAJ,V24 ZAJ	EA	1
5120-00-288-6574	HANDLE: mattock-pick, railroad or clay pick (19207) 11677021	All Except DAL,V18, ZAL	EA	1
6545-00-922-1200	KIT: first aid (19207) 11677011	All Except DAL,V18, ZAL	EA	1
6150-01-022-6004	Intervehicle power cable, NATO slave, 24-volt, 20-ft long (61951) 11682336-1	A	EA	1
6240-00-044-6914	LAMP: incandescent, S8 bulk, S contact, bayonet base, 28V (61951) 1683	DAL,V18, ZAL	EA	1
5120-00-243-2395	MATTOCK: pick type, 5-lb, w/o handle (19207) 11677022	All Except DAL,V18, ZAL	EA	1
2530-01-461-2473	MODULE, DIAGNOSTIC, CTIS manual control (19207) 12470092	ZAA,ZAB ZAC,ZAD ZAE,ZAF ZAG,ZAH ZAJ,ZAK ZAL,ZAM ZAN,ZAQ	EA	1
3940-01-449-2385	NET, DRAFT COVER (098P0) B9154-090-168-2R	DAW,DAX ZAA,ZAB	KT	1
5340-00-682-1505	PADLOCK SET: keyed alike, 1-3/4-in., w/clevis and chain, composed of 5 padlocks and 7 keys (96906) MS21313-52	All Except DAL,V18, ZAL	EA	1
4020-00-231-2581	ROPE: manila, 3 strand, 3/8-in. diameter, 1-1/8-in. circumference, 50-ft long, 325-lb capacity [in compartment no. 2 (pg B-24)] (81348) TR605	DAL,V18, ZAL	EA	1

Section II. ADDITIONAL AUTHORIZATION LIST (Cont'd)

(1) NATIONAL STOCK NUMBER	(2) DESCRIPTION CAGEC & PART NUMBER	USABLE ON CODE	(3) U/M	(4) QTY AUTH
4020-00-238-7734	ROPE: manila, 3 strand, 3/4-in. diameter, 2-1/4-in. circumference, 50-ft long, 1,350-lb capacity [in compartment no. 2 (pg B-24)] (81348) TR605	DAL,V18, ZAL	EA	1
4020-00-231-9014	ROPE: manila, 3 strand, 1-in. diameter, 3-in. circumference, 300-ft long, 2,250-lb capacity [in compartment no. 2 (pg B-24)] (81348) TR605	DAL,V18, ZAL	EA	1
5110-00-242-7147	SAW: crosscut, 1-man, 4-1/2-ft blade, 5-ft long w/supplementary handle (pg B-24) (81348) GGG-S-64	DAL,V18, ZAL	EA	1
5120-00-293-3336	SHOVEL: hand, round point, D handle, short size (19207) 11655784	All Except DAL,V18, ZAL	EA	1
4030-01-477-0508	SNAP LINK, CARGO: 14 ea per kit (098PO) NEI 40 WGB	DAW,DAX ZAA,ZAB	KT	1
7240-00-177-6154	SPOUT, can gasoline, flex, w/filter cap assembly screen (7420-00-152-6433), 2-1/4-in. outside diameter, 16-in. long (19207) 11677020	DAL,V18, ZAL	EA	1
1670-00-725-1487	TIEDOWN STRAP: Ratchet with integral hooks, 20 ft long (81349) MIL-T-272 60 TYPECGUIB	All Except V18,V19, V20,V21, V22	EA	1
6220-01-377-9133	TOW LIGHT BAR (19207) 12375702	DAL,V18	EA	1
6150-01-379-7272	TOW LIGHT CABLE (19207) 12375703	DAL,V18	EA	1
2590-01-436-9145	TOW LIGHT AND CABLE ASSY (19207) 12450235		KT	1
9905-00-148-9546	WARNING DEVICE KIT, HIGHWAY REFLECTIVE TRIANGLE: (80244) RR-W-1817, set no. 3	All Except DAL,V18, ZAL	KT	1
2540-00-912-1848	WHEEL CHOCK BLOCK: Aluminum fixed-welded plates, 15 x 12 x 9.375 inches (97403) 13211E3357	A	EA	1

APPENDIX D
EXPENDABLE/DURABLE SUPPLIES AND MATERIALS LIST

Section I. INTRODUCTION

D-1. SCOPE

This appendix lists expendable/durable supplies and materials you will need to operate and maintain M939, M939A1, and M939A2 series vehicles.

D-2. EXPLANATION OF COLUMNS

a. Column (1) – Item Number. This number is assigned to the entry in the listing.

b. Column (2) – Level. This column identifies the lowest level of maintenance that requires the listed item. Codes used in this column are:

C — Operator/Crew
O — Unit Maintenance

c. Column (3) - National Stock Number. This is the national stock number assigned an item. Use this number to request or requisition that item.

d. Column (4) - Description, CAGEC, and Part Number. Indicates the federal item name and, if required, a description to identify the item. The last line for each item indicates the Commercial and Government Entity Code (CAGEC) (in parentheses) followed by the part number. Refer to appendix B for CAGE codes used.

e. Column (5) - Unit of Measure (U/M). Indicates the measure used in performing the actual maintenance function. This measure is expressed by an abbreviation (such as ea (each), oz (ounce), gal. (gallon)). If the unit of measure differs from the unit of issue, requisition the lowest unit of issue that will satisfy your requirements.

Section II. EXPENDABLE/DURABLE SUPPLIES AND MATERIALS LIST

(1) ITEM NUMBER	(2) LEVEL	(3) NATIONAL STOCK NUMBER	(4) DESCRIPTION CAGEC AND PART NUMBER	(5) U/M
1	O	6830-00-264-6751	ACETYLENE, TECHNICAL: gas filled acetylene, 225-cu ft (to be filled/refilled locally) (81348) BB-A-106	cu ft
			ANTIFREEZE: PERMANENT ETHYLENE GLYCOL [-60°F (-51°C)] INHIBITED (MIL-A-46153)	
2	C	6850-01-441-3218	1-GALLON CONTAINER	gal.
3	C	6850-00-181-7933	5-GALLON CONTAINER	gal.
4	C	6850-01-441-3223	55-GALLON DRUM	gal.
			ANTIFREEZE: PERMANENT TYPE; ARCTIC GRADE [-90°F (-68°C)] (MIL-A-11755)	
5	C	6850-01-441-3248	55-GALLON DRUM	gal.
6	O	6850-00-926-2275	CLEANING COMPOUND,WINDSHIELD 16-OUNCE BOTTLE, CONCENTRATED	oz
			GREASE, AUTOMOTIVE AND ARTILLERY GAA (MIL-G-10924)	
7	C	9150-00-065-0029	2-1/4-OUNCE TUBE	oz
8	C	9150-01-197-7693	14-OUNCE CARTRIDGE	oz
9	C	9150-01-197-7690	1-3/4-POUND CAN	lb
10	C	9150-01-197-7689	6-1/2-POUND CAN	lb
11	C	9150-01-197-7692	35-POUND CAN	lb
12	C	9150-01-197-7691	120-POUND DRUM	lb
			HYDRAULIC FLUID: transmission (24617) Dexron® III	
12.1	C	9150-01-353-4799	1-QUART CAN	qt
		1950-01-114-9968	55-GALLON DRUM	gal.
			INHIBITOR: CORROSION, LIQUID COOLING SYSTEM; POWDER FORM (0-I-490)	
Deleted				
14	C	9140-00-286-5296	55-GALLON DRUM, 16 GAUGE	gal.
15	C	9140-00-286-5297	55-GALLON DRUM, 18 GAUGE	gal.
16	C	9140-00-286-5294	BULK	gal.
			OIL, FUEL, DIESEL, DF-1: WINTER (VV-F-800)	
17	C	9140-00-286-5288	55-GALLON DRUM, 16 GAUGE	gal.
18	C	9140-00-286-5289	55-GALLON DRUM, 18 GAUGE	gal.

Section II. EXPENDABLE/DURABLE SUPPLIES AND MATERIALS LIST (Contd)

(1) ITEM NUMBER	(2) LEVEL	(3) NATIONAL STOCK NUMBER	(4) DESCRIPTION CAGEC AND PART NUMBER	(5) U/M
19	C	9140-00-286-5286	BULK	gal.
			OIL, FUEL, DIESEL DF-A: ARCTIC (VV-F-800)	
20	C	9140-00-286-5284	55-GALLON DRUM, 16 GAUGE	gal.
21	C	9140-00-286-5285	55-GALLON DRUM, 18 GAUGE	gal.
22	C	9140-00-286-5283	BULK	gal.
			OIL, LUBRICATING, ENGINE, ARCTIC (ICE, SUB-ZERO) OEA (SAE OW-20) (MIL-L-46167)	
23	C	9150-00-402-4478	1-QUART CAN	qt
24	C	9150-00-402-2372	5-GALLON CAN	gal.
25	C	9150-00-491-7197	55-GALLON DRUM, 16 GAUGE	gal.
			OIL, LUBRICATING, EXPOSED GEAR, CW (VV-L-751)	
26	C	9150-00-234-5197	5-POUND CAN	lb
27	C	9150-00-261-7891	35-POUND PAIL	lb
			OIL, LUBRICATING, GEAR, MULTI-PURPOSE, GO 80/90 (MIL-L-2105)	
28	C	9150-01-035-5392	1-QUART CAN	qt
29	C	9150-01-035-5393	5-GALLON DRUM	gal.
30	C	9150-01-035-5394	55-GALLON DRUM, 16 GAUGE	gal.
			OIL, LUBRICATING, GEAR MULTI-PURPOSE, GO 75 (MIL-L-2105)	
31	C	9150-01-035-5390	1-QUART CAN	qt
32	C	9150-01-035-5391	5-GALLON DRUM	gal.
33	C	9150-01-152-4119	55-GALLON DRUM, 16 GAUGE	gal.
34	C	9150-00-183-7807	BULK	gal.
			OIL, LUBRICATING, OE/HDO 10W (MIL-L-2104)	
Deleted				
36	C	9150-00-186-6668	5-GALLON CAN	gal.
37	C	9150-00-191-2772	55-GALLON DRUM, 16 GAUGE	gal.
			OIL, LUBRICATING, OE/HDO 30 (MIL-L-2104)	
38	C	9150-01-178-4726	1-QUART CAN	qt
39	C	9150-00-188-9858	5-GALLON DRUM	gal.
40	C	9150-00-189-6729	55-GALLON DRUM, 16 GAUGE	gal.

Section II. EXPENDABLE/DURABLE SUPPLIES AND MATERIALS LIST (Contd)

(1) ITEM NUMBER	(2) LEVEL	(3) NATIONAL STOCK NUMBER	(4) DESCRIPTION CAGEC AND PART NUMBER	(5) U/M
			OIL, LUBRICATING, OE/HDO 15/40 (MIL-PRF-2104)	
40.1	C	9150-01-152-4117	1-QUART CAN	qt
40.2	C	9150-01-152-4118	5-GALLON DRUM	gal.
40.3	C	9150-01-152-4119	55-GALLON DRUM, 16 GAUGE	gal.
41	C	9150-00-183-7808	BULK	gal.
			OIL, TURBINE FUEL, AVIATION Grade JP-8	
42	C	9130-01-031-5816	BULK	gal.
43	O	6830-00-292-0129	OXYGEN, TECHNICAL: gas filled oxygen, 240-cu ft (to be filled/refilled locally) (81348) BB-O-925	cu ft
			SOLVENT, DRYCLEANING SD-2	
44	C	6850-00-664-5685	1-QUART CAN	qt
45	C	6850-00-281-1985	1-GALLON CAN	gal.
			METHYL ALCOHOL, METHANOL	
46	O	6810-00-597-3608	1-GALLON CAN	gal.
47	O	6810-00-275-6010	5-GALLON CAN	gal.

APPENDIX E
STOWAGE AND SIGN GUIDE

E-1. SCOPE

This appendix shows the location for stowage of equipment and material required to be carried on M939/A1/A2 series vehicles.

E-2. GENERAL

The stowed equipment locater is designed to help inventory items required for safe and efficient operation.

E-3. STOWAGE LOCATIONS

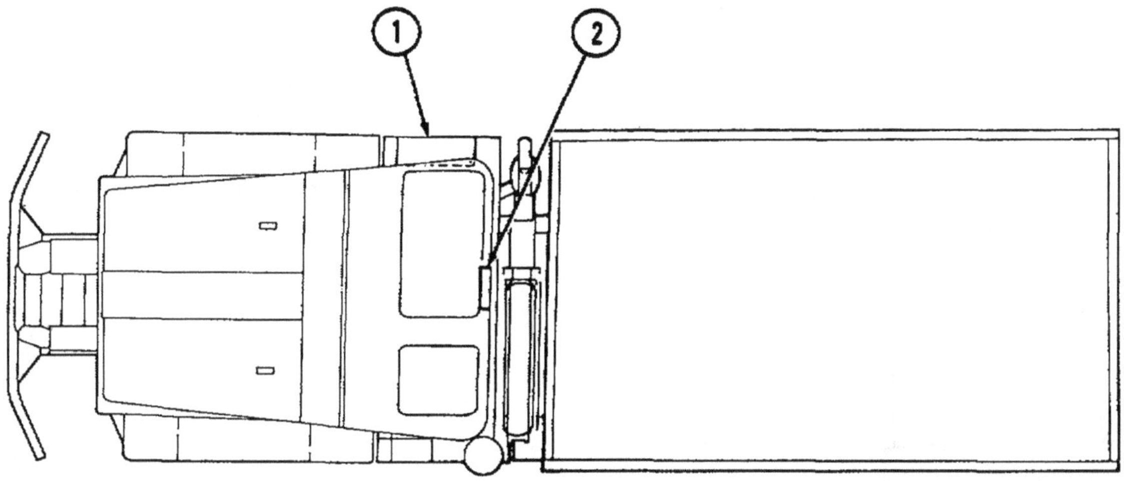

1
2

ALL (A) VEHICLES

KEY	LOCATION
1	Toolbox, right access step
2	Map compartment, inside cab rear wall

E-3. STOWAGE LOCATIONS (Cont'd)

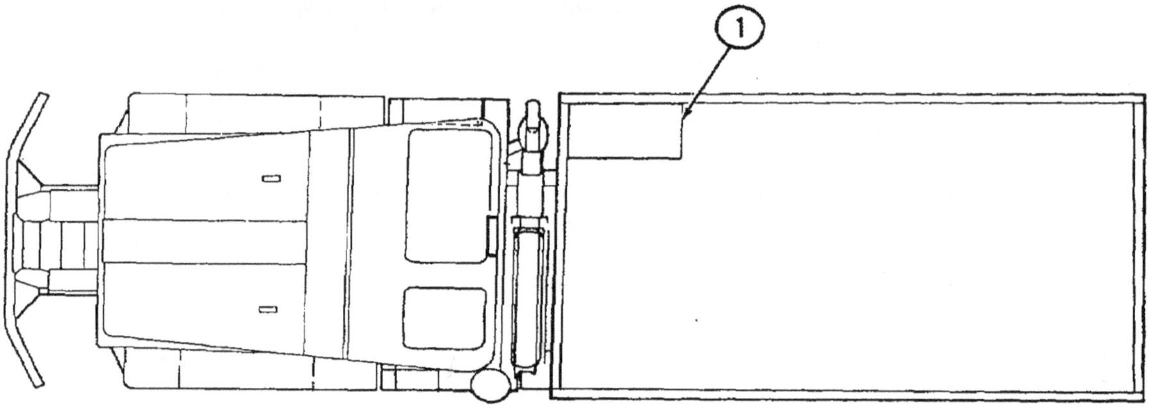
1

M923 (V15), M923A1 (DAW), M923A2 (ZAA), M925 (V14), M925A1 (DAX), M925A2 (ZAB) M927 (V17), M927A1 (DAC), M927A2 (ZAC), M928 (V16), M928A1 (DAD), M928A2 (ZAD)

KEY	LOCATION
1	Toolbox, vehicle right side frame rail

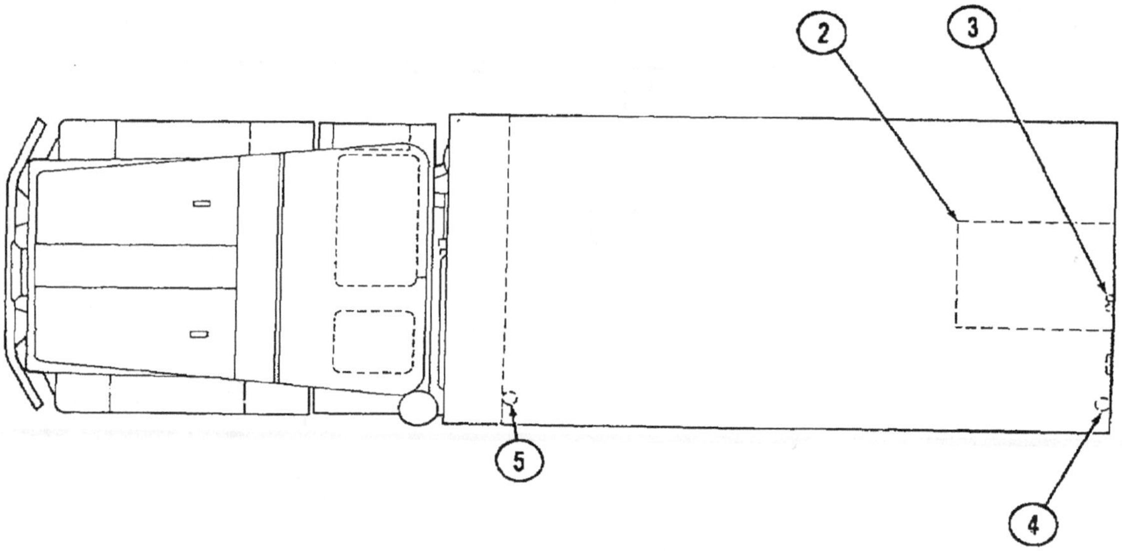
2
3
5
4

M934 (V24), M934A1 (DAJ), M934A2 (ZAJ)

KEY	LOCATION
2	Tray, rear exterior of van body (under door)
3	Clips, attached to interior left van door
4	Mount, fire extinguisher, interior left rear
5	Mount, fire extinguisher, interior left front

E-3. STOWAGE LOCATIONS (Cont'd)

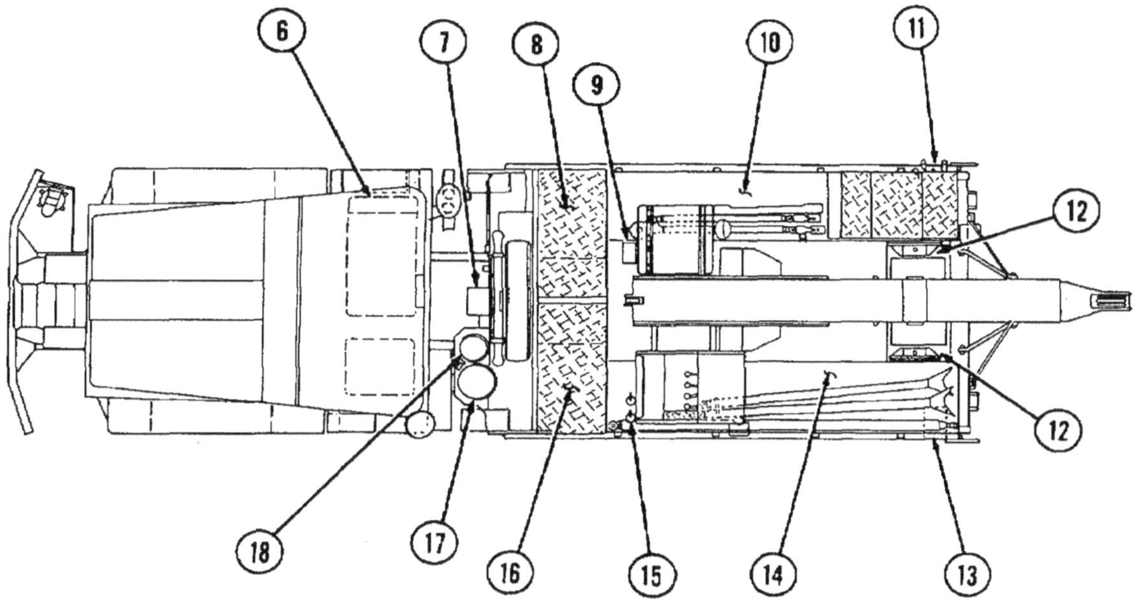
6
7
8
9
10
11
12
12
13
14
15
16
17
18

M936 (V18), M936A1 (DAL), M936A2 (ZAL)

KEY	LOCATION
6	Utility compartment, inside of cab
7	Mounting provision for gas can bracket, forward of spare tire
8	Compartment No. 1B, right front of vehicle body
9	Spare gas can bracket, exterior rear of hydraulic oil reservoir
10	Right rear deck of vehicle body
11	Compartment No. 3, right rear of vehicle
12	Boom jack base bracket, left and right of rear winch
13	Compartment No. 2, left rear of vehicle
14	Left rear deck of vehicle body
15	Fire extinguisher brackets, exterior rear of gondola cab
16	Compartment No. 1A, left front of vehicle body
17	Acetylene cylinder, behind left rear of cab
18	Oxygen cylinder, behind left rear of cab

INDEX

INDEX

A (Contd)

B

INDEX

INDEX

C (Contd)

INDEX

C (Contd)

D

INDEX

INDEX

INDEX

F (Contd)

INDEX

INDEX

INDEX

L (Contd)

INDEX

O (Contd)

INDEX

O (Contd)

INDEX

INDEX

INDEX

INDEX

INDEX

INDEX

INDEX

INDEX

T (Contd)

INDEX

INDEX

By Order of the Secretary of the Army:

DENNIS J. REIMER
General, United States Army
Chief of Staff

Official:

JOEL B. HUDSON
Administrative Assistant to the
Secretary of the Army
05676

By Order of the Secretary of the Air Force:

RONALD R. FOGLEMAN
General, United States Air Force
Chief of Staff

Official:

HENRY VICCELLIO, JR.
General, United States Air Force
Commander, Air Force Materiel Command

Distribution:

To be distributed in accordance with the initial distribution number (IDN) 380385, requirements for TM 9-2320-272-10.

THE METRIC SYSTEM AND EQUIVALENTS

LINEAR MEASURE

1 Centimeter = 10 Millimeters = 0.01 Meters = 0.3937 Inches

1 Meter = 100 Centimeters = 1,000 Millimeters = 39.37 Inches

1 Kilometer = 1,000 Meters = 0.621 Miles

SQUARE MEASURE

1 Sq Centimeter = 100 Sq Millimeters = 0.155 Sq Inches

1 Sq Meter = 10,000 Sq Centimeters = 10.76 Sq Feet

1 Sq Kilometer = 1,000,000 Sq Meters = 0.386 Sq Miles

CUBIC MEASURE

1 Cu Centimeter = 1,000 Cu Millimeters = 0.06 Cu Inches

1 Cu Meter = 1,000,000 Cu Centimeters = 35.31 Cu Feet

LIQUID MEASURE

1 Milliliter = 0.001 Liters = 0.0338 Fluid Ounces

1 Liter = 1,000 Milliliters = 33.82 Fluid Ounces

TEMPERATURE

Degrees Fahrenheit (F) = °C • 9 ÷ 5 + 32

Degrees Celsius (C) = °F - 32 • 5 ÷ 9

212° Fahrenheit is equivalent to 100° Celsius

89.96° Fahrenheit is equivalent to 32.2° Celsius

32° Fahrenheit is equivalent to 0° Celsius

WEIGHTS

1 Gram = 0.001 Kilograms = 1,000 Milligrams = 0.035 Ounces

1 Kilogram = 1,000 Grams = 2.2 Lb

1 Metric Ton = 1,000 Kilograms = 1 Megagram = 1.1 Short Tons

APPROXIMATE CONVERSION FACTORS

TO CHANGE	TO	MULTIPLY BY
Inches	Millimeters	25.400
Inches	Centimeters	2.540
Feet	Meters	0.305
Yards	Meters	0.914
Miles	Kilometers	1.609
Square Inches	Square Centimeters	6.451
Square Feet	Square Meters	0.093
Square Yards	Square Meters	0.836
Square Miles	Square Kilometers	2.590
Acres	Square Hectometers	0.405
Cubic Feet	Cubic Meters	0.028
Cubic Yards	Cubic Meters	0.765
Fluid Ounces	Milliliters	29.573
Pints	Liters	0.473
Quarts	Liters	0.946
Gallons	Liters	3.785
Ounces	Grams	28.349
Pounds	Kilograms	0.4536
Short Tons	Metric Tons	0.907
Pound-Feet	Newton-Meters	1.356
Pounds Per Square Inch	Kilopascals	6.895
Miles Per Gallon	Kilometers Per Liter	0.425
Miles Per Hour	Kilometers Per Hour	1.609

TO CHANGE	TO	MULTIPLY BY
Millimeters	Inches	0.03937
Centimeters	Inches	0.3937
Meters	Feet	3.280
Meters	Yards	1.094
Kilometers	Miles	0.621
Square Centimeters	Square Inches	0.155
Square Meters	Square Feet	10.764
Square Meters	Square Yards	1.196
Square Kilometers	Square Miles	0.386
Square Hectometers	Acres	2.471
Cubic Meters	Cubic Feet	35.315
Cubic Meters	Cubic Yards	1.308
Milliliters	Fluid Ounces	0.034
Liters	Pints	2.113
Liters	Quarts	1.057
Liters	Gallons	0.264
Grams	Ounces	0.035
Kilograms	Pounds	2.2046
Metric Tons	Short Tons	1.102
Newton-Meters	Pound-Feet	0.738
Kilopascals	Pounds Per Square Inch	0.145
Kilometers Per Liter	Miles Per Gallon	2.354
Kilometers Per Hour	Miles Per Hour	0.621

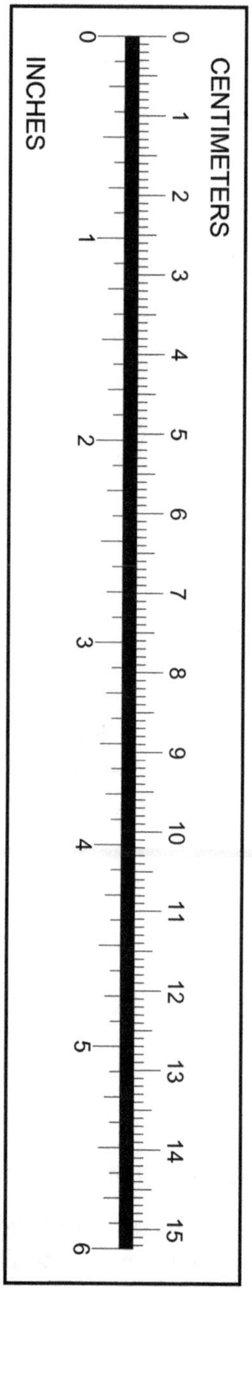
CENTIMETERS
INCHES

www.ingramcontent.com/pod-product-compliance
Lightning Source LLC
Chambersburg PA
CBHW080602270326
41928CB00016B/2899